DK COMPLETE rspb
BIRDS
OF BRITAIN AND EUROPE

Rob Hume

LONDON, NEW YORK, MUNICH,
MELBOURNE, AND DELHI

DK LONDON
Senior Art Editor Ina Stradins
Editor Miezan van Zyl
Senior Producer Alice Sykes
Pre-production Producer
Nikoleta Parasaki
Jacket Designer Mark Cavanagh
Jacket Editor Manisha Majithia
**Jacket Design Development
Manager** Sophia MTT
Managing Art Editor Michelle Baxter
Managing Editor Angeles Gavira Guerrero
Publisher Sarah Larter
Art Director Philip Ormerod
Associate Publishing Director
Liz Wheeler
Publishing Director Jonathan Metcalf

DK DELHI
Editors Susmita Dey, Himani Khatreja,
Neha Pandey, Pallavi Singh
Art Editors Sanjay Chauhan,
Suhita Dharamjit, Rakesh Khundongbam,
Upasana Sharma
Senior Art Editor Mahua Mandal
DTP Designers Arvind Kumar, Shanker
Prasad, Umesh Rawat
Managing Editor Rohan Sinha
Deputy Managing Art Editor
Sudakshina Basu
Production Manager Pankaj Sharma
Pre-production Manager Balwant Singh

First published in Great Britain in 2002
This edition published in 2013
by Dorling Kindersley Limited
80 Strand, London WC2R 0RL

2 4 6 8 10 9 7 5 3 1
001—192571—09/13

Copyright © 2002, 2007, 2009, 2013
Dorling Kindersley Limited
A Penguin Company

ISBN: 978-1-4093-3507-8

Printed and bound in China by South China
Co. Ltd.

See our complete catalogue at
www.dk.com

CD TRACK LISTING

Track	Species	Sound	Page	Track	Species	Sound	Page
1	Black-throated Diver	Display calls	104	42	Tawny Owl	i Song	261
2	Little Grebe	i Call	106			ii Call	
		ii Winter call		43	Barn Owl	Call	256
3	Fulmar	Calls at nest	112	44	Nightjar	Song	265
4	Bittern	Booming call	122	45	Swift	Call	267
5	Grey Heron	i Flight call	129	46	Hoopoe	Song	274
		ii Adult call		47	Green Woodpecker	Call	278
6	Mute Swan	i Threatened call	59	48	Great Spotted	i Drumming, calls	280
		ii Chicks call			Woodpecker		
7	Whooper Swan	Call	61			ii Flight call	
8	Bewick's Swan	Call	60	49	Skylark	i Song	320
9	Pink-footed Goose	Call	63			ii Call	
10	Greylag Goose	Call	65	50	Crested Lark	Call	317
11	Canada Goose	i Call	66	51	Woodlark	Song	319
		ii Call of a flock		52	Swallow	Song	325
12	Brent Goose	Call	68	53	Meadow Pipit	Song	406
13	Mallard	i Female quacks	73	54	Tree Pipit	Song	405
		ii Chicks call		55	White/Pied Wagtail	Call	403
14	Wigeon	i Male call	70	56	Yellow Wagtail	Call	401
		ii Female call		57	Grey Wagtail	Call	402
15	Peregrine	i Call	162	58	Wren	Song	364
		ii Alarm calls		59	Dunnock	Song	390
		at nest		60	Robin	i Song	376
16	Grey Partridge	i Male song	99			ii Call	
		ii Call		61	Nightingale	Song	378
17	Quail	Male song	97	62	Black Redstart	Song	380
18	Pheasant	Call	100	63	Stonechat	Call	385
19	Corncrake	Song	165	64	Song Thrush	Song	373
20	Moorhen	Call	166	65	Mistle Thrush	Song	375
21	Coot	i Call	167	66	Blackbird	Song	371
		ii Young call		67	Garden Warbler	Song	338
22	Oystercatcher	i Piping display	180	68	Blackcap	Song	337
		call		69	Sedge Warbler	Song	350
		ii Call		70	Reed Warbler	Song	352
23	Ringed Plover	Flight call	185	71	Great Reed Warbler	Song	353
24	Grey Plover	Call	182	72	Cetti's Warbler	Song	332
25	Golden Plover	i Call	181	73	Willow Warbler	Song	336
		ii Summer call		74	Chiffchaff	Song	335
26	Lapwing	i Male flight call	183	75	Goldcrest	Song	355
		ii Call		76	Great Tit	Song	307
27	Dunlin	i Call	198	77	Coal Tit	Song	309
		ii Adult, young		78	Blue Tit	Song	306
		calls		79	Willow Tit	Call	310
28	Common Sandpiper	Call	203	80	Marsh Tit	Call	311
29	Green Sandpiper	i Call	204	81	Long-tailed Tit	Call	313
		ii Flight call		82	Nuthatch	Call	360
30	Redshank	Flight call	209	83	Golden Oriole	Song	286
31	Greenshank	Call	206	84	Magpie	Call	296
32	Curlew	i Call	189	85	Jay	Call	297
		ii Song		86	Jackdaw	Call	299
33	Snipe	i Drumming	212	87	Rook	Call	300
		ii Flight call		88	Carrion Crow	Call	301
34	Black-headed Gull	i Call of a flock	237	89	Raven	Call	303
		ii Call		90	Starling	Song	365
35	Herring Gull	Call	242	91	House Sparrow	Call	395
36	Kittiwake	Calls at colony	236	92	Chaffinch	Song	412
37	Common Tern	i Call	233	93	Linnet	Flight call	419
		ii Adult, young		94	Goldfinch	Call	417
		calls		95	Greenfinch	Song	414
38	Stock Dove	Song	249	96	Bullfinch	Call	426
39	Woodpigeon	Song	250	97	Crossbill	Call	422
40	Collared Dove	Song	251	98	Yellowhammer	Song	431
41	Turtle Dove	Song	252	99	Corn Bunting	Song	437

For optimum use, please play the enclosed CD on an audio-CD player. If used on a PC, please ensure your internet connection is switched off so that the correct track listing displays. The publisher is not responsible for incorrect track listings that have been posted on internet sites.

CONTENTS

How This Book Works 6

INTRODUCTION
Evolution 8
Anatomy 10
Life Cycle 12
Courtship and Mating 14
Nests and Eggs 16
Plumage 18
Flight 20
Feeding 22
Song and Calls 24
Migration 26
Western Palearctic 28
Wetlands 30
Estuaries and Low-lying
    Coasts 32
Rocky Coasts, Islands,
    and the Open sea 34
The Far North 36
Northern Mountains and
    Moorlands 38
Southern Mountains
    and Crags 40
Lowland Heath and
    Mediterranean Scrub 42
Farmland and Grassland 44
Forest and Woodland 46
Gardens, Parks, and Towns 48
Watching Birds 50
Conservation 52

SPECIES GUIDE
Wildfowl 55
Gamebirds 90
Divers and Grebes 101
Petrels and Shearwaters 111
Gannets, Cormorants,
    and Pelicans 117
Bitterns and Herons 121
Flamingos 131
Storks 131
Birds of Prey 136
Rails, Crakes, and Coots 163
Cranes and Bustards 168
Waders 173
Auks 214
Skuas, Terns, and Gulls 220
Pigeons and Doves 247
Cuckoos 247
Owls 254
Nightjars 255
Swifts 266
Kingfishers, Bee-Eaters,
    Rollers, and Hoopoe 270
Woodpeckers and
    Wryneck 275

Orioles 285
Shrikes 285
Crows 291
Tits and Allies 304
Larks 314
Martins and Swallows 322
Warblers and Allies 328
Waxwings, Wallcreepers, Nuthatches,
    and Treecreepers 357
Wrens, Starlings, and
    Dippers 363
Chats and Thrushes 368
Accentors 389
Flycatchers 389
Sparrows 394
Wagtails and Pipits 399
Finches 410
Buntings 428

Rare Species 438
Vagrants 487

Glossary 500
Index 501
Acknowledgments 511

HOW THIS BOOK WORKS

This guide covers just under 800 bird species from the Western Palearctic region (Europe, the Middle East, and North Africa). The species are organized into three sections: the first profiles common European species, with each given detailed, full-page treatment; the second covers over 190 rarer birds in concise, quarter-page entries; the third section consists of a list of rare visitors (vagrants) as well as birds that live in North Africa and the Middle East.

▽ INTRODUCTION

The species are organized conventionally by order and family, with a group introduction. The birds of a family are shown together, with annotations highlighting key distinguishing features.

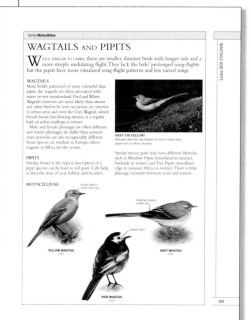

Family **Motacillidae**

WAGTAILS AND PIPITS

WHILE SIMILAR TO LARKS, these are smaller, slimmer birds with longer tails and a more steeply undulating flight. They lack the larks' prolonged song-flights but the pipits have more ritualized song-flight patterns and less varied songs.

WAGTAILS
More boldly patterned or more colourful than pipits, the wagtails are often associated with water or wet meadowland. Pied and White Wagtail, however, are more likely than almost any other bird to be seen on tarmac or concrete in urban areas and even the Grey Wagtail, which breeds beside fast-flowing streams, is a regular bird on urban rooftops in winter.
Male and female plumages are often different and winter plumages are also duller than summer ones; juveniles are more recognizably different. Some species are resident in Europe, others migrate to Africa for the winter.

PIPITS
Streaky brown is the typical description of a pipit species can be hard to tell apart. Calls help, as does the time of year, habitat, and location.

GREY OR YELLOW?
Although called the Grey Wagtail, this bird confuses many people with its yellow coloration.

Similar species pairs may have different lifestyles, such as Meadow Pipits (moorland in summer, lowlands in winter) and Tree Pipits (woodland edge in summer, Africa in winter). There is little plumage variation between sexes and seasons.

MOTACILLIDAE

YELLOW WAGTAIL
GREY WAGTAIL
PIED WAGTAIL

WAGTAILS AND PIPITS

399

MAPS

Each profile includes a map showing the range of the species, with colours reflecting seasonal movements. Migration ranges are not always mapped, as some migrants simply leave one site, turn up in another, and are not seen in between.

KEY
■ *Summer distribution*
■ *Resident all year*
■ *Winter distribution*
■ *Seen on migration*

COLOUR BAND
The information bands at the top and bottom of each entry are colourcoded for each family.

COMMON NAME

IN FLIGHT
Illustrations show the bird in flight, from above and/or below (note that differences of season, age, or sex are not always visible in flight).

DESCRIPTION
Conveys the main features and essential character of the species including:

VOICE: *a description of the species' calls and songs.*

NESTING: *the type of nest and its usual location; the number of eggs in a clutch; the number of broods in a year; the breeding season.*

FEEDING: *how, where, and what the species feeds on.*

HABITAT/BEHAVIOUR
Additional photographs show the species displaying typical behaviour in one of its preferred habitats.

SIMILAR SPECIES
Similar-looking species are identified and the key differences pointed out.
♂ *= male,* ♀ *= female*

SUBSPECIES
Panels show significant subspecies, together with distribution and distinguishing features.

▽ COMMON SPECIES

The main section of the book features the 330 most commonly seen European bird species. Each entry follows the same easy-to-access structure.

Order **Passeriformes**   Family **Motacillidae**

Yellow Wagtail 🔊 56

green

two white bars on blackish wings
bright yellow stripe over eye

MALE (SPRING)

green back

IN FLIGHT

MALE (SPRING)

white sides to black tail

long, spindly black legs

Elegant and colourful, the Yellow Wagtail, particularly the summer male, is a highly distinctive bird. However, autumn birds, especially juveniles, cause confusion with rarer species and also juvenile Pied Wagtails, which can appear strongly yellowish. The call always helps to identify it. In summer, it lives around pools and reservoirs and damp, grassy fields where horses and cattle disturb the insects it eats. In winter, it is found near big mammal herds on African plains.
VOICE Call distinctive, loud, full, flat, or rising *tsli*, or *tsweep* or *tswi-eep*; song repetition of brief, chirping phrases.
NESTING Grassy cup in vegetation on ground; 5 or 6 eggs; 2 broods; May–July.
FEEDING Forages on ground, skipping and leaping after flies in short flycatching sallies; eats insects and other invertebrates.

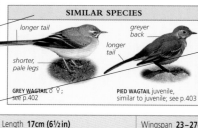

SIMILAR SPECIES

longer tail
greyer back
longer tail

shorter, pale legs

glossy crown

GREY WAGTAIL ♂ ♀; see p.402
PIED WAGTAIL juvenile, similar to juvenile; see p.403

S
M. f. f (C Euro
blue-g and ch

M. f. fe (SE Euro

Length **17cm (6½ in)**	Wingspan **23–27cm (9–10½**
Social **Small flocks**	Lifespan **Up to 5 years**

LENGTH, WINGSPAN AND WEIGHT: *length is tip of tail to tip of bill; measurements are averages or ranges.*

SOCIAL: *the social unit the species is usually found in.*

LIFESPAN: *the average or maximum life expectancy.*

STATUS: *the conservation status of the species; the symbol † means the data available can only suggest a provisional status.*

CLASSIFICATION
The top band of each entry provides the scientific order, family, and species names (see p.500 for full definitions of these terms).

GROUP NAME
The common name of the group the species belongs to is at the top of each page.

Family **Motacillidae** Species *Anthus richardi*
Richard's Pipit
This is a large pipit, Skylark-like (see p.320) in its size, bulk, and general plumage. It often stands upright, breast pushed out, on long, thick legs, its bold, strong bill quite distinct. There is no trace of a crest. The face is marked by a bold whitish area around the eye with a dark mark beneath; there is a short black line on each side of the throat (less marked on a Tawny Pipit, see p.404). The long tail is often bobbed.
OCCURRENCE Regular, but scarce, late autumn migrant in NW Europe, from Asia.
VOICE Loud, rasping *schreep* and quieter variations.

long blackish tail with white sides
very long hind claws

Length 17–20cm (6½–8in) Wingspan 29–33cm (11½–13in)

Family **Motacillidae** Species *Anthus godlewski*
Blyth's Pipit
Only recently observed with any regularity in Europe, Blyth's Pipit is difficult to identify, resembling Richard's and juvenile Tawny Pipits (see p.404). It is fractionally smaller than Richard's, with a shorter tail, a slightly richer underside colour, a slightly shorter, pointed bill, and shorter hind claws. It may look more subtly like a small pipit species and more wagtail-like than Richard's, but only close observation and several clearly heard calls can separate them for certain.
OCCURRENCE Very rare vagrant in NW Europe, from Asia.
VOICE Slightly higher than Richard's, less explosive, with fading, breathy quality, *psh-ev*.

short tail
wagtail-like shape

Length 15–17cm (6–6½in) Wingspan 26–30cm (11–12in)

Family **Motacillidae** Species *Anthus hodgsoni*
Olive-backed Pipit
Looking rather dark and uniform above or bright and streaked in front in a brief view, this pipit reveals a subtle pattern on closer examination. It is greenish, with very soft streaking above, and has a dark cap, a broad, bright cream stripe above the eye, a dark stripe through the eye, and a cream spot on the ear coverts. The underside is bright yellow-buff to buff with bold black streaks. It frequently walks in longish vegetation, bobbing its tail, but flies into trees if disturbed.
OCCURRENCE Rare vagrant in NW Europe from Asia, mostly in late autumn.
VOICE Tree Pipit-like hoarse or buzzing *spes* or *tess*.

bold dark streaks on pale underside

Length 14–15cm (5½–6in) Wingspan 24–27cm (9½–10½in)

Family **Motacillidae** Species *Anthus gustavi*
Pechora Pipit
Slim and streaky like most pipits, the Pechora Pipit has bolder pale stripes on the back (edged black), striking white wingbars, a buff breast, a white belly streaked with black, and a pinkish-based bill. It is difficult to see well and crouches when disturbed. The breast/belly contrast, wingbars, and longer wingtips help separate it from a juvenile Red-throated Pipit (see p.407).
OCCURRENCE Rare vagrant in autumn in NW Europe, from Asia.
VOICE Short, slightly buzzed, clicking *dzrp*, not often heard.

bold streaks
two wingbars

Length 14–15cm (5½–6in) Wingspan 23–25cm (9–10in)

471

◁ **RARE SPECIES**
Over 190 less common birds are presented on pp. 438–486. Arranged in the same group order used in the main section, these entries consist of one clear photograph of the species accompanied by a focused description.

PHOTOGRAPHS
These illustrate the species in different views and plumage variations. Significant differences relating to age, sex, and season are shown and the images labelled accordingly; if there is no variation, the images have no label. Unless stated otherwise, the bird shown is an adult.

FLIGHT PATTERNS
This feature illustrates and briefly describes the way the species flies. See panel below.

▷ **VAGRANTS**
Very rare visitors and peripheral species are listed at the back of the book with a brief description, including where the species is from.

MAPS
See panel on p.6

VAGRANTS
The list that follows consists of birds that occur only very rarely in Europe, known as vagrants or accidentals. It also looks a little further beyond Europe, to list those birds whose normal range is the Middle East and North America and covering a total faunal area known as the Western Palearctic.
Vagrants to Europe arrive from Asia and North America (and fewer from Africa). Western Europe, especially the UK, is well placed to receive birds that are blown off course from eastern North America and cross the Atlantic. It was thought that such birds cross the ocean on board ship, but it is now accepted that even small birds can, with a following wind, survive a flight across the Atlantic, although they probably do not survive long afterwards. Larger species, however, such as some wildfowl, may live for years in Europe and a few (that have been trapped, ringed, and released in order to follow their movements) have even returned to North America in subsequent years. These are not, in any true sense, European birds, but are included here to complete the range of species that have been recorded. Many appear again, others may not; by their nature these "accidentals" are unpredictable.

Common Name	Scientific Name	Family/Scientific Name	Description
Wildfowl			
White-faced Whistling Duck	Dendrocygna viduata	Wildfowl/Anatidae	Large, noisy duck, vagrant in North Africa, from southern Africa
Lesser Whistling Duck	Dendrocygna javanica	Wildfowl/Anatidae	Small duck from Africa
Bar-headed Goose	Anser indicus	Wildfowl/Anatidae	Pale grey goose from Asia
Spur-winged Goose	Plectropterus gambensis	Wildfowl/Anatidae	Large goose from Africa
Cotton Pygmy-goose	Nettapus coromandelianus	Wildfowl/Anatidae	Small duck, vagrant in North Africa, from southern Africa
Baikal Teal	Anas formosa	Wildfowl/Anatidae	Colourful surface-feeding duck from Asia
Cape Teal	Anas capensis	Wildfowl/Anatidae	Surface-feeding duck, vagrant in North Africa, from southern Africa
Red-billed Teal	Anas erythrorhyncha	Wildfowl/Anatidae	Surface-feeding duck, vagrant in North Africa, from southern Africa
Cape Shoveler	Anas smithii	Wildfowl/Anatidae	Surface-feeding duck, vagrant in North Africa, from southern Africa
Southern Pochard	Netta erythrophthalma	Wildfowl/Anatidae	Diving duck, vagrant in North Africa, from southern Africa
Canvasback	Aythya valisineria	Wildfowl/Anatidae	Large, pale Pochard-like duck from North America
Redhead	Aythya americana	Wildfowl/Anatidae	Pochard-like diving duck, vagrant in UK, from North America
Gamebirds			
Caucasian Grouse	Tetrao mlokosiewiczi	Grouse/Tetraonidae	Sleek black grouse, in Middle East
Caspian Snowcock	Tetraogallus caspius	Gamebirds/Phasianidae	Large mountain grouse, in Middle East

487

CLASSIFICATION *(left panel)*

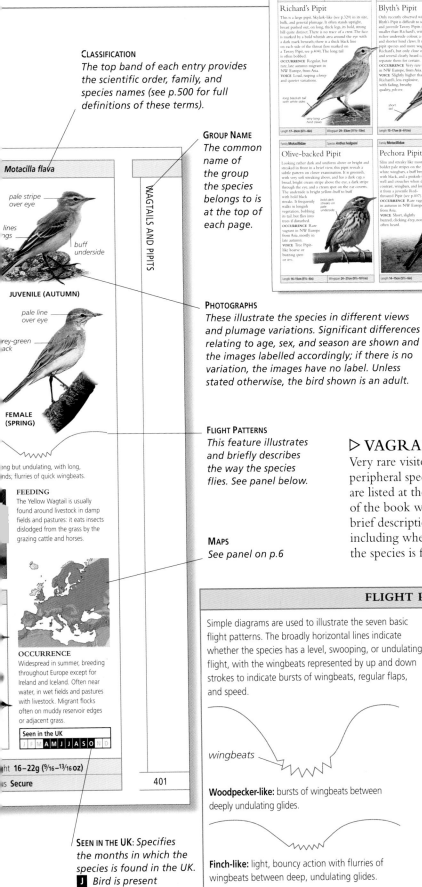

Motacilla flava

pale stripe over eye
lines ...ngs
buff underside

JUVENILE (AUTUMN)

pale line over eye
...rey-green ...ack

FEMALE (SPRING)

...ng but undulating, with long, ...nds; flurries of quick wingbeats.

FEEDING
The Yellow Wagtail is usually found around livestock in damp fields and pastures: it eats insects dislodged from the grass by the grazing cattle and horses.

OCCURRENCE
Widespread in summer, breeding throughout Europe except for Ireland and Iceland. Often near water, in wet fields and pastures with livestock. Migrant flocks often on muddy reservoir edges or adjacent grass.

Seen in the UK
| J | F | M | A | M | J | J | A | S | O | N | D |

...ht **16–22g** (9/16–13/16 oz)
...s **Secure**

401

SEEN IN THE UK: *Specifies the months in which the species is found in the UK.*
J Bird is present
J Bird not present

FLIGHT PATTERNS
Simple diagrams are used to illustrate the seven basic flight patterns. The broadly horizontal lines indicate whether the species has a level, swooping, or undulating flight, with the wingbeats represented by up and down strokes to indicate bursts of wingbeats, regular flaps, and speed.

wingbeats

Woodpecker-like: bursts of wingbeats between deeply undulating glides.

Finch-like: light, bouncy action with flurries of wingbeats between deep, undulating glides.

Sparrowhawk-like: straight, with several quick, deep beats between short, flat glides.

Gull-like: continually flapping, with slow, steady wingbeats.

Duck-like: continually flapping, with fast wingbeats.

Kite-like: deep, slow wingbeats between soaring glides.

Swallow-like: swooping, with burst of wingbeats between glides.

EVOLUTION

Most scientists believe birds evolved from dinosaurs. Whether the first birds climbed trees and began to glide back to the ground, or ran along the ground and learned to lift off, is still debated. What we do know is that, over hundreds of millions of years, birds spread over the globe and adapted to exploit every habitat except deep underwater. Some species are still evolving separate subspecies today.

GREAT AUK EXTINCTION
Extinction is forever: the total loss of a species from the world. Modern rates of extinction are exceptionally high and still accelerating. In Europe, however, we have lost only one species in historical times: the Great Auk. This large, flightless relative of the Razorbill bred in Scotland but was hunted until few remained; the final survivors were killed by collectors of stuffed birds. The last pair to be seen alive were then killed off Iceland in 1844.

THE MISSING LINK

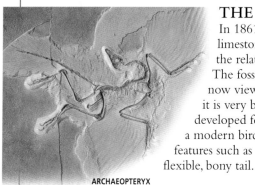

In 1861 a series of fossils were found in limestone beds in Germany that pointed to the relationship between dinosaurs and birds. The fossilized creature, named *Archaeopteryx*, is now viewed as the "missing link" because it is very bird-like in appearance, with well-developed feathers that are just like those of a modern bird, and yet still has many reptilian features such as teeth in its jaws and a long, flexible, bony tail.

ARCHAEOPTERYX

BIRD ANCESTRY

This table shows the relationships between modern birds (all within the sub-class Neornithes), and their age. Using the branches of the evolutionary tree we can see how different groups split from common ancestors. Songbirds divided off from the rest more than 100 million years ago.

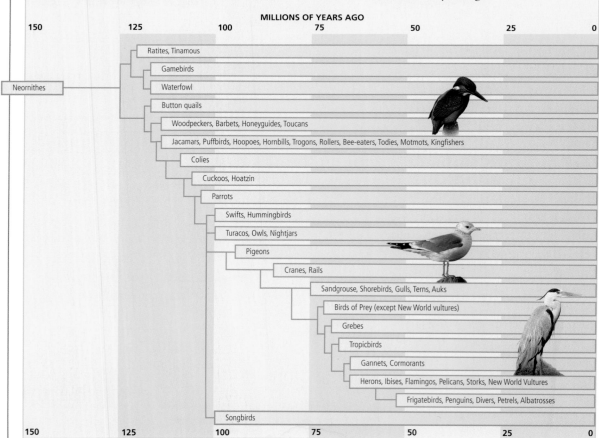

SPECIES AND VARIETIES

The birds in this book are species – not varieties or breeds. Species retain their identity over millions of years. They evolved and separated from each other to such a degree that each will only breed with a mate of its own species. Hybrids are infertile and quickly die out.

Scientists in the past looked at a bird's appearance, and studied behaviours, calls, habitat, and geographical range. Based on these criteria, they defined most of the species that we now know, such as Robin, Mallard, and Chaffinch.

Species are grouped into genera. Each genus contains close relatives, such as the Chaffinch and Brambling – which are clearly alike. There are variations within species – referred to as subspecies or races – and at what point a race becomes a species can be debatable. Modern techniques, using DNA studies, have changed many long-held opinions, but recent splits may have gone too far. Hence, the redpoll group, split into one or two species of Arctic Redpoll, and Common Redpoll and Lesser Redpoll, may be just one or two species with regional variations. There is still much that is not certain in bird classification.

CHAFFINCH

BRAMBLING

CLOSE RELATIVES
The Chaffinch and the Brambling are both species within the genus *Fringilla*. The resemblance between the two in size, structure, and pattern is remarkably obvious. Most other finches belong to the genus *Carduelis*.

NATURAL SELECTION

"Only the fittest survive." That is the theory of natural selection, and it is true that the birds best adapted to an environment have an advantage over the rest and produce offspring that perpetuate their own genetic make-up.

There are many ways in which a species may evolve in response to changing habitats or food. If such a change happens in only one part of a species' range, and the birds in that area remain separate, they may change so much that they can no longer interbreed. The result is two species instead of one. There is a halfway stage: two groups may differ in size or colour, but can still interbreed. At this point they are called subspecies or races.

PERSISTENT STARLING
The common Starling is successful through most of Europe but does not breed in Iberia and North Africa, where the Spotless Starling replaces it. This latter species must have evolved in isolation but has persisted despite an influx of common Starlings into Spain every winter.

CLASSIFYING BIRDS

The purpose of classifying birds is to indicate the relationship between them while at the same time giving each species a unique name. Birds are grouped into orders, each with one or more families. The next subdivision is the genus, which has one or more species (denoted by a two-part name). Some species are split into subspecies, indicated by a third name.

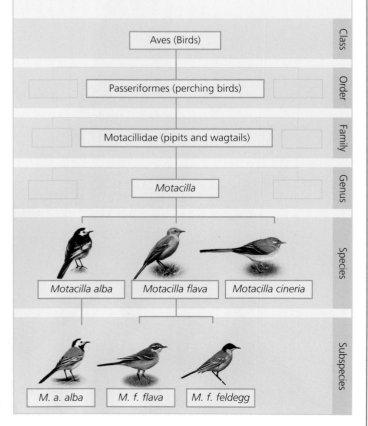

Aves (Birds) — Class

Passeriformes (perching birds) — Order

Motacillidae (pipits and wagtails) — Family

Motacilla — Genus

Motacilla alba | *Motacilla flava* | *Motacilla cineria* — Species

M. a. alba | *M. f. flava* | *M. f. feldegg* — Subspecies

ANATOMY

Although there is a huge diversity in shape, size, and outer appearance, all birds have a similar internal structure. They are vertebrates, so have a jointed internal skeleton with two forelimbs and two hindlimbs, just like fish, reptiles, and mammals – including humans.

Their bones are like ours, but the proportions are very different. For example, the wing bones are like a human arm; the "inner wing" equivalent to our forearm, and the "outer wing" like the bones of our hand. What makes them unique is that they are the only animals with feathers.

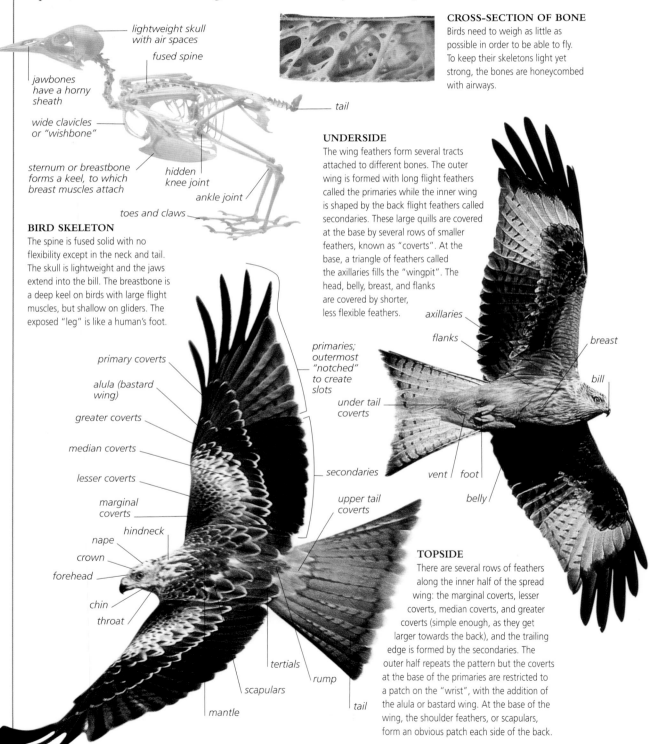

lightweight skull with air spaces

fused spine

jawbones have a horny sheath

wide clavicles or "wishbone"

sternum or breastbone forms a keel, to which breast muscles attach

hidden knee joint

ankle joint

tail

toes and claws

CROSS-SECTION OF BONE
Birds need to weigh as little as possible in order to be able to fly. To keep their skeletons light yet strong, the bones are honeycombed with airways.

BIRD SKELETON
The spine is fused solid with no flexibility except in the neck and tail. The skull is lightweight and the jaws extend into the bill. The breastbone is a deep keel on birds with large flight muscles, but shallow on gliders. The exposed "leg" is like a human's foot.

UNDERSIDE
The wing feathers form several tracts attached to different bones. The outer wing is formed with long flight feathers called the primaries while the inner wing is shaped by the back flight feathers called secondaries. These large quills are covered at the base by several rows of smaller feathers, known as "coverts". At the base, a triangle of feathers called the axillaries fills the "wingpit". The head, belly, breast, and flanks are covered by shorter, less flexible feathers.

axillaries

flanks

breast

bill

primaries; outermost "notched" to create slots

under tail coverts

secondaries

vent foot

belly

primary coverts

alula (bastard wing)

greater coverts

median coverts

lesser coverts

marginal coverts

hindneck

nape

crown

forehead

chin

throat

tertials

scapulars

mantle

upper tail coverts

rump

tail

TOPSIDE
There are several rows of feathers along the inner half of the spread wing: the marginal coverts, lesser coverts, median coverts, and greater coverts (simple enough, as they get larger towards the back), and the trailing edge is formed by the secondaries. The outer half repeats the pattern but the coverts at the base of the primaries are restricted to a patch on the "wrist", with the addition of the alula or bastard wing. At the base of the wing, the shoulder feathers, or scapulars, form an obvious patch each side of the back.

WING MARKINGS

Look at the feather tracts on a bird when perched and in flight. On some species most feathers are visible on the closed wing. On others, such as this Gull-billed Tern, the primaries and secondaries are hidden: all we can see are the primary tips. A large area between the back and the wingtips is formed by rounded, plain grey feathers called the tertials. On some birds these are large and obvious but in flight, as the wings straighten, they may slide out of sight under the scapulars. Therefore what is a prominent feature at rest may disappear in flight.

"wrist"

tertials hidden in flight

"elbow"

secondaries now visible on spread wing

full length of primaries revealed on open wing

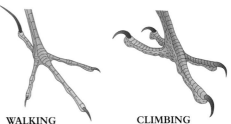

scapulars

primary tips

tertials cover half of primaries

wing coverts cover secondaries

OPEN AND CLOSED WINGS
This Gull-billed Tern has very long, tapered, pointed wings that reach well beyond its tail when folded but extend to reveal the obvious "wrist" (or carpal) joint and the elbow joint closer to the body. Only the fore edge of the inner half of the wing has any solid muscle; the rest is just feathers.

FEET

The shape of the feet indicates the lifestyle of a bird. For example, webbed feet or toes with broad lobes each side aid swimming, while feathered feet help prevent heat loss. There are variations, but below are four of the main shapes and details of the actions they enable.

WALKING
The long hind claw is typical of small birds, such as pipits, that run or walk in grassy places.

CLIMBING
Two backward-facing toes and strong claws allow climbing birds to grip onto a branch.

HUNTING
Sharp, hooked claws grasp live prey; a strong grip makes the hind claw a lethal weapon.

SWIMMING
Webbed feet provide extra thrust under the water's surface, which is ideal for swimming.

FEATHERS

Feathers not only allow flight and keep a bird warm and dry, but they also add a variety of colour, pattern, and shape. Some develop purely for decoration, while others provide cryptic patterns to help the bird avoid predators. The large, stiff quills that support a bird in the air, the "flight feathers", and the equally large tail feathers, are usually wider on one side than the other to create the aerofoil shape that gives a bird extra lift. They are overlain at the base by smaller "coverts". The feathers that smooth the shape of a bird's body are the contour feathers, while loose down feathers form an insulating underlayer.

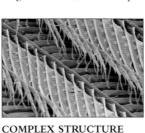

COMPLEX STRUCTURE
Feathers are amazingly complex. This close-up shows that the vanes each side of the central shaft "zip" together with minute hooks and barbs.

outer web

inner web

shaft

DOWN FEATHER

CONTOUR FEATHER

PRIMARY COVERT

TAIL FEATHER

LIFE CYCLE

A bird's appearance can vary significantly as a result of age or seasonal change. Newly hatched, chicks may be naked or downy. The down is quickly replaced by a first set of feathers, called the juvenile plumage. In autumn, some of these feathers are moulted and replaced (the wing and tail feathers are usually retained) to produce a first winter plumage. In the following spring, a partial moult produces the first summer plumage. From late summer onwards, all the feathers are replaced in a complete moult. Smaller birds may now be in their adult plumage; larger species, such as many of the gulls and birds of prey, have more intermediate (immature) stages: second winter, second summer, third winter,

third summer, and so on (as illustrated here by the Black-headed Gull, whose sequence of plumages is related to the seasons). There are variations on the theme. For example, while most birds have their brightest plumage in summer, wildfowl are at their best while pairing up in midwinter; the males become dull in summer, in an "eclipse" plumage.

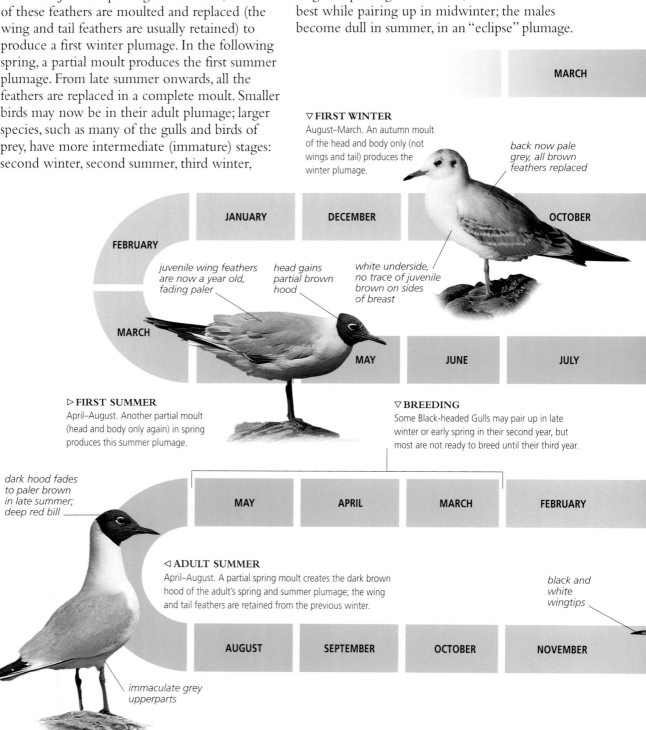

MARCH

▽ **FIRST WINTER**
August–March. An autumn moult of the head and body only (not wings and tail) produces the winter plumage.

back now pale grey, all brown feathers replaced

JANUARY **DECEMBER** **OCTOBER**

FEBRUARY

juvenile wing feathers are now a year old, fading paler

head gains partial brown hood

white underside, no trace of juvenile brown on sides of breast

MARCH

MAY **JUNE** **JULY**

▷ **FIRST SUMMER**
April–August. Another partial moult (head and body only again) in spring produces this summer plumage.

▽ **BREEDING**
Some Black-headed Gulls may pair up in late winter or early spring in their second year, but most are not ready to breed until their third year.

dark hood fades to paler brown in late summer; deep red bill

MAY **APRIL** **MARCH** **FEBRUARY**

◁ **ADULT SUMMER**
April–August. A partial spring moult creates the dark brown hood of the adult's spring and summer plumage; the wing and tail feathers are retained from the previous winter.

black and white wingtips

AUGUST **SEPTEMBER** **OCTOBER** **NOVEMBER**

immaculate grey upperparts

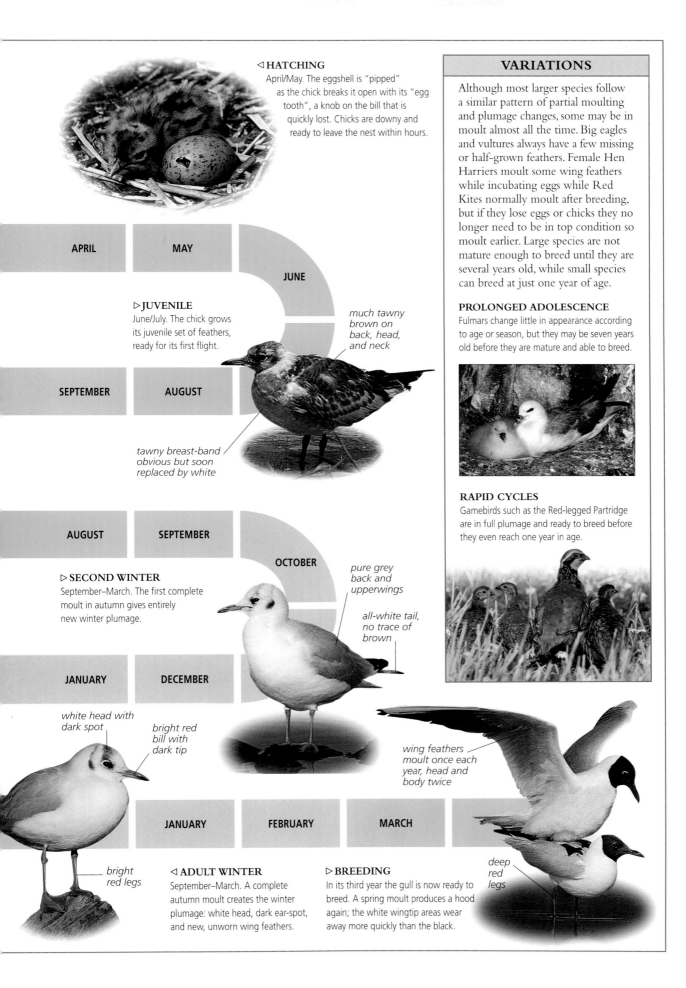

◁ **HATCHING**
April/May. The eggshell is "pipped" as the chick breaks it open with its "egg tooth", a knob on the bill that is quickly lost. Chicks are downy and ready to leave the nest within hours.

APRIL   MAY

JUNE

▷ **JUVENILE**
June/July. The chick grows its juvenile set of feathers, ready for its first flight.

much tawny brown on back, head, and neck

SEPTEMBER   AUGUST

tawny breast-band obvious but soon replaced by white

AUGUST   SEPTEMBER

OCTOBER

▷ **SECOND WINTER**
September–March. The first complete moult in autumn gives entirely new winter plumage.

pure grey back and upperwings

all-white tail, no trace of brown

JANUARY   DECEMBER

white head with dark spot

bright red bill with dark tip

bright red legs

wing feathers moult once each year, head and body twice

JANUARY   FEBRUARY   MARCH

◁ **ADULT WINTER**
September–March. A complete autumn moult creates the winter plumage: white head, dark ear-spot, and new, unworn wing feathers.

▷ **BREEDING**
In its third year the gull is now ready to breed. A spring moult produces a hood again; the white wingtip areas wear away more quickly than the black.

deep red legs

VARIATIONS

Although most larger species follow a similar pattern of partial moulting and plumage changes, some may be in moult almost all the time. Big eagles and vultures always have a few missing or half-grown feathers. Female Hen Harriers moult some wing feathers while incubating eggs while Red Kites normally moult after breeding, but if they lose eggs or chicks they no longer need to be in top condition so moult earlier. Large species are not mature enough to breed until they are several years old, while small species can breed at just one year of age.

PROLONGED ADOLESCENCE
Fulmars change little in appearance according to age or season, but they may be seven years old before they are mature and able to breed.

RAPID CYCLES
Gamebirds such as the Red-legged Partridge are in full plumage and ready to breed before they even reach one year in age.

13

COURTSHIP AND MATING

Birds spend most of their time keeping their distance from each other. To breed, they must break down barriers so that they can come into contact, if only briefly. Those that rear young together as pairs need a stronger, longer-lasting pair bond, so that they can rely on one another to risk their lives for the sake of their young. Courtship must help foster this trust.

COURTSHIP

Courtship has two functions. It bonds pairs together in a lasting partnership but, initially, it also helps the female to choose which male to mate with. A female is impressed by the size, colour, ability of a male to perform complex courtship rituals, and sometimes even his competence at fighting off other males. These attributes indicate that a male is fit, strong, and efficient. A female will invest a huge amount of time and energy into the rearing of her chicks and therefore must make the right decision when choosing which male to breed with.

MUTUAL DISPLAY
Gannet s ritualized postures reinforce their commitment to their nest and to each other. They fence with raised bills in greeting and bow with open wings to show ownership of the nest. Fencing develops into nape nibbling, mutual preening, and mating.

CHASE AND DISPLAY
A male Redshank is determined to mate with the hen of his choice and must impress her with his persistence and colourful displays. She will eventually give in to him if she is suitably interested. Courtship displays continue for some weeks as the pair learn to trust each other and accept close contact in order to mate and share parental duties.

FIT TO BREED
Wildfowl have ritualized calls and displays specific to their species. This male Ruddy Duck is literally blowing bubbles to impress: he rattles his bill against his breast feathers, forcing air from between them into the water.

THE LEK
A lek is a communal display ground where males of some species, such as these Black Grouse, have mock battles. The outcome is serious, though: females choose the strongest, most dominant males to mate with.

TOP MALE
A male Pheasant calls and thrashes his wings, raising his tail to make himself look as big as he can. He does this in order to dominate other males and attract a hen. Once he has her attention, he will tilt towards her, drooping his nearside wing and spreading his tail, to show himself off to best advantage.

BREEDING

When it comes to reproducing, there isn't just one favoured strategy. Even within some species there is variation. The Dunnock may form a simple pair that stays together all summer, but some males have more than one mate and, indeed, so do some females. Even within apparently monogamous species, fidelity is not always the norm, and should one of the pair die, the survivor usually has little difficulty finding a new mate. Unless the population is in decline, there is usually a healthy surplus.

MATING

The act of mating is brief, but can be frequent. Ospreys mate scores of times during the egg-laying period, but a single mating is enough to fertilize a whole clutch of eggs. Most birds mate on the ground or a perch. However, Swifts may mate in the air whereas ducks, such as these Goldeneyes, mate on water.

MONOGAMOUS BONDS

Most birds are monogamous, although many are quick to seize the chance to mate with a passing stranger. It seems that an extra mate is viewed as insurance against possible failure, as it doubles the chance of finding a fit and successful partner. Some species, such as the Mute Swan, however, pair for life and maintain a year-round bond that is rarely broken.

MUTUAL PREENING

Strengthening the bond between a male and female bird takes many forms. These Guillemots are preening each other; such intimate contact means that all the usual barriers that keep individuals apart have been broken down while they are rearing their offspring.

PROMISCUITY

The male Capercaillie mates with many hens. This gives him as many chances as possible to sire healthy offspring. Afterwards, he has nothing more to do with the hens.

REVERSAL OF ROLES

In a few species, such as the Dotterel, the female is larger and brighter than the male. She lays a clutch of eggs for him to incubate, then goes off to find another male.

NESTS AND EGGS

A bird's life, behaviour, and appearance revolves around finding a mate and producing eggs in order to ensure the survival of its genes. The nest is a safe place to lay those eggs, incubate them, and raise the resulting brood, so individuals will spend time locating and building the perfect nest for their situation.

BUILDING A NEST

The birds within a species will create nests that are remarkably constant in terms of size, shape, structure, and the materials from which they are made; building such nests seems to be instinctive. However, different birds use an extraordinary variety of techniques to create a whole range of structures. Some of these are little more than scrapes in the ground with a few pebbles or shells as lining. Others are hugely complex, and some are masterpieces of construction and effective camouflage.

COLLECTING MATERIAL
Puffins line their burrows with grass and scraps of vegetation from nearby slopes

CUP NEST
Most small birds make an open, cup-shaped nest that has a rough base, a neat superstructure, and a fine, soft, warm lining for the eggs and chicks. This kind of nest can take a week or more to build.

DUPING

Some species habitually lay their eggs in other birds' nests and leave them to rear their young. Not all of these birds remain parasites at all times. Many ducks, and even Swallows and Starlings, lay eggs in other nests while still incubating a clutch in their own.

CUCKOO
The Cuckoo never makes its own nest. Once hatched in a foster-parent's nest, its chick throws all other eggs out to gain their sole attention.

EGGS AND HATCHING

TYPES OF EGGS
While most eggs are oval and have a camouflage pattern, there are variations (some examples pictured). Eggs laid out of sight in dark holes, for example, are white. Gamebirds and owls lay spherical eggs, while wading birds lay pear-shaped eggs. Aerial species such as Swifts have narrow bodies so they lay longitudinal eggs. The pear shape of the Guillemot's egg prevents it from rolling off the narrow cliff ledge it is laid on.

ELLIPTICAL   LONGITUDINAL

PEAR SHAPED

OVAL   SPHERICAL   CONICAL

NEAT ARRANGEMENT
Waders lay four eggs that fit neatly under the sitting birds body. The pointed ends also accommodate the long, folded legs of chicks that can run within hours of hatching.

THE HATCHING PROCESS
Chicks call to each other and to their parents from within the egg, helping to co-ordinate their hatching. They use a tiny "egg tooth" on the bill tip to break the shell and then struggle until they push the two ends apart.

CRACKING   FORCING   BREAKING OUT

NEST SITES FOR ALL PURPOSES

Most nests are vulnerable to predators, which may eat the eggs, chicks, and sometimes even the adult that is within the nest. Even wooden nestboxes may be raided by woodpeckers. So birds think about the safest place to locate their nests. Small birds tend to hide their nests away in thick bushes, or suspend them beneath the long branches of conifers. Many species nest inside holes that they either stumble across or excavate themselves in trees or earth banks. Larger species may rely on inaccessibility and make large nests of sticks in plain sight at the tops of trees.

TUNNELLING INTO SAND
Sand Martins dig a metre into a solid earth or soft sandstone cliff with their feet. The inner end of the tunnel then broadens out into a nest chamber that will house four or five chicks.

NO NEST
The Little Ringed Plover makes a shallow scrape in sand for its eggs, giving it little or no lining. When disturbed, the bird runs off and relies on the eggs' camouflage pattern to save them. While hawks and eagles make large nests, falcons never do: they lay their eggs straight onto a bare ledge.

PLASTERWORK
The Nuthatch uses a woodpecker's hole or a natural hole in a tree. It plasters the entrance with mud to get a perfect fit – just big enough for it to squeeze inside – which should protect the nest from predators.

TREE HOLE
Woodpeckers, such as this Lesser Spotted Woodpecker, excavate holes in living trees, using their chisel-like bills. The nest chamber is left unlined apart from a few chippings.

NESTING ON BUILDINGS
White Storks have long nested on buildings, especially on church towers, although some still use trees. In parts of Europe they use telegraph poles. In Spain, poles with cartwheels are provided specially for them.

FLOATING NEST
Black-necked Grebes build floating heaps of weed that are anchored to the bottom. If they have to leave the site, they cover their eggs with a few scraps of weed in order to hide them from predators.

COMPLEX STRUCTURE
The Long-tailed Tit's nest is a masterpiece of spiders' webs, moss, lichens, and feathers. It is hard to see, and stretches as the chicks grow bigger.

PLUMAGE

Feathers are unique to birds. They keep them warm, are lightweight, and add aerodynamics that allow flight. They also provide colours and patterns that are used for camouflage, display, and communication. Feathers are renewed at least every year by a process of moult, but older ones can look slightly worn and faded. All birds keep their plumage in as good a condition as possible by frequent preening and bathing in water or sometimes in dust.

MARKINGS

Feathers create complex patterns but these usually have regular structures. On the head, the crown may have a central stripe and darker sides; above the eye may be a superciliary stripe, while through the eye there could be an eye-stripe. Wings may have wingbars across the tips of the coverts or along the base of the flight feathers. There may be streaks, spots, or bars on the body. Each of these marks helps us to distinguish individual species.

superciliary stripe

eye-stripe

CIRL BUNTING

rump colour

hood or cap

wingbar

tail pattern

CHAFFINCH

DUNNOCK

lengthwise streaks

MISTLE THRUSH

spots

eye-ring

bars

RED-LEGGED PARTRIDGE

MALE AND FEMALE

Plumage differences between sexes may just be focused on the colour or pattern. However, they are often structural too as a lot of males are bigger than the females.

fanned tail to impress females

CAPERCAILLIE

Female

camouflage pattern protects from predators

Male

JUVENILE AND ADULT

Many young birds don't look like their parents. This is because young have no need to impress possible mates. It is more important for them to be camouflaged from predators and to avoid being mistaken for an intruding adult.

Juvenile

HERRING GULL

Adult

dark plumage prevents mistaken attack by territorial parent

white body gives long-distance visual contact when feeding

SUMMER AND WINTER

Summer plumage is intended to look good, to show off to other males and females. In winter this is less important – camouflage is a better option to keep safe from predators.

camouflaged, not needing to display

Summer

SNOW BUNTING

male at his most striking in summer

Winter

SUBSPECIES

Subspecies or races occupy isolated areas and may look subtly different. Birds tend to be larger in cold areas to reduce heat loss. The Wheatear found in Greenland, for example, is bigger than the European race.

browner above, brighter below

clean grey and pale buff

Greenland race

European race

WHEATEAR

MOULTING

Moult proceeds in a regular order, according to species. Small feathers are often replaced twice a year, large ones annually. Each old feather is pushed out when a new one grows at its base. The new feather begins as a tiny bump – a "goose bump" – on the skin and grows as a shiny sheath that bursts open at the tip to reveal the soft webs. On most birds moult is not easy to see, but on some larger ones, such as gulls, it is possible to spot gaps where feathers are missing. Old feathers usually become paler and pure colours, such as grey and green, tend to turn browner as the feathers age. These contrasts can sometimes be seen on a bird at close range, or in a photograph, but detailed studies of moult rely on trained bird ringers, who catch birds and examine their stage of moult before releasing them.

grey wing coverts of winter

boldly patterned new feathers give breeding colours

BLACK-TAILED GODWIT IN MOULT
This Godwit is midway between grey winter and red-brown summer plumages. Look at the old, worn, greyish feathers, which are being replaced by new red ones with dark centres and fresh white tips.

WOOD PIGEON
Look carefully to find dull, old feathers with frayed edges amongst new, fresh, grey ones with neat white fringes.

fresh grey next to old browner feathers

new wingtip feathers are dark, old ones faded

ABERRATIONS AND VARIATIONS

Not all birds are perfect examples of their species. Some are "aberrant", because of some genetic deficiency. All-white "albinos" rarely survive long. "Leucistic" birds are white, or patchy white, through lack of pigment, while others ("greying") become white with age. "Melanistic" individuals have too much dark pigment, and look dark brown or black. Other variations are a normal part of everyday life. For example, feathers naturally bleach and fade in the sun and wear away at the tips during the course of months of wear.

WHITE BLACKBIRD
The normally black Blackbird quite often throws up partly white variants. Such birds lack inherited pigment, or lose it as they grow older.

WEAR AND TEAR

Feathers are tough, but a year of hard wear takes its toll. Like our own fabrics, feathers are bleached paler by the sun, while their inner edges, usually tucked out of sight, stay fresh. New feathers have sharp, neat, and tidy fringes, but gradually fade and become a little ragged. On some birds, such as finches and some wading birds, pale edges crumble away in spring to reveal brighter breeding season colours beneath, with no need for moult.

CAMOUFLAGE

A major function of plumage pattern is to break up the shape of a bird so it is hard to see. Both predators and prey use camouflage: one to get close to its meal, the other to avoid detection. For example, Oystercatcher flocks confuse predators with a mass of dazzling white patches.

HIDING AWAY
Stripes on the front of a Bittern make it extremely hard to see when it stretches upright in the dead reed stems of winter. It is much easier to spot in the summertime.

FLIGHT

All the birds that are found within Europe can fly. It is this ability that allows them to travel the globe, moving far and wide to exploit seasonal abundances of food while escaping any shortages. It is fascinating that these small creatures are able to fly across oceans, mountains, and deserts, although such journeys often tax them to the limit. Their most significant features, which provide them with the means to fly and control the direction of that flight, are their wings and tail.

HOVERING LESSER KESTREL
Birds of prey can hover while they search for food. To do this, they position their head into the wind, flicker their wings, and fan their tail.

WING AND TAIL SHAPE

The shape of wings varies from species to species (as shown below), and is largely dependent on the type of flight used. Generally, long, narrow wings (such as a Swallow's) are most efficient for sustained, fluent, manoeuvrable flight. Long, broad wings enable many birds to ride the winds using little energy. Short, round wings allow for shorter bursts of rapid, whirring beats and provide fast acceleration – useful for ground-dwelling birds escaping predators. Tails are used for balance, steering, and braking; a long, stiff tail ensures good balance while a forked tail or long tail that can open and close like a fan acts as a rudder and a brake.

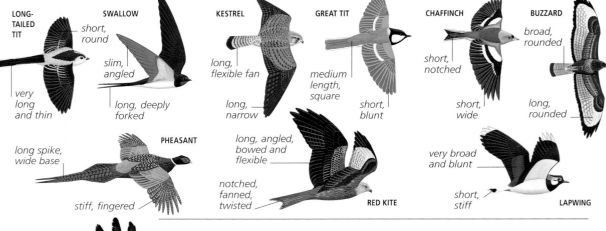

LONG-TAILED TIT — very long and thin

SWALLOW — short, round — slim, angled — long, deeply forked

PHEASANT — long spike, wide base — stiff, fingered

KESTREL — long, flexible fan — long, narrow

long, angled, bowed and flexible — notched, fanned, twisted — RED KITE

GREAT TIT — medium length, square — short, blunt

CHAFFINCH — short, notched — short, wide

very broad and blunt — short, stiff — LAPWING

BUZZARD — broad, rounded — long, rounded

EXPERT EAGLE
A White-tailed Eagle raises its wings, separates the wingtip feathers to allow air to slip through, fans its tail as a brake, then swings forward to strike with its feet. It uses powerful wing flaps to climb away from the water with its load.

BALANCE AND CONTROL

Large birds such as eagles and buzzards have a delicacy in the air that belies their size and shape. Their fingered primary feathers are "notched" on each side, creating slots at the wingtip to increase stability within flight and reduce turbulence. Their remarkable balance and precise control in flight allow them to home in on and catch their prey.

TAKING OFF

Getting airborne uses a lot of energy. Many birds take off into the wind from dry land, flapping their wings hard in a figure of eight pattern to create lift and forward propulsion; others jump from a clifftop or tree, moving forwards into the air. Most water birds need to run along the surface of the water in order to build up the necessary momentum.

TAKING OFF FROM LAND
This Grey Heron stretches forwards, pushes down as hard as it can with its wings, and leaps up with a powerful spring of its legs to rise into the air.

LANDING

Birds fly surprisingly fast and so have to reduce their speed quickly before landing. Most birds swing their bodies backwards and fan their tails to assist braking, flapping their wings against the direction of flight. Just before impact they thrust their feet forward to act as shock absorbers.

HEAVYWEIGHT SWAN
The Mute Swan is close to the upper weight limit for flight. It needs to run along the water to get aloft: its legs are too short to give much of a leap from dry land.

TAIL BRAKE
This Woodpigeon (below) has its head up, feet ready to push down, and its wings well back. Its broad tail, when fully spread, acts as an air brake and enables a safe landing.

CO-ORDINATION AND SPEED

Travelling in flocks requires great co-ordination. Each bird takes its cue from the bird ahead or to one side of it, so the decision of the leading bird to turn, rise, or fall carries fluidly through the flock. Many birds can fly at speed, but sustaining this is costly in energy.

SHORT BURSTS OF SPEED
Grey Partridges have deep breast muscles and short wings, which allow bursts of low, fast flight.

SETTLING ON WATER
Water cushions this Mallard's landing, while its webbed feet act as skis. Its wings beat forward and back to reduce the overall speed.

CO-ORDINATION
A mixed flock of Oystercatchers and Knots makes a remarkable sight in the air as it turns and twists like smoke, without a single collision. Such birds have super-quick reactions and tight control.

FEEDING

Birds eat a wide range of items from a variety of sources; some, for example, perch to eat berries while others dive into water to fish. The overall shape of a bird is a strong indication of what it eats and how it obtains its food. This is most obviously revealed in the shape of its bill, but there are also clues in its head and neck shape, and the length and shape of its legs and feet.

BLACKBIRD forages for worms and berries

BLUE TIT pecks at tiny insects and seeds

REED WARBLER picks up insects

CHAFFINCH cracks seeds and picks up caterpillars

BILL SHAPE

There are many subtle variations in bill shape, but there are a few basic forms that perform specific functions. For example, long, thin bills probe into soft mud and sand to grab worms, and thick beaks crack seeds or pluck grass. Hooked bills, on the other hand, can tear flesh, while saw-toothed bills are able to grasp and hold on to slippery fish.

CURLEW probes for worms in deep mud

GREYLAG GOOSE tears grass and roots

AVOCET sweeps sideways for tiny shrimps in water

MALLARD dabbles for seeds from water

GREY HERON grasps fish

GOLDEN EAGLE tears meat

DABBLING
"Dabbling" involves opening the beak while skimming it across the surface of the water. Dabblers filter water through a fine mesh at the sides of their bill in order to trap tiny seeds and organisms that they then swallow.

WATER FEEDERS

Birds use various methods to obtain food from water. Some wade or swim in order to pick insects from the surface and shore line. Others dive headlong into the water to catch fish. Razorbills actually dive as deep down as 100m (330ft) underwater, using their wings to propel them downwards. Grebes, divers, cormorants, and diving ducks are able to dive underwater from their sitting position on the surface, while other birds, such as swans, reach the bottom by "upending" and using their long necks.

UPENDING
This Shelduck is unable to reach the bottom by just dipping its head under the water's surface, so it gets a little deeper by swinging its whole body over and stretching its neck.

FISHING
A Kingfisher catches fish by grabbing them, not stabbing, despite its sharp bill. It usually drops from a perch, then flies back up with a fish in its beak. It will then beat the fish against a branch before swallowing it.

PROBING
The long bill of a Godwit is the ideal tool for pushing deep into soft mud in order to probe for worms and molluscs. However, if the ground is too hard it cannot feed.

LAND FEEDERS

Birds of all shapes and sizes are land feeders. Many, such as pigeons and finches, feed on the ground as well as in trees, depending on the time of year. Geese, some waders, and pipits are, on the other hand, exclusively ground birds; some birds, such as bustards and cranes, are so big that they have no option but to stay on the ground. Others, such as woodpeckers, prefer trees, and chip away at bark to locate insects. The type of food that they eat affects the way that bird communities feed. Some take in food that is spread over a wide area but is not very abundant, so they disperse themselves in order to avoid competing with each other. Others eat food, such as seeds, that is only located in a few places. It is often in plentiful supply, however, which allows them to feed in sociable flocks.

BERRY EATERS
Redwings and other thrushes gorge themselves in the autumn and early winter if the berry crop is good. If the crop fails then they must turn to worms and other food, or fly great distances in search of berries and fruit elsewhere.

TEARING FLESH
Birds of prey catch food with their feet, which have incredibly sharp talons. However, they often kill the prey and rip it apart using their hooked bills.

TEARING GRASS
Geese use their broad bills to tear at grass, shoots, and roots. This type of food is easy to find, so they are able to feed together without needing to defend their territories to gain sole use of the food supply.

PROBING IN THE GROUND
The Hoopoe uses its slim, faintly curved bill to probe into loose soil and under clumps of earth or roots to reach worms, grubs, and a variety of insects.

AERIAL FEEDERS

Catching insects in flight is a skilful operation, and is undertaken in different ways. Nightjars have tiny bills but wide mouths that are fringed with bristles – these allow them to catch flying moths after dark. Swifts catch tiny insects high in the sky by day while swallows chase bigger flies low down over meadows. Hobbies also catch bigger insects, as well as small birds, but use their feet to do so. All of these examples eat their prey in mid-air. Flycatchers and many warblers, however, return to a perch once they have snapped up their prey.

FLY-CATCHING EXPERT
The Spotted Flycatcher sits alert and upright on a perch until it spots a small moth or a fly. It darts out, twisting and turning with great agility, to catch the prey in its bill with a loud "snap", then returns to the same perch to eat its meal at leisure.

KLEPTOPARASITISM

Many birds are quick to take advantage of smaller, weaker species by stealing their food. This is called kleptoparasitism. For example, Great Black-backed Gulls harry Puffins returning to their burrows with fish while Black-headed Gulls chase lapwings in fields, forcing them to drop juicy worms. Sometimes even one bird of prey will rob another.

POWERFUL SKUAS
Great Skuas not only kill birds but also force down and steal food from other seabirds such as Kittiwakes, Fulmars, and even Gannets.

SONG AND CALLS

Birds' voices are remarkably far-carrying and can convey a variety of messages to other birds. A bird will use many vocal sounds but each type is normally constant within a species. This helps us to use song and calls to identify individual birds. Usually it is the males that sing to attract females for mating and to repel other males from their nesting area.

UNIQUE INSTRUMENTS

Birds have no voice box or larynx, but a muscular organ called the syrinx at the base of the windpipe. A series of membranes are stretched and relaxed by bunches of muscles, and vibrate as air passes across them. Some birds have a simple syrinx so produce little variety of song. But complex ones produce great variations in pitch and quality.

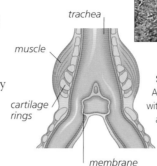

trachea

muscle

cartilage rings

membrane

SYRINX
A bird can use the muscles within the syrinx, which are attached to rings of cartilage, to change the sound that is produced.

KEEPING IN TOUCH
Many calls are contact notes, which are used by birds as they go about their everyday lives. Such calls help to keep flocks and family groups together as they feed or move about, even when they are within thick cover.

VARYING FUNCTIONS

It can be difficult to distinguish between a song and a call but basically they have different functions. A song is primarily used to attract a mate or to defend a territory. It can be varied and intricate, made up of a complex set of notes. Calls are usually simpler and are used to pass on information – such as an alarm call warning of a predator. Such calls are high and thin, to penetrate through dense woods. Birds also call in flight, purely to keep in touch.

AGGRESSIVE DEFENCE
Loud, harsh cries are given by terns, gulls, and skuas, such as this Long-tailed Skua, when they chase intruders that venture too close to their nests. Their alarm notes have an obvious urgency, sounding hysterical if their chicks are threatened.

CONSTANT REPETITION
The Song Thrush is easy to identify when in song. It sings a few notes – mellow or loud and challenging – in a short phrase. Each quick burst of notes is repeated two or three times before another theme is introduced.

UNMUSICAL PERFORMANCE
The Fulmar sits on its nesting ledge and greets its mate as it flies by, or settles alongside it, with a burst of raucous, throaty cackling. To us the calls are coarse and unmusical, but they probably help Fulmars to identify each other and are an important part of courtship.

SCREAMING DISPLAY

Swifts form high-speed "screaming parties" that dash around in the sky calling loudly. This practice seems to have a social significance within the breeding colony. They call as they approach the nest, too, alerting their partner to their imminent arrival.

HUNGER CALL

Baby birds stimulate their parents to feed them by calling loudly, just like a human baby cries for attention. They risk attracting a predator, so the parent is forced to provide food to keep them quiet.

WHERE ARE YOU?

Young birds in large colonies may wander away from the nest and can easily be lost. Only their own parents feed them. Loud, whining calls help these young gulls to keep in touch with their parents.

MECHANICAL SOUNDS

Not all sounds that birds create are vocal. Some species produce distinctive mechanical sounds during their display that have the same function as song, communicating with other birds within their species over long distances. The Snipe combines its bleating sound with a visual display, using a steeply undulating, switchback flight. Some pigeons clap their wings together in display but also use this same movement to raise an alarm. Certain owls and nightjars do a similar thing, clapping their wings beneath their bodies in display flights. The woodpecker uses a purely mechanical sound (see below).

DISPLAY DRUMMING

Snipe dive through the air with tail feathers outspread, creating a vibrant bleating noise known as "drumming".

DRUM ROLL

A different mechanical sound, also called "drumming", is made by a woodpecker. It creates a sudden burst of sound by rapidly vibrating its bill against a resonant branch. The abrupt drum roll carries well through a dense forest.

DAWN CHORUS

No-one is really sure why so many birds sing most persistently at dawn. It is a wonderful experience, especially in a large wood just as the sky brightens on a spring morning. Suddenly all the territorial birds for miles around sing loudly together; but the performance is short-lived and the song becomes erratic.

EARLY PERFORMER

In most areas the Robin is one of the dominant songsters in the dawn chorus. It also sings under street lights at night, apparently fooled by the artificial lighting.

MIGRATION

Northern Europe is full of insect food in summer, when long days allow birds time to feed both themselves and their young. Such resources are too good to miss, but few birds can survive in the far north in winter when the days become short and cold. Only by huge movements of millions of birds can they exploit feeding opportunities to the full. The regular rhythms of migration are an essential part of many species' existence.

INSTINCTIVE MOVE

Birds migrate by instinct. In many species adults and young birds migrate separately, but the youngsters are somehow programmed to move at the right time and to follow the right routes. Some gather into large flocks and migrate together. Others simply slip away one night alone, embarking on a journey that remains one of nature's most magical and dramatic undertakings. Before proceeding with such a journey a bird needs to ensure that it is in tiptop condition – this often requires some careful preparation.

MAKING THEIR WAY

Birds navigate in a number of ways. They certainly use the sun and the stars and may also see polarized light, which allows them to judge where the sun is even on cloudy days. They probably have a magnetic sense too, and may even be able to detect the position of the poles as they refer to the sky, which would indicate their latitude. What we still don't understand is how they know which way they should fly.

NIGHT MIGRANTS
Many songbirds, such as Goldcrests, migrate at night, relying on the stars to find their way. A sudden onset of cloud and fog may "ground" thousands of them along a coast, making exciting birdwatching the next morning.

A SIGN OF THE SEASONS
Migrating geese make a marvellous spectacle and sound. In parts of northern Europe they are a visible sign of the changing seasons as they migrate south in autumn and north in spring.

Canadian Arctic Islands

North America

PUTTING ON WEIGHT
Small warblers such as the Sedge Warbler double their weight before they migrate. Sedge Warblers eat aphids in reedbeds before flying across the Sahara in one flight that may last four days. Others feast on rich, sugary berries before they migrate. Such birds quickly put on layers of fat: essential fuel for their journeys.

WHEN TO DEPART
Changing day length in spring and autumn is more of a clue that the time is right to migrate than changes in temperature. A bird's internal clock takes note of the seasonal changes and stimulates a restlessness at migration time. It also starts off the hormonal changes that make physical adjustments, such as the accumulation of extra fat for long-distance journeys.

OVERLAND FLIGHTS
Broad-winged birds such as White Storks use up too much energy in flapping flight so must glide over long distances. To do this, they must be able to gain height, which they do by riding thermals or "bubbles" of warm, rising air. These only form over land, so the birds must cross the sea at the narrowest points, such as Gibraltar and Istanbul.

READY TO GO
Swallows and House Martins migrate by day, feeding on insects as they go. They gather in large flocks in autumn, before making a move together towards Africa.

GLOBETROTTERS

Waders and wildfowl are amazing travellers but so are some small, familiar birds. For example, Swallows from the United Kingdom travel to the far south of Africa in winter. Barnacle Geese fly north in spring to breed around the Arctic. Such birds that breed in the far north remain in the south till late in May, when the Arctic snows begin to melt. The map below shows three examples of long-distance migration.

DASH FOR THE NORTH

The Arctic Tern is one of the greatest globetrotters. It breeds in 24 hours of daylight in the Arctic, then spends the northern winter in 24-hour daylight, flying over southern oceans before making the journey north again.

ARCTIC OCEAN

Franz Joseph Island

Greenland

Svalbard

Novaya Zemlya

Jan Mayan

KARA SEA

BARENTS SEA

Iceland

NORTH
ATLANTIC

Europe

BLACK SEA

Azores

CASPIAN
SEA

Madeira

MEDITERRANEAN SEA

ATLANTIC
OCEAN

Canary Islands

Sahara Desert

Cape Verde
Islands

Africa

KEY

— Arctic Terns
— Knots
— Fieldfares

South
America

SOUTH
ATLANTIC

2,000km

2,000 miles

SWEEPING SOUTH

West Europe is on the great East Atlantic flyway, which is a migration route for birds from vast areas of the Arctic and northern Europe. The path of waders such as Knots makes a huge sweep southwards, emptying regions on both sides of the Atlantic in autumn.

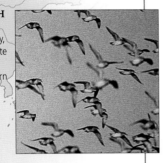

WINTER WANDERERS

Fieldfares move south and west in winter to avoid cold weather. In midwinter they may be forced to travel further still if there is severe cold or snow, but they return as soon as conditions improve. Several species are nomadic, wandering wherever there is food during the winter months.

PARTIAL MIGRANTS

Some species are resident, remaining in the same area all year round. Others are what are called partial migrants. This means that part of the the species population is resident, inhabiting an area that can sustain them year round, while the other part lives in less hospitable areas and so needs to migrate south during the winter months to find food.

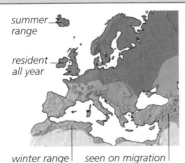

summer range

resident all year

winter range | seen on migration

MEADOW PIPIT

Meadow Pipits are summer visitors in some areas, resident in others. Those that breed in the north go south for winter (see map).

WESTERN PALEARCTIC

Europe, North Africa, and Asia (north of the Himalayas) form one large entity, called the Palearctic. It is identified by a characteristic set of plants and animals, which adhere to natural rather than political boundaries. This book focuses on the western part of this region – west of the Ural Mountains – giving an overview of the birdlife in an area extending slightly beyond Europe.

BEWICK'S SWAN
A number of species breed widely around the Arctic and move to western Europe in winter to escape severe weather conditions. The Bewick's Swan is typical of these.

SONG THRUSH
Thrushes are found almost worldwide, but form a distinctive part of Western Palearctic birdlife. They include some of the best known songbirds in Europe, such as the Blackbird, and widely travelled migrants, such as this Song Thrush.

ROBIN
Robins are related to thrushes and evolved in Western Palearctic forests. They adapted to forest life, finding worms and insects in earth turned up by moles and foraging wild boars. More recently they have turned to garden habitats.

DIFFERENT ENVIRONMENTS

"Palearctic" refers to the arctic conditions endured by much of the region in the Ice Ages; the legacy of this remains, with several species still moving back to areas from which they were driven by the ice. The Western Palearctic is claearly delimited by the sea in the west and the Sahara in the south, but its eastern boundary is harder to draw. Within this region, the habitats available to birds are amazingly varied. Arctic and Siberian habitats provide tundra and a belt of dense forest. There are also grasslands and steppes and a mixed European zone with temperate forests, wetlands, and mountains. The Mediterranean has distinctive hot, dry summers and cooler winters, while North Africa and the Middle East are hot and arid but also have snow-capped mountain peaks. Birds have adapted in many ways to this great range of challenges and opportunities.

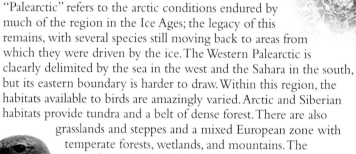

WHEATEAR
The Wheatear is an exceptionally widespread representative of its family, breeding from Africa to the Arctic. Other wheatears prefer to stay in more southerly regions, in hot, often semi-arid, habitats.

DARTFORD WARBLER
Warblers include many typical Palearctic forms. The Dartford Warbler is one of several centred on Mediterranean heathland.

DISTRIBUTION

Various distribution patterns occur in the region. Birds such as the Kestrel breed across Europe, Africa, and Asia. Others, such as the Long-eared Owl, breed in North America, Europe, and Asia. Species found across Europe and Asia are labelled "Palearctic", while the Robin is solely "European", and the Dartford Warbler "Mediterranean".

THE REGION AND ITS BIRDS
The Western Palearctic includes Europe as well as countries around its edges (the region is outlined in pink here). The Canary Islands, Azores, and North Africa are Palearctic in their birdlife. In the Middle East, the mix includes African and Oriental species.

PALEARCTIC BIRDS

Most Palearctic birds are unique to the region. Of the hundreds of Palearctic songbirds, for example, only 16 also occur in the Americas. Of the rest of the bird species in the region, just 100 are found in the Americas. There is more of an overlap with the Orient and Africa, but there are still 65 genera that are only ever found in the Palearctic. However, the Western Palearctic has only half as many species as the Eastern Palearctic.

A SELECT FEW
While there are just three species of kingfisher in the Western Palearctic, and only one occurs in Europe, there are 88 worldwide. In some ways the Palearctic is more easily defined by what it lacks than by what it has. What it does have, nonetheless, is a unique combination of birdlife.

HABITAT RANGE
Western Palearctic habitats create a variety of bird communities. Those living in deserts, which have hot days and cold nights, face quite different challenges to those in northern forests and on high ground.

WETLANDS

Wetlands include a range of watery habitats, from the edge of the sea to lakes, reservoirs and lagoons, rivers, marshes, and seasonal floods. Water means just one thing to birds – abundant food, in the form of fish, invertebrates, and plant material. Wetland birds exploit these food stocks in innumerable ways: swimming on or under water, wading into it, flying over it, or living in the dense, upright stems of plants found alongside the edge. Unfortunately, many of these habitats have been reduced by centuries of drainage and some wetland habitats and their birds are under serious threat today.

SALINE SPECIALIST
Greater Flamingos require salty water full of tiny invertebrates and algae. They sweep their odd bills upside down through the water to filter out food.

LOWLAND FLOODS AND WASHES

Wet grasslands with scattered pools are great places for breeding waders in spring. In winter, the areas flood and become magnets for waterfowl. Many of the best areas are managed as nature reserves; by controlling water levels and grazing livestock the best conditions for birds can be produced.

SWANS ON WET PASTURE
Bewick's and Whooper Swans are wild, migrant swans from the north that travel to wet grassland areas in northwest Europe each winter.

WETLAND FEAST
Flood plains and low-lying river valleys flood during heavy rainfall. Worms, insects, seeds, and other vegetable matter float to the surface or are washed up at the water's edge for waterfowl to feed on.

MARSH HARRIER
This bird of prey flies low over the reeds looking for prey, such as young waterbirds.

FRESHWATER MARSHES

Bitterns, Marsh Harriers, Reed Warblers, and Bearded Tits all depend on areas of reed growing up from the shallow water found within freshwater marshes. They build their nests in the safety of dense reeds, using the stems and leaves as nest materials. Bitterns need wet reedbeds, so they can catch fish without going into the open. Bearded Tits feed on both insects and seeds in the reeds.

REED WARBLER
Small patches of reed are perfect for this songbird. Its strong feet provide a good grip on upright perches. It weaves its deep, cup-shaped nest around several stems.

RESERVOIRS AND LAKES

Huge numbers of water birds penetrate far inland to take advantage of lakes, which add variety to birdlife in otherwise dry areas. Reservoirs with natural shores are excellent, especially if the water level falls, as the mud attracts migrant waders. Flooded gravel workings have steep shores and so few waders, but are ideal for ducks. The dragonflies found there are food for Hobbies.

WILDFOWL BONANZA
Lowland reservoirs have great concentrations of wildfowl. In winter, huge roosts of gulls appear as lakes freeze over.

LITTLE EGRET
With a changing climate, this egret has spread north in western Europe. It is attracted to lakes and open shores.

ADAPTATIONS

■ **Fine mesh** located at the sides of their bills allow dabbling ducks to sieve seeds from water. They skim their opened beaks across the surface to take in water before filtering it.

■ **The distinctively long toes** of Egrets, Moorhens, and Bitterns spread their weight and support them as they walk on floating vegetation.

DABBLING TEAL

Coots and grebes have long, lobed toes, which aid propulsion when they are in water.

■ **The uniquely flattened and round-tipped bill** of the Spoonbill sweeps sideways, halfopen, through the shallow water, until it touches a fish. It then quickly snaps its bill shut in order to feed.

flattened bill tip

SPOONBILL

WHERE TO WATCH

❶ LAKE MYVATN, ICELAND

This northern lake sees an abundance of birds including Common Scoters, Long-tailed Ducks, Wigeon, Barrow's Goldeneyes, Harlequin Ducks, Red-necked Phalaropes, Whooper Swans, and Ptarmigan.

❷ OUSE WASHES, UK

This is made up of pools and low-lying grassy fields that often flood. Breeding birds include Snipe and Black-tailed Godwit; migrant waders and terns are frequent; and in winter there are Wigeon, Pintail, and Bewick's and Whooper Swans.

❸ COTO DOÑANA, SPAIN

This vast coastal wetland sees Spanish Imperial Eagles, Black Kites, White Storks, Little and Cattle Egrets, Purple and Night Herons, Black-Winged Stilts, Avocets, Ruffs, Spoonbills, and Whiskered Terns as well as an abundance of waders, ducks, and geese throughout the winter months.

❹ CAMARGUE, FRANCE

A huge southern delta complex full of rice fields and lagoons– breeding sites for egrets, herons, Greater Flamingos, and Avocets.

GREATER FLAMINGOS

❺ ALBUFERA MARSHES, MAJORCA

This large reedbed with small open lagoons and ditches is an excellent place in summer for Great Reed and Moustached Warblers and Eleonora's Falcons.

❻ NEUSIEDLER SEE, AUSTRIA

A lake and reedy marsh complex, this area has a great variety of egrets, herons, Little Bitterns, Whiskered Terns, Ferruginous Ducks, and River Warblers.

❼ DANUBE DELTA, ROMANIA

This magnificent, huge wetland complex, leading to the Black Sea, is a vital habitat for Dalmatian and White Pelicans and Pygmy Cormorants. There is also an abundance of herons, egrets, spoonbills, Glossy Ibises, Black- Tailed Godwits, Whiskered Terns, Penduline Tits, Red-footed Falcons, and White-tailed Eagles here.

LITTLE EGRET

❽ PORTO LAGO, GREECE

This lake and coastal marsh attracts Dalmatian and White Pelicans, Great White and Little Egrets, and a host of other wetland birds.

■ *Wetland areas*

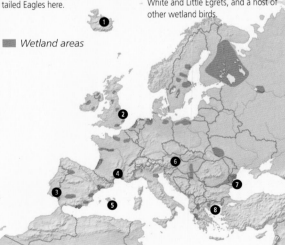

ESTUARIES and LOW-LYING COASTS

A river broadening towards the sea deposits mud and silt over vast areas that are exposed at low tide. The sides of such an estuary and other stretches of soft coast consolidate into salt marshes, where muddy creeks wind through green swards of salt-tolerant vegetation. Sand dunes, shingle spits, shell banks, and sand or pebble beaches all provide habitats for birds on low-lying shores.

ESTUARIES

A muddy estuary is an excellent source of food for numerous birds as fish, tiny snails, worms, shellfish, and other invertebrates are very abundant. The twice-daily flow of the tides also enriches the inter-tidal mud and sand with nutrients. Birds from vast areas of northern Europe and Asia rely on such estuaries from autumn right through to late spring as they rarely freeze over. During those seasons swimmers, divers, waders, probers, and aerial feeders all exploit the food that is to be found in the mud, sand, and shallow water. In the summertime, gulls, terns, ducks, larks, and pipits also breed on the firmer salt marshes that are situated all the way around the edge of an estuary.

SLAVONIAN GREBE

Grebes breed on freshwater lakes but move to the sea in the autumn. Slavonian Grebes, such as this one in its winter plumage, can often be seen drifting into an estuary with the rising tide.

DUNES AND LAGOONS

Sand consolidates into tall, grassy dunes with damp hollows ideal for waders, Skylarks, and Meadow Pipits. Shallow lagoons above high tides offer feeding areas for gulls, Shelducks, and Ringed Plovers, while dry sand spits are nest sites for terns and waders.

FLOCKING TO THE BEACH
Black-headed Gulls find safe refuge on offshore bars and beaches, resting between feeding sessions.

SAND AND SHINGLE

Harder beaches, which are made up of sand and gravel, are not as good as soft mud for long-billed, probing waders. However, these beaches are used by short-billed waders, such as plovers, that can pick food from the surface or from between stones. Seaweed and other debris washed up at high tide form a "tidewrack" or strandline, along which other birds, including some land birds such as Starlings, can feed on tiny sand hoppers and other small creatures.

SWEEPING AVOCET
An Avocet sweeps its upcurved bill sideways through shallow water, catching tiny crustaceans.

SALTMARSH RICHES
Large marshes have an abundance of salt-tolerant plants that attract insects, so are full of food for birds. Redshanks, Black-headed Gulls, and Mallards nest in these places in summer.

ADAPTATIONS

■ **The curved bill** of the Curlew is strong enough to catch and break up crabs.

■ **With flattened, chisel-shaped bills**, Oystercatchers prise shellfish off rocks and break into them.

■ **The camouflage patterning of the Little Tern's eggs** allows it to lay them straight onto sand or shingle without building a nest. They are perfectly camouflaged against the beach so are usually safe even when unattended.

LITTLE TERN

■ **Broad, flattened bills with rough edges** are swept in semi-circular directions across soft mud and shallow water by Shelducks. This enables them to gather up tiny snails, which they then feed on.

■ **Long legs** allow waders, such as Knots, to stand upright in water, while their long toes stop them sinking into the sticky mud. Their sensitive bill tips allow them to probe for worms. The length of the bill is a good clue to the depth to which a species will probe and the size of food they take.

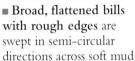

SHELDUCK

KNOTS

WHERE TO WATCH

WADERS FLOCK TO SAFE ROOSTS AT HIGH TIDE

❶ MORECAMBE BAY, ENGLAND

This is a vast estuary complex with mostly sandy flats fringed with extensive salt marshes and sand dunes. Firm banks above the high tide mark offer refuges for wildfowl and waders to roost at high tide. The area attracts Oystercatchers, Bar-tailed Godwits, Knots, Dunlins, Curlews, Shelducks, Redshanks, Ringed Plovers, and Turnstones.

❷ THE WASH, ENGLAND

The wash is a huge enclosed estuary that opens out onto the North Sea. Most of it is muddy, but there are sand banks at the mouth and extensive saltmarshes around the edges. It sees big flocks of Knots, Dunlins, Bar-tailed Godwits, and Grey Plovers. In winter it attracts tens of thousands of Pink-footed Geese, Brent Geese, and Wigeon.

PINK-FOOTED GEESE

❸ IJSSELMEER, NETHERLANDS

This enclosed and part-reclaimed estuary sees migrant gulls, terns, and waders in spring and autumn while in winter it has great numbers of Tufted Ducks, Pochards, Scaup, Wigeon, Goosanders, and Smews.

❹ WATTENMEER, GERMANY

This is the eastern end of the Netherlands' Waddenzee – a vast, shallow area of estuarine character where over a million water birds appear in spring and autumn.

AVOCETS

❺ EBRO DELTA, SPAIN

This area is half reclaimed but is still great for birds; in summer it has the world's biggest colony of Audouin's Gulls as well as Gull-billed Terns, Kentish Plovers, and Pratincoles. In autumn and winter Greater Flamingos and many wildfowl and waders can be seen.

▬ *Low-lying coasts*

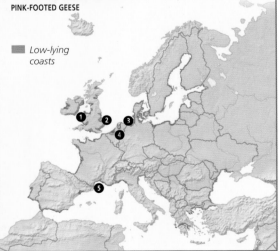

ROCKY COASTS, ISLANDS, AND THE OPEN SEA

Most of the European coastline is a "hard" coast of some kind, with rocks or cliffs and stony beaches. The majority of these shores are less attractive to birds than soft estuaries and marshes, but seabirds must still come to land on cliffs and islands in order to breed.

ROCKY AND STONY SHORES

Stony beaches and wave-washed rocks do see some waders from autumn to spring. Typical rocky shore waders are short-billed species such as Turnstones and Purple Sandpipers, which feed on invertebrate food found among weedy, barnacle-encrusted rocks right at the edge of the waves. Knots, Dunlins, Curlews, even migrant Common Sandpipers, however, often feed around flatter rocks by the shoreline. Nearby higher rocks and isolated islets out of reach of predators offer solid sites for nesting gulls and terns.

SHELLFISH EATER
Oystercatchers hammer open mussels on seaweedy rocks.

HIGH TIDE REFUGE
Even though they feed on muddy estuaries, waders may move to nearby higher rocks when the tide covers the mudflats. The Dunlins shown here may be joined by Redshanks, Knots, Bar-tailed Godwits, and Curlews.

ROCKY ISLANDS, OFFSHORE STACKS, AND CLIFFS

Within western Europe, some of the rocky islands, offshore stacks, and mainland coasts that have sheer cliffs see birds in huge numbers. Most seabirds nest in colonies and so choose the best of these sites, at which Gannets, Fulmars, Herring Gulls, Kittiwakes, Guillemots, Razorbills, and Puffins create some of Europe's finest bird spectacles. Different island areas see particular seabirds coming in to nest in burrows created in crevices in rocks or cliffs, old rabbit burrows in soil, or burrows they dig out themselves. Mediterranean and Cory's Shearwaters, for example, prefer the Mediterranean, while Manx Shearwaters nest around Britain and Ireland; Black Guillemots and Arctic Terns prefer to nest on low, rocky islets, while Shags and Cormorants like broad ledges on larger cliffs.

NOCTURNAL VISITOR
Storm Petrels spend most of their time at sea but must come to land to nest. As they are easily captured by gulls, they will only land after dark.

CLIFF NEST
Cliff ledges are out of reach of predators, so attract seabirds. They are difficult places though: seabirds' chicks would fall to their deaths if they didn't have the natural instinct to keep still on a tiny ledge. Shags build big nests of sticks and weed, Kittiwakes create a nest on the tiniest outcrop, while Guillemots do not make nests at all.

OPEN SEA

Many birds live at sea outside the breeding season. Gannets, Fulmars, Puffins, petrels, and shearwaters stay way out in the middle of the Atlantic. Migrants from much farther south pass European coasts on their ocean wanderings, including shearwaters from the southern hemisphere that appear west of Europe in late summer and autumn. All of these birds occasionally come close inshore during gales but prefer life on the open sea, where there is an abundance of food such as fish, jellyfish, and plankton. But this food source is not uniformly distributed: ocean currents and seasonal changes mean the birds have to travel great distances in order to locate food, often far from the cliffs that offer them nest sites.

SEA RESIDENT
The Kittiwake spends months in the middle of the ocean. It is attracted to fishing boats, where it feeds on discarded, undersized fish.

GUILLEMOTS
Guillemots dive from the water's surface for fish such as sandeels; they rarely feed close to land.

ADAPTATIONS

■ **A cushioned skull and air sacs** on the head and neck help the Gannet to survive its spectacular 30m (100ft) plunges into the sea for fish.

■ **Tubular nostrils** on shearwaters and petrels – known as "tubenoses" – help them to excrete excess salt. (necessary because they mainly drink salt water from the sea).

■ **The pyramidal shape of Guillemots' eggs** helps them to incubate one egg under a wing and also stops the eggs from rolling off a narrow ledge.

GANNETS

■ **Narrow wings** form stiff "paddles" that Razorbills and Puffins use to "fly" acrobatically deep underwater when they are in pursuit of fish.

■ **A special bill** with a fleshy "rosette" at the base helps the Puffin to keep the edges of its mandibles parallel as it opens them, so that it can carry a line of fish neatly held tight with its tongue.

PUFFIN

WHERE TO WATCH

❶ ROST, NORWAY

This island group houses many seabird colonies, including hundreds of thousands of Puffins as well as guillemots, Razorbills, Storm Petrels, and Leach's Petrels.

❷ ORKNEY ISLANDS, UK

A spectacular archipelago with Fulmars, Guillemots, Razorbills, Kittiwakes, Arctic Terns, Arctic Skuas, and Great Skuas.

❸ BASS ROCK, UK

This has a stunning Gannet colony that reaches more than 40,000 pairs. Also sees guillemots, Puffins, Kittiwakes, and other gulls and is easily reached by boat.

BASS ROCK

❹ BEMPTON CLIFFS, UK

These cliffs house an unusual mainland Gannet colony (most are found on islands), and one of the easiest large seabird colonies to see in the UK, with safe viewing platforms above sheer cliffs. There are many Kittiwakes and guillemots, Razorbills, Puffins, Herring Gulls, Fulmars, and some Shags and Cormorants. It is a good place to view offshore seabird migration.

❺ BERLENGO, PORTUGAL

This island attracts thousands of Yellow-legged Gulls, Cory's Shearwaters, Shags, and guillemots.

❻ GREEK ISLANDS

Many Greek islands have Yellow-legged Gulls, Cory's Shearwaters, and Mediterranean Shearwaters. Some have breeding Eleonora's Falcons and Audouin's and Mediterranean Gulls, as well as an excellent selection of land birds such as shrikes and warblers, including migrants in autumn.

❼ MAJORCA, SPAIN

The Balearic Islands in general are rocky and have many cliffs where seabirds can nest. They are good places to view Cory's and Mediterranean Shearwaters, often close inshore, as well as Yellow-legged Gulls. In late summer Eleonora's Falcons nest locally, preying on migrant songbirds. Some islets and headlands have the rare and local Audouin's Gull.

CORY'S SHEARWATERS

Rocky coasts

THE FAR NORTH

In the winter, the far north is a dark, frozen, and barren place. However, by late spring the snow is melting, the days are starting to get very long, and insects and their larvae abound in myriad pools. These far northern regions, with their summer food bonanza, attract millions of birds. They are all dependent on the Arctic during the summertime but are unable to survive there at other times of the year. Only such hardy species as the Ptarmigan can remain in the far north all year round, the rest will appear farther south as migrants or winter visitors.

PTARMIGAN
In winter the Ptarmigan turns white to match its surroundings. Its feet are densely feathered to help it to avoid heat loss; they also act as "snow shoes".

THE COAST AND OPEN SEA

The sea is rich in invertebrate food and fish so seabirds and ducks abound on many of the northern coasts. This is an exciting area, as birds that are rare in Europe head north and east into Siberia in spring, and others that are essentially Arctic species appear in a handful of sheltered bays on the fringes of their normal range. In Europe, this is the one small patch of land and sea that has the character of the Arctic, so it sees some of its great birds.

FULMAR
Fulmars nest on cliffs, occupying small, earthy ledges. They need access to the open sea, where they often feed around fishing boats.

ARCTIC TUNDRA

Open tundra is bleak and exposed as it is beyond the northern limits of tree growth. It is a very tough environment for birds but its dwarf shrubs provide seeds and shoots as food for Ptarmigan, which are widespread on broad, rocky ridges. These ridges are also nesting places for Snowy Owls. Small cliffs may have nesting birds of prey, which survive on rodents and birds, but they usually have to move south in winter. Skuas that nest around the Arctic, feeding mostly on lemmings, are purely summer visitors. It is the ability to feed in perpetual summer daylight, and the close proximity of many pools, that draws geese and waders to nest on the higher ridges.

ARCTIC TERN
The broad, stony tundra slopes of larger islands and headlands are ideal for nesting Arctic Terns.

WHOOPER SWANS
This is the swan of northern lakes, bogs, and river deltas. The Bewick's Swan, on the other hand, breeds around more open, exposed tundra pools.

NORTHERN FJORDS
Sheltered bays and cold water full of fish offer security and food for seabirds and wildfowl in what is otherwise the bleak and windswept far northern tip of Europe.

POOLS

From autumn until May or early June the pools of the tundra are frozen and covered with snow. Waders, ducks, and geese that breed up here are still in their wintering areas until May; slightly south, larger lakes on the tundra fringe see birds in early spring. Once the short summer is under way, the pools are alive with insects and their larvae. Long-tailed Ducks and Red-necked Phalaropes gorge themselves on the thick rime of insect life found on some northern lakes. Downy chicks of waders and ducks can feed themselves within hours of hatching and find life easy with so much food; but they also face a barrage of hungry predators.

STELLER'S EIDER

In spring and summer. a few hundred spectacular Steller's Eiders form dense flocks offshore in food-rich seas off the extreme north of Norway.

ADAPTATIONS

■ **Thickly feathered legs** like those of the Rough-legged Buzzard provide invaluable added warmth for life in northern areas.

feathered legs

ROUGH-LEGGED BUZZARD

■ **The thick, strong bill** of the Long-tailed Skua is also hard and hooked at the tip. This allows it to catch small rodents and some small birds more easily as it only uses its bill, not its feet, when hunting.

■ **White plumage** helps vulnerable birds such as Willow Grouse stay hidden, but also allows predators such as the Snowy Owl to get close to them unseen.

SNOWY OWL

WHERE TO WATCH

❶ ICELAND

Harlequin Ducks and Barrow's Goldeneyes are not found anywhere else in Europe, while Red-necked and Grey Phalaropes, Brunnich s Guillemots, Glaucous Gulls, Little Auks, Puffins, White-tailed Eagles, and Gyr Falcons are also specialities of Iceland. Thousands of pairs of Pink-Footed Geese and Whooper Swans, Black-tailed Godwits, Long-tailed Ducks, and Common Scoters breed in the area, and Iceland Gulls are regular visitors to the region during the winter months.

WHOOPER SWANS

❷ VARANGER FJORD, NORWAY

The area around Varanger in Norway is exceptionally rich in birds during spring and summer. Offshore migrants include Long-tailed and Pomarine Skuas and White-billed Divers, while Steller s and King Eiders are regular visitors. A large range of exciting breeding birds includes Red-necked Phalaropes and Black-throated Divers.

BLACK-THROATED DIVER

■ *Tundra areas*

NORTHERN MOUNTAINS AND MOORLANDS

These are tough habitats for wildlife, and many birds are only summer visitors to the uplands, breeding when insects or other small birds are available as prey. In winter, snow and exposure to gales and rain make life impossible for all but a few hardy species. Some birds survive on a very restricted diet, in areas that offer little variety. Nevertheless, these are imposing and often inspiring places that do see exciting birds.

NORTHERN MOUNTAINS

These mountains reproduce the cold, windswept, and barren conditions of far northern lowlands, and so winter sees most birds journeying south or moving to lowland areas. In summer, however, these same habitats offer greater possibilities for migrants, as insect and plant life has a brief period of abundance and small rodents multiply quickly in the grasslands, providing food for predators. The rocky gullies and crags provide nest sites for birds and good feeding opportunities.

DOTTEREL
Dotterels like broad, rolling ridges and stony plateaux at high altitude, where they feed on insect life in summer.

EXPOSED MOUNTAIN REGIONS
On high, exposed ground the tundra-like conditions, which include sparse vegetation, cliffs, and screes, are exploited by only a small selection of hardy sub-Arctic birds.

MOORLAND

Moorland forms on peaty ground where conditions are cold and wet for much of the year. This means dead plant material is slow to rot, so it builds up into thick layers of peat. Where this becomes waterlogged, extensive blanket bog forms. This is a rare habitat worldwide, but is well represented in northwest Europe. Such places are impoverished in wildlife terms, but still attract a small selection of birds. Some of these birds, such as Skylarks, are widespread elsewhere, while others are upland moor and bog specialists. For example, Dunlins prefer wet bogs, while Golden Plovers choose more open spaces of grassy moors or heather for feeding and nesting.

RED GROUSE
Red Grouse can be scarce on heathery moors. Only where the habitat is managed for them do they appear in large numbers. They feed almost exclusively on heather shoots and seeds.

WHINCHAT
The Whinchat likes gentle slopes with bracken and heather, or young conifer plantations.

CONIFER PLANTATIONS

Many moorlands have been planted with alien conifers. These provide temporary homes for Black Grouse, Hen Harriers, and Whinchats until they grow too tall and dense. Chaffinches, Robins, and Coal Tits may then move in.

ADAPTATIONS

■ **Camouflage** helps Dotterels, Golden Plovers, and other breeding birds to blend in to barren surroundings.

■ **Ptarmigan turn white in winter**, helping them to hide in the snow. Their feet are feathered to reduce winter heat loss.

feathered legs

PTARMIGAN

■ **The long, fine-tipped bills** of Dunlin are ideal for picking insects and small worms from moss-covered bogs in summer, and for probing in mud for worms during winter.

■ **Open moors** have few perches so birds such as Golden Plovers and Skylarks sing in flight to claim their territorial rights and attract females.

■ **Muscular gizzards and elongated intestines** allow grouse to swallow grit to help grind up the shoots of tough, heathery plants.

SKYLARK

WINTER MOORLAND

Overgrazed moors have little heather and can degenerate into swathes of acid grassland. Few birds appear, but Meadow Pipits and Skylarks will feed on grass seeds and insects while Curlews take worms from boggy places.

WHERE TO WATCH

THE CAIRNGORMS

❶ VATNA JOKULL, ICELAND

The high central plateau of Iceland is bleak and forbidding. Even in summertime, it has a thick central ice cap that is almost devoid of birdlife. Around it is a spectacular landscape that sees a small range of birds. Bare ground provides habitat for Ptarmigans, Snow Buntings, and rare Gyr Falcons and Snowy Owls. Locally, Pink-footed Geese and Purple Sandpipers breed.

PURPLE SANDPIPER

❷ DOVREFJELL, NORWAY

This national park has a mix of bogs, lakes, woodland, snowfields, and mountain peaks and plateaux. Here there are Rough-legged Buzzards, Cranes, Dotterels, Temminck's Stints, Red-necked Phalaropes, Shore Larks, Bramblings, and Lapland and Snow Buntings in summer.

❸ CAIRNGORMS, SCOTLAND

A unique area in the UK with extensive boulder fields, screes, and cliffs in a very high plateau, where Dotterels, Ptarmigan, a few Snow Buntings, Ravens, and Golden Eagles breed.

❹ NORTH PENNINES, ENGLAND

This is rolling moorland with heather and limestone grassland, dissected by wooded valleys and rocky gulleys with tumbling streams. Many Golden Plovers, Curlews, Wheatears, Snipe, Meadow Pipits, and Skylarks breed, as well as a few Merlins, Short-eared Owls, Dunlins, Dippers, Twites, Whinchats, Stonechats, and Ring Ouzels.

CURLEW

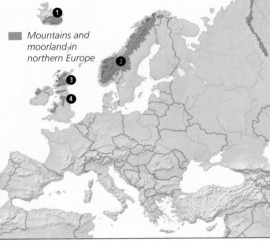

Mountains and moorland in northern Europe

SOUTHERN MOUNTAINS AND CRAGS

The highest, harshest, snowiest peaks of southern European mountains are little different from their counterparts in the north, but many others are snow-free and characterized by big, bare crags, dizzying cliffs, and deep gorges with rushing rivers. Forests of beech, oak, and pine climb high on the mountain slopes, creating conditions that attract woodland species more characteristic of northern Europe. The southern influence, however, remains strong, with the presence of birds of prey hunting overhead, and, in the valleys, many other species that are not found farther north.

HABITAT MIXTURE

With their peaks, high pastures, and valleys, mountainous areas bring together a variety of habitats within a small area. From one spot it is possible to see riverside birds, woodland birds, and those characteristic of grassy pastures and hedgerows, as well as species that prefer cliffs and crags. Slightly further up the hillside, alpine species can be heard or seen flying across the valleys or over the highest peaks. Superimposed upon this natural diversity there are regional differences. For example, some species, including various eagles and other birds of prey, are more easterly and so are found in the Balkans but not within Iberia. Birds of prey have also been wiped out of some areas due to centuries of persecution, although they remain numerous in Spain. Some species – such as eagles, vultures, buzzards, and falcons – occupy different "niches", exploiting particular foods and nest sites. This may allow them to live close together without competing against each other. Alternatively, it may mean that species remain apart, each inhabiting an area that meets its particular requirements.

COASTAL CLIFFS
Several mountain birds also breed on crags by the sea that offer safe nesting sites.

BLACK WHEATEAR
Warmer, south-facing stony slopes and crags in Iberia attract this eyecatching wheatear, as well as Black-eared Wheatears and Stonechats.

CLIFFTOP THRUSH
Limestone crags are the favoured haunt of the Blackbird-sized Blue Rock Thrush, which ventures anywhere from sea level to high in the mountains.

STUNNING PEAKS
High mountain peaks with deep gorges and shady valleys have a great range of birds of prey, many of which nest lower down but move up each day to hunt over high ground. Cliff faces have their own unique birds – some move out to feed on high-altitude grassy meadows.

WHERE TO WATCH

❶ MASSIF CENTRAL, FRANCE

This famous mountainous region in France has seen the reintroduction of Griffon and Black Vultures. It also has natural populations of Red Kites, Egyptian Vultures, Short-toed, Booted, and Golden Eagles, Eagle Owls, and Alpine Accentors.

TYPICAL LIMESTONE CRAGS

❷ PYRENEES

Straddling France and Spain, the Pyrenees offer wonderful birds in spectacular settings. There is a great range of eagles, vultures, and Kites, as well as alpine and cliff birds such as Ptarmigan, Wallcreepers, Alpine Choughs, Alpine Swifts, Alpine Accentors, Snowfinches, and other birds that thrive in mountains and forests.

❸ SIERRA DE GUADARAMA, SPAIN

This is a superb area for Black, Griffon, and Egyptian Vultures, as well as eagles and a variety of upland species such as the Rock Thrush, Blue Rock Thrush, Black Redstart, Rock Bunting and Citril Finch. The local Black Stork also breeds in a few locations within the Sierra de Guadarama.

GOLDEN EAGLE

❹ SIERRA NEVADA, SPAIN

Although not so rich in birds of prey as some northern ranges, this big range of hills still sees Griffon and Egyptian Vultures, Golden, Booted, and Bonelli's Eagles, and Peregrines. It is a good place to spot Alpine Accentors, Black and Black-eared Wheatears, Rock Thrushes, and Alpine Swifts.

❺ PUIG MAYOR, MAJORCA

The mountains and valleys of Majorca have some remote and spectacular sites that attract Black Vultures, which are rare and local. In summer, many cliffs have Eleonora's Falcons, Red Kites, Ospreys, Blue Rock Thrushes, Black-eared Wheatears, Rock Buntings, Alpine and Pallid Swifts, and Crag Martins. Alpine Accentors only visit in the wintertime.

❻ EVROS MOUNTAINS, GREECE

Greece no longer has the rich populations of birds of prey that used to be there, but in the extreme northeast the mountains attract specially conserved Black Vultures as well as Imperial, Lesser Spotted, and White-tailed Eagles, Lanners, Long-legged Buzzards, and Levant Sparrowhawks.

Mountainous areas in southern Europe

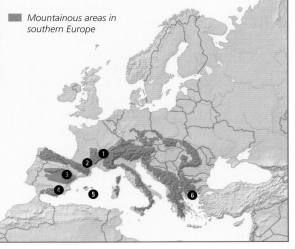

GRIFFON VULTURE IN SOARING FLIGHT

ADAPTATIONS

- **Big, soaring birds** are attracted to more southerly areas rather than northwest Europe, which lacks the warm, rising air currents that offer them so much lift. These species have evolved to make the best use of air currents, including cold winds on the high tops. This enables them to travel long distances and search for widely scattered food (such as animal carcasses) over vast areas – with the least expenditure of energy. To aid them in this, they also have exceptional eyesight.
- **The fine bill** of the Wallcreeper probes for insect food in damp recesses, under dark overhangs on cliff faces, and deep inside shady gorges.
- **The strong, curved bill** of the Chough can overturn animal droppings and prise cushions of grass and lichens from rocks to expose invertebrates.
- **The exceptionally tough oesophagus** of the Lammergeier has developed in order to cope with sharp-ended fragments of shattered bone, which it feeds on. The bird drops large bones onto rocks to break them into pieces small enough to swallow.

CHOUGH

LOWLAND HEATH AND MEDITERRANEAN SCRUB

In Northwest Europe lowland heath is a rare and restricted habitat. It is often found in a thin strip along a rocky coast but also in special regions with sandy or acid soils, such as the New Forest and Brecklands in England and a few areas of the Netherlands and northern France. In southern Europe, however, much larger areas of rough ground with short, aromatic shrubs and evergreen bushes cover sunny slopes.

MEDITERRANEAN SIZZLER
A typical bird found in Mediterranean scrub is the Serin, whose jingling, sizzling song is a familiar summer sound.

LOWLAND HEATHS

These are habitats for a few specialist species that require the mixture of open space and patchy scrub, which is often maintained by grazing animals and periodic fires. Dartford Warblers are resident in such areas while Nightjars are summer visitors to them. In the summertime a variety of birds take advantage of the insect food and nesting opportunities found in dense heather and gorse. Some thickets have Nightingales that prefer dense vegetation down to ground level, while open spaces see nesting Stone-curlews and Woodlarks, which need bare earth to pick up insects from. Dragonflies breed where there are pools in boggy valleys and attract hunting Hobbies.

HEATHLAND
Coastal heath bordering an estuary provides a unique habitat.

COASTAL DUNES
A lack of water and erosion by the wind create harsh conditions for most birds. However, Crested Larks feed in the open spaces.

MEDITERRANEAN SCRUB

Mediterranean slopes have bare rocks, patches of short grass, cushions of short, spiny herbs, and taller, thick, evergreen bushes. This is the "garrigue" or "maquis" habitat, wonderful for birds such as Rollers, Great Grey Shrikes, and Lesser Kestrels that take large insects and lizards from the ground. Overhead, many kinds of birds of prey may be seen, while on the bush tops are shrikes, which are small bird- and insect eaters. Ground feeders as varied as Red-legged Partridges, Hoopoes, larks, pipits, and buntings are attracted to the scrubland while the bushes have an exciting array of small warblers, mostly restricted to the Mediterranean region in Europe. There are some habitats that are found in just a few scattered locations and are home to specific species. For example, patches of introduced cactus attract Rufous Bush Robins and rocky gullies are perfect nesting sites for Rock Sparrows.

RUFOUS BUSH ROBIN
This large chat is one of an essentially African family that prefers dry, sandy gullies, small cliffs, and patches of prickly pear cactus.

ADAPTATIONS

■ **Enormous pupils** allow the Stone-curlew to feed at dawn and dusk; its large eyes are ideal for maximum light gathering. They shrink to pinpoints by day.

■ **The long, slender tail** of the Sardinian Warbler helps it to balance as it tips up to dive head first into a bush; the tail is also used to warn others of potential danger.

■ **Big eyes and very short, wide bills fringed with bristles** characterize the Rednecked Nightjar and Nightjar. Both these adaptations allow them to feed after dark, catching flying moths in their mouths in mid-air.

SARDINIAN WARBLER

NIGHTJAR

■ **The hooked bills of Shrikes** allows the birds to tear at their prey, but their feet are not especially strong. Instead, they impale large prey such as lizards, small birds, and beetles on thorns, so that they can tear at them easily.

BOOTED EAGLE
A small eagle, this species hunts other birds and small mammals that are plentiful on bushy slopes and heaths.

BIRD-RICH SCRUB
This bushy habitat is full of flowers and insects, so offers up an abundance of food for birds that eat seeds or large and small insects. The undergrowth also provides plenty of secure nesting places.

WHERE TO WATCH

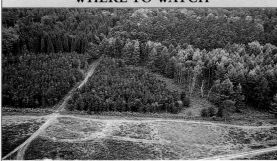

HEATH AND WOODLAND IN THE NEW FOREST

❶ NEW FOREST, UK

This is a large, rolling tract of heath and forest. The grass is kept close-cropped by ponies, which makes the area ideal for Woodlarks. The gorse and heather areas attract Dartford Warblers. Snipe and Curlews nest in damp patches, while Hobbies feed over open heath.

DARTFORD WARBLER

❷ BRECKLAND, UK

Woodlarks and Nightjars breed in felled conifer plantations on the dry, sandy soil in this area, while Stone-curlews nest on the grassy heaths and nearby fields. The grassland also attracts Wheatears, Stonechats, and Curlews.

❸ LA CRAU, FRANCE

An area of very dry, stony plains, with Little Bustards, Pin-tailed Sandgrouse, Rollers, Stone-curlews, and various larks.

❹ MAJORCA

This has many areas of heathland worth exploring to see local Marmora s and Spectacled Warblers as well as Subalpine and Sardinian Warblers. It is also good for Thekla Larks and Short-toed Larks. Some areas also attract Stone-curlews, Rock Sparrows, Woodchat Shrikes, and Rock Thrushes.

❺ ALGARVE, PORTUGAL

This part of Iberia has a mixture of sand dunes, open woodland, wetland, and rich Mediterranean scrub. Birds to be found here include Short-toed, Crested, and Thekla Larks, Great Spotted Cuckoo, and Sardinian, Subalpine, and Spectacled Warblers.

SUBALPINE WARBLER

 Heath and scrub areas

FARMLAND AND GRASSLAND

Natural grassland habitats have been almost eliminated from most of Europe by millennia of agricultural development. Those remaining are now rare and fragile. They host a collection of wildlife, including birds whose requirements are very specific, and which are often shy and easily disturbed by human activity. Change of any kind, including irrigation, ploughing, over-grazing, and encroachment by trees, spells disaster for them.

GREAT BUSTARDS
These birds are among Europe's most threatened, as they face pressure from agriculture.

FOLLOWING THE PLOUGH
Not all birds avoid farming activities: Lapwings, Black-headed Gulls, Jackdaws, and Rooks find food on ploughed fields.

FARMLAND

Agricultural land is naturally rich and full of birds but intensification, especially change from mixed farming to a dry, unvarying monoculture, removes birds from vast areas of countryside. However, arable land that is ploughed for growing a variety of crops does offer feeding opportunities – a wealth of invertebrate food when the earth is turned over and insect food in growing crops. Old, unimproved grassland, which is grazed by animals, has a wide range of plants and insects and many safe nest sites.

GRASSLAND

Dry grassy steppes have species that are adapted to living within semi-arid environments, including bustards, sandgrouse, and larks. They survive in such places by foraging within the short vegetation, but irrigation and development schemes have ruined a lot of these habitats and threaten many more, including the last great steppe lands in Eastern Europe. Other grassy habitats, such as the cold, wet northern moorlands through to hot Mediterranean scrub are "farmed" by being grazed by millions of sheep. They would not look the same nor have the same birds if such usage changed as the areas would quickly be invaded by scrub. On the other hand, too many sheep and goats reduce the variety of plants and destroy the structure of the vegetation, so grazing at just the right intensity is vital to the survival of birds and other wildlife.

LESSER KESTREL
This handsome falcon has suffered a long-term decline. This may be due to the pesticides that have reduced its insect prey in both Europe and Africa, where it spends the winter.

WHITE STORK
White Storks survive on farmland unless it is intensively cultivated and drained. They eat frogs and small rodents, which require marshy ground or tall, rich grassland.

ABUNDANT FOOD SUPPLIES
Grassland in northern and western Europe has periodic increases in vole populations, which attracts Short-eared Owls and Kestrels in large numbers.

NORTHERN VISITORS
Huge numbers of geese from far northern breeding areas, including Pink-footed, Barnacle, and White-fronted Geese, have come to rely on European farmland in winter – feeding on grass and waste crops.

ADAPTATION

- **Skylarks and Calandra Larks** have no perches to sing from in open fields: they pour out prolonged songs from high-level song flights instead.
- **Lapwings and Stone-curlews** are happy to adapt to farmland habitats so long as there is a mixture of pasture on which they can situate their nests, and bare earth where their growing chicks can forage for insects.
- **Sandgrouse** are able to survive on one drink a day and they fly huge distances to get it. They also carry water to their nestlings, holding it in their belly feathers.

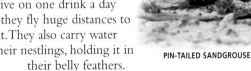

PIN-TAILED SANDGROUSE

- **Geese** have changed their eating habits and now enjoy feasting on waste crops, such as sugar beet tops and carrots.
- **Choughs** feed in old pastures, probing for ants with their long, curved bills.

CHOUGH

WHERE TO WATCH

❶ ISLAY, SCOTLAND

In summer, the damp fields here see breeding Snipe, Lapwings, Redshanks, and Curlews. Choughs, Twites, and Rock Doves feed in the fields and along the field margins. In winter, huge flocks of Barnacle and Greenland White-fronted Geese feed on grassy pastures.

❷ SALISBURY PLAIN, ENGLAND

This is an exceptional area of chalk downland, which has especially large numbers of Skylarks as well as other widespread farm birds such as Yellowhammers and Linnets. It also attracts some Stone-curlew and Grey Partridges.

❸ TEXEL, NETHERLANDS

A wonderful island to visit all year round, it sees breeding Black-tailed Godwits, Ruffs, and Redshanks on wetter nature reserves in summer. In winter, large numbers of Lapwings, Snipe, Oystercatchers, Golden Plovers, Brent, Bean, and White-fronted Geese, Wigeon, Mallards, Shovelers, other wildfowl, birds of prey, and finches, including Twites, visit the area.

TWITES

❹ BELCHITE, SPAIN

These semi-arid northern Spanish steppes have a nature reserve area in which can be found Dupont s, Crested, Lesser Short-toed, Short-toed, and Thekla Larks as well as Stone-curlews, Pin-tailed Sandgrouse, various warblers, Hobbies, Lesser Kestrels, and other birds of prey.

CRESTED LARK

❺ BIEBRZA MARSHES, POLAND

This farmed wetland and grassland region has rare Aquatic Warblers and Great Snipes as well as a great variety of other exciting species – from Black and White Storks to Cranes, Corncrakes, Greater Spotted Eagles, and White-winged Black Terns.

❻ HORTOBAGY, HUNGARY

This is a huge area of farmland and semi-natural steppe, which attracts Red-footed Falcons, Great Bustards, Aquatic Warblers, Rollers, Bee-Eaters, and Lesser Grey Shrikes in summer. Tens of thousands of Cranes pass by on their migration route and White-tailed Eagles visit in winter.

▥ *Grassland areas*

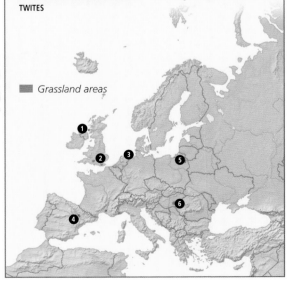

FOREST AND WOODLAND

Most of Europe would naturally be covered with forest, but human activities have destroyed many areas and left others impoverished. Even so, Europe's forests and woodlands still include such varied types as Mediterranean evergreen oak and ancient Scots pine forest.

OAK AND BEECH WOODS
Oak woods (below) let in more light than beech (right), so have a thicker, more varied shrub layer beneath and consequently a greater variety of birds.

DECIDUOUS WOODS

The character of deciduous woods changes greatly with the seasons, so offers different opportunities and challenges to birds. Deciduous trees lose their leaves in winter to survive the cold. Insects are abundant and active in summer, but many of them survive the winter as cocoons or eggs, or hibernate, so insect-eating birds are forced to migrate unless they can turn to seeds in winter. In contrast, many seed-eating species are year-round woodland residents. The gnarled trunks of deciduous trees have many crevices for birds to nest in.

PIED FLYCATCHER
Pied Flycatchers need holes for nesting, most easily found in mature deciduous trees.

EVERGREEN FORESTS

Evergreen trees – mostly conifers – keep their leaves year-round. Conifers tend to be smoother and straighter than many deciduous trees, and have fewer holes for nesting birds. Their insect food is more uniform through the seasons but their seed production tends to fluctuate, resulting in a few good years between several poor ones. Some birds breed well in such forests during the good years but move out in a nomadic search for food when the seeds run out. Conifers have tough needle-leaves and their dense foliage casts a deep shade, so few plants can grow underneath them. This reduces the variety of feeding and nesting possibilities for birds.

CROSSBILLS
Cones have nutritious seeds but are protected by tough, overlapping scales. Crossbills have evolved a special, cross-tipped bill that can prise these apart. They reach the seeds inside using their tongues.

CONIFERS
Conifers suit a wide variety of birds as the trees mature. Old trees and open glades at the edge of a forest create a light, accessible area.

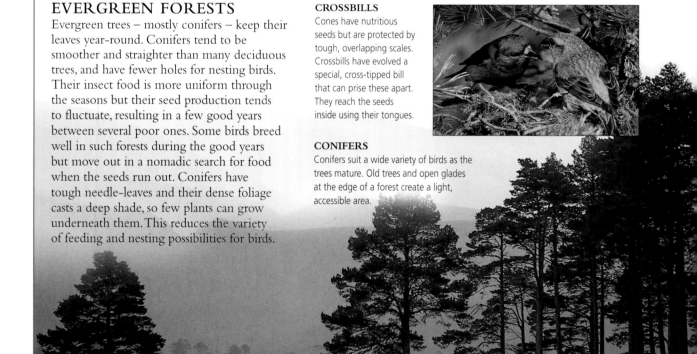

TREECREEPER
The Treecreeper spends its whole life carefully searching tree bark for insects.

ADAPTATIONS

■ **An elongated, backward-pointing outer toe and a stiff tail** provide the Green Woodpecker with a sure grip as it perches on broad branches.

GREEN WOODPECKER

■ **The stout, dagger-like bill** of the Nuthatch is used for breaking into nuts and seeds, which it wedges into crevices in bark.

■ **A Woodcock's dead-leaf camouflage** is perfect for nesting on the forest floor.

■ **A Sparrowhawk's long tail** helps it twist and turn quickly in tight spaces between trees when chasing prey.

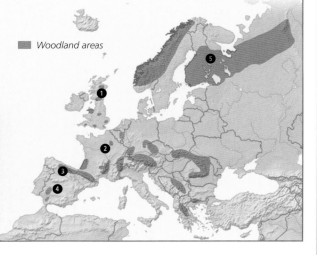

SPARROWHAWK

WHERE TO WATCH

❶ ABERNETHY, SCOTLAND

A magnificent area of rolling moor and ancient pine forest (the largest remaining tract of native pine forest in Britain) with dense bilberry, crowberry, heather, and juniper – ideal for rare Capercaillies. The mature trees are good for Crested Tits as well as Parrot, Scottish, and Two-barred Crossbills. Younger plantations have Black Grouse.

YOUNG CONIFER PLANTATION

❷ FONTAINEBLEAU, FRANCE

This is an excellent area of deciduous and mixed forest and parkland, which has great possibilities for many woodland species such as interesting birds of prey, six species of woodpeckers – including Black, Middle Spotted, and Grey-headed – as well as a good selection of warblers in summer.

❸ PICOS DE EUROPA, SPAIN

A fine mountain range in northwest Spain, often cloudy and wet but with spectacular forested peaks and gorges and some extensive hillside oak and pine woods, where there are Griffon Vultures, Bonelli s, Short-toed, Booted, and Golden Eagles, Black Woodpeckers, Crested Tits, Goshawks, and Bonelli s Warblers.

CRESTED TIT

❹ EXTREMADURA, SPAIN

Woodlands in this region are mixed or of open cork oak. It sees a superb selection of forest and woodland-glade birds exploiting the nesting and feeding opportunities, including Great Spotted Cuckoos, Azure-winged Magpies, and Rollers.

❺ KUUSAMO, FINLAND

In a heavily forested country, this is a specialized area of mostly coniferous forest. Rare birds include Red-flanked Bluetails, Arctic Warblers, and Three-toed Woodpeckers. The habitat also attracts birds such as Waxwings, Siberian Jays, and Parrot Crossbills.

■ *Woodland areas*

GARDENS, PARKS, AND TOWNS

Gardens vary greatly according to their location: some are little more than enclosed pieces of Mediterranean scrub while others represent woodland habitats in miniature. Urban gardens bring birds right into cities and tend to have a selection (though limited) of woodland birds but lack most of the bigger, shyer, or more demanding species. Artificial food in gardens is a lifeline to huge numbers of birds – and not just in winter. Spring is an important period when birds are trying to build up energy levels to begin nesting but many natural foods are running short. Town parks provide birds with shrubberies, lawns, and mature trees, while a park lake will attract wetland species.

PUBLIC SPACES

Town and city centres attract few species but spilled food and scraps and the abundance of ledges and cavities in buildings offer food and nest sites for an increasing number. Town and city parks are often much better areas for birds, especially if there is a lake or pond, and some shy species become surprisingly tame if they are not disturbed. Town or feral pigeons, which are derived from wild Rock Doves, are abundant in towns, while large lakes attract various ducks, coots, and gulls that forage for scraps. Few birds feed on roads, paved areas, and rooftops but wagtails often do so, finding insects trapped in rainwater puddles. The large numbers of town pigeons are increasingly attracting predatory peregrines into city centres and onto tall buildings in industrial areas. As towns are warmer than the surrounding countryside in winter, some species fly in specially to roost in city trees or on buildings.

BLACK REDSTART
In much of Europe Black Redstarts nest in cavities in buildings, which take the place of natural cliff habitats.

URBAN SECRETS
Dunnocks display to one another in town parks and gardens. Their sober appearance belies a complex social life: males and females often have more than one mate and can frequently be seen in "wing-waving displays".

ROOSTING WAGTAILS
Pied Wagtails feed on roofs, tarmac-covered footpaths, car parks, and town centre riverbanks, and each evening come together in large roosts for warmth and protection. They often choose ornamental trees for this, but equally seek out glasshouses and industrial sites in towns.

SUBURBAN SURPRISES
Industrial areas and roads on urban outskirts often see birds, such as these Lapwings, flocking in large numbers.

PRIVATE GARDENS

Ordinary suburban and rural gardens are becoming increasingly important wildlife havens. If there are thick shrubberies, flowerbeds with freshly turned earth, patches of lawn, and perhaps a few garden ponds, a suburban area can be almost as rich as a piece of natural woodland. These small habitats offer a range of feeding opportunities for all kinds of birds, some of which live permanently in the gardens while others just visit from nearby woods and streams. Even the shy Kingfisher and Grey Heron may make an early morning raid on a goldfish pond. Gardens are far poorer for birds, however, if they are too regimented and over-tidy.

GARDEN FEEDER
Robins are among many woodland birds that have become garden inhabitants. They like thick hedges and shrubs, close to pieces of bare ground and short grass where they can forage for worms and insects.

BERRY EATERS
Many birds eat berries, especially in autumn when they provide much needed energy. To attract birds like the Song Thrush, plant shrubs such as elderberry and cotoneaster.

EXPLOITING THE GARDENER
Blackbirds and Robins are well known for their bold and fearless behaviour, especially when a gardener is turning over soil and they get a good chance of grabbing a worm.

BIRD-TABLES
By placing food on a bird-table each day we are giving birds regular meals when natural food may be lacking. As an added bonus, we can watch them as they eat.

ADAPTATIONS

The most successful urban birds are species that can find habitats and food that resemble their more natural requirements out of town.

■ **Starlings** have strong muscles that can open their bills when they probe into a lawn for a leatherjacket. Their eyes are able to swivel backwards, so that they can watch for predators without having to stop searching for food.

■ **House Martins** naturally nest on cliff faces, but long ago learned to come into town and nest under the eaves of buildings. They do have problems when dry weather makes finding mud difficult.

HOUSE MARTIN

■ **Old buildings** provide all kinds of holes and cavities for birds to nest in, and Kestrels through to House Sparrows and Starlings exploit these. Swifts are now almost entirely reliant on buildings. However, as old buildings are being knocked down and new houses and office blocks go up that offer no nest sites, such birds may find life more difficult.

■ **Woodpigeons and town pigeons** can see a broader range of light than humans can. Grains and seeds, which we find difficult to find on gravel, stand out to their eyes as different colours and are quickly pecked up with great precision.

WOODPIGEON

WATCHING BIRDS

There are a vast number of opportunities to birdwatch – whether you visit a different country specifically for that purpose, go to a nature reserve, or simply look at the species that can be found every day in your garden or on your journey to work. You will find that there are a huge range of birds to see and the more you look, the more you will want to learn about the differences there are between them so that you can start to identify individual species.

WINTER ROBIN
In winter, putting out food or leaving apples unpicked on a tree will attract Robins and thrushes to your garden.

BIRD BOXES
Most gardens have food but no natural nest sites for hole-nesting birds such as sparrows, tits, and Starlings. Artificial boxes give them a chance to nest where you can watch them.

GETTING STARTED

It can be difficult to get close enough to birds to identify them as they tend to perch high on trees or far out on lakes. Don't let this put you off, however, as with practice and experience you will learn how to pick out unique characteristics from afar. The better you get, the more you will enjoy birdwatching. To help with this process, buy yourself an identification guide, a pair of binoculars, and a notebook – a knowledgeable friend is a great asset too.

BINOCULARS AND TELESCOPES

Binoculars are essential for birdwatching. They are described by figures, such as 10 × 40 or 8 × 30. The first is the magnification while the second is the diameter of the large lens. Higher magnification means larger images, but the lenses are also larger, making it harder to hold the binoculars steady; the field of view is also narrower and the image duller. You will need to compromise. There are top-quality binoculars that give really bright images but they are still heavy and expensive. It is best to opt for a magnification between 7 and 10. If you want to look at birds far away and need more power, consider buying a telescope. It will magnify 20 or 30 times but you will need a 60 or 80mm wide lens to let in enough light.

THE ESSENTIALS
Choose binoculars that you can handle easily, and buy the best you can afford. A camera and telescope can also be useful in the field.

BINOCULARS

focusing wheel

the "object lens" described by its diameter in millimetres

focus wheel

adjustable angle

objective lens

tripod

TELESCOPE

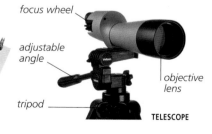

use a camera with a long lens to capture detail

SLR CAMERA

USING A NOTEBOOK

Notebooks are really invaluable for novice and experienced birdwatchers alike. Note down what you see if you want to keep records at home: it is much better than trying to rely on shaky memories later. If you come across a bird that is new to you, or has an unusual plumage, try to take a description – you will need to look closely at the bird in order to do this. It is a useful discipline, for example, to look at the colour of the bill, legs, and any special markings. Also write down the calls you hear and try to make a few sketches, however rough they are.

CODE OF CONDUCT

There is one simple rule for birdwatchers to follow: the welfare of the bird comes first. It is easy to get carried away, and to inch forward while trying to see a bird that you think might be rare or is one that you have not seen before. It may be tempting to chase a bird if it starts to move away, or to try flushing a bird out one last time, but all you will do is scare it away. Most birds are not bothered too much by everyday disturbance, but use your common sense. Don't disturb birds while they are at their nests, as you may put them in danger, and never be tempted to handle eggs. Keep all types of disturbance to a minimum – for the sake of the birds, other people, and your own view.

VIEWING CONDITIONS

Viewing conditions are an important consideration. It may seem obvious but many people forget that a white bird will look dark against a bright sky, while a dark brown bird may look remarkably pale against a ploughed field or winter hedgerow. White gulls lit by low, evening light may look orange on the sunlit side and quite blue on the shaded side; in strong sunlight they simply look brilliant white and dark grey. Dull light in rain or snow may reveal the subtleties of their greys and whites much better.

DAWN
Low, weak, misty light with an orange glow, which is not good for revealing detail.

MISTY MORNING
Mist may make a bird look large but it hides detail of colour and pattern, or even shape.

MORNING LIGHT
Low light may exaggerate contrast, but the brightness is perfect for seeing detail.

MIDDAY
Overhead sun can cast surprisingly deep shadows and "washes out" the colours on the upperside.

INTO THE SUN
Looking into bright light makes it difficult to see much more than a silhouette, even on a pale bird.

EVENING LIGHT
Lovely soft, warm light, but beware a rather orange cast to many colours.

USING BINOCULARS
If you see a bird that you want to view close-up, keep your eyes on it as you lift your binoculars.

USING A TRIPOD
Modern telescopes are short and need support, preferably on a good, firm tripod. This is a clumsy, heavy combination but the reward is unbelievably close views.

CONSERVATION

Conservation is simply wise management, which aims to maintain or enhance the numbers and variety of wild species and the habitats on which they depend. The birdlife in most parts of Europe has undergone dramatic changes almost everywhere over several centuries: little of Europe is now "natural", apart from the far north, some forests, mountains, and seas. It is therefore difficult to say that the present birdlife is "as it should be", and that all change must be resisted. Most people would, however, agree that we should at least try to maintain the variety of birds that we do have now. Every effort should be made to ensure that we do not lose any species from the European avifauna.

GREAT BUSTARD
The loss of grassland habitat, particularly to intensive farming, now threatens the last few thousand Great Bustards more severely than ever before.

CROWDED COASTS
Waders such as Redshanks use estuaries and other coastal habitats. Unfortunately for them, coasts are also ideal for various developments, from ports and industry to sport and leisure. Even people using beaches may disturb feeding or roosting flocks. Estuary birds face many threats.

THREATENED HABITATS

Conserving birds is not usually a case of preventing persecution or over-hunting. Most threats arise because birds' habitats are being damaged or destroyed. No wild creature can survive unless its needs for food, breeding sites, and safe refuges from predators are met. Protecting habitats usually involves planning and managing human activity, especially farming, forestry, drainage of marshes, and other large-scale changes. Local developments such as building roads and airports or urban housing expansion also affect the habitats. Food and habitat protection is often a political matter. A clear example of this is the farming and fishery policies, which affect countryside birds and seabirds across the whole of Europe.

DEGRADED RIVERS
Many rivers are dredged and straightened, which removes the habitat for riverside birds that prefer muddy shallows, sheltered bays, or beds of reed and sedges. Erosion eats into banks used by Kingfishers, while pollution, especially the acidification of rivers, affects the food supply – from tiny aquatic invertebrates to fish. An increased frequency of summer drought also affects river birds.

GLOBAL WARMING
Wading birds, wildfowl, gulls, and terns that use low-lying sea coasts and nearby lagoons for feeding and nesting face severe threats from the effects of global warming and sea-level rise. This presents many challenges to coastal conservation, not least finding sensitive ways to protect the coast.

FINDING SOLUTIONS

Conservation is complicated, and to achieve a political solution to some of the more intractable problems conservationists must have hard facts and figures. There is no substitute for good, reliable research to give a solid and well-respected backing to any conservation argument. Once the problems and their causes are understood, it is possible to have a more organized and objective approach to finding practical and economic solutions which other people, such as farmers, foresters, and gamekeepers, then put into practice.

RADIO TAGGING

This Black Grouse is being fitted with a minute radio transmitter so that its movements can be followed. Black Grouse have declined rapidly so finding out more about them and their problems is essential if solutions are to be found.

CORNCRAKE

In some countries, farmers are paid to maintain uncut field corners that provide early spring cover for Corncrakes, and to continue growing hay and cutting it late in the season. Such schemes are essential for the bird's survival.

GANNET PROBLEMS

Pollution and discarded waste, especially at sea, may cause unexpected problems for birds. Gannets build their nests out of discarded nylon twine and other debris that can entangle their growing chicks. Greater awareness and improved rules for the disposal of waste at ports rather than overboard at sea would help solve such problems.

RECORD KEEPING

Conservation in Europe relies on collating facts and figures, following trends, and identifying sudden changes in numbers and distributions of birds. This has a proud tradition of extensive amateur involvement, as thousands of people add their own records to the national and international data upon which our collective knowledge depends.

CONSERVATION GROUPS

Bird and habitat conservation groups worldwide work together in a global federation called BirdLife International. Each country has a BirdLife "partner". The largest is the RSPB in the UK. The RSPB works for a healthy environment that is rich in birds and other wildlife. It relies on the support and generosity of others to make a difference. Without members and other supporters it would lack the finance to do its work. When it does find and advocate solutions to problems faced by birds in the countryside, it relies on the willingness of decision-makers, landowners, and others to take notice, and act upon its advice. There are many other bird conservation bodies, bird clubs, and local groups that also play a vital role in conservation. Contact your country's BirdLife partner for further details of societies and groups near you if you would like to get involved.

THREATENED LESSER KESTREL

SPECIES GUIDE

WILDFOWL

THIS LARGE GROUP OF water birds splits into clear groups, with a few "in between" species. They have webbed feet, rather short but specialized bills, and swim well, although many spend much of their time on dry land.

DUCKS
The two main groups are surface feeders (or dabbling ducks) and diving ducks. The former feed on land or from the water surface (sieving food through their bills, gleaning grain, plucking vegetation), sometimes tipping over ("upending") but rarely diving. The diving ducks feed beneath the surface, diving as they swim (not plunging from the air): some are animal feeders, others vegetarian; some species feed at night.

SHELDUCKS
A few large species fall between ducks and geese, with an easy walk and rather long legs and neck, but in other respects they are more like the true ducks.

ANATIDAE

SEA DUCK
The Eider is a marine bird in the north and west of Europe, feeding largely on seabed shellfish.

GEESE
Geese are mostly large and terrestrial, feeding on dry land or marshes but returning to water (a lake or the sea) each night for a safe roost. They are social, flying in large flocks with loud, evocative calls – among the finest sights in Europe in winter.

SWANS
European swans are all-white when adult (elsewhere there are black and white species). They are longer-necked than the geese, tending to be more aquatic, but also feed extensively on drier ground, often in mixed flocks.

black facial knob

long yellow wedge on side of bill extends beyond nostrils

MUTE SWAN
p.59

rounded yellow bill patch falls short of nostrils

BEWICK'S SWAN
p.60

WHOOPER SWAN
p.61

ANATIDAE *continued*

yellow-orange legs

BEAN GOOSE
p.62

very dark, round head contrasts with pale breast

PINK-FOOTED GOOSE
p.63

white forehead blaze

WHITE-FRONTED GOOSE
p.64

large orange bill

GREYLAG GOOSE
p.65

white chinstrap

CANADA GOOSE
p.66

blue-grey back with black and white bars (irregular on juvenile)

BARNACLE GOOSE
p.67

bold white stern

BRENT GOOSE
p.68

red knob on forehead

♂

♀

no knob on bill

SHELDUCK
p.69

chestnut head and neck

mottled greyish to rust-brown body; dark legs (orange on Mallard)

♀

♂

black stern

WIGEON
p.70

♂

white patch

♀

GADWALL
p.71

bright green patch on hindwings

♂

♀

green head

brown, grey, or olive bill; small orange marks

♂

♀

streaked brown body with grey legs

TEAL
p.72

MALLARD
p.73

long, black tail spike

♂

pointed tail

♀

bold white stripe over eye

♂

pale, mottled brown plumage

♀

PINTAIL
p.74

GARGANEY
p.75

green-black head

♂

streaked, pale brown body

♀

rich red head

♂

brown head

♀

SHOVELER
p.76

POCHARD
p.77

long, wispy tuft on nape

♂

short crest on nape

♀

black head with green gloss

♂

rich brown head

♀

TUFTED DUCK
p.78

SCAUP
p.79

ANATIDAE *continued*

long, wedge-shaped head with green patch

closely barred brown body

rounded orange shield and pink-red bill

barred rusty-brown

♂

♀

EIDER
p.80

♂

♀

KING EIDER
p.81

long, flexible tail point

thick dark bill

all-black body with duller or paler wings (browner in summer)

greyish white lower face

♂

♀

LONG-TAILED DUCK
p.82

♂

♀

COMMON SCOTER
p.83

yellow sides of bill

dark bill tapers to long tip

green-glossed black head with bold white spot

dark brown head

♂

♀

VELVET SCOTER
p.84

♂

♀

GOLDENEYE
p.85

white crest droops over black nape

brown cap

wispy crest on green-black head

fuzzy crest

♂

♀

SMEW
p.86

♂

♀

RED-BREASTED MERGANSER
p.87

green-black head

dark brown head

blue bill

dull dark-grey brown body

♂

♀

GOOSANDER
p.88

♂

♀

RUDDY DUCK
p.89

Order **Anseriformes**	Family **Anatidae**	Species **Cygnus olor**

Mute Swan 🔊 6.I, 6.II

grey-brown plumage becomes blotched white

grey bill

JUVENILE

all-white plumage

black facial knob

reddish orange bill, angled down

outstretched neck

ADULT

IN FLIGHT

relatively long, pointed tail

ADULT

long neck, often curved or upright

A large, familiar bird, strikingly white and obvious even at great range, the Mute Swan is generally quite tame, even semi-domesticated in its behaviour and choice of habitat. Territorial pairs are aggressive, even to people or their dogs, using impressive displays of arched wings and loud, hissing calls. In some floodplains, small groups regularly feed on dry land, a habit that is more consistent with the two "wild swans", Bewick's and Whooper.

VOICE Strangled trumpeting and hissing notes.
NESTING Huge pile of vegetation at water's edge; up to 8 eggs; 1 brood; March–June.
FEEDING Plucks vegetable matter from short grass in fields and salt marshes, pulls the same from shallow water, or upends in deeper water.

FLIGHT: heavy but powerful, direct, with neck outstretched; strong, regular wingbeats produce throbbing sound.

THREAT DISPLAY
When threatened, the Mute Swan raises its wings like sails, lowers its head, and makes a loud, rough hiss.

SHELTERED YOUNG
Small cygnets seek shelter and refuge between the wings of a parent swan.

NEST
The nest is a large mound of vegetation built at the water's edge. The female lays up to 8 eggs and incubates them.

OCCURRENCE
Many breed on park lakes and other small pools, others on natural lakes, reservoirs, and rivers, almost throughout Europe; in W Europe, commonly seen on sheltered sea coasts and marshes. May join other swans feeding on flat, open fields.

Seen in the UK
J F M A M J J A S O N D

SIMILAR SPECIES

BEWICK'S SWAN see p.60
black and yellow bill
short tail

wedge-shaped head
black and yellow bill
smaller

WHOOPER SWAN see p.61

Length **1.4–1.6m (4½–5¼ft)**	Wingspan **2.08–2.38m (6¾–7¾ft)**	Weight **10–12kg (22–26lb)**
Social **Small flocks**	Lifespan **15–20 years**	Status **Secure**

Order **Anseriformes**	Family **Anatidae**	Species ***Cygnus columbianus***

Bewick's Swan  8

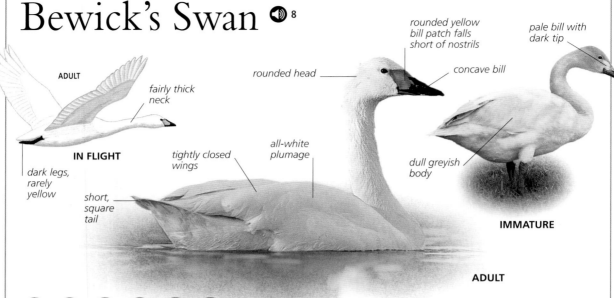

ADULT

fairly thick neck

rounded yellow bill patch falls short of nostrils

pale bill with dark tip

rounded head

concave bill

IN FLIGHT

dark legs, rarely yellow

short, square tail

tightly closed wings

all-white plumage

dull greyish body

IMMATURE

ADULT

FLIGHT: direct, strong; regular wingbeats, with simple quiet whistle from feathers at close range, no loud, throbbing sound.

The smallest of the swans, Bewick's Swan is typically rather stocky, although it can look surprisingly thin-necked at times. Unlike the Mute Swan, this is a thoroughly wild swan, although in certain places where it is fed in winter, it has become much more approachable. It is a very vocal bird, especially in flocks, its conversational chorus often penetrating the winter gloom over long distances.

VOICE Loud, bugling notes, less strident or trumpeting than Whooper Swan; often soft, conversational chorus from flocks.

NESTING Pile of grass stems and similar vegetation at edge of pool in tundra; 3–5 eggs; 1 brood; May–June.

FEEDING Often grazes on grass or cereal crops, or eats root crops in ploughed fields; feeds less often in water.

REMARK Subspecies *C. c. columbianus* (North America, very rare in Europe) has tiny yellow spot on black bill.

MIXED FLOCK
Bewick's Swans, mixed with other wildfowl, including Mute and Whooper Swans, can be watched closely from hides at several reserves. Elsewhere they are shy and wild.

YELLOW BILL
Any swan with yellow on the base of the bill will be a wild migrant to western Europe: a Bewick's or a Whooper Swan.

OCCURRENCE
Breeds in N Siberia. In winter, moves to lowlands of W Europe, mostly agricultural land, usually in traditional areas, occupied year after year. In some places, concentrates on reserves where it is fed.

Seen in the UK
J F M A M J J A S O N D

SIMILAR SPECIES

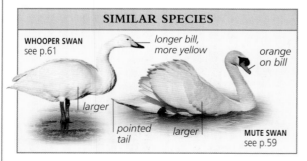

WHOOPER SWAN see p.61

longer bill, more yellow

orange on bill

larger

larger

pointed tail

MUTE SWAN see p.59

Length **1.15–1.27m (3¾–4¼ft)**	Wingspan **1.8–2.1m (6–7ft)**	Weight **5–6.5kg (11–14¼lb)**
Social **Flocks**	Lifespan **Up to 10 years**	Status **Localized**

Order **Anseriformes**	Family **Anatidae**	Species **Cygnus cygnus**

Whooper Swan 🔊 7

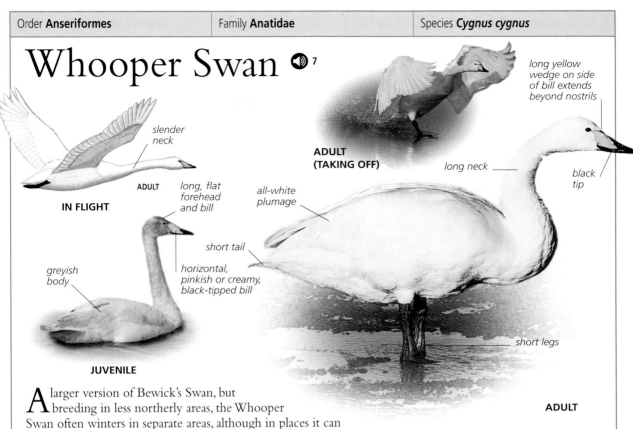

slender neck

ADULT

IN FLIGHT

long yellow wedge on side of bill extends beyond nostrils

ADULT (TAKING OFF)

long neck

black tip

long, flat forehead and bill

all-white plumage

short tail

greyish body

horizontal, pinkish or creamy, black-tipped bill

short legs

JUVENILE

ADULT

A larger version of Bewick's Swan, but breeding in less northerly areas, the Whooper Swan often winters in separate areas, although in places it can be found together with the Mute Swan and Bewick's Swan. The Whooper Swan is a wild, usually shy bird, far less approachable than the Mute Swan and, like Bewick's, usually more terrestrial. Although equally large, it is a more agile bird than the Mute Swan; however, the Whooper Swan lacks the arch-necked elegance (it holds its neck bolt upright and head horizontal) and rich bill colours of the latter.
VOICE Loud trumpeting call, slightly lower-pitched; simpler bugling than Bewick's Swan, often three or four syllables instead of two.
NESTING Big, domed structure of grass and reed stems at water's edge or built up from bottom of shallow lake; 5–8 eggs; 1 brood; April–June.
FEEDING Plucks leaves and stems from short vegetation on dry ground, or digs roots and waste crops from ploughed earth; feeds on aquatic plants mostly in summer.

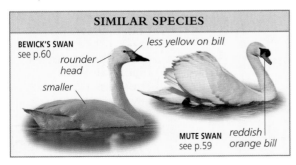

FLIGHT: powerful and direct; regular wingbeats with little wing noise.

UPRIGHT NECKS
Whooper Swans have long, slim necks, held upright, with head horizontal, when alert.

DENSE FLOCKS
Whooper Swans are found in dense flocks on a few nature reserves where they are fed.

SIMILAR SPECIES

BEWICK'S SWAN see p.60

less yellow on bill

rounder head

smaller

MUTE SWAN see p.59

reddish orange bill

OCCURRENCE
Breeds in Scandinavia and Iceland on remote pools; in winter locally across NW and C to SE Europe on large lakes and marshes. Increasing numbers winter on reserves with Bewick's Swans, taking advantage of artificial feeding.

Seen in the UK
J F M A M J J A S O N D

Length **1.4–1.6 m (4½–5¼ft)**	Wingspan **2.05–2.35m (6¾–7¾ft)**	Weight **9–11kg (20–24lb)**
Social **Flocks**	Lifespan **Up to 10 years**	Status **Secure**

| Order **Anseriformes** | Family **Anatidae** | Species *Anser fabalis* |

Bean Goose

broad dark tail-band

dark grey wings

fine white tail tip

extended neck

IN FLIGHT

ADULT

longer head-bill profile than Pink-footed Goose

dark brown head

orange band on black bill

cream bars across brown back (less regular bars on juvenile)

pale brown breast

yellow-orange legs

ADULT

This large, social goose returns each winter to traditional areas, feeding in the same fields and roosting on the same lakes year after year. A dark brown goose, it has two basic forms, one long-necked and long-billed, the other shorter-necked and more like a Pink-footed Goose in appearance. Except in the Low Countries it is a scarce bird, not usually seen in large numbers. Its long, dark head and neck and cleanly barred back help to identify it in flocks of White-fronted Geese, with which it sometimes mingles.

VOICE Deep, two- or three-syllable trumpeting, *ung-ung* or *unk-uk-uk*.

NESTING Down- and feather-lined hollow on ground near bog pools, in open tundra or in forest clearings; 4–6 eggs; 1 brood; June.

FEEDING Grazes on short grass; picks up grain and root crops from stubble or ploughed fields, often in traditional areas used for decades.

FLIGHT: strong and direct, long neck obvious; in lines of "V"s.

FLOCKS
Bean Geese feed in flocks, at times mixing with other geese, often on arable land.

OCCURRENCE
Breeds in N Scandinavia on bogs and tundra pools. In winter, mostly around S Baltic/North Sea and E Europe; one flock in England, one in Scotland. These are in traditional sites, threatened by disturbance and habitat change.

Seen in the UK
J F M A M J J A S O N D

SIMILAR SPECIES

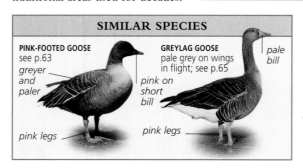

PINK-FOOTED GOOSE see p.63
greyer and paler
pink legs

GREYLAG GOOSE pale grey on wings in flight; see p.65
pale bill
pink on short bill
pink legs

SUBSPECIES

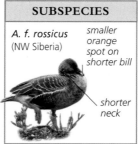

A. f. rossicus (NW Siberia)
smaller orange spot on shorter bill
shorter neck

| Length **66–84cm (26–33in)** | Wingspan **1.47–1.75m (4¾–5¾ft)** | Weight **2.6–3.2kg (5¾–7lb)** |
| Social **Flocks** | Lifespan **Up to 10 years** | Status **Secure** |

Order **Anseriformes**	Family **Anatidae**	Species *Anser brachyrhynchus*

Pink-footed Goose 🔊 9

narrow dark tail-band

broad white tail tip

pale grey wings

IN FLIGHT

ADULT

white-barred, pale grey back (browner, less neatly barred on juvenile)

very dark, round head contrasts with pale breast

small bill with pink band

darker bars on flanks

pale to rich pink legs

dark underwings

ADULT

ADULT

With large population increases in recent decades, the Pink-footed Goose is found in tens of thousands in favoured places, with regular daily feeding and roosting regimes. Evening flights to the roosts, especially, are spectacular, unless the feeding geese are encouraged to stay away all night under the light of a full moon. By day, they feed in dense flocks on fields, and are sometimes surprisingly difficult to locate. They are usually wary and difficult to approach.

VOICE Like Bean Goose but less deep, resonant *ahng-unk* and frequent higher *wink-wink*.

NESTING Down-lined nest on ground in open tundra and exposed rocky slopes; 4–6 eggs; 1 brood; June–July.

FEEDING Large flocks eat grass, waste grain, sugar beet tops, carrots, and potatoes.

FLIGHT: direct, strong flight, head and neck short; forms long lines and "V" shapes.

FEEDING IN FLOCKS
Pink-footed Geese generally feed in dense groups with much noise and activity. They frequently mix with other geese species.

SIMILAR SPECIES

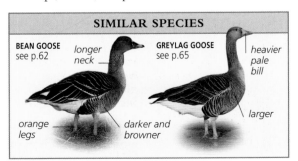

BEAN GOOSE see p.62

longer neck

orange legs

darker and browner

GREYLAG GOOSE see p.65

heavier pale bill

larger

OCCURRENCE
Breeds in Greenland, Iceland, and Svalbard. Moves to Great Britain and Low Countries in winter, roosting on large lakes, estuaries, and low-lying islands offshore; feeding on marshes, pasture, and arable land, close to coast.

Seen in the UK
J F M A M J J A S O N D

Length **64–76cm (25–30in)**	Wingspan **1.37–1.61m (4½–5¼ft)**	Weight **2.5–2.7kg (5½–6lb)**
Social **Large flocks**	Lifespan **10–20 years**	Status **Secure**

Order **Anseriformes**	Family **Anatidae**	Species *Anser albifrons*

White-fronted Goose

broad dark tail-band

ADULT

outstretched head and neck

grey on wings

IN FLIGHT

white forehead blaze

pink bill

oily-brown body looks greyer in strong sun

pale bars on back

no black bars below

orange legs

black bars on belly

pink bill with dark tip

JUVENILE

ADULT

vivid orange legs

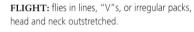

FLIGHT: flies in lines, "V"s, or irregular packs, head and neck outstretched.

One of the most colourful and lively of the grey geese, the White-fronted Goose returns each winter to regular locations. Its flocks often attract stragglers of other species, and hence are particularly significant for bird-watchers. Knowledge of this, more common, species and its variable appearance in different lighting conditions is valuable when trying to identify other geese.

VOICE High, yodelling, yelping notes, *kyu-yu, ku-yu-yu* or *lo-lyok*.
NESTING Down-filled nest on ground; 5 or 6 eggs; 1 brood; June.
FEEDING Grazes on firm ground during steady forward walk, taking grass, roots, some winter wheat, and grain.

SUBSPECIES

A. a. flavirostris
(Greenland)

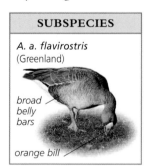

broad belly bars

orange bill

WATERSIDE GOOSE
Water is essential for drinking and also as a safe refuge for night-time roosts. Floods, broad rivers, and estuaries are preferred.

SIMILAR SPECIES

GREYLAG GOOSE
see p.65

larger

pink legs

LESSER WHITE-FRONTED GOOSE
see p.444

yellow eye-ring

small, bright pink bill

OCCURRENCE
Breeds in Greenland and far N Russia. Winters on pastures and coastal marshes in Great Britain, Ireland, Low Countries, S France, and E Europe. Often with, or near, other geese in winter, sometimes in huge flocks.

Seen in the UK
J F M A M J J A S O N D

Length **65–78cm (26–31in)**	Wingspan **1.3–1.65m (4¼–5½ft)**	Weight **1.9–2.5kg (4¼–5½lb)**
Social **Flocks**	Lifespan **15–20 years**	Status **Secure**

| Order **Anseriformes** | Family **Anatidae** | Species **Anser anser** |

Greylag Goose  10

very pale grey upperwings

ADULT

outstretched head

IN FLIGHT

pale underwings

ADULT

pale, plain head

large orange bill

brown-grey plumage (less neatly barred on juvenile)

large pale body

pink legs (rarely orange)

ADULT

Of all grey geese, the Greylag (the direct ancestor of the domestic goose) most resembles the farmyard goose. It is also the most easily seen grey goose in the UK because it has been introduced in many lowland areas where sizeable flocks are now semi-tame and resident all year round. Winter visitors are still wild and shy. In much of eastern and central Europe, the Greylag is naturally resident in extensive marshes.

VOICE Loud, clattering, and honking notes like farmyard bird, *ahng-ahng-ahng, kang-ank.*

NESTING Sparsely lined ground nest, often on island; 4–6 eggs; 1 brood; May–June.

FEEDING Grazes; plucks grass and cereal shoots, digs for roots and waste grain.

FLIGHT: powerful, head outstretched; in "V"s and long lines.

SLOW FLIERS
Flocks of Greylag Geese in flight are rather slower and heavier than other grey geese but become agile when losing height before landing.

OCCURRENCE
Breeds widely in Europe. Large numbers from Iceland visit NW Europe in winter, coming to coastal marshes and pastures. Introduced (feral) birds common in some places, including S England and Scotland where they are resident and far less wild than most.

Seen in the UK
| J | F | M | A | M | J | J | A | S | O | N | D |

SIMILAR SPECIES

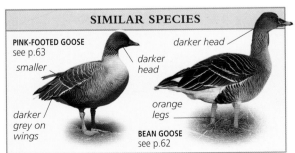

PINK-FOOTED GOOSE
see p.63

smaller

darker head

darker grey on wings

darker head

orange legs

BEAN GOOSE
see p.62

FEEDING
The Greylag Goose, like all grey geese, has a horizontal, head-down posture when feeding and shows a striking white rear.

| Length **74–84cm (29–33in)** | Wingspan **1.49–1.68m (5–5½ft)** | Weight **2.9–3.7kg (6½–8¼lb)** |
| Social **Flocks** | Lifespan **15–20 years** | Status **Secure** |

| Order **Anseriformes** | Family **Anatidae** | Species *Branta canadensis* |

Canada Goose 🔊 11.I, 11.II

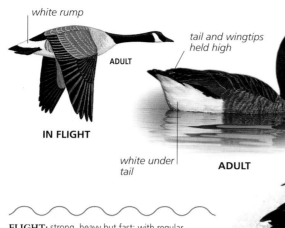

white rump

ADULT

IN FLIGHT

tail and wingtips held high

white under tail

ADULT

black bill

white chinstrap

black head and neck

swan-shaped, brown body (juvenile duller)

pale breast

black legs

ADULT

FLIGHT: strong, heavy but fast; with regular wingbeats; often in "V"s.

Originally imported from North America as an ornamental bird, the Canada Goose is now well established in many areas as a largely resident and rather tame bird, with little of the romance of "wild geese". Most have lost their migratory instinct. A remarkably successful and adaptable bird, it is sometimes considered a messy and aggressive pest in town parks. It often mingles with equally "artificial" groups of introduced Greylags. The Canada Goose is, nevertheless, a handsome bird.

VOICE Deep, loud, two-syllable, rising *ah-ronk!*

NESTING Down-lined scrape on ground, often on small island; loosely colonial; 5 or 6 eggs; 1 brood; April–June.

FEEDING Grazes on grass and cereals, takes some aquatic plants.

FLOCK ON WATER
Big flocks of Canada Geese are often seen on ornamental lakes in undisturbed estates.

HONKING PARTIES
Flocks of Canada Geese usually draw attention to themselves by their loud, honking calls.

FAMILY GROUP
The male and the female Canada Geese are alike and the goslings soon look like duller versions of their parents.

OCCURRENCE
Mostly in UK, Scandinavia, and Low Countries, on marshes, reservoirs, and flooded pits or surrounding grassland. Very few wild vagrants from North America are seen with other geese in W Europe in winter.

Seen in the UK
J F M A M J J A S O N D

SIMILAR SPECIES

BARNACLE GOOSE
see p.67

white face

black breast

smaller and greyer

GREYLAG GOOSE X CANADA GOOSE HYBRID
• duller head/face patterns
• orange bill
• often pinkish legs

| Length **90–110cm (35–43in)** | Wingspan **1.5–1.8m (5–6ft)** | Weight **4.3–5kg (9½–11lb)** |
| Social **Flocks** | Lifespan **20–25 years** | Status **Localized** |

Barnacle Goose

cream-tinged white face
(pure white on juvenile)

small, stubby
black bill

black eye
patch

glossy
black neck
and chest

pale grey
wings

ADULT

IN FLIGHT

ADULT

blue-grey back with
black and white bars
(irregular on juvenile)

white
underside

strongly contrasted
wings and underside

ADULT

ADULT

~~~

FLIGHT: strong, quick; flocks form irregular packs.

Few birds so numerous as this are restricted to such localized, traditional wintering sites. Barnacle Geese are predictably found, mostly on protected areas, from autumn until late spring, and are generally rather shy and unapproachable except when viewed from hides. They are not nearly so widespread as Canada Geese and, except for a handful of escapees and semi-tame groups, not so "suburbanized". Adult birds have strongly contrasted upper- and underparts, while the juveniles are duller and lack the cream tinge on the white face, which is so evident in the adults.

VOICE Harsh, short bark, creating chattering, yapping, unmusical chorus from flocks.
NESTING Feather-lined nest on ground or on cliff ledges in Arctic tundra; 4–6 eggs; 1 brood; May–June.
FEEDING Large flocks graze on grass, clover, and similar vegetation.

CONTRASTED COLOURS
Barnacle Geese, like other goose species, are especially prone to showing strong contrasts in low winter sun.

IRREGULAR FLOCK
Flocks of Barnacle Geese form irregular packs in flight, not "V"s or chevrons.

OCCURRENCE
Breeds in Greenland and Svalbard with populations remaining separate all year. Mostly in Iceland, W Scotland, Ireland, and Low Countries in winter, with migrants through Baltic, on grassy pastures and salt marshes.

Seen in the UK
J F M A M J J A S O N D

SIMILAR SPECIES

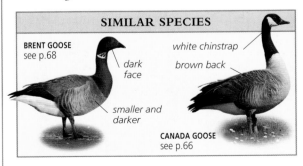

BRENT GOOSE
see p.68

dark
face

smaller and
darker

white chinstrap

brown back

CANADA GOOSE
see p.66

Length **58–70cm (23–28in)**	Wingspan **1.32–1.45m (4¼–4¾ft)**	Weight **1.5–2kg (3¼–4½lb)**
Social **Flocks**	Lifespan **Up to 18 years**	Status **Localized**

Order **Anseriformes**	Family **Anatidae**	Species *Branta bernicla*

Brent Goose 🔊 12

- **black head**
- **black bill**
- **white patch high on neck (absent on juvenile)**
- **dark grey-brown upperparts**
- **black chest**
- **brown underside**
- **black legs**

JUVENILE — pale bars across wings

ADULT (DARK-BELLIED) IN FLIGHT — uniform wings

bold white stern

ADULT (DARK-BELLIED)

In winter, this goose of low-lying, often muddy, coasts is common and increasingly tame in many areas, coming into estuaries and harbours and even feeding on roadside fields at high tide. Flocks are often on water, upending like ducks in order to feed. At low tide most are scattered over estuary mud or in the remaining narrow channels. Their pleasant, growling calls are far-carrying and typical of many estuaries from October to March. Although little longer than a Mallard, the Brent Goose usually looks much bigger.

VOICE Rhythmic, deep, throaty *rronk rronk*, creating loud, murmuring chorus from large flocks.

NESTING Feather-lined nest on ground near shallow pool; 4–6 eggs; 1 brood; May–June.

FEEDING Eats eelgrass and algae on mudflats, increasingly cereals and grass on fields.

FLIGHT: fast, strong; deep, quick wingbeats; in irregular masses or long lines.

LOOSE FLOCK
Loose flocks of Brent Geese rise from creeks and muddy channels.

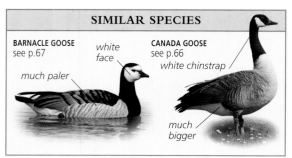

ESTUARY BIRD
Brent Geese are commonly seen scattered over estuary mud and the adjacent arable land at low tide (pale-bellied birds pictured).

SUBSPECIES

B. b. hrota
(Ireland, NE England)

whitish underparts

B. b. nigricans
(vagrant from North America)

white flanks

white in front

blackish belly

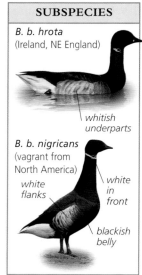

SIMILAR SPECIES

BARNACLE GOOSE see p.67 — white face, much paler

CANADA GOOSE see p.66 — white chinstrap, much bigger

OCCURRENCE
Breeds on Arctic tundra. Mostly winters in Great Britain, Ireland, and Low Countries, with large migrations through Baltic. Pale-bellied birds from Greenland winter separately from dark-bellied Siberian breeders.

Seen in the UK
J F M A M J J A S O N D

Length **56–61cm (22–24in)**	Wingspan **1.1–1.21m (3½–4ft)**	Weight **1.3–1.6kg (2¾–3½lb)**
Social **Flocks**	Lifespan **12–15 years**	Status **Localized**

| Order **Anseriformes** | Family **Anatidae** | Species *Tadorna tadorna* |

Shelduck

bold black wingtips

MALE

IN FLIGHT

red knob on forehead

black head

bright red bill

white body

broad orange band around chest

pink legs

brown-black cap

white underside

pink or grey bill

grey legs

IMMATURE

paler patch on cheek

no knob on bill

FEMALE

tawny orange band around foreparts

MALE

Mostly but not exclusively coastal, the Shelduck is widespread and easily identified. Family groups gather together in late summer when most adults fly to the Helgoland Bight (Germany) to moult; at other times, pairs or small, loose flocks are usual. The bright white plumage is easily visible at great range across dark estuary mud. With the ever-increasing number of gravel workings in lowland areas, Shelducks have been able to spread inland to take advantage of newly flooded, worked-out pits.

VOICE Goose-like *a-ank* and growled *grah grah*; various whistling notes from male and rhythmic *gagagagaga* from female in spring.

NESTING In holes on ground, between straw bales, in old buildings, under brambles, and also in trees; 8–10 eggs; 1 brood; February–August.

FEEDING Typically sweeps bill from side to side over wet mud to find algae, snails, and small crustaceans; also grazes and upends in shallow water.

FLIGHT: strong, fast-flying, but rather heavy, goose-like action.

UPENDING
Shelducks often upend to feed on submerged plants and animals.

OCCURRENCE
Widespread as breeding and wintering bird on coasts but only locally in Mediterranean. Mostly found on sandy or muddy shores, especially sheltered estuaries, with some on freshwater lakes, reservoirs, or flooded pits well inland.

| Seen in the UK |
| J F M A M J J A S O N D |

GRAZING DUCKS
Pairs of Shelducks may sometimes be found feeding around the shores of lakes and reservoirs.

SIMILAR SPECIES

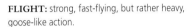

MALLARD ♂ similar to ♂ ♀; see p.73 greyer body

yellow bill

dark green head

| Length **58–65cm (23–26in)** | Wingspan **1.1–1.33m (3½–4¼ft)** | Weight **0.85–1.4kg (1¾–3lb)** |
| Social **Flocks** | Lifespan **5–15 years** | Status **Secure** |

| Order **Anseriformes** | Family **Anatidae** | Species ***Anas penelope*** |

Wigeon 🔊 14. I, 14. II

dull grey wings

FEMALE

white belly

pointed tail

bold white patch on wings (adult)

IN FLIGHT

MALE (SUMMER)

redder than female; retains white on wings

round grey body, paler than Teal

black and white stern

chestnut head and neck

yellow forehead

short, black-tipped greyish bill

pink breast

white band on wings

white belly

MALE (WINTER)

mottled greyish to rust-brown body; dark legs (orange on Mallard)

round head

black-tipped bluish bill

FEMALE

L ike most ducks, the Wigeon forms close flocks on water while feeding, advancing across a salt marsh or meadow in a tight-packed mass. Such a flock looks richly colourful and adds to the effect with constant loud calls. Wigeon are generally shy and fly off when approached, large numbers forming wheeling flocks circling above a marsh or heading for the safety of a reservoir. They have good reason to be wary, being the prime target of many fowlers.

VOICE Male has loud, explosive, musical whistle, *whee-oo*; female has deep, rough, abrupt growl.
NESTING In long vegetation on ground, near water; 8 or 9 eggs; 1 brood; April–July.
FEEDING Grazes on short grass, often in dense flocks; also feeds in shallow water, taking seeds, shoots, and roots.

FLIGHT: fast, wader-like, head protruding, wings swept back and pointed, tail pointed.

GRAZING
Dense Wigeon flocks feed on short grass near water. The entire flock usually faces one way while feeding.

OCCURRENCE
Breeds in N Europe and N UK, on edges of moorland pools and lakes in forests. Much more widespread in winter on estuaries and freshwater marshes, and on grassy areas surrounding reservoirs and water-filled pits.

| Seen in the UK |
| J F M A M J J A S O N D |

SIMILAR SPECIES

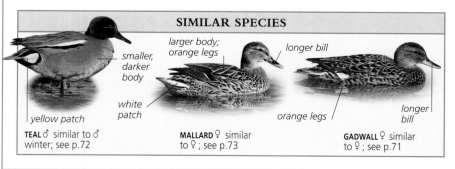

smaller, darker body

yellow patch

TEAL ♂ similar to ♂ winter; see p.72

larger body; orange legs

white patch

MALLARD ♀ similar to ♀; see p.73

longer bill

orange legs

GADWALL ♀ similar to ♀; see p.71

longer bill

| Length **45–51cm (18–20in)** | Wingspan **75–86cm (30–34in)** | Weight **500–900g (18–32oz)** |
| Social **Dense flocks** | Lifespan **Up to 15 years** | Status **Secure** |

| Order **Anseriformes** | Family **Anatidae** | Species *Anas strepera* |

Gadwall

white patch near base of wings

protruding head

white patch

FEMALE

white belly in flight

head squarer, paler than Mallard's

dark bill with orange sides

mottled brown body

orange-sided bill

white belly

MALE (WINTER)

IN FLIGHT

white patch on wings

MALE (SUMMER)

pale brown head

pale area

grey body

black stern

steep forehead

narrow, straight black bill

orange legs like Mallard (unlike Pintail, Wigeon and Teal)

MALE (WINTER)

A large, elegant, surface-feeding duck, with a smaller, squarer head than the Mallard, the Gadwall is usually far less abundant. Dull colours at a distance reveal exquisite patterns at close range. Pairs are often seen flying over territories in spring, with characteristic calls attracting attention. In autumn and winter, Gadwalls often flock on reservoirs and pits, frequently scattered through flocks of Coots (see p.168), exploiting the food that the latter bring up from the bottom when they dive.

VOICE Male has high, nasal *pee* and croaked *ahrk*, female has loud quack.

NESTING Down-lined hollow on ground near water; 8–12 eggs; 1 brood; April–June.

FEEDING Mostly feeds in shallow water, dabbling and upending for seeds, insects, roots, and shoots of aquatic plants.

FLIGHT: strong, quick, direct, head protruding; frequently in pairs.

FEEDING GADWALL PAIR
The male (left) is "upending" to reach deeper food, a typical surface-feeding duck technique.

SIMILAR SPECIES

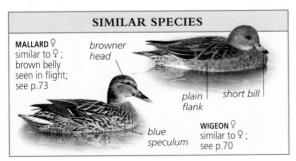

MALLARD ♀ similar to ♀; brown belly seen in flight; see p.73

browner head

plain flank

short bill

blue speculum

WIGEON ♀ similar to ♀; see p.70

OCCURRENCE
Mostly breeds in mid- and W Europe, on lakes and rivers with reeds or wooded islands. In winter, more westerly, on open waters such as big reservoirs and flooded pits, but prefers some shelter. Scarce on salt marshes and quiet estuaries.

Seen in the UK
J F M A M J J A S O N D

| Length **46–56cm (18–22in)** | Wingspan **84–95cm (33–37in)** | Weight **650–900g (23–32oz)** |
| Social **Flocks** | Lifespan **10–15 years** | Status **Vulnerable** |

| Order **Anseriformes** | Family **Anatidae** | Species *Anas crecca* |

Teal

white central bar

pale streak beside tail

MALE (WINTER)

greyish head, sometimes with dark-capped effect

grey bill

pale streak beside tail

MALE (SUMMER)

FEMALE/JUVENILE

streaked brown body with grey legs

green band on brown head

thin horizontal white line along side

bright green patch on hindwings

thick midwing bar

pale leading edge

green patch

FEMALE

IN FLIGHT

black-edged yellow triangle under tail

grey body (plumage similar to female's in summer)

MALE (WINTER)

The smallest common surface-feeding duck, the Teal is agile and quick in flight, its movements recalling those of a wader. Nervous groups may often swoop down to a sheltered bay, only to dart over the water surface and wheel up and away once more; sometimes several such approaches precede their eventual settling. In places, Teal occur in hundreds but 20–40 are more typical, scattered along well-vegetated or muddy shores, or in wet marshes.

VOICE Male has loud, ringing, high-pitched *crik crik* that can be easily heard at long range across marshes or estuaries; female has high quack.

NESTING Down-lined hollow near water; 8–11 eggs; 1 brood; April–June.

FEEDING Mostly in water or on muddy shores, taking plants and seeds.

FLIGHT: quick, active, twisting; swooping in to settle like waders.

DRAB DUCK
An absence of bright colours on the bill and legs helps to identify the female Teal.

OCCURRENCE
Breeds in N and E Europe on freshwater marshes and wet moors and heaths, including high moorland pools. Winters more widely in S and W Europe, mostly on fresh waters with muddy edges and around estuaries.

Seen in the UK
J F M A M J J A S O N D

SIMILAR SPECIES

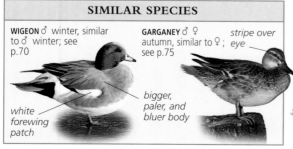

WIGEON ♂ winter, similar to ♂ winter; see p.70

white forewing patch

GARGANEY ♂ ♀ autumn, similar to ♀; see p.75

stripe over eye

bigger, paler, and bluer body

A. carolinensis

Green-winged Teal
Closely related N American species, rare in W Europe

vertical white line near chest

| Length **34–38cm (13½–15in)** | Wingspan **58–64cm (23–25in)** | Weight **250–400g (9–14oz)** |
| Social **Small flocks** | Lifespan **10–15 years** | Status **Secure** |

Order **Anseriformes**	Family **Anatidae**	Species *Anas platyrhynchos*

Mallard ◀)) 13.I, 13.II

MALE

white underwings

white tail

blue speculum

brown, grey, or olive bill; small orange marks

FEMALE

streaked brown body

browner head

bill remains yellow

MALE (SUMMER)

becomes browner

FEMALE

curly central tail feathers

purple-blue, white-edged speculum

green head

yellow bill

white neck-ring

brown breast

dark belly

IN FLIGHT

MALE (WINTER)

The most widespread and familiar of all the ducks, this is the one most often seen in town parks, coming for bread on lakes or on the riverside: but there are also many that are truly wild and shy. The different breeds of the farmyard duck are the domestic forms of the Mallard. They vary from very dark brown to all-white. The status of the wild form is obscured by the release of thousands bred for shooting.

VOICE Male whistles quietly; female has loud, raucous *quark quark*.
NESTING Down-lined nest on ground or in raised sites; 9–13 eggs; 1 brood; January–August.
FEEDING Takes small aquatic invertebrates, seeds, roots, shoots, and grain from shallows while upending or dabbling, or from dry ground.

FLIGHT: strong, fast; wingbeats mostly below body level; often in groups.

DUCKLINGS
Mallard ducklings follow the mother to water. They stay under her care until they are able to take care of themselves.

OCCURRENCE
Nests practically anywhere within reach of almost any kind of water, from towns to remote moorland pools and northern lakes, almost throughout Europe. In winter, more western, often on estuarine salt marshes but less so on open sea. Mostly feeds on arable fields and muddy margins of lakes.

Seen in the UK
| J | F | M | A | M | J | J | A | S | O | N | D |

SIMILAR SPECIES

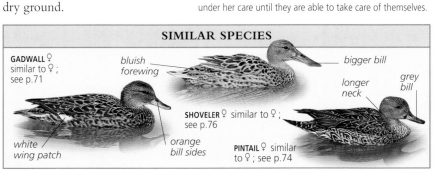

GADWALL ♀
similar to ♀;
see p.71

bluish forewing

white wing patch

orange bill sides

SHOVELER ♀ similar to ♀;
see p.76

PINTAIL ♀ similar to ♀; see p.74

bigger bill

longer neck

grey bill

Length **50–65cm (20–26in)**	Wingspan **81–95cm (32–37in)**	Weight **0.75–1.5kg (1¾–3¼lb)**
Social **Flocks**	Lifespan **15–25 years**	Status **Secure**

| Order **Anseriformes** | Family **Anatidae** | Species *Anas acuta* |

Pintail

long black tail spike

MALE (WINTER)

outstretched head and neck

white trailing edge

FEMALE

dark patch

pale belly

IN FLIGHT

yellowish patch

brown head

white neck-stripe and breast

grey bill with black lines

MALE (WINTER)

dull greyish body

grey bill

pale breast

MALE (SUMMER)

plain, pale tawny head

mottled grey-brown body with grey legs

grey bill

pointed tail

FEMALE

L arge and slim, the Pintail is perhaps the most elegant of all the surface-feeding ducks. It is numerous in a few traditional wintering areas, both salt- and freshwater, but otherwise rather scarce, usually in ones and twos among commoner waterfowl. Individuals in autumn flocks are always a bit of a challenge for the birdwatcher, before the striking winter plumage of the males has fully developed.

VOICE Male has low, short whistle; female's quack like Mallard's but quieter.
NESTING Leaf- and down-lined hollow on ground; 7–9 eggs; 1 brood; April–June.
FEEDING Mostly dabbles and upends in water but also grazes on grass and marsh and visits cereal fields for spilt grain.

FLIGHT: quick, straight, with head and neck outstretched, tail long.

UPENDING PINTAILS
Pintail drakes tip up to feed, revealing the long tail, black vent, and white belly.

OCCURRENCE
Breeds mostly in N and E Europe, sporadically in W, nesting on moorland and coastal marshes. In winter, flocks concentrate on traditional areas on estuaries and fresh marshes south to Mediterranean, with very small numbers turning up elsewhere.

Seen in the UK
J F M A M J J A S O N D

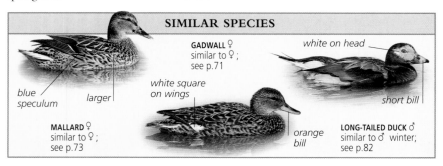

SIMILAR SPECIES

blue speculum

larger

MALLARD ♀
similar to ♀;
see p.73

GADWALL ♀
similar to ♀;
see p.71

white square on wings

orange bill

white on head

short bill

LONG-TAILED DUCK ♂
similar to ♂ winter;
see p.82

| Length **53–70cm (21–28in)** | Wingspan **80–95cm (32–37in)** | Weight **550–1,200g (20–43oz)** |
| Social **Flocks** | Lifespan **15–25 years** | Status **Vulnerable** |

| Order **Anseriformes** | Family **Anatidae** | Species ***Anas querquedula*** |

Garganey

pale forewings

MALE (SPRING)

FEMALE

dark hindwings with two equal white bars

IN FLIGHT

broad pale line almost broken over eye

dull wing patch

dark eyestripe

JUVENILE

juvenile almost identical to female, pale spot near bill and pale line on cheek separate them from Teal

white spot near bill

pale, mottled brown plumage

pale line over eye

FEMALE

pale stripes above and below eye

blotched, dark brown plumage

MALE (AUTUMN)

bold white stripe over eye

pinkish brown back

blue-grey flanks

MALE (SPRING)

Small and colourful, the Garganey is unusual in Europe as a summer visiting duck that spends the winter in Africa. It is mostly scarce and thinly spread but spring flocks in the eastern Mediterranean can be substantial. In autumn, pairs or small groups can be found swimming with other wildfowl. They tend to associate with Teal and Shovelers and picking one or two autumn migrants out from a flock of mixed ducks is a challenge.

VOICE Male makes short, dry rattle; female rather silent but has short, high-pitched quack.

NESTING Down-lined hollow in vegetation near water; 8–11 eggs; 1 brood; May–June.

FEEDING Dabbles and upends in water, taking tiny invertebrates, roots, and seeds.

FLIGHT: quick, twisting and turning easily; flocks almost wader-like.

WATERSIDE DUCK
Garganeys like wet, grassy marshes and shallow floods with reeds and sedges in spring.

OCCURRENCE
Common in Mediterranean in spring, and scarce breeder in N and W Europe, mostly on wet, grassy, freshwater marshes. Small numbers of migrants join other ducks on lakes and reservoirs in autumn.

| Seen in the UK |
| J F **M A M J J A S O** N D |

SIMILAR SPECIES

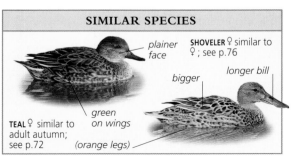

plainer face

SHOVELER ♀ similar to ♀; see p.76

bigger

longer bill

TEAL ♀ similar to adult autumn; see p.72

green on wings

(orange legs)

| Length **37–41cm (14½–16in)** | Wingspan **63–69cm (25–27in)** | Weight **250–500g (9–18oz)** |
| Social **Family groups** | Lifespan **Up to 10 years** | Status **Vulnerable** |

Order **Anseriformes**	Family **Anatidae**	Species *Anas clypeata*

Shoveler

grey forewings

FEMALE

dark belly

streaked, pale brown body

heavy bill with orange sides

FEMALE

pale crescent on face

dark head

rufous-tinged flanks

MALE (SUMMER)

pale blue forewings

MALE

green-black head

dark rufous flanks

striking yellow eye

long, heavy, shovel-like black bill

IN FLIGHT

MALE (WINTER)

bright white breast

M ale Shovelers in breeding plumage are obvious and easily identified by their green heads, white breasts, and chestnut sides, while females are distinctive at close range but best identified by shape at a distance. They have the typical plumage of female dabbling ducks: streaked and pale brown overall. While taking flight, Shovelers make a characteristic "woofing" noise with their wings.

VOICE Male has deep *took took*; female has deep, quiet quack.

NESTING Down- or leaf-lined hollow near water; 8–12 eggs; 1 brood; March–June.

FEEDING Dabbles, often in tight, circular flocks, for seeds and invertebrates, with bill thrust forward on water surface or underwater and shoulders practically awash.

FLIGHT: quick, agile; strong, deep wingbeats, short-tailed effect.

FEEDING IN WATER
To reach slightly deeper food, Shovelers tip up on end, when the long, pointed wingtips become especially conspicuous.

SIMILAR SPECIES

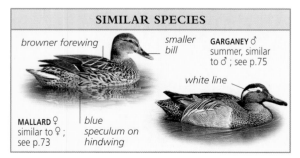

browner forewing

smaller bill

GARGANEY ♂ summer, similar to ♂; see p.75

white line

MALLARD ♀ similar to ♀; see p.73

blue speculum on hindwing

OCCURRENCE
Breeds mostly in E Europe on reedy pools, generally in lowland areas. In winter, more widespread in W, on fresh water, marshes, and sheltered estuaries with grassy salt marshes crossed by creeks. Some reservoirs attract autumn flocks.

Seen in the UK

J F M A M J J A S O N D

Length **44–52cm (17½–20½ in)**	Wingspan **70–84cm (28–33in)**	Weight **400–1,000g (14–36oz)**
Social **Flocks**	Lifespan **10–20 years**	Status **Secure**

| Order **Anseriformes** | Family **Anatidae** | Species **Aythya ferina** |

Pochard

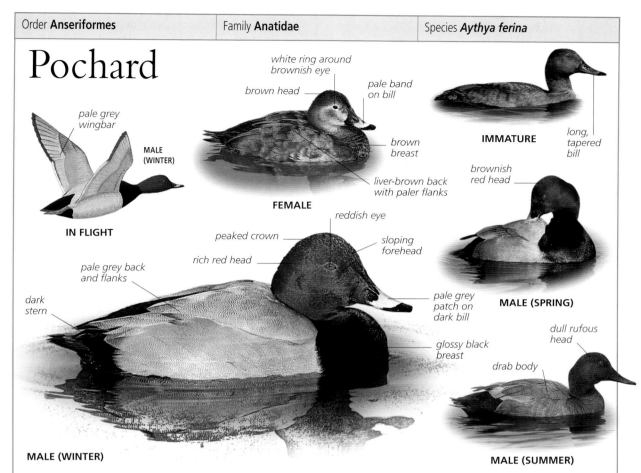

pale grey
wingbar

**MALE
(WINTER)**

white ring around
brownish eye

brown head

pale band
on bill

brown
breast

liver-brown back
with paler flanks

FEMALE

IN FLIGHT

IMMATURE

long,
tapered
bill

brownish
red head

reddish eye

peaked crown

sloping
forehead

rich red head

pale grey back
and flanks

dark
stern

pale grey
patch on
dark bill

MALE (SPRING)

dull rufous
head

glossy black
breast

drab body

MALE (SUMMER)

MALE (WINTER)

Together with the Tufted Duck, which often associates with it, the Pochard is one of the common inland diving ducks. Flocks are generally less active by day than Tufted Ducks, often sleeping for long periods. They are typically tightly packed and often made up largely of one sex, mostly males in the UK. In late autumn, hundreds of migrants may appear on a lake for a day or two and then move on at night. In summer, Pochards are dispersed and generally rare breeding birds in western Europe, including the UK.

VOICE Wheezing rise-and-fall call from displaying male; purring growl from female.

NESTING Large pad of leaves and down in reeds near water; 8–10 eggs; 1 brood; April–July.

FEEDING Dives from surface, taking seeds, shoots, and roots; often feeds by night.

FLIGHT: fast and direct, with fast, deep wingbeats; flies in loose flocks.

FLAPPING WINGS
Pochards rise up and flap their wings in a "comfort" movement used by many waterfowl species.

SIMILAR SPECIES

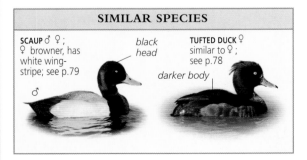

SCAUP ♂ ♀;
♀ browner, has white wing-stripe; see p.79

♂

black
head

TUFTED DUCK ♀;
similar to ♀;
see p.78

darker body

OCCURRENCE
Widespread breeder on reedy lakes in E Europe, rather scarce in W. Common non-breeder on fresh water in W Europe; big numbers on migration in late autumn and widespread smaller flocks on flooded pits and similar waters.

| Seen in the UK |
| J F M A M J J A S O N D |

| Length **42–49cm (16½–19½in)** | Wingspan **72–82cm (28–32in)** | Weight **700–1,000g (25–36oz)** |
| Social **Large flocks** | Lifespan **8–10 years** | Status **Secure** |

| Order **Anseriformes** | Family **Anatidae** | Species *Aythya fuligula* |

Tufted Duck

bold white wingbar

MALE

FEMALE

IN FLIGHT

slight tuft

dark, dull brown body

MALE (SUMMER)

dark brown body with paler flanks

short crest on nape

may have white on tail

FEMALE

yellow eye

long, wispy tuft on nape

bluish bill with large black tip

black body with white flanks

MALE (WINTER)

This is a common diving duck, swimming and feeding in flocks, with individuals disappearing under in search of food. Males are strongly contrasted except in mid-summer, while females are dark and dull. Flocks are often mixed with Pochards and are worth scanning through for individuals of rarer species that are naturally attracted to their company. Tufted Ducks may be semi-tame and come for food on ponds in town parks. Even flocks on lakes and reservoirs often tolerate a quite close approach, or simply swim away from disturbance.

VOICE Deep, grating growl; male calls with nasal whistles during courtship.

NESTING Down-lined hollow in long vegetation close to water; 8–11 eggs; 1 brood; May–June.

FEEDING Dives underwater from surface to find molluscs and insects.

FLIGHT: fast but not manoeuvrable; flocks make tight, irregular packs.

WINTER FLOCK
Tufted Ducks, with Pochards, form quiet, sleepy flocks on inland waters; these flocks are sometimes quite large.

SIMILAR SPECIES

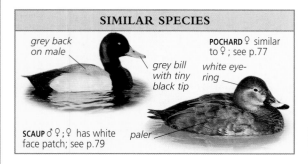

grey back on male

grey bill with tiny black tip

SCAUP ♂ ♀; ♀ has white face patch; see p.79

POCHARD ♀ similar to ♀; see p.77

white eye-ring

paler

OCCURRENCE
Widespread; breeds in long grass around fresh waters such as flooded pits. In winter, widespread and often abundant, with huge numbers on sheltered coastal waters, and small flocks on gravel pits, reservoirs, and sheltered coasts.

| Seen in the UK |
| J F M A M J J A S O N D |

| Length **40–47cm (16–18½in)** | Wingspan **67–73cm (26–29in)** | Weight **450–1,000g (16–36oz)** |
| Social **Large flocks** | Lifespan **10–15 years** | Status **Secure** |

Order **Anseriformes**	Family **Anatidae**	Species ***Aythya marila***

Scaup

broad white wingbar

MALE (WINTER)

IN FLIGHT

smaller, more blurred white face than adult

pale cheek

JUVENILE

white blaze on face

rich brown head

FEMALE (WINTER)

pale cheeks

FEMALE (SUMMER)

black head with green gloss

steeper forehead than Tufted Duck

yellow eye

bigger bill than Tufted Duck

broader body than Tufted Duck

pale grey back

white flanks (grey on Pochard)

black around tail

round nape

broad, blue-grey bill with small black tip

all-dark head

grey bars

MALE (SUMMER)

MALE (WINTER)

Typically sea ducks, Scaup regularly appear inland in small numbers, usually associated with Tufted Ducks and Pochards, but similar-looking hybrids must be ruled out. Flocks of Scaup on the sea are attractive, if less lively than scoters or Long-tailed Ducks, tending to fly less and to seem much more settled and sedate on the water. They tend to prefer more sheltered parts of outer estuaries than the broad, open, windswept bays loved by the more energetic scoters. Flocks are generally quite approachable.

VOICE Male lets out low whistles in display, mostly silent otherwise; female has deep growl.

NESTING Nest lined with feathers and down, on ground, near water; 8–11 eggs; 1 brood; April–June.

FEEDING Dives from surface to find invertebrates, waste grain, and aquatic plants.

FLIGHT: fast and direct; looks thickset and powerful; fast wingbeats.

FLOCK ON SEA
Wintering Scaup are gregarious and swim in flocks of a few score on sheltered seas, with the white flanks of adult males showing up well.

SIMILAR SPECIES

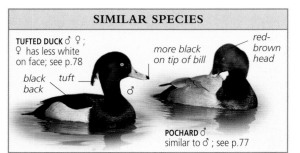

TUFTED DUCK ♂ ♀;
♀ has less white on face; see p.78

black back   tuft

red-brown head

more black on tip of bill

POCHARD ♂
similar to ♂; see p.77

OCCURRENCE
Northern breeder, on wild moors and tundra in Scandinavia and Iceland. In winter, flocks found in regular, traditional places, mostly in S Baltic and North Sea, isolated groups and individuals turning up with Tufted Ducks inland.

Seen in the UK
J F M A M J J A S O N D

Length **42–51cm (16½–20in)**	Wingspan **67–73cm (26–29in)**	Weight **0.8–1.3kg (1¾–2¾lb)**
Social **Flocks**	Lifespan **10–12 years**	Status **Localized**

| Order **Anseriformes** | Family **Anatidae** | Species *Somateria mollissima* |

Eider

white patch on rear flank

MALE

dark hind-wings

FEMALE

IN FLIGHT

IMMATURE

wedge-shaped head and bill

closely barred brown body

dark belly

FEMALE

black crown

long, wedge-shaped head with green patch

white upperparts

wedge-shaped bill

pinkish breast

black underparts

unbarred dark body

MALE (SUMMER)

MALE (WINTER)

An entirely marine bird, the Eider is highly sociable and often seen in large rafts offshore. It is, however, equally familiar as a tame duck around coastal rocks and bays. Female Eiders with young in summer are easy to identify and so are spring males, calling and displaying, in northern harbours. In a few areas, flocks of Eiders remain all year but make no attempt to nest. These groups show remarkable fidelity to certain sites.

VOICE Male has sensuous, cooing *aa-ahooh*; female deep growls and *kok-kok-kok*.
NESTING Down-lined hollow on ground, exposed or well hidden; 4–6 eggs; 1 brood; April–June.
FEEDING Dives from surface to take crabs, shellfish, shrimps, and starfish.

FLIGHT: heavy, low, straight but fast, with deep, steady wingbeats.

MALE FLOCK
Large flocks of male Eiders gather offshore in estuaries and bays.

FEMALE AT NEST
The Eider makes a down-lined nest near water, often openly, on which the female sits tight to incubate.

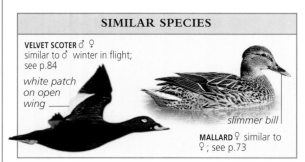

SIMILAR SPECIES

VELVET SCOTER ♂ ♀ similar to ♂ winter in flight; see p.84

white patch on open wing

MALLARD ♀ similar to ♀; see p.73

slimmer bill

OCCURRENCE
Breeds in N Great Britain, Iceland, and Scandinavia, on low-lying coasts and islands with rocky shores and weedy bays. Winters on sea south to W France, with large flocks in sandy bays and over mussel beds. Very rare inland.

| Seen in the UK |
| J F M A M J J A S O N D |

| Length **50–71cm (20–28in)** | Wingspan **80–108cm (32–43in)** | Weight **1.2–2.8kg (2¾–6¼lb)** |
| Social **Large flocks** | Lifespan **10–15 years** | Status **Secure** |

| Order **Anseriformes** | Family **Anatidae** | Species ***Somateria spectabilis*** |

King Eider

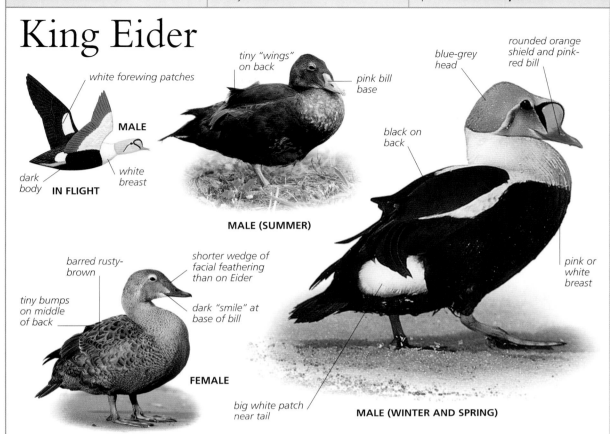

white forewing patches

MALE

dark body **IN FLIGHT**

white breast

tiny "wings" on back

pink bill base

MALE (SUMMER)

blue-grey head

rounded orange shield and pink-red bill

black on back

pink or white breast

MALE (WINTER AND SPRING)

barred rusty-brown

shorter wedge of facial feathering than on Eider

tiny bumps on middle of back

dark "smile" at base of bill

FEMALE

big white patch near tail

This is a colourful eider, closely similar in many respects to the common Eider but a more northerly and rarer bird. It is usually seen as a rare stray in flocks of Eiders far to the south of its usual range, occasionally remaining for several weeks and sometimes returning in subsequent years. Males in full plumage are easy to tell from common Eiders, moulting males in summer less so, and females and young birds difficult, except with close scrutiny.

VOICE Male has cooing calls in spring; female has harsh, croaking note.

NESTING Nest lined with down from female's body, on ground near water, often on small island; 4–5 eggs; 1 brood; May–July.

FEEDING Dives for molluscs, crustaceans, and insect larvae from tundra pools in summer.

FLIGHT: direct, rather heavy but fast with steady wingbeats.

STRIKING APPEARANCE
The male is colourful and easy to tell in winter and spring, but females are much more like Eiders, requiring careful identification.

SIMILAR SPECIES

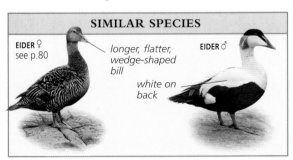

EIDER ♀
see p.80

longer, flatter, wedge-shaped bill

EIDER ♂

white on back

OCCURRENCE
Rare and local breeder in N Iceland and extreme N Scandinavia; otherwise rare, usually with Eiders, on shores of Scandinavia, Baltic, and northern North Sea, mostly autumn to spring.

| Seen in the UK |
| J F **M A M** J J A **S O N D** |

| Length **60cm (24in)** | Wingspan **90–100cm (36–40in)** | Weight **1.5–2kg (3¼–4½lb)** |
| Social **Small flocks** | Lifespan **10–15 years** | Status **Secure** |

| Order **Anseriformes** | Family **Anatidae** | Species *Clangula hyemalis* |

Long-tailed Duck

MALE (WINTER)

dark wings

narrrow, dark centre and broad white sides to rump

dark cheek patch

thick dark bill

pale flanks

FEMALE (WINTER)

smudgy white band around eye

JUVENILE (WINTER)

white neck

FEMALE (SUMMER)

FEMALE (WINTER)

IN FLIGHT

long, flexible tail point

white and pale grey body

dark cheek patch

pink band on stubby dark bill

white rear flanks

rich brown body

white face patch

MALE (SUMMER)

MALE (WINTER)

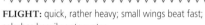

While occasional individuals appear inland briefly, these ducks are marine birds, living offshore in flocks, often mixed with scoters. Lively and active, they often fly low over the waves, splashing down, then flying again. When feeding, they spend long periods underwater. The irregular and complicated patterns, especially of males, may confuse novice birdwatchers, although they are quite distinctive. These birds are identified by their dark cap and cheeks, stubby upturned bill, white beside the tail, and dark wings.

FLIGHT: quick, rather heavy; small wings beat fast; splashes heavily onto water.

VOICE Male makes loud, rhythmic, yodelling calls, *a-ahulee*; female growls.

NESTING Down-lined hollow on ground near water; 4–6 eggs; 1 brood; May–June.

FEEDING Dives from surface to take molluscs and crustaceans.

FAMILY GROUP
Breeding Long-tailed Ducks are restricted to the far north, but are often numerous on suitable lakes.

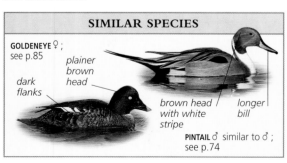

SIMILAR SPECIES

GOLDENEYE ♀; see p.85

plainer brown head

dark flanks

brown head with white stripe

longer bill

PINTAIL ♂ similar to ♂ ; see p.74

OCCURRENCE
Breeds in Iceland and Scandinavia on bleak moorland. Winters at sea off N Great Britain and in North Sea and Baltic, typically well offshore but drifting into bays and estuaries with tide, especially in early spring.

| Seen in the UK |
| J F M A M J J A S O N D |

| Length **38–60cm (15–23½in)** | Wingspan **73–79cm (29–31in)** | Weight **520–950g (19–34oz)** |
| Social **Flocks** | Lifespan **Up to 10 years** | Status **Secure** |

Common Scoter

paler wingtips

slim neck

MALE

IN FLIGHT

blackish cap

grey bill

dark brown body

greyish white lower face

FEMALE

long, pointed tail often raised

all-black body with duller or paler wings (browner in summer)

round head

pointed bill with yellow patch

thin neck

MALE (WINTER)

O̲n a calm day, if Common Scoters are close inshore, their musical calls can be heard, but usually they are seen as distant dots appearing intermittently on the swell, or flying in long lines along the horizon. Migrants appear for very short stays inland in mid- and late summer. Common Scoters favour large, shallow, sheltered bays, although they are able to ride out storms with ease. Traditional sites may have thousands of birds, which are vulnerable to oil pollution. These sites may be inhabited throughout the year, with large numbers of moulting birds in bays with plentiful food in late summer and autumn.

VOICE Male has musical, piping whistle; female makes deep growls.

NESTING Down- and leaf-lined hollow near water, often on island; 6–8 eggs; 1 brood; March–June.

FEEDING Dives from surface to find shellfish, crustaceans, and worms.

FLIGHT: fast, low; in long, wavering lines and packs; rapid, deep wingbeats and sideways rolling action.

SOCIAL SEA DUCK
Large flocks of the very social Common Scoter often swim on heaving swell or fly low over the waves well offshore.

SIMILAR SPECIES

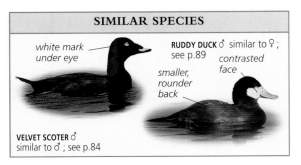

white mark under eye

VELVET SCOTER ♂
similar to ♂; see p.84

RUDDY DUCK ♂ similar to ♀; see p.89

contrasted face

smaller, rounder back

OCCURRENCE
Breeds on moorland pools in Iceland, Scandinavia, and N Great Britain. Winters on coasts around UK, North Sea, and Baltic, fewer south to Mediterranean. Flocks found in regular sites almost throughout year.

Seen in the UK
J F M A M J J A S O N D

Length **45–54cm (18–21½in)**	Wingspan **79–90 cm (31–35 in)**	Weight **1.2–1.4 kg (2¾–3lb)**
Social **Large flocks**	Lifespan **10–15 years**	Status **Secure**

| Order **Anseriformes** | Family **Anatidae** | Species ***Melanitta fusca*** |

Velvet Scoter

MALE

IN FLIGHT

broad white wing patch

dark brown body

white spots on face

thick neck

dark bill tapers to long tip

white eye-spot

wedge-shaped face

yellow sides of bill

red legs may show as bird dives (dark on Common Scoter)

FEMALE

black body (browner in summer)

red legs

MALE (WINTER)

A large, almost Eider-like diving duck, spending almost all of its time at sea, the Velvet Scoter is usually seen in small numbers in larger Common Scoter flocks. On the water, it is difficult to pick out from the smaller species, but the wing patches revealed in flight are immediately obvious. Females are dark-faced in fresh plumage but soon develop white face spots as the dark feather tips wear off. Individual variation adds to the identification problem. If a Velvet Scoter turns up inland, it may stay for a few days, giving a rare chance to get a good look at this surprisingly elegant duck.
VOICE Male whistles; female growls, but generally quiet, especially in winter.
NESTING Down-lined hollow near water; 6–8 eggs; 1 brood; May–July.
FEEDING Dives from surface, to find shellfish, shrimps, crabs, and marine worms.

FLIGHT: fast, low, heavy but strong; usually direct or in wide arc over sea.

WHITE WING PATCHES
Bold white wing patches distinguish the Velvet Scoter from the Common Scoter in flight or when it flaps its wings on the water.

SIMILAR SPECIES

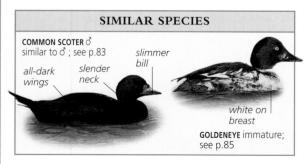

COMMON SCOTER ♂ similar to ♂; see p.83

all-dark wings

slender neck

slimmer bill

white on breast

GOLDENEYE immature; see p.85

OCCURRENCE
Breeds in Scandinavia, along coasts and on tundra pools. Winters in North Sea and Baltic on sheltered coasts. Small numbers join big Common Scoter flocks in summer and autumn.

| Seen in the UK |
| J F M A M J J A S O N D |

| Length **52–59cm (20½–23 in)** | Wingspan **90–99cm (35–39 in)** | Weight **1.1–2kg (2½–4½lb)** |
| Social **Flocks** | Lifespan **10–12 years** | Status **Localized** |

| Order **Anseriformes** | Family **Anatidae** | Species **Bucephala clangula** |

Goldeneye

large white belly

MALE (WINTER)

yellow eye

green-glossed black head with bold white spot

triangular dark bill

black marks on sparkling white body

MALE (WINTER)

extensive white on wing

FEMALE

IN FLIGHT

body becomes white with age

JUVENILE (MALE)

face develops white spot in winter

grey body can look very dark; rounded back

eyes pale yellow or white

dark brown head

FEMALE

yellow patch on grey bill

There are usually more Goldeneyes in a group than is usually apparent until they fly off: typically, at least half of these expert divers are underwater at any one time. Usually shy, they are easily disturbed by people on the shore or by boats, flying off in tight-packed groups with a loud whistling from their wings. Flocks are typically predominantly females and immatures, although more adult males may appear in spring (summer males look like females).

VOICE Frequent nasal, mechanical *ze-zeee* in display; female has grating double note.

NESTING Down-lined tree hole and nest box; 8–11 eggs; 1 brood; April–June.

FEEDING Dives constantly from surface to find molluscs and crustaceans.

FLIGHT: very quick and strong, with rather short wings; fast, deep wingbeats create loud whistle.

FLYING
Goldeneyes are shy and quick to fly off. Their wings make a loud whistling sound.

RESTING FLOCK
After feeding well, Goldeneyes rest in flocks, head withdrawn and tails cocked.

SIMILAR SPECIES

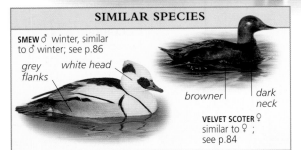

SMEW ♂ winter, similar to ♂ winter; see p.86

grey flanks   *white head*

browner   *dark neck*

VELVET SCOTER ♀ similar to ♀; see p.84

OCCURRENCE
Breeds in N Europe (but only rarely in Scotland), in wooded areas beside cold freshwater lakes. Numbers often increased by provision of nest boxes. In winter, widespread on lakes, reservoirs, and estuaries.

| Seen in the UK |
| J F M A M J J A S O N D |

| Length **42–50cm (16½–20in)** | Wingspan **65–80cm (26–32in)** | Weight **600–1,200g (21–43oz)** |
| Social **Small flocks** | Lifespan **Up to 8 years** | Status **Secure** |

| Order **Anseriformes** | Family **Anatidae** | Species ***Mergellus albellus*** |

Smew

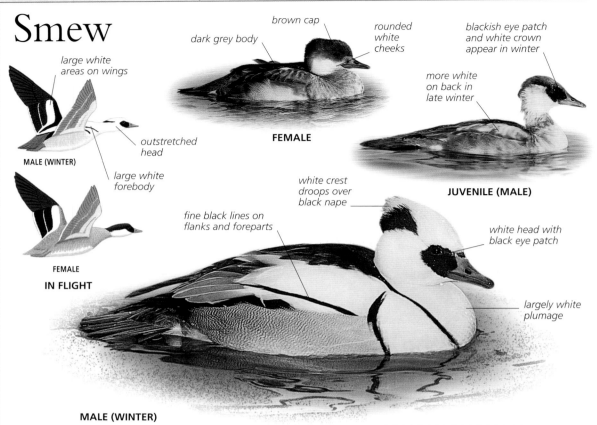

large white areas on wings

MALE (WINTER)

outstretched head

large white forebody

FEMALE
IN FLIGHT

brown cap

dark grey body

rounded white cheeks

FEMALE

blackish eye patch and white crown appear in winter

more white on back in late winter

JUVENILE (MALE)

white crest droops over black nape

fine black lines on flanks and foreparts

white head with black eye patch

largely white plumage

MALE (WINTER)

In the UK, this is a scarce bird, found in small groups at most, but in the Low Countries and the Baltic, hundreds sometimes feed together in very active, busy flocks. The white males are usually much scarcer than females and immatures, collectively known as "redheads". Even where there are one or two males in a flock on a complex of gravel pits, they can be quite hard to spot as the birds tend to fly around a lot. They particularly associate with Goldeneyes and may also be found with Tufted Ducks.

VOICE Silent in winter.

NESTING Tree holes, often of Black Woodpecker, near water, or in nest boxes; 4–6 eggs; 1 brood; April–June.

FEEDING Dives often from surface, eating small fish and insect larvae.

FLIGHT: flies fast and low; broad white foreparts and outstretched neck very obvious.

STUNNING DRAKE
The winter male is one of the most attractive of European wildfowl. In summer, the male Smew looks like the female.

SIMILAR SPECIES

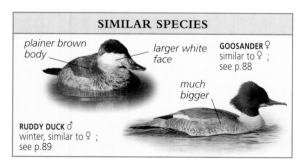

plainer brown body

larger white face

GOOSANDER ♀
similar to ♀ ;
see p.88

much bigger

RUDDY DUCK ♂
winter, similar to ♀ ;
see p.89

OCCURRENCE
Breeds in far NE Europe by lakes and rivers in forested areas. In winter, common in large, mobile flocks on Baltic and Low Country coasts, scarce on inland waters west to Great Britain, when a few turn up on pits and reservoirs.

Seen in the UK

| J | F | M | A | M | J | J | A | S | O | N | D |

| Length **36–44cm (14–17½in)** | Wingspan **55–69cm (22–27in)** | Weight **500–800g (18–29oz)** |
| Social **Flocks** | Lifespan **Up to 8 years** | Status **Vulnerable** |

Order **Anseriformes**	Family **Anatidae**	Species ***Mergus serrator***

Red-breasted Merganser

fuzzy crest

brownish grey body

brownish head

MALE (WINTER)

bold white wing patches

FEMALE

ginger-brown head

FEMALE

red legs **IN FLIGHT**

MALE (SUMMER)

smudgy foreneck

wispy crest on green-black head

broad white line between black back and grey flank

slim, slightly upcurved red bill

white collar

brown, black-edged breast

MALE (WINTER)

A "sawbill" duck, the Red-breasted Merganser is found both on fast, clean rivers in hilly regions and at the coast in summer, and usually on the sea at other times. It often stands out on sandy shores or rocks. Males display to females in winter and spring with ritualized, jerky, curtseying movements, fanning their spiky crests and opening their bills. In winter, typically a male or two will be seen with a handful of "redheads" (females and immatures) but in places a few hundred might flock together.

VOICE Quiet, sometimes low, rolling croak or growl.

NESTING In long grass on ground or among rocks; 8–11 eggs; 1 brood; April–June.

FEEDING Dives from surface to find small fish and invertebrates.

FLIGHT: fast, direct, with long head and neck and tail giving marked cross-shape.

DISPLAYING MALES
Male Red-breasted Mergansers stretch forward and jerk their heads forward while raising the rear body, in energetic displays.

SIMILAR SPECIES

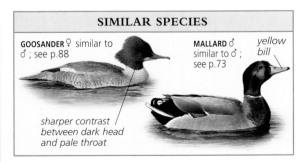

GOOSANDER ♀ similar to ♂; see p.88

MALLARD ♂ similar to ♂; see p.73

yellow bill

sharper contrast between dark head and pale throat

OCCURRENCE
Breeds by coasts or along fast rivers in N Great Britain, Iceland, Scandinavia, and Baltic region. Winters south to Greece and N France, mostly on coasts. Big moulting groups can be seen off sandy and rocky shores in late summer.

Seen in the UK
J F M A M J J A S O N D

Length **51–62cm (20–24in)**	Wingspan **70–85cm (28–34in)**	Weight **0.85–1.25kg (1¾–2¾lb)**
Social **Family groups/Flocks**	Lifespan **Up to 8 years**	Status **Secure**

| Order **Anseriformes** | Family **Anatidae** | Species ***Mergus merganser*** |

Goosander

large white
wing patch

MALE (WINTER)

FEMALE
IN FLIGHT

striped face

greyish body

JUVENILE

dark brown
head

smooth, downward-
pointing crest

blue-grey
body

red legs

FEMALE

sharply defined white throat and
dark collar (blurred on Merganser)

thick-based,
hooked
plum-red bill

green-black
head

salmon-pink to
white body

long tail

MALE (WINTER)

FLIGHT: often low, fast, and direct, with elongated
but heavy shape.

The largest "sawbill", with a long, serrated bill
for grasping fish, the Goosander is more of a
freshwater bird than the Red-breasted Merganser,
especially outside the breeding season. It is found in
small groups in winter, often looking remarkably large
on small inland lakes and pits on still, misty days. In
summer, breeding pairs prefer upland reservoirs and
shallower, fast-flowing, clear streams with plenty of
boulders and stony shores. Usually a shy bird, the
Goosander is easily scared off even at long range.
VOICE Harsh *karrr* and cackling notes.
NESTING Hole in trees near water; 8–11 eggs;
1 brood; April–July.
FEEDING Dives from surface, moving long distances
underwater in larger lakes, to take fish.

LAZY MALES
In winter, male Goosanders swim about on open water, often drifting
inactive for long periods. Males look like the females in summer.

SIMILAR SPECIES

**RED-BREASTED
MERGANSER** ♂♀;
♂ has dark breast;
see p.87

blurred
face

dark breast

MALLARD ♂
similar to ♂♀;
see p.73

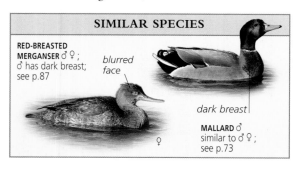

OCCURRENCE
Breeds beside rivers and lakes in
Iceland, Scandinavia, and N Great
Britain. Winters south to Balkans
and France, mostly on fresh water.
Bigger flocks mostly in regular
sites on larger reservoirs, smaller
numbers on pits or rivers.

| Seen in the UK |
| J F M A M J J A S O N D |

| Length **57–69cm (22½–27in)** | Wingspan **82–98cm (32–39in)** | Weight **1–1.6kg (2¼–3½lb)** |
| Social **Small flocks** | Lifespan **Up to 8 years** | Status **Secure** |

| Order **Anseriformes** | Family **Anatidae** | Species *Oxyura jamaicensis* |

Ruddy Duck

all-dark wings

MALE (SUMMER)

white face

IN FLIGHT

dull dark grey-brown body

cheek stripe

dark cap on pale head

dark grey bill

FEMALE

dull grey-brown body

blackish bill

MALE (WINTER)

large, round head

black cap and nape

pure white cheeks

rounded back

rufous body

blue bill

stiff tail, laid flat or angled upwards

MALE (SUMMER)

/\/\/\/\/\/\/\/\/\/\/\/\/\/\/\/\

FLIGHT: fast, low, weak, with whirring wingbeats; direct, with little agility.

An accidental introduction to Europe, the Ruddy Duck escaped from collections in the 1950s and has since become established in several countries. It is a freshwater bird, family parties pottering about reedy shores. It moves to larger lakes and reservoirs in winter, flocks numbering several hundreds in the most favoured places which have become traditional moulting and wintering areas.

VOICE Mostly silent; odd grunts, also slaps bill against chest in display.

NESTING Large, floating pile of vegetation in tall reeds, often "roofed" over by meshed stems; 6–10 eggs; 1 brood; April–June.

FEEDING Dives from surface, reappearing like a cork; takes insect larvae and seeds.

DISPLAYING MALE
The male Ruddy Duck vibrates his bill against the breast, pushing out air in a flurry of bubbles from the feathers.

OCCURRENCE
Breeds on reedy pools and flooded pits in Great Britain and less commonly, adjacent areas of continent. In larger reservoirs and more open waters in autumn, mostly in a few flocks at regular sites, ones and twos elsewhere.

| Seen in the UK |
| J F M A M J J A S O N D |

SIMILAR SPECIES

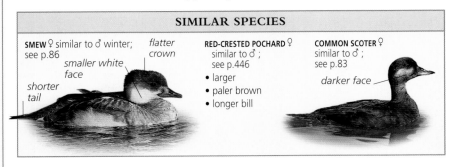

SMEW ♀ similar to ♂ winter; see p.86

flatter crown

smaller white face

shorter tail

RED-CRESTED POCHARD ♀ similar to ♂; see p.446
• larger
• paler brown
• longer bill

COMMON SCOTER ♀ similar to ♂; see p.83

darker face

| Length **35–43cm (14–17in)** | Wingspan **53–62cm (21–24in)** | Weight **350–800g (13–29oz)** |
| Social **Winter flocks** | Lifespan **Up to 8 years** | Status **Secure†** |

GAMEBIRDS

A MIXED GROUP of ground birds, these are all short-legged, short-billed species that feed on vegetable matter, some with a very restricted diet. Their chicks, however, require an abundance of energy-giving insects. Several have ritualized social behaviour, typified by the "lekking" of Black Grouse. Males display in order to get the best chance of being selected by a female, but take no part in incubating eggs or rearing the young.

FACIAL WATTLES
Several species of gamebirds, such as this Pheasant, have fleshy red appendages on the head. Pheasants have been introduced for shooting in many countries.

GROUSE
Round-bodied species with feathered legs and feet, and typically cryptic plumages, grouse live in demanding conditions, including bleak heather moors and high mountain tops.

GIANT GROUSE
Biggest of all grouse is the male Capercaillie, a threatened species of pine forest habitats. It is a social bird where numbers remain high, but often solitary in summer.

PHEASANTS
Long-tailed male pheasants are gorgeous birds, while females are generally smaller and duller. Some are secretive and hard to find.

PARTRIDGES
Small and rotund, partridges have far less difference between male and female than the grouse and pheasants, and the males take a greater part in caring for the family. They are social birds outside the breeding season. The Quail is a smaller species and a long-distance migrant, wintering in Africa.

TETRAONIDAE

white, black, and rufous mottles on flanks

red over eye

"salt and pepper" barring on grey upperparts

HAZEL GROUSE
p.92

WILLOW GROUSE
p.93

PTARMIGAN
p.94

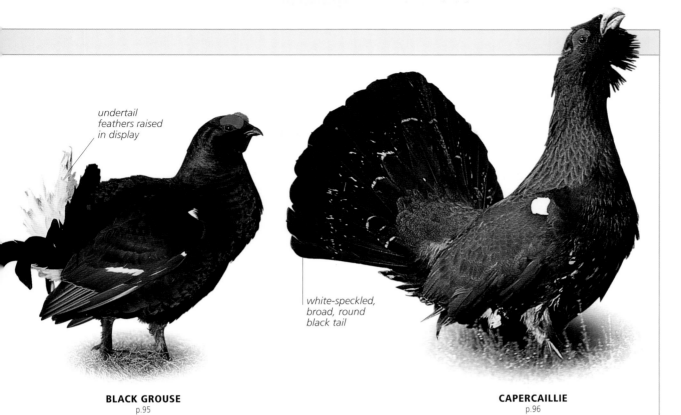

undertail feathers raised in display

white-speckled, broad, round black tail

BLACK GROUSE
p.95

CAPERCAILLIE
p.96

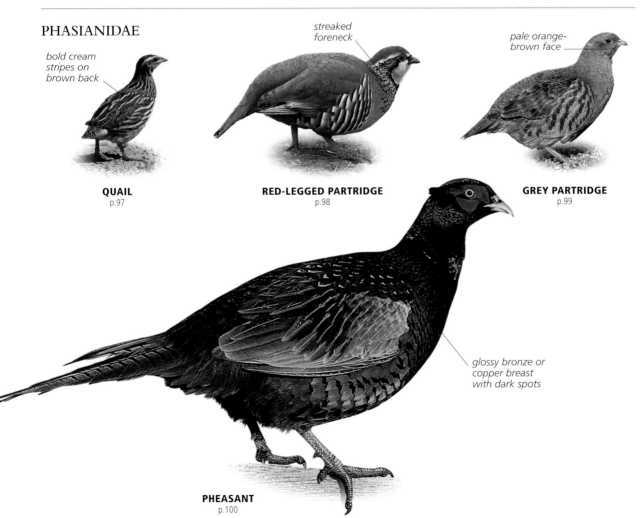

PHASIANIDAE

bold cream stripes on brown back

streaked foreneck

pale orange-brown face

QUAIL
p.97

RED-LEGGED PARTRIDGE
p.98

GREY PARTRIDGE
p.99

glossy bronze or copper breast with dark spots

PHEASANT
p.100

| Order **Galliformes** | Family **Tetraonidae** | Species ***Bonasa bonasia*** |

Hazel Grouse

arched, pale grey-brown wings

whitish band on each side of back

IN FLIGHT

pale grey rump

black chin edged white

black throat

grey above, pale below with large dark spots

small, round head

boldly spotted flanks

dumpy body

FEMALE

broad grey tail with dark tip

white, black, and rufous mottles on flanks

MALE

A woodland grouse, this is a rare and elusive bird of northeast, central, and eastern Europe. It prefers deep, dark, coniferous forests, which makes it hard to see because it is shy, quiet, and inconspicuous. It often feeds on the ground in the shade of deep undergrowth and moves away out of sight as people approach.

VOICE Various liquid, slightly ticking notes and a curious song of fine, very high notes, falling in pitch, unexpected for such a large bird.

NESTING Shallow, lined depression well hidden on ground; 7–11 eggs; 1 brood; April–June.

FEEDING Eats various shoots, leaves, berries, and insects, mostly from the ground or in thick conifer sprays.

FLIGHT: fast, with quick wingbeats producing a distinctive whirring sound.

FOREST DWELLER
Hazel Grouse live secretive, quiet lives deep in forests and dense undergrowth; rare close views reveal beautifully marked and subtly coloured plumages.

SIMILAR SPECIES

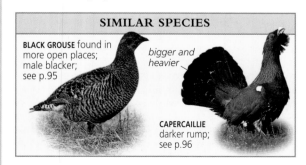

BLACK GROUSE found in more open places; male blacker; see p.95

bigger and heavier

CAPERCAILLIE darker rump; see p.96

OCCURRENCE
Widespread in central European mountain forests, and farther north and east, but rare and local almost everywhere. Does not move outside normal breeding areas.

Seen in the UK
J F M A M J J A S O N D

| Length **34–39cm (12–15in)** | Wingspan **48–54cm (19–22in)** | Weight **350–490g (13–18oz)** |
| Social **Family groups** | Lifespan **Up to 7 years** | Status **Secure** |

| Order **Galliformes** | Family **Tetraonidae** | Species *Lagopus lagopus* |

Willow Grouse

MALE (SUMMER)

WINTER

IN FLIGHT

white wings

black tail

white wings

thick bill

all-white body

WINTER

red over eye

thick black bill

rich red-brown body (female yellow-brown, more marbled)

dark scaly bars

white belly

MALE (SUMMER)

A thickset, small-headed, round-bodied, chicken-like bird of moor and heath, the Willow Grouse is extensively shot for sport but, unlike the Pheasant, not reared and released. The British and Irish race, commonly known as the Red Grouse, is distinctive, retaining an essentially unchanged plumage pattern all year round. Grouse have declined over many decades in most areas, struggling to maintain high numbers in a relatively artificial environment.

VOICE Remarkable, deep, staccato calls echo across moors, *kau-kau-kau-ka-ka-karrr-rrr-g'bak, g'bak, bak*.

NESTING Sparsely lined scrape on ground in heather; 6–9 eggs; 1 brood; April–May.

FEEDING Plucks shoots and seeds from heather while standing or walking slowly; also takes variety of berries and seeds.

FLIGHT: explosive escape, then long, fast, low flight with whirring beats and long glides on arched wings.

PATCHY TRANSITION
Moulting birds have a contrasted look; the dark head is the last to go white in autumn, and first to turn brown in spring.

SUBSPECIES

Red Grouse
L. l. scoticus
(Great Britain, Ireland)

dark reddish brown

dark wings

yellow-brown

MALE

FEMALE

SIMILAR SPECIES

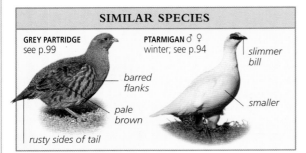

GREY PARTRIDGE
see p.99

PTARMIGAN ♂ ♀
winter; see p.94

slimmer bill

barred flanks

pale brown

smaller

rusty sides of tail

OCCURRENCE
On heather moorland in Great Britain and Ireland, most on places managed for shooting, rarely moving from breeding areas even in hard winter weather. In northern forests and clearings in Scandinavia and extreme NE Europe.

| Seen in the UK |
| J F M A M J J A S O N D |

| Length **37–42cm (14½–16½in)** | Wingspan **55–66cm (22–26in)** | Weight **650–750g (23–27oz)** |
| Social **Small flocks** | Lifespan **Up to 7 years** | Status **Secure** |

Order **Galliformes**	Family **Tetraonidae**	Species *Lagopus muta*

Ptarmigan

MALE (SUMMER)

black tail

neck extended

MALE (WINTER)

white wings

IN FLIGHT

yellow speckling on brown body

white wings

small, round head

red over eye

FEMALE (SUMMER)

small, delicate bill

slim neck

white plumage

small bill

"salt and pepper" barring on grey upperparts

black line between bill and eye

round, bantam-shaped body

white plumage; piebald in spring and autumn

MALE (WINTER)

white belly

FEMALE (WINTER)

MALE (SUMMER)

A high-altitude grouse in the south of its range, on lower, barren ground farther north (even within Scotland), the Ptarmigan is a smaller, more delicate version of the Willow Grouse. In the UK, it is found only on the highest Scottish peaks and extreme northern moors. It is difficult to separate from Willow Grouse in winter, and females require care to separate from Red Grouse in summer.

VOICE Low, dry, croaking notes, especially four-syllable *arr-kar-ka-karrrr*; also cackling "belch".

NESTING Scrape on ground, lightly lined with grass; 5–9 eggs; 1 brood; May–July.

FEEDING Shoots, leaves, buds, seeds, and berries of variety of low-growing shrubs; also takes insects, which are important food for chicks.

FLIGHT: flies up at close range with powerful, rising flight on stiff, fast-beating wings; long downhill glides.

SEASONAL CHANGE
Various parti-coloured plumage patterns can be seen on the Ptarmigan in spring and autumn.

SIMILAR SPECIES

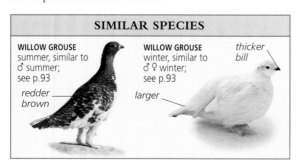

WILLOW GROUSE summer, similar to ♂ summer; see p.93

redder brown

WILLOW GROUSE winter, similar to ♂ ♀ winter; see p.93

thicker bill

larger

OCCURRENCE
Breeds widely in Iceland, N and W Scandinavia, and very locally in Scotland, Pyrenees, and Alps, on open tundra or rocky shores, and boulder fields. In S Europe, only on highest stony peaks that mimic tundra environment.

Seen in the UK
J F M A M J J A S O N D

Length **34–36cm (13¹/₂–14in)**	Wingspan **54–60cm (21¹/₂–23¹/₂in)**	Weight **400–600g (14–21oz)**
Social **Small flocks**	Lifespan **Up to 7 years**	Status **Secure**

| Order **Galliformes** | Family **Tetraonidae** | Species ***Tetrao tetrix*** |

Black Grouse

broad white wingbar

elongated shape

MALE

blue sheen on neck

bold white shoulder spot

undertail feathers raised in display

curved, broad-tipped outer feathers on tail

fine pale bar on mid-brown wings

slightly notched tail (lacks contrast of Red Grouse)

FEMALE

IN FLIGHT

white patch under tail

glossy black plumage

large, heavy, cockerel-like body

dark-barred, yellow-brown or grey plumage

FEMALE

MALE

A large grouse of moorland edges and forest clearings, the Black Grouse has declined over most of its European range. At the leks (the spring display sites), the males display with mock fights, to impress the reclusive females that watch from hidden vantage points nearby. They are subject to disturbance at leks, and the males usually fly off at long range if approached.

VOICE Female has gruff bark; displaying male has far-carrying, dove-like, rolling coo with regular rhythm and explosive "sneeze".

NESTING Hollow on ground beneath heather or bracken, with little or no lining; 6–10 eggs; 1 brood; April–July.

FEEDING Wide variety of seeds, berries, buds, shoots, leaves, and flowers of many shrubs, sedges, and trees; chicks eat insects.

FLIGHT: strong, often high, direct, over long distances with regular wingbeats; occasional glides.

DRAMATIC MALE
Males on open ground are easily visible at great distances.

COURTSHIP DISPLAY
In spring, males gather at dawn in open places to display with drooped wings and lyre-shaped tails.

SIMILAR SPECIES

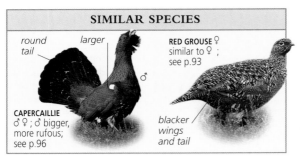

round tail　*larger*

CAPERCAILLIE
♂♀; ♂ bigger, more rufous; see p.96

RED GROUSE ♀
similar to ♀; see p.93

blacker wings and tail

OCCURRENCE
Widespread in N and W Britain, Scandinavia, Alps, and NE Europe, very local in Low Countries with long-term decline. Varied habitats including woodland, pastures, heaths, moors, or new plantations on heather moor.

Seen in the UK
J F M A M J J A S O N D

| Length **40–55cm (16–22in)** | Wingspan **65–80cm (26–32in)** | Weight **0.75–1.4kg (1³/₄–3lb)** |
| Social **Flocks** | Lifespan **Up to 5 years** | Status **Vulnerable** |

| Order **Galliformes** | Family **Tetraonidae** | Species *Tetrao urogallus* |

Capercaillie

brown wings

MALE

white-speckled, broad, round black tail

red wattle

short, hooked thick bill

spiky beard

huge, blackish grey body

bold white shoulder spot

FEMALE

dark bars on orange tail

IN FLIGHT

dark bars on rufous-ginger body

broad orange chest

white patches on flanks and belly

FEMALE

MALE

The largest grouse and a bird of pine forest and boggy forest clearings with thick undergrowth, the Capercaillie is sensitive to disturbance and typically shy and secretive. It may burst almost from underfoot in a forest, but usually flies up at long range from an open clearing. It is nowhere common and in some areas, including Scotland, seriously threatened.

VOICE Pheasant-like crow; male in spring has prolonged, croaking "song" ending with cork-popping and gurgling notes.

NESTING Hollow on ground, often at bases of trees, lined with grass, pine needles, and twigs; 5–8 eggs; 1 brood; March–July.

FEEDING Pine needles, buds of several shrubs and trees, shoots and leaves, and berries of various herbs and shrubs, especially bilberry; feeds in trees in winter.

FLIGHT: often flies up far away and goes off in long, low, fast flight with heavy wingbeats; also bursts up at close range with great clatter.

SPRING DISPLAY
Where they are common, females may gather around displaying males in early spring.

OCCURRENCE
Breeds in Scotland and N Spain, widely in Scandinavia and from Alps eastwards, declining in many areas. Prefers ancient, natural pine forest; more sparsely found in pine plantations. Feeds in nearby boggy clearings with bilberry, juniper, and heather, and on treetops in winter, but remains within such areas all year.

| Seen in the UK |
| J F M A M J J A S O N D |

SIMILAR SPECIES

BLACK GROUSE ♀
similar to ♀;
see p.95

notched tail

greyer and smaller

PINE FOREST BIRD
The Capercaillie inhabits pine forest, feeding on the treetops in winter, and in marshy clearings nearby during the rest of the year.

| Length **60–85cm (23¹/₂–34in)** | Wingspan **0.87–1.25m (2³/₄–4ft)** | Weight **1.5–4.4kg (3¹/₄–9³/₄lb)** |
| Social **Family groups** | Lifespan **Up to 10 years** | Status **Secure†** |

Order **Galliformes**	Family **Phasianidae**	Species ***Coturnix coturnix***

Quail 🔊 17

IN FLIGHT

MALE
- dark, pointed tail
- narrow, long, dark wings

FEMALE
- striped head
- pale throat

MALE
- striped crown
- small bill
- black on throat
- bold cream stripes on brown back
- small, rotund body
- dark stripes on flanks

This is a bird that is heard but rarely seen: it seldom flies and, living as it does in long grass or cereals, it is almost impossible to see on the ground unless it ventures onto an open track. Migrants occasionally appear in more exposed places and can then be watched more easily, although they remain secretive and skulking. When Quails do fly, they look unexpectedly long-winged, and may be confused with other species, such as young Partridges which can fly well before they are full grown. Quails usually fly in a short, fast, low arc before dropping down out of sight, and are unlikely to be flushed a second time. They are generally much more common and widespread in warmer, southern parts of Europe than farther north, but have declined in numbers in the face of modernized agriculture even there, and future prospects are not very bright.

VOICE Unique loud, far-carrying song, full, liquid, rhythmic *quick-we-wik*; also quiet mewing notes.
NESTING Slight hollow lined with vegetation and well hidden in crops or grass; up to 12 eggs; 1 brood; May–June.
FEEDING Walks slowly forwards, picking up seeds and shoots and snatching small insects from ground or foliage.

FLIGHT: low, quite quick; fast wingbeats and short glides, almost Snipe-like but drops quickly into cover.

EXPOSED MIGRANT
Migrants occasionally rest in fields with sparse crops and can sometimes be seen in the open.

SIMILAR SPECIES

GREY PARTRIDGE
see p.99
- rufous tail

CORNCRAKE
see p.166
- rufous wings
- longer legs

OCCURRENCE
Widespread north to Baltic, but erratic at northern edge of its range. Breeds in extensive tracts of long grass or cereal fields, mostly in warm, dry areas. Increased numbers appear in some "Quail years".

Seen in the UK
J F M **A M J J A S O** N D

Length **16–18cm (6¹⁄₂–7in)**	Wingspan **32–35cm (12¹⁄₂–14in)**	Weight **70–135g (2¹⁄₂–5oz)**	
Social **Family groups**	Lifespan **Up to 8 years**	Status **Vulnerable**	97

| Order **Galliformes** | Family **Phasianidae** | Species *Alectoris rufa* |

Red-legged Partridge

straight, stiff wings

white chin

rufous tail

IN FLIGHT

white stripe below cap

grey-brown cap

streaked foreneck

plain, pale brown upperparts

red bill

black line around white face, speckled breast

dark red-brown tail

FLIGHT: runs more than flies; wings straight, stiff, beating rapidly between flat-winged glides.

red legs

black, brown, and blue-grey barred flanks

This neat, colourful, and attractive partridge is seriously threatened by irresponsible introductions of other species within its range, producing a mixed hybrid population. Close views are required to rule out Red-legged Chukar hybrids which are widespread in England. In their native range, Red-legged Partridges are quite elusive birds of warm, open, stony slopes and fields; in the UK, they prefer light soils in arable areas. Red-legged Partridges can often be seen perched on haystacks, barns, and farm buildings.

VOICE Deep, gobbling and hissing or chuckling mechanical calls, *chuk-uk-ar, k'chuk-ar, k'chuk-ar.*

NESTING Grass-lined hollow scrape on ground beneath low vegetation; 7–20 eggs; 1 brood; April–June.

FEEDING Eats leaves, shoots, berries, acorns, and seeds, including beech-mast picked up from ground; chicks eat insects.

STAYING CLOSE
Family groups of Red-legged Partridges walk slowly over open ground or sit together inconspicuously in short stubble.

DRINKING PARTY
Temporary pools attract family groups to drink and bathe after rain in otherwise dry areas.

OCCURRENCE
Resident in Portugal, Spain, France, N Italy, and introduced in UK. Breeds on ground in places that have open slopes with scattered aromatic shrubs and much bare, stony, or sandy ground; in arable areas with dry, sandy fields; and less commonly on grassy heaths and coastal dunes.

| Seen in the UK |
| J F M A M J J A S O N D |

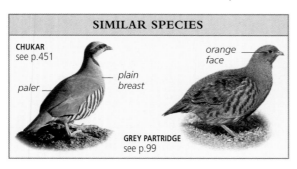

SIMILAR SPECIES

CHUKAR see p.451

paler

plain breast

orange face

GREY PARTRIDGE see p.99

CAMOUFLAGE
Despite the bright patterns of this, the prettiest partridge, it is well camouflaged in most situations.

| Length **32–34cm (12¹⁄₂–13¹⁄₂in)** | Wingspan **45–50cm (18–20in)** | Weight **400–550g (14–20oz)** |
| Social **Family groups** | Lifespan **Up to 6 years** | Status **Vulnerable** |

Order **Galliformes**	Family **Phasianidae**	Species *Perdix perdix*

Grey Partridge 🔊 16.I, 16.II

slightly arched, pale brown wings

rusty orange tail sides

IN FLIGHT

pale orange-brown face

small head

dull brown bill

streaked back

dumpy, pale brown body

pale streaks on grey breast

broad, horseshoe-shaped, dark brown patch on belly

dull brown legs

FLIGHT: low, fast, on bowed wings, with quick wingbeats and short glides.

This small, neat, grouse-like bird is typical of old-fashioned farmland with meadows, arable crops, and plenty of hedges; extensive cereal prairies suit it far less well. Its territorial call on summer evenings draws attention to it where it manages to survive in modern intensively farmed landscapes. It moves secretively through grassy habitats, often pausing to raise its head and look around. Family groups gather together in tight flocks, called "coveys" and sometimes fly off together in such groups if disturbed.

VOICE Distinctive, mechanical, creaky, low, rhythmic note, *kieeer-ik* or *ki-yik*.

NESTING Shallow scrape on ground, lined with some grass and leaves, well hidden under long grass; 10–20 eggs; 1 brood; April–June.

FEEDING Feeds as it walks over ground, taking seeds, leaves, and shoots; chicks feed on insects.

STUBBLE BIRD
Winter corn stubbles provide good habitat but are now rarely left for long. Intensive farming has led to large declines.

FAST FLIGHT
A covey dashes past low and fast on whirring wings, with frequent glides.

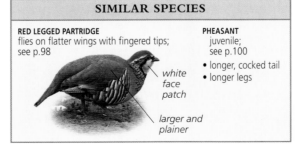

SIMILAR SPECIES

RED LEGGED PARTRIDGE
flies on flatter wings with fingered tips; see p.98

PHEASANT
juvenile; see p.100
• longer, cocked tail
• longer legs

white face patch

larger and plainer

OCCURRENCE
Widespread from UK, France, extreme N Spain, east across Europe, and north to Finland. In farmland, heaths, dunes, and especially traditional grassy meadows with abundant insect food; in reduced numbers in modern farmland with grassy field margins and hedges.

Seen in the UK
J F M A M J J A S O N D

Length **29–31cm (11½–12in)**	Wingspan **45–48cm (18–19in)**	Weight **350–450g (13–16oz)**
Social **Family groups**	Lifespan **Up to 5 years**	Status **Vulnerable**

| Order **Galliformes** | Family **Phasianidae** | Species *Phasianus colchicus* |

Pheasant 🔊 18

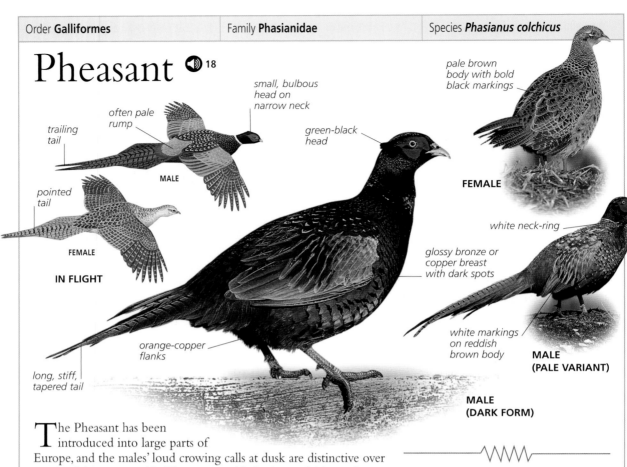

- trailing tail
- often pale rump
- MALE
- pointed tail
- FEMALE
- **IN FLIGHT**
- small, bulbous head on narrow neck
- green-black head
- glossy bronze or copper breast with dark spots
- orange-copper flanks
- long, stiff, tapered tail
- pale brown body with bold black markings
- **FEMALE**
- white neck-ring
- white markings on reddish brown body
- **MALE (PALE VARIANT)**
- **MALE (DARK FORM)**

The Pheasant has been introduced into large parts of Europe, and the males' loud crowing calls at dusk are distinctive over much of the countryside. The status and behaviour of this species are difficult to specify because of the frequent presence of young birds reared and released, unprepared for life in the wild, to be shot. "Wild" birds frequently resort to marshy, reedy places, as well as woodland edges where they are most familiar and characteristic.

VOICE Loud, explosive *corr-kok!* with sudden whirr of wings; also loud clucking in flight.

NESTING Hollow on ground, under overhanging cover such as brambles, unlined or with thin scattering of grass stems; 8–15 eggs; 1 brood; April–July.

FEEDING Takes all kinds of food, from seeds and berries to insects and lizards, from ground in its powerful bill.

FLIGHT: low, short bursts with whirring wings, trailing tail; explosive escape from underfoot.

WHIRRING WINGS
The call of the richly patterned male Pheasant is followed by a sudden burst of wingbeats that create a very brief, loud whirring sound.

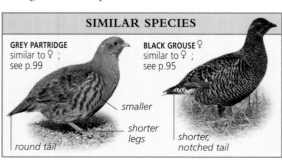

SIMILAR SPECIES

GREY PARTRIDGE similar to ♀; see p.99

BLACK GROUSE ♀ similar to ♀; see p.95

- smaller
- shorter legs
- round tail
- shorter, notched tail

OCCURRENCE
Very local in Spain, Portugal, and S Scandinavia but widespread through mid- and W Europe. Found widely in varied habitats, chiefly in very mixed countryside, in arable fields, woods, reedbeds, heaths, and moorland edges.

Seen in the UK
J F M A M J J A S O N D

| Length **52–90cm (20½–35in)** | Wingspan **70–90cm (28–35in)** | Weight **0.9–1.4kg (2–3lb)** |
| Social **Small flocks** | Lifespan **Up to 7 years** | Status **Secure** |

DIVERS AND GREBES

Entirely water birds except when nesting, these sleek, dagger-billed birds have thick plumage, narrow wings, short tails, and legs set very far back on the body. This makes it difficult for them to move on land: they merely shuffle on their bellies, pushing with their feet, onto the nest and off again into water. Their feet are not webbed, but have broad lobes along each toe (the hind toe is very small). These lobes fold together as the foot is pushed forwards through water, reducing drag, but open out when pushed against it to give greater propulsion.

RED THROAT IN SUMMER
Divers have distinctive plumages when breeding, such as this Red-throated Diver, but in winter they become more anonymous in dark brown and white.

DIVERS

These are bigger and longer-bodied than grebes, flying more often (indeed, the Red-throated Diver flies from its nesting lake to feed in the sea) and more northerly in their distribution in summer. They have loud, wailing calls.

GREBES

Round-bodied but long-necked, grebes are widespread in Europe, breeding on rivers, marshes, and larger lakes. They have barking, croaking, or trilling calls, less wild and dramatic than the divers' vocalizations. They characteristically cover their eggs with waterweed whenever they leave the nest, in an effort to deter predators.

GAVIIDAE

red throat, uptilted bill

RED-THROATED DIVER
p.103

grey head, black throat

BLACK-THROATED DIVER
p.104

black head

GREAT NORTHERN DIVER
p.105

DIVERS AND GREBES

PODICIPEDIDAE

pale yellow spot at base of bill

black head plumes

grey and white face

LITTLE GREBE
p.106

GREAT CRESTED GREBE
p.107

RED-NECKED GREBE
p.108

stiff golden-yellow wedge on black head

fan-shaped yellow or bronze ear tufts

SLAVONIAN GREBE
p.109

BLACK-NECKED GREBE
p.110

GRACEFUL DISPLAY
Black-necked Grebes show off their dramatic plumage in courtship displays; the sexes look alike.

| Order **Gaviiformes** | Family **Gaviidae** | Species *Gavia stellata* |

Red-throated Diver

hump-backed, dark-winged; larger, longer than grebes

all-brown back

upturned grey bill

striped grey nape

tapering, dark red throat patch

ADULT (WINTER)

outstretched head

ADULT (SUMMER)

uptilted bill (other divers' horizontal)

white face extends above eye

pale dusky face

white-speckled back

ADULT (SUMMER)

IN FLIGHT

JUVENILE

ADULT (WINTER)

A low-profile, swimming bird, not found on land except at the nest, this diver is distinguished by its slim, tapered bill held angled upwards. Red-throated Divers nest on small freshwater pools but fly off to feed at the coast. They are typically found flying high overhead to and from the sea in summer, calling loudly. In winter, they are mostly marine birds. Brown-backed and grey-headed with a dark red throat patch in summer, the plumage of the Red-throated Diver is less distinctive in winter.

VOICE Loud, high wail and fast, staccato quacking (in flight) in summer; quiet in winter.
NESTING Scrape almost on shoreline, in danger of flooding; 2 eggs; 1 brood; April–July.
FEEDING Dives for fish and other aquatic creatures; underwater for long periods, reappears at some distance from point of dive.

FLIGHT: low and straight over sea, head outstretched, legs slightly drooped; steady, strong wingbeats.

CALL POSTURE
Red-throated Divers, like other divers, use several strange, ritualized postures while calling on breeding pools in summer.

SIMILAR SPECIES

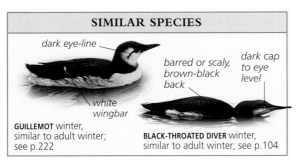

dark eye-line

barred or scaly, brown-black back

dark cap to eye level

white wingbar

GUILLEMOT winter, similar to adult winter; see p.222

BLACK-THROATED DIVER winter, similar to adult winter; see p.104

OCCURRENCE
Breeds on small, remote moorland pools and lakes in north, but feeds on sea in N Scotland, Iceland, and Scandinavia. In winter, more widespread around W European shores on open coasts and estuaries, but very rare inland.

Seen in the UK
J F M A M J J A S O N D

| Length **50–60cm (20–23½ in)** | Wingspan **1.06–1.16m (3½–3¾ ft)** | Weight **1.2–1.6kg (2¾–3½lb)** |
| Social **Winter flocks** | Lifespan **Up to 20 years** | Status **Vulnerable** |

Order **Gaviiformes**	Family **Gaviidae**	Species *Gavia arctica*

Black-throated Diver 🔊 1

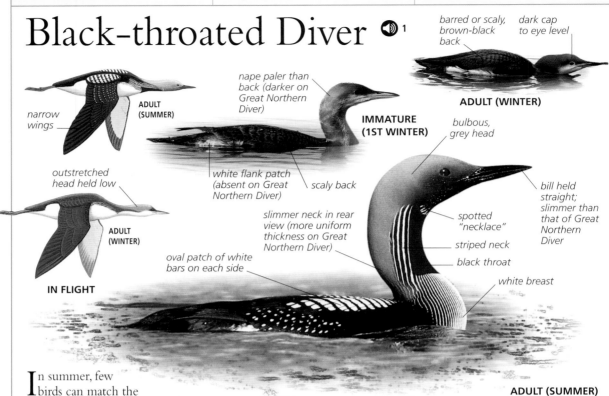

barred or scaly, brown-black back

dark cap to eye level

ADULT (WINTER)

narrow wings

ADULT (SUMMER)

nape paler than back (darker on Great Northern Diver)

IMMATURE (1ST WINTER)

bulbous, grey head

outstretched head held low

white flank patch (absent on Great Northern Diver)

scaly back

bill held straight; slimmer than that of Great Northern Diver

ADULT (WINTER)

slimmer neck in rear view (more uniform thickness on Great Northern Diver)

spotted "necklace"

oval patch of white bars on each side

striped neck

black throat

IN FLIGHT

white breast

ADULT (SUMMER)

In summer, few birds can match the exquisite patterning of the Black-throated Diver. In winter, it is duller and harder to tell from a Great Northern or Red-throated Diver, but the slightly bulbous head, slim, straight bill, and narrow body are distinctive, along with the greyish nape, paler than the back. Small parties of Black-throated Divers gather in coastal bays in summer. However, this bird is generally solitary, swimming low on the sea.

VOICE Wild, loud wailing notes in summer; silent in winter.

NESTING Shallow scoop near water's edge on island (or raft) in lake; 2 eggs; 1 brood; April–July.

FEEDING Dives long and deep for fish, reappearing far away; slides evenly head-first into water with barely a ripple.

FLIGHT: head low, outstretched, legs trailed; narrow wings have slightly whip-like action.

SUMMER GATHERING
Black-throated Divers gather in impressive groups, swimming close together with heads raised.

NESTING ON RAFT
Black-throated Diver nests are often subject to flooding or left high and dry; artificial nest rafts ensure greater breeding success.

SIMILAR SPECIES

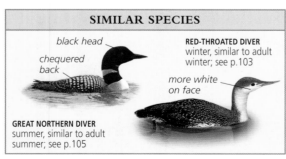

black head

chequered back

RED-THROATED DIVER winter, similar to adult winter; see p.103

more white on face

GREAT NORTHERN DIVER summer, similar to adult summer; see p.105

OCCURRENCE
Breeds on large lakes with small islands in remote areas of N Scotland and Scandinavia. In winter, more widespread but scarce on coasts of NW Europe. Often in larger estuaries or broad sandy bays; rare inland.

Seen in the UK
J F M A M J J A S O N D

Length **60–70cm (23½–28in)**	Wingspan **1.1–1.3m (3½–4¼ft)**	Weight **2–3kg (4½–6½lb)**
Social **Small summer flocks**	Lifespan **Up to 20 years**	Status **Vulnerable**

Order **Gaviiformes**	Family **Gaviidae**	Species *Gavia immer*

Great Northern Diver

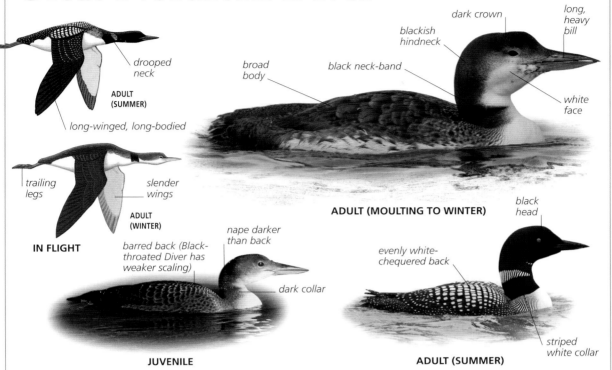

ADULT (SUMMER)
drooped neck

long-winged, long-bodied

IN FLIGHT

ADULT (WINTER)
trailing legs
slender wings

dark crown
blackish hindneck
broad body
black neck-band
long, heavy bill
white face

ADULT (MOULTING TO WINTER)

JUVENILE
nape darker than back
barred back (Black-throated Diver has weaker scaling)
dark collar

ADULT (SUMMER)
black head
evenly white-chequered back
striped white collar

One of the largest divers, the Great Northern Diver has a heavy, dagger-like bill, and often an angular head shape with a "bump" on the forehead. It is a splendidly chequered bird in breeding plumage. In winter, its large size, very broad body, heavy bill, and dark nape (while the barred back is paler) are distinguishing features. A young Cormorant might be taken for it, or vice versa, in a poor view. The Great Northern Diver also swims very low in the water, but never stands out on a perch.

VOICE Wolf-like wailing and tremulous laughing notes in summer; silent in winter.
NESTING Shallow scrape on water's edge; 2 eggs; 1 brood; April–June.
FEEDING Feeds on large fish, crabs, and other aquatic life, often bringing big flatfish to surface after long dive.

FLIGHT: flies low and direct; typical long, slender wings of diver, legs trailing, neck drooped.

FLAPPING WINGS
The Great Northern Diver, like most other divers, often sits up on the water and flaps its wings, revealing its white underparts.

SIMILAR SPECIES

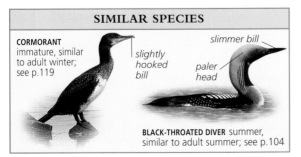

CORMORANT immature, similar to adult winter; see p.119
slightly hooked bill

slimmer bill
paler head

BLACK-THROATED DIVER summer, similar to adult summer; see p.104

OCCURRENCE
Breeds on larger lakes in Iceland. Scarce in winter but widespread on wide estuaries, in broad, sandy coastal bays, and also on wilder, open water in W Europe; rare inland on bigger reservoirs or flooded pits.

Seen in the UK

| J | F | M | A | M | J | J | A | S | O | N | D |

Length **70–80cm (28–32in)**	Wingspan **1.27–1.47m (4¼–4¾ft)**	Weight **3–4kg (6½–8¾lb)**
Social **Solitary**	Lifespan **10–20 years**	Status **Secure†**

| Order **Podicipediformes** | Family **Podicipedidae** | Species **Tachybaptus ruficollis** |

Little Grebe 🔊 2.I, 2.II

trailing feet

ADULT (WINTER)

dark wings

IN FLIGHT

buffish face

ADULT (WINTER)

buff foreneck

JUVENILE

rounded body

rufous face

blackish cap

straight, pointed bill

pale yellow spot at base of bill

blunt buff tail end

ADULT (SUMMER)

Little Grebes are small, dark, short-billed, and round as a ball, their near lack of tail rendering them especially buoyant on freshwater rivers, lakes, and ponds. Their loud, trilling or whinnying calls ring around marshes in summer. In winter, they often move to larger waters, less prone to freeze, and more rarely to the coast. They may then mix loosely with other species of waterfowl but tend to keep in little groups, slightly dispersed, in their own corner of the lake.

VOICE High, loud, rapid trill that fades away; silent in winter.

NESTING Floating mound of wet weed, anchored to stem or branch; 4–6 eggs, which the bird covers if it leaves the nest; 1 brood; April–June.

FEEDING Dives for small fish, aquatic insects, and molluscs, often diving under with a little leap and reappearing like a cork.

FLIGHT: flies little, usually very low, skittering over water on small wings.

ALERT ON WATER
When alarmed, the Little Grebe looks longer-necked and less dumpy, and in winter, can look surprisingly like a Black-necked Grebe.

SIMILAR SPECIES

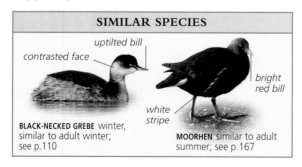

uptilted bill

contrasted face

BLACK-NECKED GREBE winter, similar to adult winter; see p.110

white stripe

bright red bill

MOORHEN similar to adult summer; see p.167

OCCURRENCE
Widespread in summer except in N Europe, breeding on broad rivers and canals, freshwater pools, and flooded pits. Dispersal to larger waters in W Europe in autumn also takes them onto sea at times, in sheltered areas.

Seen in the UK
J F M A M J J A S O N D

| Length **25–29cm (10–11½in)** | Wingspan **40–45cm (16–18in)** | Weight **100–120g (3⅝–4oz)** |
| Social **Small flocks** | Lifespan **10–15 years** | Status **Secure†** |

Order **Podicipediformes**	Family **Podicipedidae**	Species **Podiceps cristatus**

Great Crested Grebe

IN FLIGHT

drooping neck

legs below body level

bold white wing patches

ADULT (WINTER)

ADULT (SUMMER)

black head plumes

white face

unique ruff

slender neck

dull, dark back

dagger-like pink bill

white neck and breast

pale greyish body

striped head

JUVENILE

tailless shape

dark back

white over eye

pink bill

white breast

ADULT (WINTER)

The upright, slender neck with a silky white front and bright white breast are always characteristic of this large, dagger-billed grebe. In summer, its black cap extends into a double, backward-facing tuft and a frill of chestnut appears on the face, used in face-to-face head-wagging displays by breeding pairs. Small groups breed close together and larger flocks form on more open water, such as large reservoirs, in winter.

VOICE Various loud barks and growling notes in summer; juveniles make loud, fluty whistles.

NESTING Pile of weed on water, anchored to vegetation; 3 or 4 white eggs; 1 brood; February–June.

FEEDING Dives from surface, staying under for lengthy periods in search of fish and large aquatic invertebrates.

FLIGHT: low, direct; head and trailing legs drooped below body level.

OCCURRENCE
Widespread except in far N Europe. Mostly breeds on flooded gravel pits, large lakes, and reservoirs; also on larger rivers. In winter, mostly on bigger reservoirs and sheltered coastal waters in W Europe; autumn migrants often on sea.

Seen in the UK
J F M A M J J A S O N D

SIMILAR SPECIES

stockier and darker

stubbier bill

RED-NECKED GREBE winter, similar to adult winter; see p.108

COURTING
Pairs of Great Crested Grebes perform complex courting rituals, diving underwater and surfacing with weeds that they offer each other.

Length **46–51cm (18–20in)**	Wingspan **85–90cm (34–35in)**	Weight **800–1000g (29–36oz)**
Social **Flocks**	Lifespan **10–15 years**	Status **Secure**

| Order **Podicipediformes** | Family **Podicipedidae** | Species **Podiceps grisegena** |

Red-necked Grebe

ADULT (WINTER)

slightly drooped neck

IN FLIGHT

dumpy shape

grey and white face

thick neck

dark cap extends below eye (white, above eye on Great Crested Grebe)

yellow base to stout, daggerlike bill

dark chestnut neck and breast

ADULT (SUMMER)

round head

whitish cheeks hooks up behind cap

dark around eye

dusky grey on foreneck

ADULT (WINTER)

striped cheeks

yellow on bill

JUVENILE

FLIGHT: low, straight, heavy, with slightly drooped neck and legs.

In much of northeast Europe, this is a typical bird of large, well-vegetated lakes; in Great Britain, it is mostly a late summer and winter visitor and never common. Like other grebes it can look dumpy, squat, and thickset, or rather slim, according to conditions and intent: an active, feeding bird looks much more slim and alert than a resting one. Such changes in shape can make judgement of size, especially at a distance on open water, very difficult, but this is a large grebe, along with the Great Crested Grebe, compared with the small grebes such as Slavonian and Black-necked Grebes.

VOICE Various growling notes; silent in winter.

NESTING Typical grebe nest: heap of waterweed with shallow depression on top, semi-floating on fresh water amongst vegetation; 3 or 4 eggs; 1 brood; April–June.

FEEDING Dives for food, mostly fish, but also feeds on various crustaceans and insects in summer.

A STRUGGLE ON LAND
Grebes are unable to walk on land, and use an ungainly shuffle to leave the nest. They are adept swimmers.

SIMILAR SPECIES

slim bill, never yellow; white over eye

longer neck

whiter foreneck

smaller bill

GREAT CRESTED GREBE winter, similar to adult winter; see p.107

SLAVONIAN GREBE winter, similar to adult winter; see p.109

OCCURRENCE

Breeds on reedy lakes and broad rivers in E Europe. In autumn and winter, moves west, mostly on quiet estuaries and sheltered coastal bays; scarce on inland waters such as flooded gravel pits and reservoirs.

Seen in the UK
| J | F | M | A | M | J | J | A | S | O | N | D |

| Length **40–46cm (16–18in)** | Wingspan **77–85cm (30–34in)** | Weight **700–900g (25–32oz)** |
| Social **Small flocks** | Lifespan **Up to 10 years** | Status **Secure** |

| Order **Podicipediformes** | Family **Podicipedidae** | Species *Podiceps auritus* |

Slavonian Grebe

IN FLIGHT

- white patch on small wings
- **ADULT (SUMMER)**

- flat crown
- stiff golden-yellow wedge on black head
- rust-red flanks
- dumpy shape
- straight, short, pale-tipped bill
- rust-red neck
- **ADULT (SUMMER)**

- well-defined white cheeks
- pale bill tip
- white foreneck
- **IMMATURE (1ST WINTER)**

- white flanks
- large white cheeks
- straight division between black cap and white cheeks
- small whitish spot in front of eye
- white foreneck and breast
- **ADULT (WINTER)**

A bird of remote, upland, northern pools in summer, the Slavonian Grebe is then easy to identify. Its complex head pattern and plumes are used in display: like other grebes, the sexes are alike. In winter, in stark black and white plumage, it is much more like some other grebes. It typically breeds in loose groups of three or four pairs, and spends the winter in twos and threes at most. These may be seen near Black-necked Grebes in favoured spots.

VOICE High, fast, whistling trills in summer; usually silent in winter.

NESTING Pile of weeds anchored to reed or sedge stem; 4 or 5 eggs; 1 brood; April–July.

FEEDING Dives to find small fish; in summer, mostly feeds on insects and various aquatic crustaceans.

FLIGHT: low, quick; may patter across water surface.

NEST BUILDING
Slavonian Grebes build soggy heaps of water weeds in which to lay their eggs, covering them for safety when they leave the nest.

SIMILAR SPECIES

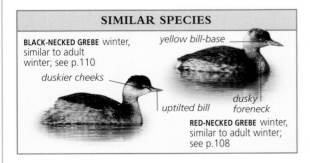

BLACK-NECKED GREBE winter, similar to adult winter; see p.110
- duskier cheeks
- uptilted bill

- yellow bill-base
- dusky foreneck
- **RED-NECKED GREBE** winter, similar to adult winter; see p.108

OCCURRENCE
Breeds in cool, wild pools with some fringing vegetation in Iceland, N Scotland, and N and E Europe. In winter, mostly on sea in NW Europe, especially on muddy estuaries, more rarely on reservoirs and pits inland.

| Seen in the UK |
| J F M A M J J A S O N D |

| Length **31–38cm (12–15in)** | Wingspan **59–65cm (23–26in)** | Weight **375–450g (13–16oz)** |
| Social **Pairs/Small flocks** | Lifespan **Up to 10 years** | Status **Secure†** |

| Order **Podicipediformes** | Family **Podicipedidae** | Species *Podiceps nigricollis* |

Black-necked Grebe

IN FLIGHT

trailing legs

white patch on slim wings

ADULT (SUMMER)

peaked crown

red eye

fine, slightly uptilted bill

black neck

fan-shaped yellow or bronze ear tufts

coppery red flanks

ADULT (SUMMER)

dusky cheeks

steep forehead

grey foreneck

JUVENILE

pale "hook" on rear cheeks

blurred cap

uptilted bill

ADULT (WINTER)

More restricted to richer, lower-lying lakes than the Slavonian Grebe as a breeding bird, the Black-necked Grebe is more frequent on fresh water in winter. It is characterized by a slim, slightly uptilted bill and round head with a peaked crown. Although it is one of the smallest grebes, barely bigger than a Little Grebe, in breeding plumage, with head erect, it can look quite large out on a still, gleaming lake. Ones and twos may be seen on estuaries in winter, swimming very buoyantly, drifting in and out with the tide. Black-necked Grebes often mix loosely with other grebe species, usually outnumbering Slavonian Grebes.

VOICE Chattering and high-pitched whistling notes; silent in winter.

NESTING Pile of wet water weeds; 3 or 4 eggs; 1 brood; March–July.

FEEDING Catches insects, molluscs, and a few fish, in lengthy dives underwater.

FLIGHT: low, weak, fluttery; head extended, legs trail.

NEST OF WEEDS
The Black-necked Grebe builds a typical grebe nest by piling up water weeds. The only time it is not on water is when it is at the nest.

SIMILAR SPECIES

sharper contrast on face

SLAVONIAN GREBE winter, similar to adult winter; see p.109

straight bill

LITTLE GREBE winter, similar to adult winter; see p.106

browner face

OCCURRENCE
Widespread but very local; breeds on pools with large amounts of reeds and other vegetation. In winter, on estuaries and in coastal bays, reservoirs, and flooded pits; in spring, often on sea in Mediterranean.

Seen in the UK
J F M A M J J A S O N D

| Length **28–34cm (11–13½ in)** | Wingspan **56–60cm (22–23½ in)** | Weight **250–350g (9–13oz)** |
| Social **Small flocks** | Lifespan **Up to 10 years** | Status **Secure** |

PETRELS AND SHEARWATERS

S O SUPREMELY ADAPTED to life at sea that they are cumbersome and vulnerable on land, petrels and shearwaters (with the exception of the cliff-nesting Fulmar) come to land only to breed and then do so only under the cover of darkness. By far the best chance of seeing most of them is from a ship at sea.

PETRELS

Like shearwaters and albatrosses, petrels' tubular nostrils excrete excess salt, and these birds are known collectively as "tubenoses". They breed in burrows or cavities, staying out of sight all day. Returning birds follow calls from their mates on the nests and use scent to find the right burrow in pitch dark.

STORM PETREL
Swallow-like in its actions, the Storm Petrel feeds on tiny plankton and oily waste out at sea.

  Petrels are mostly small and insignificant over the open sea, but are dainty fliers, coping with the most ferocious gales as they skim the wave crests. They are sometimes driven close inshore and may then find it difficult to get back out to sea, sometimes ending up, exhausted, inland. Several species, especially the Storm Petrel, follow ships to feed on organisms that are disturbed in the wake.

The Fulmar is a larger bird, easily seen on its open cliff ledge nest or flying beside cliff tops during the day.

SHEARWATERS

Superb fliers, using air currents over the waves, shearwaters are almost helpless ashore and in danger from predatory gulls and skuas when returning to their colonies. They fly with stiffly outstretched wings and long glides between brief periods of wingbeats. In still air, they look rather heavy but with a wind become wonderfully capable, banking steeply over onto one wingtip then to the other, showing alternately dark upperparts and light underparts as they fly past far offshore. Young birds may be exhausted and blown inland by autumn gales.

GREAT SHEARWATER
An ocean-going migrant, this species breeds in the northern winter on islands in the South Atlantic.

PROCELLARIIDAE

"tubed" nostrils

yellow bill

weak legs, cannot stand

FULMAR
p.112

CORY'S SHEARWATER
p.113

MANX SHEARWATER
p.114

HYDROBATIDAE

large white rump

forked tail

STORM PETREL
p.115

LEACH'S PETREL
p.116

Order **Procellariiformes**	Family **Procellariidae**	Species *Fulmarus glacialis*

Fulmar 🔊³

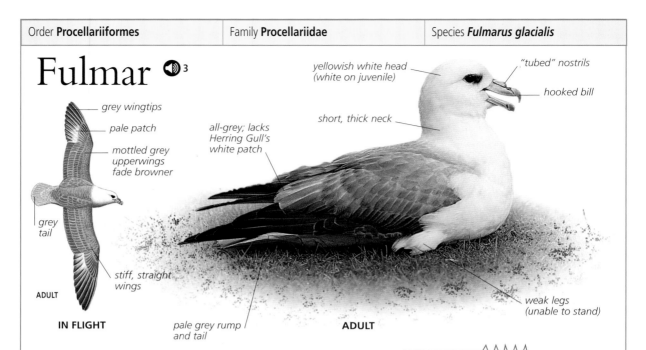

grey wingtips

pale patch

mottled grey upperwings fade browner

grey tail

stiff, straight wings

ADULT

IN FLIGHT

pale grey rump and tail

yellowish white head (white on juvenile)

"tubed" nostrils

hooked bill

short, thick neck

all-grey; lacks Herring Gull's white patch

weak legs (unable to stand)

ADULT

Gliding very low over the open sea or along clifftops, the Fulmar is a strong flier. Superficially gull-like, it is a "tubenose" (having large, raised nasal tubes) more closely related to albatrosses. Some Fulmars spend much of the year visiting breeding cliffs, even in winter, and can be viewed at close range as they sail by on the wind currents. Their real home, however, is over the wildest, windswept seas of the North Atlantic. Large flocks, often mixed with Gannets, gulls, and skuas, follow fishing fleets: Fulmar numbers have increased in the past century with far more food made available from these vessels.

VOICE Loud, harsh, throaty cackling.

NESTING On rocky or earth ledge, rarely ledges on buildings; 1 egg; 1 brood; April–June.

FEEDING Feeds mostly on fish offal from trawlers, small fish, jellyfish, squid, and other marine organisms.

FLIGHT: direct, on narrow, stiff, flat wings; gliding in wind, flapping heavily in calm weather low over sea.

SWIMMING IN SEA
Drinking saltwater, while swimming in the sea, is normal for "tube-noses" such as the Fulmar. Excess salt is excreted through the nostrils.

BREEDING
The Fulmar breeds in loose colonies on ledges on steep coastal cliffs or in burrows on inaccessible slopes, but also locally on buildings.

SIMILAR SPECIES

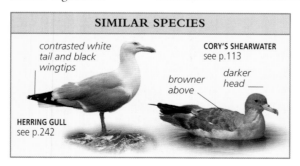

contrasted white tail and black wingtips

CORY'S SHEARWATER
see p.113

browner above

darker head

HERRING GULL
see p.242

OCCURRENCE
Breeds in NW Europe on cliffs, earth ledges, and even buildings or, where abundant, on grassy banks, usually close to sea. Out on open sea, sometimes mouths of estuaries or smaller bays.

Seen in the UK
J F M A M J J A S O N D

Length **45–50cm (18–20in)**	Wingspan **1–1.12m (3¼–3¾ft)**	Weight **700–900g (25–32oz)**
Social **Flocks**	Lifespan **20–30 years**	Status **Secure**

Order **Procellariiformes**	Family **Procellariidae**	Species *Calonectris diomedea*

Cory's Shearwater

long, tapered, slightly rounded wings

pale bill

hint of dark "W" across wings

IN FLIGHT

white underwings

dark brown above

all-white underparts

dull greyish head with no sharp cap

pale bill

slightly bowed wings

This large shearwater flies lazily by – low and rather heavily – sometimes quite close inshore, in small groups; however, strong winds allow it to show great mastery in the air. In southern Europe and off northwest Africa, this is an abundant seabird. In the Mediterranean, it is quite frequently seen in summer from many islands and headlands. Near nesting areas, birds flying to their burrows after dark make loud, strange calls, even over towns, such as Funchal in Madeira.

VOICE Loud, varied, wailing sounds near breeding sites; mostly silent at sea.

NESTING Hole among rocks in scree and on cliff, or burrow on steep slope; used only at night; 1 egg; 1 brood; March–July.

FEEDING Takes fish, squid, shrimps, jellyfish, and waste from fishing vessels in shallow dives from surface of sea.

FLIGHT: low, swerving slowly in long arcs; in wind, banks steeply and rises to great height.

LIGHT AND SHADE
Strong light gives these swimming Cory's Shearwaters a pale-naped appearance, with unusually dark faces and contrasting upperparts.

SIMILAR SPECIES

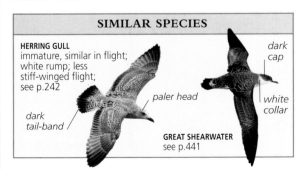

HERRING GULL
immature, similar in flight; white rump; less stiff-winged flight; see p.242

dark tail-band

paler head

dark cap

white collar

GREAT SHEARWATER
see p.441

OCCURRENCE
Mostly out at sea, but sometimes in broad bays or close inshore off headlands and islands. Outside breeding season, in western approaches off Ireland, N France, and SW England.

Seen in the UK
J F M A M J **J A S** O N D

Length **45–56cm (18–22in)**	Wingspan **1–1.25m (3¼–4ft)**	Weight **700–800g (25–29oz)**
Social **Flocks**	Lifespan **Up to 20 years**	Status **Vulnerable†**

| Order **Procellariiformes** | Family **Procellariidae** | Species **Puffinus puffinus** |

Manx Shearwater

stiff wings

IN FLIGHT

black cap

white throat

silvery white below

blackish upperparts; looks browner in strong sun

dark cap

thin dark bill

white flank bulges up each side of rump

weak legs, cannot stand

Nesting sites of Manx Shearwaters are relatively localized, nearly all on islands; these seabirds are commonly seen on surrounding seas and are sometimes very numerous, especially in the evenings as they gather prior to going ashore. In autumn, large numbers are seen close to the shore during gales and a few are blown inland each year. Like other shearwaters and storm petrels, they only go to the nest in the cover of darkness, but many are still caught and killed by gulls. They are ungainly on land, moving with a shuffle, using legs, wings, and bill to scramble over rough ground.

VOICE Loud, strangled wailing and chortling sounds at night around breeding colony.

NESTING Uses rabbit or Puffin burrow or similar tunnel, or hole in scree; 1 egg; 1 brood; April–July.

FEEDING Flocks gather over fish or small squid, diving from surface or after short plunge from air.

FLIGHT: fast, especially downwind, with long, swerving glides, banking on one wingtip then wheeling over onto other; wingbeats fast, stiff, flickering.

FLAP AND GLIDE
In calm air, shearwaters fly low over the water with many more deep wingbeats and short, flat glides.

SIMILAR SPECIES

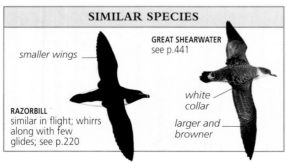

GREAT SHEARWATER
see p.441

smaller wings

RAZORBILL
similar in flight; whirrs along with few glides; see p.220

white collar

larger and browner

OCCURRENCE

Over open sea, except when breeding. Large colonies on islands, sometimes high up on rocky mountain peaks, more often in burrows in turf on lower slopes, in NW Europe. In autumn, widespread off coasts.

| Seen in the UK |
| J F **M A M J J A S O** N D |

| Length **30–38cm (12–15in)** | Wingspan **76–82cm (30–32in)** | Weight **350–450g (13–16oz)** |
| Social **Flocks** | Lifespan **25–30 years** | Status **Localized†** |

Order **Procellariiformes**	Family **Hydrobatidae**	Species *Hydrobates pelagicus*

Storm Petrel

broad-based wings taper and sweep back

white line along underwings

IN FLIGHT

settles on water with wings raised

large white rump

round head with small, "tubed" bill

sooty black body

broad, rounded tail

all-dark upperwings

short legs (unable to stand)

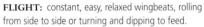

FLIGHT: constant, easy, relaxed wingbeats, rolling from side to side or turning and dipping to feed.

It is quite remarkable that such tiny, delicate creatures as Storm Petrels spend months on end far out at sea, surviving the most difficult weather. They return to islands, more rarely headlands, to breed, coming to land only at night for fear of predation by gulls and skuas. At sea, they fly slightly erratically but quite strongly, low over waves, pattering at times, or swooping like Swallows over the surface, but they are often able to overtake ferries and other ships at surprising speed.

VOICE Soft purring trill with abrupt ending, at nest.

NESTING Hole amongst rocks or in old wall, or in small burrow; 1 egg; 1 brood; April–July.

FEEDING Picks up tiny pieces of offal, fish oil, and marine invertebrates from surface of sea in flight.

HOMING IN ON FOOD
Petrels fly very low, scattered over vast areas of open sea. Their sharp sense of smell locates oily food and small flocks gather to feed.

SIMILAR SPECIES

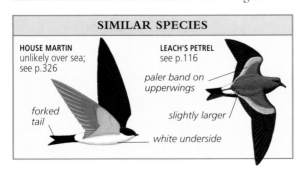

HOUSE MARTIN unlikely over sea; see p.326

forked tail

LEACH'S PETREL see p.116

paler band on upperwings

slightly larger

white underside

OCCURRENCE
Most breed in NW Europe but colonies also in Mediterranean; otherwise, lives out at sea. Difficult to see from most coasts but can be frequent in S Ireland just offshore; elsewhere, not so frequent as Leach's Petrel, even during storms.

Seen in the UK
J F M **A M J J A S O** N D

Length **14–17cm (5½–6½in)**	Wingspan **36–39cm (14–15½in)**	Weight **23–29g (¹³⁄₁₆–1¹⁄₁₆oz)**
Social **Small flocks**	Lifespan **Up to 20 years**	Status **Localized†**

| Order **Procellariiformes** | Family **Hydrobatidae** | Species *Oceanodroma leucorhoa* |

Leach's Petrel

"V"-shaped white rump with dark central line

pale panel on upperwings

IN FLIGHT

long, angled, arched wings

angular shape

sharp-winged, tern-like shape

dark underwings

notch in tail hard to see

sooty brown back

forked tail

FLIGHT: quick, strong, tern-like, with strong beats, erratic twists, turns, leaps, and changes of speed.

Slightly larger than the Storm Petrel, Leach's Petrel is still a tiny bird to live out at sea all its life, buffeted by Atlantic gales. Like other petrels and shearwaters, it ventures to land only to breed and only at night (unless it is deep inside its burrow), but may be seen offshore from suitable headlands in autumn gales. It may then also be driven inland, appearing over reservoirs rather than being stranded in unlikely places like some other storm-driven birds.

VOICE Rattling, chattering coo at nest.

NESTING Burrow or cavity among rocks; one bird incubates, while other is at sea, returning only at night; 1 egg; 1 brood; April–July.

FEEDING Picks up tiny, floating pieces of offal, fish oil, jellyfish, and marine invertebrates, from surface in flight.

BLOWN ONTO BEACHES
Tired by their efforts to fly out of coastal bays against a gale, Leach's Petrels may briefly patter across the tideline or even over a beach.

SIMILAR SPECIES

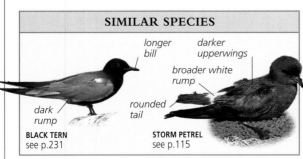

longer bill

darker upperwings

broader white rump

rounded tail

dark rump

BLACK TERN
see p.231

STORM PETREL
see p.115

OCCURRENCE
Breeds on few islands in NW Europe; more widespread in autumn in North Atlantic, but scarce in North Sea. Appears during gales off NW England and N Wales and is regular but rare inland in autumn gales.

| Seen in the UK |
| J F M **A M J J A S O N** D |

| Length **18–21cm (7–8½in)** | Wingspan **43–48cm (17–19in)** | Weight **40–50g (1⁷⁄₁₆–1¾oz)** |
| Social **Small flocks** | Lifespan **Up to 24 years** | Status **Localized†** |

GANNETS, CORMORANTS, AND PELICANS

AT FIRST SIGHT a mixed bunch, this group of water birds shares several characteristics, including a long inner wing with an obvious, backward-pointed "elbow" joint near the body, and broad webbing across all four toes (wildfowl have webs only between the front three). They all feed on fish, some being restricted to the sea, others coming inland too.

DENSE FLOCKS
White Pelicans feed and rest together in tightly packed flocks, which look wonderful when they take flight. Members of this family are rarely seen in the UK and Europe.

GANNETS
These are magnificent plunge-divers, seeing fish from high up and diving headlong, or fishing from a lower altitude if the fish are close to the surface, spearing into the water at an angle. They are often seen from the coast but spend winter well out at sea. They breed in a small number of large colonies, mostly on offshore islands. Gannets are among the most spectacular sights in European birdwatching.

CORMORANTS
The Cormorant is a generalist, able to feed in the open sea, in quiet estuaries, and in lakes or rivers. It nests in trees, as well as on cliffs. The Shag nests only on cliffs and feeds in saltwater. The Pygmy Cormorant is a freshwater bird in summer, nesting in marshes, but may move to the coast at other times.

PELICANS
Familiar, huge, ponderous birds on the water, known for the remarkable "pouch" beneath the bill, pelicans are brilliant fliers, often in beautifully coordinated lines, V-shaped flocks, or swirling packs. Among the world's biggest flying birds, they are always impressive.

SULIDAE

yellow-buff head

PHALACROCORACIDAE

small hook at tip of bill; thicker bill than Shag

short crest

GANNET
p.118

CORMORANT
p.119

SHAG
p.120

| Order **Pelecaniformes** | Family **Sulidae** | Species ***Morus bassanus*** |

Gannet

ADULT

long, narrow wings

protruding head

narrow, pointed tail

IMMATURE (1ST WINTER)

white band above tail

IN FLIGHT

black wingtips

dagger-like bill

yellow-buff head

blackish body with white spots

white plumage

ADULT

IMMATURE

piebald plumage turns white with age

IMMATURE

The biggest of the European seabirds, related to the even larger pelicans (see p.440), the Gannet is typically seen as a brilliant white bird offshore, circling and diving for fish, or flying singly or in groups. It is highly social at its nesting colonies and loosely so at sea. The Gannet's forward-facing eyes and cushioned head and neck equip it to pinpoint fast-moving fish and dive headlong to catch them.
VOICE Regular, rhythmic, throaty chorus at nest; silent at sea.
NESTING Pile of seaweed and debris on broad ledge high above sea; 1 egg; 1 brood; April–July.
FEEDING Catches fish such as mackerel and pollack underwater in shallow and sloping dive from air, or vertical dives from greater height.

FLIGHT: in strong winds, banks and veers like a giant shearwater; in light winds, steady, powerful flight with regular beats of straight wings.

DENSE COLONIES
Gannet colonies, usually along cliff ledges and steep slopes, are dense and often very large.

SIMILAR SPECIES

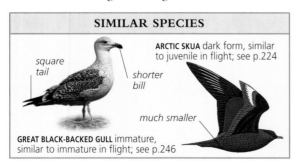

square tail

shorter bill

ARCTIC SKUA dark form, similar to juvenile in flight; see p.224

much smaller

GREAT BLACK-BACKED GULL immature, similar to immature in flight; see p.246

OCCURRENCE
Forms colonies on rocky islands north from NW France, occupying them from early spring until late autumn. Widespread in Atlantic and North Sea while feeding and migrating, some entering W Mediterranean; scarce in winter.

Seen in the UK
J F M A M J J A S O N D

| Length **85–89cm (34–35in)** | Wingspan **1.65–1.8m (5½–6ft)** | Weight **2.8–3.2kg (6¼–7lb)** |
| Social **Flocks** | Lifespan **16–20 years** | Status **Localized** |

Cormorant

ADULT

longer tailed than geese

neck kinked in flight

outstretched head

broad wings

white underside

JUVENILE

IN FLIGHT

lower, longer forehead profile than Shag

small hook at tip of thick bill; thicker bill than Shag

orange-yellow near bill

white on face

black underside

short legs with web across all toes

ADULT (SUMMER)

long, broad tail

wedge-shaped head and bill

brown back

mostly white below

IMMATURE

bill tilted upwards

upright neck

low body

ADULT (SUMMER)

FLIGHT: strong, often high; head outstretched, long tail, rather broad wings with regular beats; long glides.

In summer, Cormorants are unmistakable, with long, pale head plumes and bold facial colours; they have a round white thigh patch in spring. They are duller in winter, but retain a distinct character, typified by their habit of perching with half-open wings, or swimming with back almost awash, neck upright, and bill angled upwards. Equally at home on salt- or fresh water, Cormorants are widespread and familiar in much of Europe.

VOICE Growling and cackling at nests and communal roosts, otherwise a quiet bird.

NESTING Bulky nest of sticks in tree and on cliff ledge, with white splashings beneath; 3 or 4 eggs; 1 brood; April–May.

FEEDING Catches fish in long underwater dive from surface; brings larger ones to surface before swallowing them.

SUBSPECIES

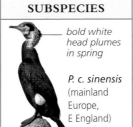

bold white head plumes in spring

P. c. sinensis
(mainland Europe, E England)

PERCHING
Cormorants have a distinctive perching stance, with half-open wings, upright neck, and bill angled upwards.

SIMILAR SPECIES

dagger-like bill

SHAG see p.120

slim bill

shorter neck

slightly smaller

GREAT NORTHERN DIVER winter, similar to immature; see p.105

OCCURRENCE
Breeds widely but very locally through Europe. Prefers sheltered estuaries and bays at coast but breeds on cliffs; inland, on reservoirs, flooded pits, and even small pools. Often found in and around harbours and marinas.

Seen in the UK

| J | F | M | A | M | J | J | A | S | O | N | D |

Length **80–100cm (32–39in)**	Wingspan **1.3–1.6m (4¼–5¼ft)**	Weight **2–2.5kg (4½–5½lb)**
Social **Flocks**	Lifespan **15–20 years**	Status **Secure**

| Order **Pelecaniformes** | Family **Phalacrocoracidae** | Species ***Phalacrocorax aristotelis*** |

Shag

short crest
steep forehead
white spot on chin
slim, slightly hooked bill
dark brown plumage
brown below
snaky neck
long, slim body
oily green-black plumage
rounded crown

JUVENILE
narrow wings
long tail
ADULT
slender head and neck

IN FLIGHT

ADULT (SUMMER)

IMMATURE

ADULT (WINTER)

Although sometimes solitary, Shags tend to breed in sizeable groups and sometimes feed in tight flocks of hundreds where they are common. They prefer the fast tide races and rough water under rocks and cliffs, and swim in groups braving the most dangerous-looking conditions; they are rare inland. The Shag has an obvious close relationship with the Cormorant and the two birds can be difficult to tell apart, but a certain slim snakiness gives the Shag a different character.

VOICE Coarse, frenzied rattling at nests; silent at sea.
NESTING Heap of grass, sticks, and seaweed on broad cliff ledge or inside cave; 3 or 4 eggs; 1 brood; May.
FEEDING Catches fish underwater, after dive from surface, often with quick, arching forward leap.

FLIGHT: direct with quick beats of narrow wings; usually keeps very low over sea, even where Cormorants fly high.

SUBSPECIES

P. a. desmaresti
juvenile
(Mediterranean)

much whiter below

STANDING
The Shag stands with wings outstretched like the Cormorant, possibly an aid to digestion after a heavy meal of fish.

SIMILAR SPECIES

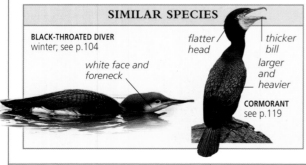

BLACK-THROATED DIVER
winter; see p.104

white face and foreneck

flatter head

thicker bill

larger and heavier

CORMORANT
see p.119

OCCURRENCE
Widespread on European coasts although scarcer in Mediterranean. Breeds on coastal cliffs and feeds off rocky coasts and islands. Rather scarce around harbours and estuaries and generally rare inland.

Seen in the UK
J F M A M J J A S O N D

| Length **65–80cm (26–32in)** | Wingspan **90–105cm (35–41in)** | Weight **1.75–2.25kg (3¾–5lb)** |
| Social **Large flocks** | Lifespan **Up to 15 years** | Status **Secure** |

BITTERNS AND HERONS

EUROPEAN HERONS AND bitterns are waterside birds, apart from the Cattle Egret, but elsewhere in the world many feed in drier places. They share characteristics such as long legs, dagger-like bills, a kinked neck (thicker in bitterns) that gives them a sudden, fast lunge to grasp prey, and binocular vision that allows them to pinpoint prey with precision. They are mostly ground-living birds, but fly capably.

REEDBED SKULKER
Bitterns need water within reeds so that they can ambush unsuspecting eels and other fish without coming out into the open.

HERONS

The larger herons are long-necked and elegant, using skill and patience when searching for prey. Although primarily fish-eaters, they eat almost anything they can catch. Egrets are generally smaller (although the Great White Egret is very big) and mostly white. Some have long, elegant plumes in the breeding season. Herons and egrets develop bright bill and leg colours for short periods in spring and the facial skin may even "blush" brightly during courtship. Most are social, breeding in colonies, but often fishing alone. The Cattle Egret feeds around live-stock and on rubbish tips, as well as on marshes, and flies to roost each evening in spectacular, flickering white flocks.

BITTERNS

Two groups are widespread worldwide, each represented by one species in Europe. The Bittern is large, heavy-bodied, and closely patterned with black on sandy buff. The Little Bittern is small, sexually dimorphic, with a large, pale, oval patch on each otherwise blackish wing. Both are shy and elusive.

ARDEIDAE

plumage provides exceptional camouflage in winter reeds

black cap and back

thin white plumes

neat, tapered oval shape

BITTERN
p.122

LITTLE BITTERN
p.123

NIGHT HERON
p.124

SQUACCO HERON
p.125

yellow bill, reddish in spring

black bill; yellow feet

grey back; black crest

snaky head and neck

very long neck

CATTLE EGRET
p.126

LITTLE EGRET
p.127

GREAT WHITE EGRET
p.128

GREY HERON
p.129

PURPLE HERON
p.130

Order **Ciconiiformes**	Family **Ardeidae**	Species **Botaurus stellaris**

Bittern 🔊 4

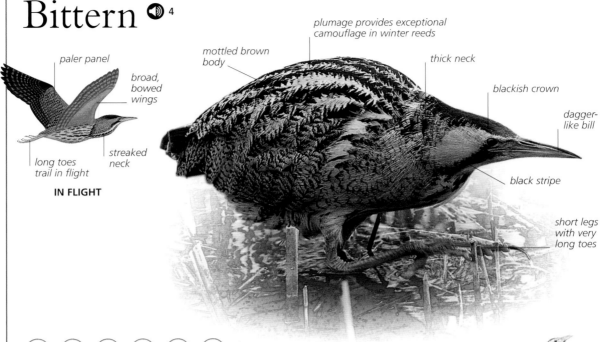

paler panel

broad, bowed wings

long toes trail in flight

streaked neck

IN FLIGHT

plumage provides exceptional camouflage in winter reeds

mottled brown body

thick neck

blackish crown

dagger-like bill

black stripe

short legs with very long toes

FLIGHT: heavy, low, slow, somewhat unsteady; on bowed, rounded wings, legs trailing.

Few birds are so restricted to a single habitat as the Bittern is to wet reedbeds. Even drier reedbeds are of no use to it: it must have deeper water, so that it can find fish in the shelter of the reeds, at the edge of secret pools and ditches. To maintain suitable conditions, in a habitat that naturally dries out over time, requires a great deal of expensive management work: Bitterns have gradually been lost from many past sites.
VOICE Deep, hollow, rhythmic boom, *ker-whooomp!*
NESTING Broad, damp nest of reed stems well out of sight in thick reedbed; 4–6 eggs; 1 brood; April–May.
FEEDING Catches fish, especially eels, in sudden grab of outstretched bill.

FORAGING FOR FOOD
The Bittern mostly feeds at the edge of thick reeds, moving slowly, with side-to-side shaking of the whole body. It may be driven into more open areas by freezing conditions.

STEALTHY FISHER
Bitterns rely on access to fish while remaining inside thick cover, typically in wet reedbeds.

OCCURRENCE
Rare bird of larger, wetter reedbeds, very local and scattered through Europe. In winter, more widespread in W Europe, forced out by frost into smaller patches of reed or more open water where it can reach small fish.

Seen in the UK
J F M A M J J A S O N D

"BITTERNING" STANCE
When alarmed, the Bittern adopts a camouflage posture with an upward-pointing bill.

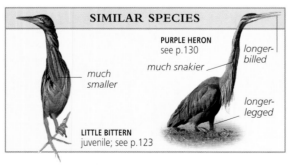

SIMILAR SPECIES

much smaller

LITTLE BITTERN
juvenile; see p.123

PURPLE HERON
see p.130

much snakier

longer-billed

longer-legged

Length **69–81cm (27–32in)**	Wingspan **1.25–1.35m (4–4½ft)**	Weight **0.9–1.1kg (2–2½lb)**
Social **Solitary**	Lifespan **10–12 years**	Status **Vulnerable†**

| Order **Ciconiiformes** | Family **Ardeidae** | Species *Ixobrychus minutus* |

Little Bittern

large pale wing patch

greenish gloss on black upperparts

black cap and back

bright buff neck

MALE

IN FLIGHT

MALE

narrowly streaked neck

light brown body

JUVENILE

streaked brown back

streaked below

pale wing patch

FEMALE

This tiny heron is secretive and usually difficult to see, except in brief, fast flights low over marshy areas when the large, oval, pale wing patches catch the eye. Occasionally, one will perch in view near the top of a stem or at the edge of an overhanging willow, and reveal its exquisite pattern. Males, especially, are beautifully and delicately coloured, with very subtle striping on the neck and a slight greenish gloss on the black upperparts.

VOICE Short, nasal call; nocturnal "song" is monotonously repeated single croak.

NESTING Small nest of stems in dense reeds or bush; 2–7 eggs; 1 brood; May–July.

FEEDING Hunts fish, frogs, shrimps, and big aquatic insects, using stealth and patience and sudden forward lunges to catch prey.

FLIGHT: quick, low, fast wingbeats, in sudden flurry; low over reedbed.

AQUATIC HUNTER
Shallow water with dense cover is ideal hunting territory for Little Bitterns, making them hard to see until they fly.

SIMILAR SPECIES

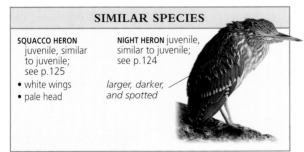

SQUACCO HERON juvenile, similar to juvenile; see p.125
• white wings
• pale head

NIGHT HERON juvenile, similar to juvenile; see p.124
larger, darker, and spotted

OCCURRENCE
Found from April to October in S and C Europe. Occupies reedbeds by rivers and marshes and also much smaller pools and flooded willow thickets. Rare spring migrant in UK.

| Seen in the UK |
| J F M **A M** J J A S O N D |

| Length **33–38cm (13–15in)** | Wingspan **49–58cm (19½–23in)** | Weight **140–150g (5–5½oz)** |
| Social **Solitary** | Lifespan **Up to 10 years** | Status **Vulnerable†** |

Order **Ciconiiformes**	Family **Ardeidae**	Species ***Nycticorax nycticorax***

Night Heron

ADULT

grey wings

IN FLIGHT

dark brown wings and back

JUVENILE

thin white plumes

black back

broad, rounded wings

black cap

JUVENILE

white forehead

short, thick bill

yellow legs (red in spring)

ADULT

pale spots on back and wings

streaked chest

JUVENILE

Night Herons are mostly active at dusk and dawn, but can be seen quite well by day if discovered at a roost. They stand on branches within trees and thickets near water, typically looking like motionless, pale spots from a distance. Once located, usually several are seen, but only when they choose to fly do the real numbers become apparent, as quite sizeable parties sometimes leave the trees. They feed in near-dark conditions, searching for fish, in typical heron style, at the water's edge.

VOICE Deep, low, short, crow-like croak.
NESTING Small stick nest in tree or bush; 3–5 eggs; 1 brood; April–July.
FEEDING Mostly nocturnal, taking small fish and large insects from water's edge.

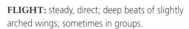

FLIGHT: steady, direct; deep beats of slightly arched wings; sometimes in groups.

SHALLOW WADER
Night Herons fish at the edge of ponds and rivers, frequently being most active at dusk but fishing all day when they have young to feed.

SIMILAR SPECIES

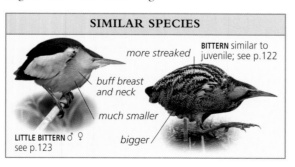

more streaked

BITTERN similar to juvenile; see p.122

buff breast and neck

much smaller

LITTLE BITTERN ♂ ♀ see p.123

bigger

OCCURRENCE
In waterside habitats, from reedbeds to tall trees by rivers and lakes, mostly from March to October in S and C Europe, but very localized in most areas. May appear farther north in spring; wintering birds may be escapees.

Seen in the UK
J F M A M J J A S O N D

Length **58–65cm (23–26in)**	Wingspan **90–100cm (35–39in)**	Weight **600–800g (21–29oz)**
Social **Roosts and breeds in flocks**	Lifespan **10–15 years**	Status **Declining**

Order **Ciconiiformes**	Family **Ardeidae**	Species *Ardeola ralloides*

Squacco Heron

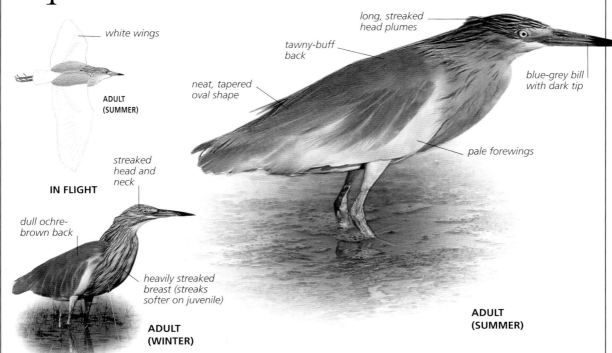

white wings

ADULT (SUMMER)

IN FLIGHT

streaked head and neck

dull ochre-brown back

heavily streaked breast (streaks softer on juvenile)

ADULT (WINTER)

long, streaked head plumes

tawny-buff back

neat, tapered oval shape

blue-grey bill with dark tip

pale forewings

ADULT (SUMMER)

In flight, the Squacco Heron is obvious because its pure white wings catch the eye at any angle. On the ground, however, it is an inconspicuous bird, the white all but hidden, the head withdrawn into squat, rounded shoulders. Typically found in overgrown ditches or streams, or on floating weeds in a larger river or marsh, it is easily overlooked unless disturbed. Very much a southern bird in Europe, it is found, very rarely, as a vagrant farther north.

VOICE Hoarse, nasal croaking call, but mostly silent.
NESTING Small nest of grass and reeds, low down in reeds; 4–6 eggs; 1 brood; April–June.
FEEDING Stands on floating weeds or in thick cover, hunting fish, frogs, and insects.

FLIGHT: low, quite quick, with fast beats of slightly arched wings; feet trail beyond tail.

WING STRETCH
This heron stands with an exaggerated forward neck stretch and open wings after preening.

WHITE IN FLIGHT
The bright white wings of the Squacco Heron are obvious only in flight, low over a marsh.

SIMILAR SPECIES

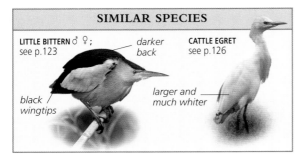

LITTLE BITTERN ♂ ♀; see p.123

darker back

black wingtips

CATTLE EGRET see p.126

larger and much whiter

OCCURRENCE
Mostly in Mediterranean region in summer, in all kinds of freshwater swamps, from weedy riversides to reedbeds and extensive floods and marshes. Only rare spring vagrant farther north.

Seen in the UK
J F M A M J J A S O N D

Length **40–49cm (16–19½in)**	Wingspan **71–86 cm (28–34in)**	Weight **230–350g (8–13oz)**
Social **Loose flocks**	Lifespan **5–10 years**	Status **Vulnerable**

Order **Ciconiiformes**	Family **Ardeidae**	Species ***Bubulcus ibis***

Cattle Egret

rich buff on back

SUMMER

looks all-white in flight at long range

dark feet

SUMMER

IN FLIGHT

rich buff crown

yellow bill, reddish in spring

white body and wings

rich buff on breast in spring

yellow bill

heavy chin and throat

all-white body

dull yellowish to brown legs

WINTER

SUMMER

FLIGHT: quick, direct, often in flocks; wingbeats quite fast and deep.

Most herons and egrets feed on fish, but Cattle Egrets have a specialist lifestyle, following large animals and picking up insects disturbed by their hooves. They also feed on freshly ploughed earth, finding small prey turned up in the furrows, and forage on refuse tips as well. Towards evening, large flocks fly in distinctive, shapeless flurries of white, gathering to roost in trees (sometimes close to buildings), often in thousands.

VOICE Occasional short, croaking or creaking notes.

NESTING Shallow nest of sticks and reeds in tree; 4 or 5 eggs; 1 brood; April–June.

FEEDING Catches insects disturbed by cattle, sheep, and goats; also eats frogs, reptiles, and mice.

CATTLE FOLLOWER
Cattle Egrets typically follow cattle in Europe; in Africa, they gather around great herds of antelopes, buffaloes, and elephants.

SPARKLING WHITE
Flocks in flight present a flickering, tight-packed effect, often looking much whiter than they may appear on the ground.

SIMILAR SPECIES

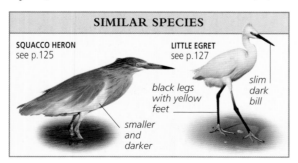

SQUACCO HERON
see p.125

LITTLE EGRET
see p.127

black legs with yellow feet

slim dark bill

smaller and darker

OCCURRENCE
Confined largely to S Spain and Portugal and extreme S France; rare vagrant farther north. Resident in coastal regions or centred on lakes with thickets in which it roosts. Feeds in fields and at refuse tips, often associated with livestock.

Seen in the UK
J F M A M J J A S O N D

Length **45–50cm (18–20in)**	Wingspan **82–95cm (32–37in)**	Weight **300–400g (11–14oz)**
Social **Feeds and roosts in flocks**	Lifespan **Up to 10 years**	Status **Secure**

| Order **Ciconiiformes** | Family **Ardeidae** | Species **Egretta garzetta** |

Little Egret

withdrawn head

slightly bowed wings

SUMMER

IN FLIGHT

all-white plumage

bluish on face

slim blackish bill

snaky neck

fan of plumes over tail

pointed breast plumes

black legs

yellow feet (may be discoloured by mud)

angular shape

lacks head and back plumes

WINTER

SUMMER

Steadily spreading northwards in western Europe, the Little Egret is a sparkling white heron of marshes, poolsides, and coasts. It is often lively, dashing about muddy creeks or even over seaweed-covered rocks in a frenzied fashion; however, like other herons, it spends much time standing still or wading in shallows, looking for prey. Little Egrets are often found in small, loose feeding flocks; they tend to form evening roosts in traditional spots, and these attract birds from far afield.

VOICE Generally silent.

NESTING Stick nest in tree, often mixed with those of other heron-like species; 3 or 4 eggs; 1 brood; April–July.

FEEDING Catches small fish, frogs, snails, and other wetland animals; may run around with flapping wings, or move more stealthily.

FLIGHT: direct, quick, with head withdrawn, legs trailing, wings only slightly bowed.

SOCIAL FEEDING
In areas where Little Egrets are common, small groups of these birds often feed together, or spread more loosely along a shore.

OCCURRENCE
Found in S Europe, north to S Great Britain, on watersides from rocky coasts to reedy lakes, but especially open, muddy or sandy shores. Breeds in treetop colonies, often with other herons and egrets.

Seen in the UK
| J | F | M | A | M | J | J | A | S | O | N | D |

SIMILAR SPECIES

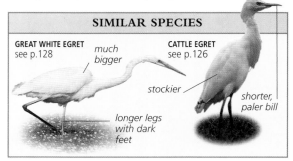

GREAT WHITE EGRET see p.128

much bigger

longer legs with dark feet

CATTLE EGRET see p.126

stockier

shorter, paler bill

| Length **55–65cm (22–26in)** | Wingspan **88–106cm (35–42in)** | Weight **400–600g (14–21oz)** |
| Social **Small flocks** | Lifespan **Up to 10 years** | Status **Secure** |

Order **Ciconiiformes**	Family **Ardeidae**	Species *Egretta alba*

Great White Egret

arched wings

all-white plumage (long back plumes in summer)

very long neck

yellow bill, often dark at tip

similar in size to Grey Heron

SUMMER

IN FLIGHT

yellowish or blackish legs with dark feet

WINTER

When it stands next to a Grey Heron, the actual size of the Great White Egret is apparent: it is truly a giant egret, as tall as the heron, if not standing a little higher. Slim, angular, but elegant, this stunningly white bird is easily seen at long range. It develops long back plumes in the breeding season, at which time the bill turns blackish. The Great White Egret often feeds in grassy, relatively dry, places, leaning forward, sometimes with a fast, shimmering tremor through the whole body as it homes in on its prey.

WINTER

FLIGHT: slow, direct, on arched wings; heavier, slower than Little Egret.

VOICE Mostly silent.
NESTING Shallow plat-form of sticks in tree; 2–5 eggs; 1 brood; April–July.
FEEDING Catches fish, amphibians, and small mammals in wet places.

NESTING ON TREES
Great White Egrets often perch on tall trees overlooking a marsh. They are sociable in summer, nesting in colonies in trees.

SIMILAR SPECIES

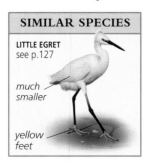

LITTLE EGRET
see p.127

much smaller

yellow feet

COURTSHIP DISPLAY
In spring, the Great White Egret spreads its long body plumes in a spectacular courtship display at the nest.

OCCURRENCE
Summer visitor, migrant, or winter visitor within SE Europe, rare at all times in W Europe. In reedbeds and extensive floods, or beside big, more open lakes, or on tall trees near marshes.

Seen in the UK
J F M A M J J A S O N D

Length **85–100cm (34–39in)**	Wingspan **1.45–1.7m (4¾–5½ft)**	Weight **1–1.5kg (2¼–3¼lb)**
Social **Small flocks**	Lifespan **10–15 years**	Status **Secure**

| Order **Ciconiiformes** | Family **Ardeidae** | Species **Ardea cinerea** |

Grey Heron 🔊 5.I, 5.II

pale grey on head

no crest

spotted foreneck

JUVENILE

white forehead

long, narrow, black head plume

dagger-like, yellow, orange, or pink bill (dull except in spring)

black spots on white foreneck

pale grey body

long legs

ADULT

withdrawn head

broad, strongly arched wings

grey-black wingtips

ADULT

IN FLIGHT

FLIGHT: direct, on broad, strongly arched wings, head withdrawn, feet trailing; also higher, or diving down from height with aerobatic twists and turns.

Usually moving slowly or standing quite still beside a pool or river, this large, pale grey heron is unmistakable. It may, however, look tall and slim with a long, erect neck, or round-shouldered and hunched with its head and neck withdrawn into the shoulders; it may also stand high in a treetop, or fly with surprising agility high overhead. It is typically shy, but in towns becomes remarkably bold and many now visit garden fish ponds on early-morning raids.

VOICE Short, harsh *fraink*; rattling and croaking sounds at nest.

NESTING Large nest of thick sticks in treetop (or bush where no tree is available); 4 or 5 eggs; 1 brood; January–May.

FEEDING Catches fish, frogs, rats, and other prey in its bill, usually after long, patient stalk before sudden strike.

SIMILAR SPECIES

PURPLE HERON
see p.130

slimmer and browner

longer bill

thin neck

SLOW, SILENT MOVER
This heron strides silently through shallows or long vegetation. It stands still for long spells.

PERCHING
The Grey Heron may sometimes be found perched on a treetop, usually in a hunched posture, with its head and neck drawn into the shoulders.

OCCURRENCE
Frequents both freshwater and saltwater habitats, from salt marsh and rocky coasts to floods and fish ponds almost throughout Europe. Some regularly visit garden ponds or town lakes, especially in winter when much habitat is frozen.

Seen in the UK
| J | F | M | A | M | J | J | A | S | O | N | D |

| Length **90–98cm (35–39in)** | Wingspan **1.75–1.95m (5¾–6½ft)** | Weight **1.6–2kg (3½–4½lb)** |
| Social **Solitary/Small flocks** | Lifespan **Up to 25 years** | Status **Secure** |

| Order **Ciconiiformes** | Family **Ardeidae** | Species ***Ardea purpurea*** |

Purple Heron

steel-grey midwing patch

arched wings narrow at base

reddish underwing

ADULT

IN FLIGHT

longer toes than Grey Heron

deeper, rounder neck bulge than Grey Heron

dark back with paler, buff plumes

ADULT

long, slim, spear-like bill

snaky head and neck

dark stripes on bright ginger neck

dark reddish shoulder patch

rich tawny underparts

thin stripes on face

brown body

paler neck

much slimmer than Bittern

JUVENILE

FLIGHT: strong, steady, on arched wings with more curved trailing edge than Grey Heron's, neck creating deeper bulge, larger feet sometimes obvious.

Much more of a reedbed bird than the Grey Heron, the Purple Heron is consequently more difficult to see unless it flies over the reeds or chooses to feed at the edge of a reedy pool. It breeds in small groups, but is generally a less social bird than most herons. It has often been predicted that the Purple Heron will spread north and breed in the UK, but this has so far failed to develop: it remains a rather rare visitor north of its usual range.

VOICE Short, simple, harsh *krekk*.

NESTING Large pile of reed stems and other vegetation, often in reeds, sometimes in trees; 4 or 5 eggs; 1 brood; February–June.

FEEDING Catches small fish, frogs, and aquatic invertebrates in its long, slender bill.

FISHING
A secretive bird, the Purple Heron typically waits patiently at the water's edge or strides through reeds in search of prey.

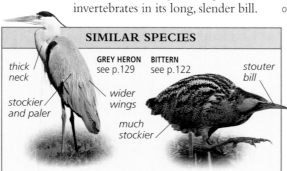

SIMILAR SPECIES

thick neck

GREY HERON see p.129

BITTERN see p.122

stouter bill

stockier and paler

wider wings

much stockier

OCCURRENCE
Generally more southerly bird than Grey Heron, absent from N Europe. Typically in reedy marshes, tall sedge beds, and wet meadows. Migrants appear north of breeding range in reedbeds and overgrown flooded areas.

Seen in the UK
J F M A M J J A S O N D

| Length **70–90cm (28–35in)** | Wingspan **1.1–1.45m (3½–4¾ft)** | Weight **1–1.5kg (2¼–3¼lb)** |
| Social **Solitary** | Lifespan **Up to 23 years** | Status **Vulnerable** |

FLAMINGOS

Tℋ**IS FAMILY OF EXTRAORDINARY** birds includes very few species, scattered world-wide, and only one, the Greater Flamingo, is found in Europe. It breeds in just a few large colonies, including one at the Camargue in southern France and one in southern Spain, but is regularly seen in sizeable flocks at a number of other shallow, salty lagoons around the Mediterranean, its only habitat type.

Nesting colonies are on very low islands, each nest being built up from the mud: nests are vulnerable to flooding, or to falls in water level that allow predators to reach the colony. The flamingos may rear very few young for several years but periodically have a "boom" year that maintains their numbers. Young birds congregate in dense "crèches" and later tend to gather around the outer fringes of large flocks.

PHOENICOPTERIDAE

extremely long neck

GREATER FLAMINGO
p.132

STORKS

O**F THE TWO SPECIES OF STORK** in Europe, one species, the White Stork, is able to live close to people, exploiting buildings and electricity pylons as nesting places and feeding extensively on refuse tips. It is, however, under pressure from the destruction of freshwater habitats and the spread of intensive agriculture. The Black Stork is, in contrast, a forest bird and often nests on remote cliffs.

Both are migrants, spending the winter far south in Africa. White Storks migrate in huge flocks, creating exciting birdwatching as they cross the Mediterranean at the narrowest point, as they must glide and soar to save energy and can only gain the benefit of rising air over land. Black Storks may migrate alone or in smaller groups, but can be seen with other large migrants, such as eagles and kites, over the sea, or through mountain passes. They may be forced down temporarily by poor weather.

CICONIIDAE

dagger-like bright red bill

white body, often soiled

BLACK STORK
p.133

WHITE STORK
p.134

THRESKIORNITHIDAE

spoon-shaped black bill, tipped yellow

SPOONBILL
p.135

Order **Phoenicopteriformes**	Family **Phoenicopteridae**	Species *Phoenicopterus ruber*

Greater Flamingo

black wingtips

red patch on narrow wings

outstretched neck

ADULT

IN FLIGHT

bright pink bill with black tip

whitish to pale pink plumage

long, pale pink legs

extremely long neck

greyish body

grey bill

dark grey legs

IMMATURE

ADULT

Flamingos, of which there are a handful of species worldwide, are instantly recognizable. The downcurved bill, long neck, long legs, and red-splashed wings of the Greater Flamingo create a dramatic and easily identified bird, one of the more exotic in appearance in Europe. Although small groups occur, most live and nest in large flocks. Single birds found away from the main range are usually escapees.

VOICE Loud, deep honking; cackling notes give goose-like chorus.

NESTING Small pillar of mud in shallow water, safe from land-based predators; 1 egg; 1 brood; April–May.

FEEDING Sweeps its bill, upside down, through shallow water, picking up tiny crustaceans; often in very shallow water but also while the bird is belly-deep, or swimming like a swan.

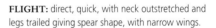

FLIGHT: direct, quick, with neck outstretched and legs trailed giving spear shape, with narrow wings.

WADING FLOCK
Flocks of Greater Flamingos typically stand in the shallows, or wade more deeply into the water in long lines. Sometimes they swim like swans.

FLYING FLOCK
Flocks form irregular shapes or long, trailing lines while flying.

SIMILAR SPECIES

CHILEAN FLAMINGO
escapee
- grey legs with pink "knees"
- may be found on lakes of northwest Europe

OCCURRENCE
Breeds in a few places on large salty lakes in Spain, Portugal, S France, Sardinia, and Turkey; more widespread in Mediterranean as non-breeding bird, both on salt pans and natural lakes, usually in very open, exposed areas.

Seen in the UK
J F M A M J J A S O N D

Length **1.2–1.45m (4–4¾ft)**	Wingspan **1.4–1.7m (4½–5½ft)**	Weight **3–4kg (6½–8¾lb)**
Social **Large flocks**	Lifespan **Up to 20 years**	Status **Localized**

Order **Ciconiiformes**	Family **Ciconiidae**	Species *Ciconia nigra*

Black Stork

long, fingered wings

darker underwing than White Stork

white wingpits

ADULT

black chest

IN FLIGHT

dull greenish black back

dull greenish bill

paler legs

JUVENILE

red around eye

long black neck

dagger-like, bright red bill

blackish plumage with green and purple gloss

black chest

white belly

long red legs

ADULT

FLIGHT: powerful, direct; on long, fingered, flat wings with steady beats and long glides; master soarer in upcurrents.

flat wings while gliding

ADULT

AT THE NEST
Black Storks nest in remote areas, not in towns or villages as do White Storks.

Much less familiar than the stockier White Stork, the Black Stork is a shier species which prefers wilder country with extensive forests, marshes, and isolated crags. It is a rare bird throughout its range, wintering in Africa and returning quite late in spring. Small numbers can be seen on migration over the Pyrenees and a few "overshoot" to unexpected places in spring.
VOICE Generally silent except for rasping notes at nest.
NESTING Big, bulky structure of sticks on rock ledge or high up in canopy of large trees; 2–4 eggs; 1 brood; May–July.
FEEDING Hunts frogs, newts, toads, and aquatic insects in wet places, and picks them up in its long, slim bill.

SIMILAR SPECIES

WHITE STORK similar in flight; see p.134

white head and breast

CORMORANT juvenile superficially similar in flight; see p.119

long tail

white chest

OCCURRENCE
Occupies extensive forests, marshes, and rocky areas mostly in Spain, Portugal, and E Europe, in summer. Migrates to Africa each autumn and very few are seen outside regular range.

Seen in the UK
J F M A M J J A S O N D

Length **90–105cm (35–41in)**	Wingspan **1.1–1.45m (3½–4¾ft)**	Weight **2.5–3kg (5½–6½lb)**
Social **Small flocks**	Lifespan **Up to 20 years**	Status **Rare**

| Order **Ciconiiformes** | Family **Ciconiidae** | Species *Ciconia ciconia* |

White Stork

broad, fingered wings

outstretched head

long red bill (dark grey on juvenile)

flat wings when gliding

white body, often soiled

black rear wings

IN FLIGHT

ADULT

ADULT

trailing legs

ADULT

long, stout red legs; majestic walk

FLIGHT: direct flight low, on long, flat, fingered wings, neck outstretched; soars magnificently, often in swirling flocks.

One of Europe's largest and most boldly patterned birds, the White Stork creates a spectacular sight when it gathers in thousands on migration to make the narrowest sea crossings possible, at Gibraltar and Istanbul. Although still widespread, it is declining in much of its range as farming becomes ever more intense and wetlands drier or more polluted. Reintroduction schemes have helped in northwest Europe but leave the origin of some western vagrants in some doubt.

VOICE Silent, but rattles bill loudly at nest.

NESTING Large, bulky nest of sticks on pole, tower, or roof of tall house, or in tree; 2–4 eggs; 1 brood; April–June.

FEEDING Catches various aquatic insects, small rodents, frogs and toads, and small fish in damp places or shallow water.

PERCHING AT NEST
White Storks often build their nest on rooftops, the size of the nests rendering them visible at great range. The birds are typically seen standing erect at their rooftop nests.

SIMILAR SPECIES

grey (black and white in strong sun)

GREY HERON bowed wings in flight; see p.129

neck withdrawn in flight

WHITE PELICAN short legs; see p.440

OCCURRENCE
Breeds in mainland Europe other than far north, and migrates to Africa in winter. Feeds on open land near marshes, rivers, and lakes, including edges of towns and villages where it often nests on high perches.

Seen in the UK
J F M A M J J A S O N D

| Length **0.95–1.1m (3–3½ft)** | Wingspan **1.8–2.18m (6–7¼ft)** | Weight **2.5–4.5kg (5½–10lb)** |
| Social **Migrates in flocks** | Lifespan **Up to 25 years** | Status **Vulnerable** |

Order **Ciconiiformes**	Family **Threskiornithidae**	Species *Platalea leucorodia*

Spoonbill

outstretched head

ADULT — JUVENILE

IN FLIGHT

black wingtips

bushy crest

neck thicker than that of egrets

all-white body

orange patch under chin

thick black legs

spoon-shaped black bill, tipped yellow

black legs thicker than those of egrets

ADULT

ADULT

all-white underwings

pink bill turns black with age

IMMATURE

Equipped with a flattened, broad-tipped bill, the Spoonbill is heron-like, but as white as an egret. It is a bird of marshes and lakes with extensive shallow water, but in winter it may also be found on coastal estuaries, striding through the shallows with its rather human-like walk. This, and its tall, upstanding presence, make it an unmistakable bird in Europe.

VOICE Silent.

NESTING Shallow platform of sticks and reed stems in reeds or tree; in colonies, rarely mixed with other species; 3 or 4 eggs; 1 brood; April–July.

FEEDING Holds bill slightly open, partially submerged, and sweeps it through water from side to side, to catch fish, molluscs, and crustaceans.

FLIGHT: strong, direct, swan-like, with head outstretched; regular wingbeats.

FLYING FLOCK
Lines and chevrons of Spoonbills tend to coordinate short glides between spells of steady wingbeats.

SIDEWAYS SWEEP
The Spoonbill wades slowly forwards in shallow water, sweeping its partly open bill sideways until it detects food; the bill is then snapped shut on the prey.

SIMILAR SPECIES

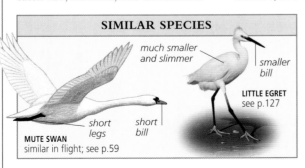

much smaller and slimmer

smaller bill

LITTLE EGRET
see p.127

MUTE SWAN
similar in flight; see p.59

short legs

short bill

OCCURRENCE
Mostly found in E Europe, locally in W, breeding around reedy lakes with surrounding bushes, but feeds on extensive salt pans, coastal marshes, and other areas of shallow water. Very few winter in W Europe.

Seen in the UK
J F **M A M J J A S** O N D

Length **80–93cm (32–37in)**	Wingspan **1.2–1.35m (4–4½ft)**	Weight **1–1.5kg (2¼–3¼lb)**
Social **Small flocks**	Lifespan **25–30 years**	Status **Endangered**

BIRDS OF PREY

THIS IS A MIXED GROUP, including birds that eat tiny insects, others that eat dead animals, and some that catch their own prey up to the size of small deer. Mostly splendid in flight, many are likely to remain perched for hours on end between feeding forays, while others are much more aerial, spending much of the day aloft.

VULTURES

Including some of Europe's biggest birds, the vultures soar high up, using superb eyesight to spot potential food on the ground: they eat meat, preferably freshly dead animals. They need warm, rising air or updraughts and live in southern Europe and mountainous areas.

EAGLES AND HAWKS

Eagles are powerful hunters with large eyes, powerful bills, and strong feet. Most have feathered legs. Buzzards are smaller, less strong, in particular smaller-billed, birds but also majestic fliers. Harriers are hunters over open ground, flying low as they try to surprise prey at close range. Bird-eating hawks such as the Sparrowhawk catch prey by surprise in a short, fast dash. Kites have long, notched tails that are swivelled, acting as rudders in their elegant flight. The Osprey is a fish-eating specialist, diving onto its prey from the air, while the Short-toed Eagle is a "snake-eagle", one of a largely African

POWER
The Golden Eagle has a powerful and charismatic presence, with a keen eye and strong hooked bill.

group with large, yellow eyes, a big head, and unfeathered legs.

FALCONS

Big falcons catch large prey and eat infrequently, spending long periods perched, but look impressive in the air when they do fly; smaller falcons may be much more active. Some eat small mammals, others mostly insects or birds.

ACCIPITRIDAE

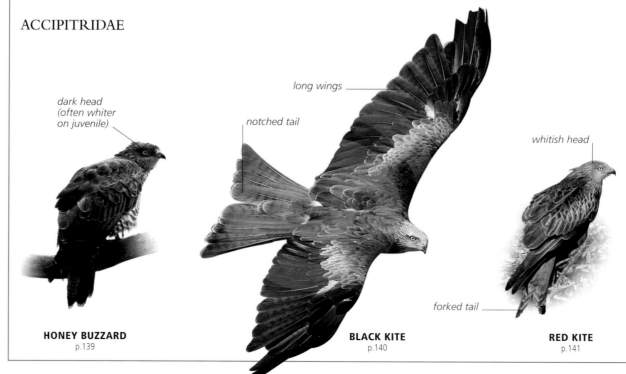

dark head
(often whiter
on juvenile)

long wings

notched tail

whitish head

forked tail

HONEY BUZZARD
p.139

BLACK KITE
p.140

RED KITE
p.141

creamy-buff head and neck

GRIFFON VULTURE
p.142

massive, dark brown body (juvenile paler)

BLACK VULTURE
p.143

huge span of flat, deeply fingered wings

WHITE-TAILED EAGLE
p.144

untidy ruff of spiky feathers

EGYPTIAN VULTURE
p.145

bulky, round head

SHORT-TOED EAGLE
p.146

cream patch on wings

MARSH HARRIER
p.147

pale grey body

HEN HARRIER
p.148

black bar across wing

MONTAGU'S HARRIER
p.149

whitish underside, with fine grey barring

GOSHAWK
p.150

ACCIPITRIDAE *continued*

orange on face

round shoulders; hunched shape

"frosty" pale feather edges

SPARROWHAWK
p.151

BUZZARD
p.152

ROUGH-LEGGED BUZZARD
p.153

pale tawny to yellowish head

dark upperparts with white patch on back

white underside

GOLDEN EAGLE
p.154

BONELLI'S EAGLE
p.155

BOOTED EAGLE
p.156

PANDIONIDAE

black band on head

OSPREY
p.157

FALCONIDAE

unspotted, rich red-brown back

pale rufous back spotted with black

small, square head with little pattern

LESSER KESTREL
p.158

KESTREL
p.159

MERLIN
p.160

large, solid body with broad shoulders

thick black streaks on pale underparts

HOBBY
p.161

PEREGRINE
p.162

Order **Accipitriformes**	Family **Accipitridae**	Species *Pernis apivorus*

Honey Buzzard

wings held flat or only slightly raised; flutters wings high over back in unique display flight

dark head (often whiter on juvenile)

yellow eye (dark on juvenile)

long wings, angled on leading edge

three dark bands on tail

ADULT

ADULT

IN FLIGHT

tiger-striped underwings and underparts

"tiger-striped" underparts

ADULT

dark bands on base of tail

narrow, Cuckoo-like head with pointed bill

in flight, tail twisted like kites

long tail, widest in centre

ADULT

FLIGHT: direct flight easy with elastic, deep wing-beats; soars with wings bowed or flattish, tips drooped.

Not a true buzzard at all, this is a unique raider of wasp and bee nests, even walking about on the ground and digging out wax and grubs with its feet. It is secretive when breeding and more easily seen on migration, as it concentrates on short sea crossings over the Baltic and the Mediterranean on its way to and from tropical Africa. Exceptionally variable in colour and pattern, it requires careful observation for positive identification.

VOICE Infrequent whistling *peee-haa, pee-ee-aah*.
NESTING Small platform of sticks and greenery in tree, often on old crow's nest; 1–3 eggs; 1 brood; April–June.
FEEDING Eats wasp and bee grubs, wax, honey, adult insects, ant pupae, and small mammals and reptiles.

OCCURRENCE
Widespread throughout Europe except in far N Scandinavia, and Iceland; rare breeder and migrant in UK. Occupies extensive forest or well-wooded hilly country, arriving in April, leaving in September. Migrants cross Mediterranean and mountain regions on regular routes.

Seen in the UK
J F M **A M J J A S O** N D

SIMILAR SPECIES

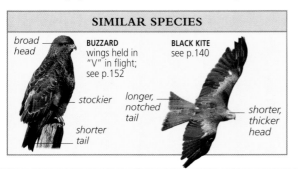

broad head

BUZZARD wings held in "V" in flight; see p.152

BLACK KITE see p.140

stockier

shorter tail

longer, notched tail

shorter, thicker head

BOLD PATTERN
The underside of most Honey Buzzards is heavily spotted and barred but the pattern varies greatly and some are much plainer.

Length **52–60cm (20½–23½in)**	Wingspan **1.35–1.5m (4½–5ft)**	Weight **600–1,100g (21–39oz)**
Social **Migrates in flocks**	Lifespan **Up to 25 years**	Status **Secure**

139

| Order **Accipitriformes** | Family **Accipitridae** | Species ***Milvus migrans*** |

Black Kite

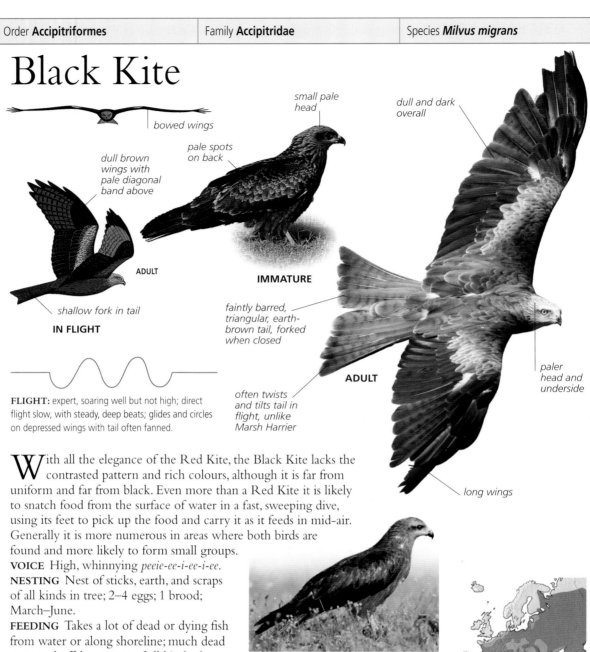

bowed wings

small pale head

dull and dark overall

dull brown wings with pale diagonal band above

pale spots on back

ADULT

IMMATURE

shallow fork in tail

IN FLIGHT

faintly barred, triangular, earth-brown tail, forked when closed

ADULT

paler head and underside

FLIGHT: expert, soaring well but not high; direct flight slow, with steady, deep beats; glides and circles on depressed wings with tail often fanned.

often twists and tilts tail in flight, unlike Marsh Harrier

long wings

With all the elegance of the Red Kite, the Black Kite lacks the contrasted pattern and rich colours, although it is far from uniform and far from black. Even more than a Red Kite it is likely to snatch food from the surface of water in a fast, sweeping dive, using its feet to pick up the food and carry it as it feeds in mid-air. Generally it is more numerous in areas where both birds are found and more likely to form small groups.

VOICE High, whinnying *peeie-ee-i-ee-i-ee*.

NESTING Nest of sticks, earth, and scraps of all kinds in tree; 2–4 eggs; 1 brood; March–June.

FEEDING Takes a lot of dead or dying fish from water or along shoreline; much dead meat and offal or scraps of all kinds; dung, small birds, reptiles, and voles.

REMARK Groups fight and chase each other over rubbish tips.

SCAVENGER
Black Kites join other birds of prey and crows at carcasses and rubbish tips, and also snatch scraps with their feet in fast, accurate swoops.

OCCURRENCE
Widespread from Spain and Portugal to Finland and south to Balkans; rare visitor to UK. Feeds around rubbish tips and over open ground, wooded slopes, coasts, and rivers; more often associated with water than Red Kite and still around towns in some places.

| Seen in the UK |
| J F M A M J J A S O N D |

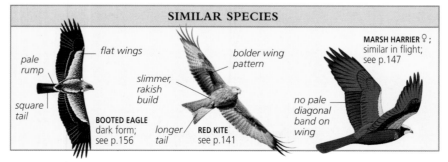

SIMILAR SPECIES

pale rump

flat wings

bolder wing pattern

MARSH HARRIER ♀; similar in flight; see p.147

slimmer, rakish build

square tail

BOOTED EAGLE dark form; see p.156

longer tail

RED KITE see p.141

no pale diagonal band on wing

| Length **48–58cm (19–23in)** | Wingspan **1.3–1.55m (4¼–5ft)** | Weight **650–1,100g (23–39oz)** |
| Social **Small flocks** | Lifespan **Up to 20 years** | Status **Vulnerable** |

Order **Accipitriformes**	Family **Accipitridae**	Species ***Milvus milvus***

Red Kite

bowed wings while soaring

pale band on upperwings

forked tail

ADULT

IN FLIGHT

bold white patch contrasts with black wrist patch

ADULT

whitish to pale red underside of tail

paler than adult

paler upperwings than adult's

pale rufous tail, deeply notched when closed

IMMATURE

whitish head

pale eye

pale tawny to rust-red body

ADULT

Agility and mastery in the air are synonymous with the kites: the Red Kite adds the appeal of colour and pattern. It is easily separated from the heavier, less elegant Buzzard, having more of the lightness of a harrier, but even greater flexibility and elasticity in its movements. Where common, it gathers in groups of ten or even up to forty wherever food is concentrated.

VOICE High, long-drawn, wailing or squealing *weieie-ee-ow*, higher-pitched than Buzzard.

NESTING Large nest of sticks, rags, earth, and rubbish in tree, usually well-hidden; 2–4 eggs; 1 brood; March–June.

FEEDING Eats dead animals, such as rabbits or sheep; catches birds up to crow or gull size in surprise dash; also feeds on insects, earthworms, and voles.

FLIGHT: direct flight slow and steady with supple, deep wingbeats; often twists tail as rudder; soars well but not usually to any great height; very aerobatic.

FLEXIBLE FLIGHT
The kite flexes its wings and twists its long tail from side to side to exploit air currents to the full; it is capable of fast stoops and twisting dives.

SIMILAR SPECIES

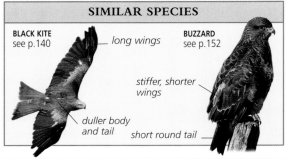

BLACK KITE see p.140

long wings

duller body and tail

BUZZARD see p.152

stiffer, shorter wings

short round tail

OCCURRENCE
Local in Great Britain, widespread in Spain, Portugal, France, parts of S Europe, and north to Baltic. In open countryside and wooded valleys, higher up in summer but in low valleys in winter, when it forages around towns and tips.

Seen in the UK
J F M A M J J A S O N D

Length **60–65cm (23½–26in)**	Wingspan **1.45–1.65m (4¾–5½ft)**	Weight **0.75–1.3kg (1¾–2¾lb)**
Social **Small flocks**	Lifespan **Up to 25 years**	Status **Secure**

| Order **Accipitriformes** | Family **Accipitridae** | Species **Gyps fulvus** |

Griffon Vulture

wings in "V" when soaring

creamy-buff head and neck

bulbous bill

pale toffee-brown back

very short dark tail

rich buff-brown wing coverts and back

darker flight feathers

narrow light bands on darker brown underwings

deeply fingered wingtips

IN FLIGHT

A massive, long-winged, short-tailed bird, the Griffon Vulture spends much time sitting quietly on cliff ledges but flies off in search of food each day. It may be in the air early on cold, windy days, using the wind to help it soar, but on calm, hot days it waits until later in the day and uses rising currents of warm air that build up over bare ground or cliffs. It soars with its wings raised and twisted at the tips, the trailing edge bulging and also slightly arched, so that, as it slowly circles, its shape appears to change. From directly below it is broad- and square-winged but from many angles the wingtips appear more pointed.

VOICE Silent apart from coarse hissing when feeding.
NESTING On bare ledge in gorge or on high cliff, in loose colonies of ten or so to several scores of pairs; 1 egg; 1 brood; April–July.
FEEDING Finds carrion (dead sheep, goats, and smaller animals); often fed at special feeding stations.

FLIGHT: heavy, deep wingbeats; mostly magnificent soaring with wings in "V".

MASTER SOARER
The Griffon can travel great distances without flapping its wings at all, using warm upcurrents to gain height.

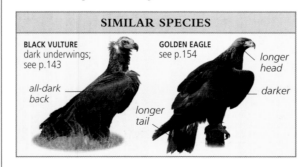

SIMILAR SPECIES

BLACK VULTURE dark underwings; see p.143

all-dark back

GOLDEN EAGLE see p.154

longer head

darker

longer tail

OCCURRENCE
Quite common in Portugal and Spain, rare in S France, Sardinia, Balkans, and Turkey. In all kinds of open areas, from lowlands to high, bleak mountain peaks, centred on a cliff or gorge where they roost and nest.

Seen in the UK
J F M A M J J A S O N D

Length **0.95–1.1m (3–3½ft)**	Wingspan **2.3–2.65m (7½–8¾ft)**	Weight **7–10kg (15–22lb)**
Social **Flocks**	Lifespan **Up to 25 years**	Status **Rare**

| Order **Accipitriformes** | Family **Accipitridae** | Species *Aegypius monachus* |

Black Vulture

very broad, square wings held flat

flat wings when soaring

pale feet

very dark forewings

ADULT

IN FLIGHT

pale head with black mask

pale brown ruff (darker on juvenile)

massive, dark brown body (juvenile paler)

fresh feathers very dark, fade paler

ADULT

FLIGHT: prolonged, expert soaring and gliding with very occasional deep, ponderous wingbeat; flaps heavily in cold, still air.

One of the world's largest flying birds, the Black Vulture's massive bulk and majestic soaring and gliding flight give it terrific impact whenever it is aloft. It exhibits great skill at utilizing every updraught or breath of wind to soar effortlessly without wingbeats. Its flat-winged flight gives the Black Vulture a very broad, rectangular shape, which is less elegant and shapely than a Griffon Vulture's. Unlike the Griffon Vulture, it nests and often perches in trees (rather than on cliffs), and also spends much time on the ground, especially near food such as a sheep or goat carcass.

VOICE Mostly silent.

NESTING Huge stick nest in flat-topped trees; 1 egg, 1 brood; April–June.

FEEDING Rarely catches live prey; mostly eats carrion; also feeds at special feeding stations.

GIANTS OF THE AIR
A very impressive bird, the Black Vulture has great presence in the sky, even among equally large Griffon Vultures; its pale head and feet may be obvious.

OCCURRENCE
Restricted to C Spain, Majorca, and very rare in NE Greece. Resident all year in these small areas and only very rare vagrant elsewhere. Rare vulture of mountainous regions and rolling uplands with mixed forest and open ground.

Seen in the UK
J F M A M J J A S O N D

SIMILAR SPECIES

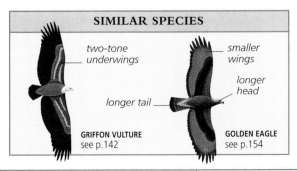

two-tone underwings

longer tail

GRIFFON VULTURE
see p.142

smaller wings

longer head

GOLDEN EAGLE
see p.154

| Length **1–1.15m (3¼–3¾ ft)** | Wingspan **2.5–2.85m (8¼–9¼ ft)** | Weight **7–11.5kg (15–25 lb)** |
| Social **Small flocks** | Lifespan **Up to 25 years** | Status **Vulnerable** |

Order **Accipitriformes**	Family **Accipitridae**	Species *Haliaeetus albicilla*

White-tailed Eagle

glides on flat wings

pale and dark brown blotches on back

short, dark tail

JUVENILE

saw-toothed trailing edge

JUVENILE

pale head

big, bright yellow bill

dark brown overall

dark tail

ADULT

short white tail

protruding head and neck

ADULT

huge span of flat, deeply fingered wings

IN FLIGHT

A huge, flat-winged bird, this eagle is now very rare over most of its former range. It survives in remote marshes and along rocky coasts and offshore islands, and appears on extensive damp coastal plains in winter. Its presence in Scotland is due to a recent reintroduction programme. Around small fishing harbours, it can become quite bold and relatively tame if unpersecuted. It typically sits upright on rocks or swoops into sheltered bays for fish.

VOICE Shrill yaps near nest in summer.
NESTING Huge pile of sticks on flat crowns of trees or cliff ledges; 2 eggs; 1 brood; March–July.
FEEDING Picks sick or dead fish and offal from water using its feet; eats dead animals and catches seabirds and hares.

FLIGHT: heavy and direct; wingbeats deep and elastic, or flexible; soars on flat wings.

SWOOPING ON PREY
This eagle dives for food with lumbering but expert agility, picking fish and offal from water with its feet.

OCCURRENCE
Most common in Scandinavia along rocky coasts, rare in central and E Europe, Balkans, Iceland, and W Scotland where reintroduced on offshore isles. Winters mostly in wide open lowlands, including farmland, with ones and twos appearing most years in some traditional areas.

Seen in the UK
J F M A M J J A S O N D

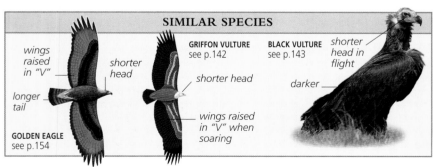

SIMILAR SPECIES

wings raised in "V"

shorter head

GRIFFON VULTURE
see p.142

BLACK VULTURE
see p.143

shorter head in flight

longer tail

shorter head

darker

wings raised in "V" when soaring

GOLDEN EAGLE
see p.154

Length **70–92cm (28–36in)**	Wingspan **2–2.45m (6½–8ft)**	Weight **3.1–7kg (6¾–15lb)**
Social **Family groups**	Lifespan **Up to 20 years**	Status **Rare**

| Order **Accipitriformes** | Family **Accipitridae** | Species **Neophron percnopterus** |

Egyptian Vulture

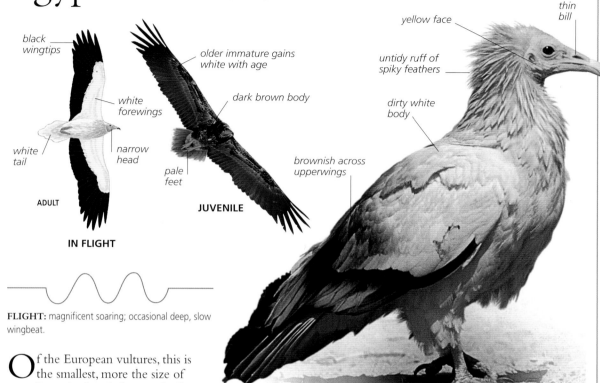

black wingtips

white forewings

white tail

narrow head

ADULT

IN FLIGHT

older immature gains white with age

dark brown body

pale feet

brownish across upperwings

JUVENILE

thin bill

yellow face

untidy ruff of spiky feathers

dirty white body

ADULT

FLIGHT: magnificent soaring; occasional deep, slow wingbeat.

Of the European vultures, this is the smallest, more the size of a medium-sized eagle but markedly larger than the buzzards. In pattern, however, adults have the black and white of White Storks (see p.134) or pelicans (see p.440), which are much bigger. Immatures are browner but spend their early years in Africa, and so are rather infrequently seen in Europe. On the ground, the plumage of an adult often looks dirty and drab, soiled by its foraging in filthy places, but against a blue sky, it is a splendid bird. Two or three Egyptian Vultures may often be seen associating with larger numbers of Griffon Vultures.

VOICE Silent.

NESTING Nest of sticks, bones, and rubbish, on cliff ledge or in small cave; 1–3 eggs; 1 brood; April–June.

FEEDING Finds all kinds of dead meat, offal, and scraps.

GLORIOUS FLIGHT
While often dirty-looking on the ground, the Egyptian Vulture against a deep blue sky in full sunlight looks stunning, almost translucent white and intense black.

SIMILAR SPECIES

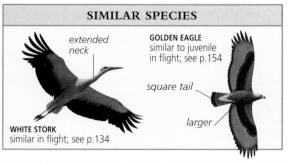

extended neck

WHITE STORK
similar in flight; see p.134

GOLDEN EAGLE
similar to juvenile in flight; see p.154

square tail

larger

OCCURRENCE
Summer visitor to Portugal and Spain, rare in S France, S Italy, and Balkans. In wooded mountainous areas, around gorges and cliffs, and also at refuse tips near small towns and villages, often with kites and larger vultures.

| Seen in the UK |
| J F M A M J J A S O N D |

| Length **55–65cm (22–26in)** | Wingspan **1.55–1.7m (5–5½ft)** | Weight **1.6–2.1kg (3½–4¾ lb)** |
| Social **Small flocks** | Lifespan **10–15 years** | Status **Secure** |

Order **Accipitriformes**	Family **Accipitridae**	Species ***Circaetus gallicus***

Short-toed Eagle

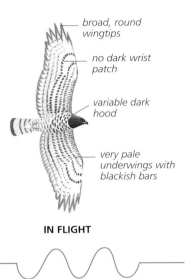

broad, round wingtips

no dark wrist patch

variable dark hood

very pale underwings with blackish bars

IN FLIGHT

bulky, round head

yellow eye

pale brown upperside

new feathers dark, old ones paler with ragged edges

grey-white underside with fine bars

bare legs

long tail

FLIGHT: glides on bowed wings pushed forward; soars with wings very long and straight, head protruding in display; hovers with heavy, floppy action.

A large, impressive eagle, the Short-toed Eagle is sometimes seen perched at close range, on a high pylon or on top of a tall tree, when its crisply barred plumage and vivid yellow eyes give it a handsome and magnificent appearance. One of the "snake-eagles", this bird of prey has strong, bare legs, a thickly feathered, rounded head, and an owl-like face. While hunting it hovers before a long, fast plunge, with its head pulled back, wings almost closed, and chest thrust out, reaching out with its feet at the last moment.

VOICE Various short, abrupt calls, *kyo, meeok*.
NESTING Bulky stick nest in crown of large tree; 1 egg; 1 brood; April–June.
FEEDING Catches snakes and lizards, typically stooping onto them at speed after hover, sometimes from great height.

PERSISTENT HUNTER
Short-toed Eagles spend hours flying over open hillsides. Their frequent hovering in search of prey is distinctive.

SIMILAR SPECIES

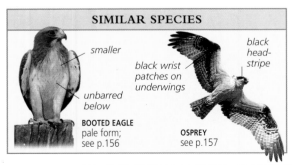

smaller

black wrist patches on underwings

black head-stripe

unbarred below

BOOTED EAGLE
pale form; see p.156

OSPREY
see p.157

OCCURRENCE
In summer, over high, open slopes, and rocky areas with short scrub, in Portugal, Spain, France, Italy, and Balkans. Prefers some woodland but mostly semi-natural vegetation such as aromatic, thorny scrub; absent from farmed areas.

Seen in the UK											

Length **62–69cm (24–27in)**	Wingspan **1.62–1.78m (5¼–5¾ft)**	Weight **1.5–2.5kg (3¼–5½lb)**
Social **Family groups**	Lifespan **Up to 15 years**	Status **Rare**

| Order **Accipitriformes** | Family **Accipitridae** | Species ***Circus aeruginosus*** |

Marsh Harrier

silvery grey midwings

broad black wingtips

MALE

square grey tail

dark brown plumage

pale head markings

IMMATURE

pale head

brown back

grey on wings

MALE

broad wings held up in "V" while gliding

**FEMALE
IN FLIGHT**

plainer, darker, longer-tailed than Buzzard

very dark brown plumage

cream patch on wings

creamy cap and throat

dark belly

MALE

FEMALE

FLIGHT: low, steady or rolling, wings raised in obvious "V" in short glides; soars well.

Long-winged and long-tailed, harriers fly low across open ground or marshland. The biggest and heaviest of them, the Marsh Harrier can be taken for a dark Buzzard or a Black Kite when soaring. It is closely associated with reedbeds, but may be seen over all kinds of open ground, especially flat pastures with ditches, when hunting or on migration. It often perches on bush tops or trees in the middle of a marsh for long spells. Males are generally smaller than females; they may breed in largely brown immature plumage.

VOICE Shrill *kee-yoo*, chattering *kyek-ek-ek-ek* or *kyi-yi-yi-yi*.
NESTING Large platform of reed stems among dense reeds over water; 4 or 5 eggs; 1 brood; April–July.
FEEDING Hunts low over marshes, diving to catch small birds, wildfowl, small mammals, and frogs.

FLIGHT PATTERN
Like all harriers, the Marsh Harrier has a slow flap and long glide flight, but with a heavier, less buoyant action.

SIMILAR SPECIES

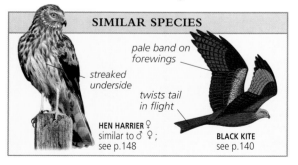

streaked underside

HEN HARRIER ♀
similar to ♂ ♀;
see p.148

pale band on forewings

twists tail in flight

BLACK KITE
see p.140

OCCURRENCE
Widespread north to Great Britain (rare) and S Scandinavia, in reedy areas or long grass in marshes. N and E breeders move south in autumn; some W European birds remain all year, over marshes and flat, open countryside near coasts.

| Seen in the UK |
| J F M A M J J A S O N D |

| Length **48–55cm (19–22in)** | Wingspan **1.1–1.25m (3½–4ft)** | Weight **400–800g (14–29oz)** |
| Social **Pairs/Family groups** | Lifespan **Up to 15 years** | Status **Secure** |

Order **Accipitriformes**	Family **Accipitridae**	Species *Circus cyaneus*

Hen Harrier

— black wingtips

wingtips wider, shorter, less tapered than Montagu's Harrier

cream and brown bars on tail

dark barring on silvery grey underwings

pale grey body

MALE

white rump

FEMALE

IN FLIGHT

wings held up in slight "V" while gliding

dark brown above

bold white rump

FEMALE

whitish line under cheeks

whitish underside

dark streaks on bright buff underside

FEMALE

grey tail

MALE

FLIGHT: deceptively quick, sailing flight with wings raised or steady, deep wingbeats; soars well.

Hen Harriers frequent heather moors in summer but are often found over coastal marshes or low-lying, rough grassland in winter. Males and females are very different in appearance and echo the plumages of the closely related Montagu's Harrier. As with most birds of prey, females are larger and broader-winged than males.

VOICE Near nest, loud, irregular *week-eek-ik-ik-ik* from female; more even *chekekekekekek* from male.

NESTING Platform of stems on ground, in rushes or heather; 4–6 eggs; 1 brood; April–June.

FEEDING Hunts low over open ground, swooping down on small birds and voles.

NESTING
This female is carrying nesting material in its bill. Hen Harriers make a bulky pad of heather stems and grass on the ground.

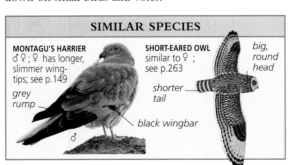

SIMILAR SPECIES

MONTAGU'S HARRIER
♂♀; ♀ has longer, slimmer wing-tips; see p.149

grey rump

♂

SHORT-EARED OWL
similar to ♀; see p.263

shorter tail

black wingbar

big, round head

OCCURRENCE
Breeds in N and E Europe, mostly on moorland or heaths, locally in C and W Europe, on moors and sometimes open farmland. Widespread over open ground with short vegetation, including marshes, in W Europe in winter.

Seen in the UK
| J | F | M | A | M | J | J | A | S | O | N | D |

Length **43–50cm (17–20in)**	Wingspan **1–1.2m (3¼–4ft)**	Weight **300–700g (11–25oz)**
Social **Roosts in groups**	Lifespan **Up to 15 years**	Status **Vulnerable**

Order **Accipitriformes**	Family **Accipitridae**	Species **_Circus pygargus_**

Montagu's Harrier

slender tapered wingtips

MALE

medium grey head

black bar across inner wings

red-brown bars on underwings

grey rump

medium grey upperparts

long, narrow black wingtips angled back

FEMALE

banded tail

white rump

pale below with bold streaks

IN FLIGHT

glides with wings raised

unmarked rufous below

JUVENILE

pale crescents above and below eye

dark brown above

FEMALE

streaked flanks

MALE

Difficult to tell from the Hen Harrier in most circumstances, Montagu's Harrier is a slimmer, thin-winged bird more closely associated with arable farmland. It is also a summer visitor to Europe, and is absent in winter. This is the most delicate of all the elegant-looking birds of the family, the swept-back, tapered wingtips often enough to aid identification even at very long range.

VOICE High, clear _yek-yek-yek_ from male, _chek-ek-ek-ek_ from female.

NESTING Nest of stems and grasses on ground in corn or heather; 4 or 5 eggs; 1 brood; April–June.

FEEDING Catches small mammals, reptiles, and small birds on or near ground, diving from low, gliding flight.

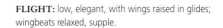

FLIGHT: low, elegant, with wings raised in glides; wingbeats relaxed, supple.

IMMATURE MALE
Young males often look dark, with a mixture of brown feathers among the adult grey.

OCCURRENCE
Widespread in N to S Great Britain (very rare) and to Baltic, from April to September, over heaths, rough grassland, marshes, and rolling cereal fields, often nesting in tall crops. Migrants typically follow low-lying coasts but also use high mountain passes.

Seen in the UK
J F M **A M J J A S O** N D

SIMILAR SPECIES

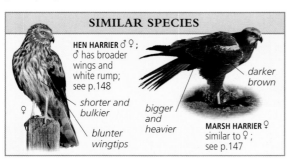

HEN HARRIER ♂ ♀; ♂ has broader wings and white rump; see p.148

shorter and bulkier

blunter wingtips

♀

darker brown

bigger and heavier

MARSH HARRIER ♀ similar to ♀; see p.147

Length **40–45cm (16–18in)**	Wingspan **1–1.2m (3¼–4ft)**	Weight **225–450g (8–16oz)**
Social **Pairs/Small flocks**	Lifespan **Up to 15 years**	Status **Secure**

| Order **Accipitriformes** | Family **Accipitridae** | Species *Accipiter gentilis* |

Goshawk

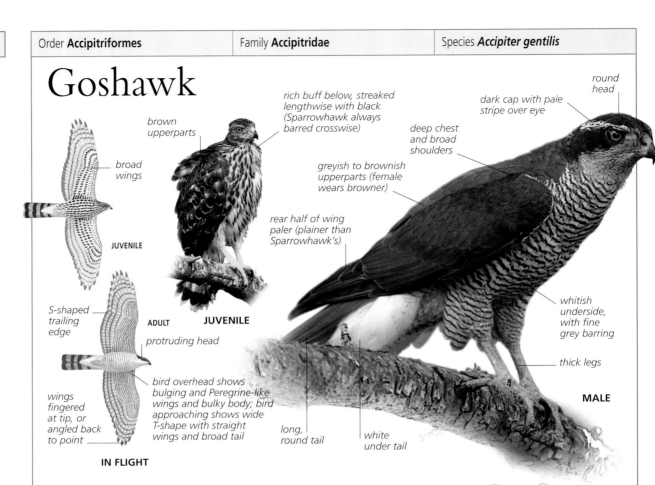

brown upperparts

broad wings

rich buff below, streaked lengthwise with black (Sparrowhawk always barred crosswise)

round head

dark cap with pale stripe over eye

deep chest and broad shoulders

greyish to brownish upperparts (female wears browner)

rear half of wing paler (plainer than Sparrowhawk's)

S-shaped trailing edge

ADULT

JUVENILE

JUVENILE

protruding head

wings fingered at tip, or angled back to point

bird overhead shows bulging and Peregrine-like wings and bulky body; bird approaching shows wide T-shape with straight wings and broad tail

long, round tail

white under tail

whitish underside, with fine grey barring

thick legs

MALE

IN FLIGHT

A powerful, awesome predator, the Goshawk is a much more substantial hawk than the Sparrowhawk; females, especially, can look remarkably large. Goshawks are scarce, through persecution, in most areas, but making a comeback in others, including the UK where they have been illegally released or simply escaped to establish wild populations. They usually sit, well-hidden, in trees and occasionally soar over forest. Goshawks are best looked for over extensive woodland in early spring, soaring over their territories.
VOICE Woodpecker-like, nasal *gek-gek-gek* and *pi-aah*.
NESTING Remarkably large, flat-topped heap of sticks and greenery close to trunk of tall tree; 2–4 eggs; 1 brood; March–June.
FEEDING Hunts boldly in forest or clearings, catching birds from thrush size to crows, gamebirds, and other birds or prey; also eats rabbits and squirrels.

FLIGHT: direct flight fast, with deep, quick wingbeats between short glides; soars with wings at full stretch, tail rounded.

BOLD PREDATOR
Goshawks eat prey where they happen to catch it, or carry smaller items to regular perches.

SIMILAR SPECIES

small head

narrower tail

SPARROWHAWK ♂ ♀; see p.151

BUZZARD slower flight; see p.152

shorter tail

OCCURRENCE
Widespread except in Iceland and Ireland, but mostly scarce. In well-wooded farmland and hills and forest, typically in mature woodland with tall conifers but plenty of space beneath trees. Few in more open country in winter.

Seen in the UK
| J | F | M | A | M | J | J | A | S | O | N | D |

| Length **48–61cm (19–24in)** | Wingspan **0.95–1.25m (3–4ft)** | Weight **500–1,350g (18–48oz)** |
| Social **Family groups** | Lifespan **Up to 20 years** | Status **Secure** |

Order **Accipitriformes**	Family **Accipitridae**	Species **Accipiter nisus**

Sparrowhawk

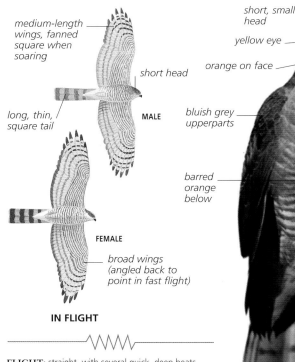

medium-length wings, fanned square when soaring

short head

MALE

long, thin, square tail

FEMALE

broad wings (angled back to point in fast flight)

IN FLIGHT

short, small head

yellow eye

orange on face

bluish grey upperparts

barred orange below

thin legs

MALE

browner above than adult

JUVENILE

brown bars below

pale line over eye

barred grey below

FEMALE

FLIGHT: straight, with several quick, deep beats between short, flat glides; soars with wings forward, tail tight closed; deep, bouncing undulations in display.

In many places still scarce after decades of accidental pesticide poisoning and centuries of persecution, the Sparrowhawk is common and familiar in other areas where its recovery has been complete. It typically soars over woods, perches inconspicuously, or dashes by, low, with a flap-flap-glide action. It is bold enough to hunt in gardens and parks but is essentially a forest-edge bird, extending its hunting range into more open country in winter. Males are much smaller than females.

VOICE Repetitive *kek-kek-kek-kek-kek*, thin, squealing *peee-ee*, but generally quiet away from nest.

NESTING Small, flat platform of thin twigs on flat branch close to trunk; 4 or 5 eggs; 1 brood; March–June.

FEEDING Hunts small birds, darting along hedges, woodland edges, or into gardens to take prey by surprise; males take tits and finches, females thrushes and pigeons.

AGILE HUNTER
Broad wings and a long tail give great manoeuvrability in tight spaces and accuracy when hunting.

OCCURRENCE
Throughout Europe, except in Iceland, in wooded farmland and hills and forest. In winter, in more open areas, including salt marshes with adjacent woodland. Hunts almost anywhere, including forays into gardens where small birds are fed.

Seen in the UK
J F M A M J J A S O N D

SIMILAR SPECIES

contrasted upperwings

more bulging wing shape

longer head

bigger

KESTREL ♂ ♀ similar in flight; more wingbeats between fewer glides, frequent hovers; see p.159

GOSHAWK ♂ ♀; see p.150

Length **28–40cm (11–16in)**	Wingspan **60–80cm (23½–32in)**	Weight **150–320g (5–12oz)**
Social **Family groups**	Lifespan **Up to 10 years**	Status **Secure**

Order **Accipitriformes**	Family **Accipitridae**	Species *Buteo buteo*

Buzzard

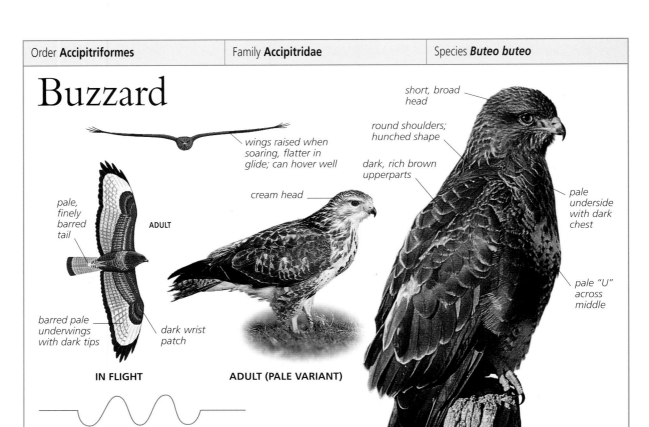

wings raised when soaring, flatter in glide; can hover well

short, broad head

round shoulders; hunched shape

dark, rich brown upperparts

pale, finely barred tail

ADULT

cream head

pale underside with dark chest

pale "U" across middle

barred pale underwings with dark tips

dark wrist patch

IN FLIGHT

ADULT (PALE VARIANT)

short, round tail

ADULT

FLIGHT: quick with slightly jerky, stiff wingbeats; soars with wings in "V", rising in broad circles.

One of the most common and most widespread of birds of prey, the Buzzard is therefore a useful yardstick by which to judge other, rarer birds. It is, however, well worth watching in its own right, too, being an impressive and exciting raptor. It is very variable, albeit around a relatively constant basic pattern. It soars in wavering, rising circles over nesting woods and perches on telegraph poles and fence posts. In some areas, such as the wooded valleys of Wales, it may be the most common bird of prey.

VOICE Noisy; frequent ringing *pee-yaah* scream or weaker *mew*; calls often while flying.

NESTING Stick nest in tree, or at base of bush on cliff ledge; 2–4 eggs; 1 brood; March–June.

FEEDING Catches small mammals, rabbits, beetles, earthworms, and some birds; eats much dead meat, including road-kill rabbits.

HEAD TO WIND
A Buzzard is able to hang motionless in the wind while searching for food; it also hovers rather heavily with deep wingbeats in calmer air.

OCCURRENCE
Widespread except in far N Europe (summer visitor in NE Europe), in wooded farmland, hills, and moors near crags and forest. Many move to Low Countries and France in winter, occupying low, flat ground with scattered woodland.

Seen in the UK
J F M A M J J A S O N D

SIMILAR SPECIES

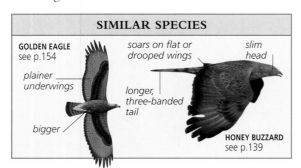

GOLDEN EAGLE see p.154

plainer underwings

bigger

soars on flat or drooped wings

slim head

longer, three-banded tail

HONEY BUZZARD see p.139

Length **50–57cm (20–22½ in)**	Wingspan **1.13–1.28m (3¾–4¼ ft)**	Weight **550–1,200g (20–43oz)**
Social **Family groups**	Lifespan **Up to 25 years**	Status **Secure**

| Order **Acccipitriformes** | Family **Accipitridae** | Species ***Buteo lagopus*** |

Rough-legged Buzzard

white tail with dark bands near tip

whitish patches on primaries

IN FLIGHT

JUVENILE

pale chest and blackish belly

dark brown above

dark trailing edge to wing (weakest on juvenile)

pale head

"frosty" pale feather edges

ADULT

FLIGHT: flight slightly more fluid and flexible than Buzzard; soars less, hovers frequently.

This northern buzzard sometimes appears in small numbers well to the south of its usual range in winter when food is short. It is clearly closely related to the Buzzard but usually sufficiently distinct to make identification straightforward. In wintering areas such as the Netherlands, there may be scores or hundreds of common Buzzards for every Rough-legged, but its regular hovering may draw attention to it. Its trademark feathered legs are often hard to see. Like some other northern species, its fortunes are closely linked to the fluctuating availability of its prey.

VOICE Loud, low, plaintive squeal, *pee-yow*.
NESTING Stick nest on cliff or in tree; 2–4 eggs; 1 brood; March–June.
FEEDING Drops onto small mammals, especially voles and small rabbits, from perch or after hover.

HEAD TO WIND
The pale head, dark belly, and pale vent show well as this individual hovers, head to wind, searching for prey on the ground.

OCCURRENCE
Breeds in Scandinavia in tundra and highland areas. Winters lower down in C Europe, few in Low Countries, very rare in UK except in years when small rodent food is scarce in north. Winter visitors often in low, expansive farmland.

| Seen in the UK |
| J F M A M J J A S O N D |

SIMILAR SPECIES

GOLDEN EAGLE juvenile, longer wings; see p.154

browner head

stockier

BUZZARD jerkier in flight; see p.152

much bigger

| Length **50–60cm (20–23½in)** | Wingspan **1.2–1.5m (4–5ft)** | Weight **600–1,300g (21–46oz)** |
| Social **Family groups** | Lifespan **Up to 10 years** | Status **Secure** |

Order **Accipitriformes**	Family **Accipitridae**	Species **Aquila chrysaetos**

Golden Eagle

wings raised in slight "V"

white on tail and wings reduces with age

IMMATURE

blacker body than adult's

pale tawny to yellowish head

dark brown plumage

bulky body and wings

IMMATURE

protruding head (less than White-tailed Eagle's)

barred dark underwings

long tail with paler area at base

IN FLIGHT   **ADULT**

ADULT

While White-tailed Eagles and the large vultures have great bulk and a massive presence, the Golden Eagle combines size with elegance and even delicacy in the air. It is often seen far off over a high peak, literally a dot in the distance, but its wide, slow circling is frequently sufficient to identify it. Close views are harder to come by, a rare chance encounter on a high peak being really memorable.

VOICE Occasional shrill yelps and whistling *twee-oo*.

NESTING Massive pile of sticks, lined with wool and greenery, on broad cliff ledge or in old pine; 1–3 eggs; 1 brood; February–June.

FEEDING Eats much dead meat, mostly sheep and deer in winter; hunts grouse, Ptarmigan, crows, hares, and rabbits.

FLIGHT: direct flight steady; frequent glides; soars with wings raised; switchback display and occasional superfast stoop or plunge with wings curved back.

STANDING BIRD
A standing Golden Eagle looks majestic. The thick, heavily feathered legs are obvious on a perched bird.

OCCURRENCE
Widespread but sparse, most frequent in Scotland, Spain, parts of Scandinavia, Italy, Balkans, and also in Alps. Mostly over remote peaks or upland forests, more rarely on steep coasts, and stays high up, away from towns and villages, and roads and other developments (unlike Buzzards).

Seen in the UK
J F M A M J J A S O N D

SIMILAR SPECIES

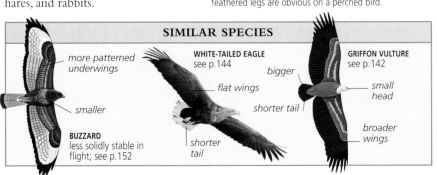

more patterned underwings

smaller

BUZZARD
less solidly stable in flight; see p.152

WHITE-TAILED EAGLE
see p.144

flat wings

shorter tail

GRIFFON VULTURE
see p.142

bigger

small head

broader wings

shorter tail

Length **75–85cm (30–34in)**	Wingspan **1.9–2.2m (6¼–7¼ft)**	Weight **3–6.7kg (6½–15lb)**
Social **Family groups**	Lifespan **Up to 25 years**	Status **Rare**

Order **Accipitriformes**	Family **Accipitridae**	Species *Hieraaetus fasciatus*

Bonelli's Eagle

dark midwing band

short head, held high

dark upperparts with white patch on back

pale orange-tawny below

rather long dark-tipped tail

IMMATURE

JUVENILE

blackish band on dark underwings

pale leading edge

streaked white underbody

ADULT

dark tail tip

pale leading edge

ADULT

long, narrow, straight-edged tail

IN FLIGHT

A strong and potent predator, Bonelli's Eagle combines the power of a large eagle with the behaviour of a fast-flying, aggressive hawk. It spends much time perching on a ledge out of sight, but periodically soars over its territory or moves off to find food; it hunts rather low down and is usually inconspicuous. It is generally scarce and very localized, with isolated pairs here and there in traditional territories, mostly in areas with steep cliffs and crags between forested slopes. Only a few hundred pairs of Bonelli's Eagle remain in Europe.

VOICE Occasional bark or shrill yelp, but mostly silent.
NESTING Large stick nest in small cave, deep ledge, or sheer cliff, rarely in tree; 2 eggs; 1 brood; February–April.
FEEDING Elegant and powerful; hunts hares, rabbits, squirrels, partridges, crows, and pigeons.

FLIGHT: glides on flat wings with angled leading edge; wingbeats shallow; soars, sometimes stoops at great speed.

STRIKING APPEARANCE
In flight, Bonelli's Eagle looks big and square with quite a long tail, long flat wings, and a small head sweeping up from the deep chest; the white body catches the light.

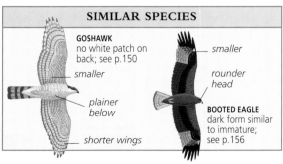

SIMILAR SPECIES

GOSHAWK
no white patch on back; see p.150
smaller
plainer below
shorter wings

smaller
rounder head

BOOTED EAGLE
dark form similar to immature; see p.156

OCCURRENCE
Occupies forested hills and crags in Spain, Portugal, S France, and Balkans; often nests and roosts in deep gorges, but not especially high up, and moves out into nearby hills and woodland to hunt. Resident.

Seen in the UK
J F M A M J J A S O N D

Length **55–65cm (22–26in)**	Wingspan **1.45–1.65m (4¾–5½ft)**	Weight **1.5–2.5kg (3¼–5½lb)**
Social **Pairs/Families**	Lifespan **Up to 15 years**	Status **Endangered**

| Order **Accipitriformes** | Family **Accipitridae** | Species *Hieraaetus pennatus* |

Booted Eagle

- diagonal pale bands on wings
- white spots
- whitish crescent on rump

BOTH FORMS FROM ABOVE

- well-fingered wingtips
- long tail with sharp corners
- pale patch
- round head
- dull brown overall

DARK FORM

IN FLIGHT

- flat or bowed wings, unlike shallow V of Buzzard

PALE FORM

- broad, round head
- white underside

PALE FORM

FLIGHT: fast, with long glides and rather flappy wingbeats; frequent soaring.

Much more frequent than Bonelli's Eagle, the Booted Eagle is a small, buzzard-sized eagle, frequently mobbed by crows and other birds of prey. It occurs in pale and dark forms but its shape and proportions help identify it. It is typical of well-wooded, hot countryside in relatively lowland areas, avoiding the moister northwestern parts of Europe. Unlike the Buzzard, it is a summer visitor to Europe.
VOICE Buzzard-like *hi-yaaah* and loud, musical, wader-like whistle, *kli-kli-kli* in display.
NESTING Bulky stick nest deep inside canopy of tree, rarely on cliff ledge; 2 eggs; 1 brood; February–April.
FEEDING Catches reptiles, small birds, and small mammals on ground, often after very fast, near-vertical stoop from considerable height.

DISTINCTIVE PATTERN
In flight overhead, the Booted Eagle shows thin, translucent trailing edges to both wings and tail, and a pale inner primary patch; head-on it reveals bright white shoulder spots.

SIMILAR SPECIES

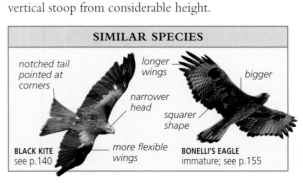

- notched tail pointed at corners
- narrower head
- more flexible wings

BLACK KITE
see p.140

- longer wings
- bigger
- squarer shape

BONELLI'S EAGLE
immature; see p.155

OCCURRENCE
In Spain, Portugal, France, and E Europe, in forests and warm, sunny, well-wooded, hilly country with mixed farmland and scrub, often close to villages. Thrives best in regions with little disturbance. From March to October.

Seen in the UK
J F M A M J J A S O N D

| Length **42–51cm (16½–20in)** | Wingspan **1.1–1.35m (3½–4½ft)** | Weight **700–1,000g (25–36oz)** |
| Social **Family groups** | Lifespan **Up to 15 years** | Status **Rare** |

Order **Accipitriformes**	Family **Pandionidae**	Species *Pandion haliaetus*

Osprey

glides on kinked wings

ADULT

long, broad wings

ADULT

black patch on underwings

ADULT

short tail with translucent pale bands

IN FLIGHT

bowed wings while soaring

blackish band along midwings

FLIGHT: gull-like but strong with long, sailing glides; soars well.

whitish crown

black stripe through eye

dark brown upperparts (bright buff feather edges on juvenile)

white underparts

ADULT

large, sharp claws

Between a Buzzard and a large eagle in size, the Osprey is impressive, and yet sometimes easily dismissed as a large immature gull at long range. Should it hover and then plunge for a fish, such uncertainty is quickly removed. Any reasonable view would reveal its unique combination of shape and pattern. The Osprey is rarely seen far from water, although it does nest and perch well away from the shore. It perches upright for hours at a time on a tree, buoy, or offshore rock.

VOICE Loud yelps and repeated, high, liquid *pyew pyew pyew* near nest.

NESTING Huge stick nest on trees or cliffs (in the past, on ruined buildings); 2 or 3 eggs; 1 brood; April–July.

FEEDING Catches fish in steep dive.

CARRYING FISH
The Osprey hovers well but heavily over water, and catches prey in a steep, headlong dive, swinging its feet forwards to grasp prey.

SIMILAR SPECIES

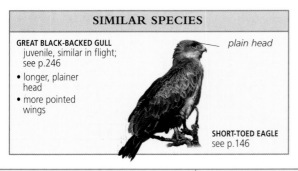

GREAT BLACK-BACKED GULL
juvenile, similar in flight; see p.246
- longer, plainer head
- more pointed wings

plain head

SHORT-TOED EAGLE
see p.146

OCCURRENCE
Breeds in wild and remote places in N Europe but much more adaptable in S. Appears in most of Europe except for Iceland, from March to October, along coasts and near large lakes and rivers.

Seen in the UK
J F M **A M J J A S O** N D

Length **52–60cm (20½–23½in)**	Wingspan **1.45–1.7m (4¾–5½ft)**	Weight **1.2–2kg (2¾–4½lb)**
Social **Family groups**	Lifespan **Up to 30 years**	Status **Rare**

Order **Falconiformes**	Family **Falconidae**	Species *Falco naumanni*

Lesser Kestrel

blue-grey panel, often hard to see

MALE

dark outer wings

FEMALE

IN FLIGHT

pale cheek spot

black-barred back

FEMALE

blue-grey head

unspotted, rich red-brown back

MALE

finely spotted, deep pink-buff chest

white claws

Sharing the basic colours and patterns of the Kestrel, the male Lesser Kestrel can be distinguished by its unspotted back, but the female can be very difficult to tell apart. While the smaller size of the Lesser Kestrel is not normally obvious, its slightly stockier shape and shorter wings help in identifying it. It is also a more social bird, often breeding in tight colonies. Lesser Kestrel numbers have declined dramatically in Europe in recent decades. It is, however, not certain whether this is caused by problems in its wintering sites in Africa or in its summer range.

VOICE Fast, raucous, triple call, *chay-chay-chay*, and nasal, high, chattering notes.

NESTING On ledges or in cavities, in loose colonies; 3–6 eggs; 1 brood; April–July.

FEEDING Catches insects in air or from ground after hovering.

blue tail with black band at tip

FLIGHT: light, quick, with shallow wingbeats; hovers like Kestrel.

bluish grey head

MALE

white underwings

brownish head

FEMALE

brown tail with black band at tip

HOVERING
The Lesser Kestrel looks much like the Kestrel while hovering; however, its chunkier body, shorter tail, and square-ended rear aid identification.

SIMILAR SPECIES

KESTREL ♂ ♀; ♀ very similar; see p.159

spotted back

♂

OCCURRENCE
Breeds socially on buildings and cliffs and feeds over hot, open ground in Spain, Portugal, S France, S Italy, and Balkans. Summer visitor declining fast in most areas; very rare out of normal range.

Seen in the UK
J F M A M J J A S O N D

Length **27–33cm (10½–13in)**	Wingspan **63–72cm (25–28in)**	Weight **90–200g (3¼–7oz)**
Social **Flocks**	Lifespan **5–7 years**	Status **Vulnerable†**

Order **Falconiformes**	Family **Falconidae**	Species *Falco tinnunculus*

Kestrel

grey tail with black band

MALE

MALE

brown-black outer wings

barred back and wings

short, round blue-grey head

FEMALE

pale rufous back spotted with black

rufous inner wings

FEMALE

pale brown inner wings

outer wings paler than on male

IN FLIGHT

black claws

slim tail (faintly barred on immature, pure grey on adult)

MALE

Usually the most familiar and easily seen bird of prey, the Kestrel has nevertheless declined in farmland areas in recent years. It is the pigeon-sized, long-winged, daytime hunter most often seen perched on telegraph poles or wires or hovering over roadsides, as if suspended on a string. Unlike Sparrowhawks, there is relatively little difference in size between the sexes, but much more in pattern: the male has a bluish grey head and red-brown upperparts while the female has a brown head and tail.

VOICE Nasal, complaining, whining *keee-eee-eeee* and variants.

NESTING On bare ledges on cliffs, in quarries, derelict buildings, high window-ledges, disused crows' nests or tree holes; 4–6 eggs; 1 brood; March–July.

FEEDING Catches small mammals, especially voles, and also beetles, lizards, earthworms, and small birds.

FLIGHT: direct flight with deep wingbeats, few glides; hovers conspicuously; soars with wings and tail fanned; aerobatic around cliffs.

POISED FOR A DIVE
The Kestrel hovers frequently, its tail often spread like a fan. The fan-shaped tail acts as a brake when the bird is about to land.

OCCURRENCE
Almost everywhere in Europe, from cities to remote mountains; common around woodland and heaths, until recently on farmland but declining as farming systems are modernized and food is restricted almost to roadside verges. Present all year but many move south in winter.

Seen in the UK
J F M A M J J A S O N D

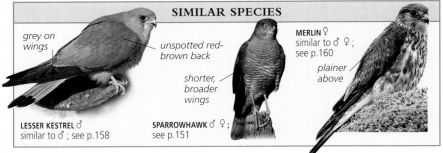

SIMILAR SPECIES

grey on wings

unspotted red-brown back

MERLIN ♀; similar to ♂ ♀; see p.160

shorter, broader wings

plainer above

LESSER KESTREL ♂; similar to ♂; see p.158

SPARROWHAWK ♂ ♀; see p.151

Length **34–39cm (13½–15½in)**	Wingspan **65–80cm (26–32in)**	Weight **190–300g (7–11oz)**
Social **Family groups**	Lifespan **Up to 15 years**	Status **Declining**

| Order **Falconiformes** | Family **Falconidae** | Species *Falco columbarius* |

Merlin

broad-based wings taper to point

pale tail with black band

MALE

mud-brown above

FEMALE

small, square head with little pattern

bluish grey upperparts

small, chunky body

cream below with streaks

brown and cream banded tail

FEMALE

IN FLIGHT

orange-buff below

MALE

FLIGHT: fast, low, direct, with almost constant wingbeats; chases prey with in-out flicked wing action and finally rapid, twisting, acrobatic chase.

A small, dynamic, chunky falcon of open ground, the Merlin usually keeps low, chasing down its prey in quick, agile flights. It spends much of its time perched on low posts, rocks, and even clods of earth, scanning the landscape. In summer, it is an elusive bird, especially if nesting on the ground, although tree-nesting pairs can be more demonstrative if approached, boldy diving at intruders.

VOICE Male has quick, sharp *kik-kik-ki-kik*; female has deeper, more nasal *kee-kee-kee-kee*; quiet away from nest and in winter.

NESTING Bare scrape on ground among heather or old crow's nest in tree; 3–6 eggs; 1 brood; April–June.

FEEDING Mostly eats small birds, caught in flight; also eats variety of large aerial insects.

LOW PERCH
A brown female Merlin perches on a low, mossy rock in a typical squat, upright pose, alert and ready to chase prey.

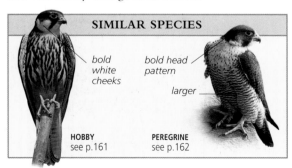

SIMILAR SPECIES

bold white cheeks

bold head pattern

larger

HOBBY
see p.161

PEREGRINE
see p.162

OCCURRENCE
Widespread but scarce breeder on moors in N Europe, sometimes using slopes with trees or edges of conifer plantations. In winter, over most of Europe in open country-side, especially pastureland and coastal marshes with open spaces.

Seen in the UK
| J | F | M | A | M | J | J | A | S | O | N | D |

| Length **25–30cm (10–12in)** | Wingspan **60–65cm (23¹/₂–26in)** | Weight **140–230g (5–8oz)** |
| Social **Family groups** | Lifespan **Up to 10 years** | Status **Secure** |

Order **Falconiformes**	Family **Falconidae**	Species *Falco subbuteo*

Hobby

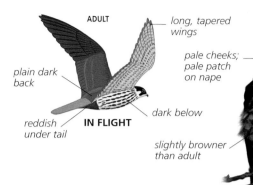

ADULT

long, tapered wings

plain dark back

reddish under tail

IN FLIGHT

dark below

black "moustache"

pale patch on cheeks and throat

pale cheeks; pale patch on nape

heavy streaks

slightly browner than adult

no red under tail

JUVENILE

thick black streaks on pale underparts

rufous thighs

short, narrow, plain tail

ADULT

FLIGHT: easy, relaxed with fluent wingbeats; hunts in gliding, swooping, patrolling flight with sudden twists and changes of height, direction, and pace.

No bird outdoes the Hobby for sheer elegance when it is hawking insects above a marsh, heath, or field on a summer evening. It floats over open space, speeding up to take prey: its seemingly effortless acceleration takes it into a dive, or a chase, or a smooth upward glide, before an extremely neat "take", using its feet to catch larger insects such as chafers and dragonflies. It is unusual among European falcons, being a summer visitor from Africa.

VOICE Clear, musical, whistled *kyu-kyu-kyu-kyu*.

NESTING No nest (like other falcons), lays eggs inside old nest of Rook or crow; 2 or 3 eggs; 1 brood; June–August.

FEEDING Catches small birds such as martins and swallows in flight and many insects such as dragonflies and large aerial beetles.

PINE TREE NURSERY
A crow's nest in a tall conifer makes an ideal place for the Hobby to nest.

DYNAMIC FLIGHT
Hobbies catch insects with deft turns, but make long, fast dives after small birds.

SIMILAR SPECIES

PEREGRINE see p.162
- stockier
- barred, not streaked, below
- broader across rump and shoulders in flight

plainer face

chunkier wings

MERLIN ♂ ♀; see p.160

OCCURRENCE
Most of Europe except far north, from April to October, breeding in clumps of trees. Hunts over open ground, heaths, farmland with trees, and marshy ground, especially around flooded pits with large flying insects.

Seen in the UK
J F M **A M J J A S O** N D

Length **28–35cm (11–14in)**	Wingspan **70–84cm (28–33in)**	Weight **130–340g (5–12oz)**
Social **Small feeding flocks**	Lifespan **Up to 10 years**	Status **Secure**

| Order **Falconiformes** | Family **Falconidae** | Species *Falco peregrinus* |

Peregrine 🔊 15.I, 15.II

ADULT

anchor-shaped wings

broad pale rump

short, square dark tail

IN FLIGHT

blue-grey upperparts

bluish eye-ring and bill-base

similar to Hobby but slimmer

yellow bill-base

black lobes on each side of face and neck

yellow eye-ring

white cheek patch

white breast

large, solid body with broad shoulders

closely barred white underside

browner than adult above with buff edges

streaked below

JUVENILE

buff tail tip

ADULT

grey bars on flanks and belly

ADULT

yellow feet

FLIGHT: direct level flight fast with quick, regular, deep, whippy wingbeats; also soars on straight, flat wings; acrobatic, with long, fast, near-vertical stoops.

One of the larger falcons, the Peregrine is a symbol of survival against the odds, these being persecution and pesticide poisoning that threatened it with extinction. Now it has made a strong comeback; in the UK, it is more common than it has ever been, although parts of its former range remain unoccupied. As with most falcons and hawks, females are considerably larger than males. Pairs often remain together for long periods, soaring over nesting cliffs.

VOICE Loud, raucous calls at nest include throaty *haak-haak-haak-haak* and whining *kee-keee-eeeeee* and *wheeee-ip*.

NESTING On broad ledge or earthy scrape on cliff, in quarry, or more rarely on building or on flatter ground; 2–4 eggs; 1 brood; March–June.

FEEDING Kills birds of sizes ranging from thrush to pigeon or grouse, sometimes larger, often rising to take them from beneath, chasing in level flight, or stooping from great height.

STRIKING ADULT
Peregrines look dark against the sky, but a close view reveals black, white, and yellow adding sharp contrast to the overall grey.

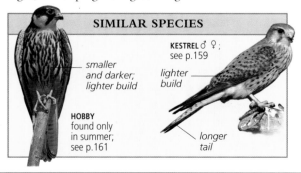

SIMILAR SPECIES

smaller and darker; lighter build

HOBBY found only in summer; see p.161

KESTREL ♂ ♀; see p.159

lighter build

longer tail

OCCURRENCE
Widespread but scarce, breeding through Scandinavia, N and W Great Britain, Spain, Portugal, Alps, Italy, and Balkans, in hills and coasts with cliffs, increasingly in cities. Present all year, some wandering in winter.

Seen in the UK
J F M A M J J A S O N D

| Length **39–50cm (15½–20in)** | Wingspan **0.95–1.15m (3–3¾ft)** | Weight **600–1,300g (21–46oz)** |
| Social **Family groups** | Lifespan **Up to 15 years** | Status **Rare** |

RAILS, CRAKES, AND COOTS

RATHER SMALL, slim birds, rails, crakes, and coots are narrow-bodied, allowing them to slip through dense vegetation; however, their deep bellies give a rounder appearance in a side view. Most live in wet places. Some are resident, others long-distance migrants. The Moorhen is common and familiar, and easy to see; the Coot is more gregarious and also easy to watch in large numbers on almost any freshwater pool. Others, however, such as the smaller crakes and the Water Rail, live in such dense vegetation that they are difficult to see at all, or come only to the edge of the reeds or sedges occasionally to give a brief glimpse. They are, however, skulking rather than shy and can sometimes be watched very closely. The Corncrake lives in dry fields of hay and clumps of irises or nettles, away from open water; it is hard to see but easy to hear its loud,

RATCHET CALL
The repeated "crek crek" of the Corncrake is an evocative sound, which is sadly declining.

repetitive "song". In most European countries it has declined severely with modernization of farming techniques.

RALLIDAE

long red bill

WATER RAIL
p.164

short, often cocked tail

SPOTTED CRAKE
p.165

tawny back with bold black streaks

CORNCRAKE
p.166

diagonal white stripe

MOORHEN
p.167

white facial shield and bill

COOT
p.168

| Order **Gruiformes** | Family **Rallidae** | Species ***Rallus aquaticus*** |

Water Rail

ADULT

slate-grey head and breast

trailing feet

IN FLIGHT

pale buff under short, cocked tail

pale to rich brown upperparts with thick dark streaks

red eye

black tip of pointed red bill

narrowly barred flanks

grey chest

pink legs

untidy bars below

dull legs

JUVENILE

ADULT

This is often a difficult bird to see, not so much because it is shy but as a result of its habit of skulking in dense waterside vegetation; occasionally a Water Rail on open mud will show itself off remarkably well. Because of its habitat requirements, it is very patchily distributed and generally scarce, but large reedbeds can have big populations, best detected by listening at dawn and dusk for their loud, squealing calls.

VOICE Loud, hard, repetitive *kipkipkipkipkip*, frequent loud, squealing and grunting (generally pig-like) notes.

NESTING Shallow dish of broad leaves and grass stems, in vegetation raised a little above water level; 6–11 eggs; 2 broods; May–August.

FEEDING Mostly feeds on insects and molluscs but very opportunistic, taking even voles and small birds, dead animals, seeds, and berries.

FLIGHT: quick, short, low flights with raised, whirring wings and dangling legs and toes.

REEDBED WADER
Water Rails typically wade through the shallows in and around reeds and swampy willow thickets, now and then appearing at the edges.

SIMILAR SPECIES

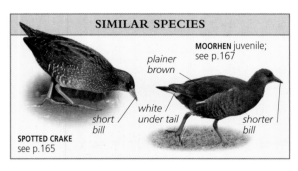

MOORHEN juvenile; see p.167

plainer brown

white under tail

short bill

shorter bill

SPOTTED CRAKE see p.165

OCCURRENCE
In most of Europe except N Scandinavia in wet reedbeds, sedges, and dense reedmace by pools; also in overgrown ditches, muddy ponds, sometimes flooded places under willows and alders, and overgrown riversides.

Seen in the UK
| J | F | M | A | M | J | J | A | S | O | N | D |

| Length **22–28cm (9–11in)** | Wingspan **38–45cm (15–18in)** | Weight **85–190g (3–7oz)** |
| Social **Family groups** | Lifespan **Up to 6 years** | Status **Secure†** |

Order **Gruiformes**	Family **Rallidae**	Species ***Porzana porzana***

Spotted Crake

ADULT

white leading edge on wing

short bill

barred underparts

buff under tail

IN FLIGHT

short, often cocked tail

rotund from side

grey-buff neck with white spots

short yellowish bill with red at base

white bars on flanks

browner on head and neck than adult

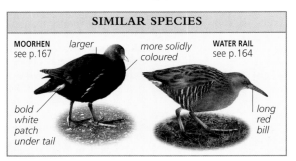

JUVENILE

ADULT

/\/\/\/\/\/\/\/\/\/\/\/\/\/\/\/\/\/\/\

FLIGHT: short, quick flights if flushed from vegetation, dropping back into cover with dangling legs.

The Spotted Crake is another "shy" bird that hides itself in thick vegetation but, at times, allows remarkably close views: careful observation reveals a beautifully patterned and almost shiny-plumaged bird. Spotted Crakes in spring are elusive but can be detected by their nocturnal whiplash calls. Most in western Europe are seen on migration in autumn, where falling water levels have exposed bands of mud along the edges of reedbeds.

VOICE Repeated, rhythmic, whipped, or dripping *hwit, hwit, hwit*, at dusk or after dark.

NESTING Small saucer of leaves and stems, placed in upright stalks raised above water or in wet marsh; 8–12 eggs; 1 brood; May–July.

FEEDING Picks various small insects and aquatic invertebrates from mud, foliage, and water.

SLIM SHAPE
Like all crakes and rails, the Spotted Crake appears deep-bodied from the side but end-on, it is slim, easily able to slip between reeds and sedges.

SIMILAR SPECIES

MOORHEN see p.167

larger

more solidly coloured

WATER RAIL see p.164

bold white patch under tail

long red bill

OCCURRENCE
Widespread except in far N Europe, but everywhere very patchy. Breeds in extensive flood-meadows; migrant in wet marsh in reedy places, often appearing at edge of muddy pools and generally elusive in dense cover rather than shy.

Seen in the UK
J F **M A M J J A S O N** D

Length **22–24cm (9–9¹/₂in)**	Wingspan **35cm (14in)**	Weight **70–80g (2¹/₂–2⁷/₈oz)**
Social **Solitary**	Lifespan **Up to 5 years**	Status **Secure**

| Order **Gruiformes** | Family **Rallidae** | Species *Crex crex* |

Corncrake 🔊 19

strong, narrow but rounded, rufous wings

pink legs

ADULT

IN FLIGHT

less grey than male

white and brown bars on flanks

FEMALE

soft grey and buff face

stout, short pink bill

grey throat and breast

tawny back with bold black streaks

MALE

FLIGHT: low, short flight with quick wingbeats and trailing feet; drops down to cover quickly.

Corncrakes skulk in tall vegetation, especially hay in summer, and require dense beds of yellow iris and nettles for early cover ion their return from Africa in spring. They are hard to see, but singing males are easy to hear especially at dusk. Migrants are generally rare and take people by surprise, suddenly rising underfoot and looking surprisingly rufous. The advent of advanced, intensive farming threatens the survival of Corncrakes in eastern Europe, where good numbers still remain; last-ditch conservation efforts have protected them in western Scotland. There they prefer hay in flowery meadows, with longer, rougher vegetation in field corners or overgrowing dry stone walls to give plenty of thick cover.

VOICE Loud, repeated double-note: at distance light, scratched or rasped *crik crik*; at close range hard, rattling, deep, vibrating *crrek crrek*.

NESTING Small, leaf- and grass-lined hollow on ground, with grass cover to conceal top; 8–12 eggs; 1 or 2 broods; May–August.

FEEDING Picks insects, seeds, leaves, and shoots from foliage and ground, in steady, springy walk.

PEERING FROM COVER
The Corncrake keeps well hidden in long grass and occasionally peers upwards into open view with its head raised.

SIMILAR SPECIES

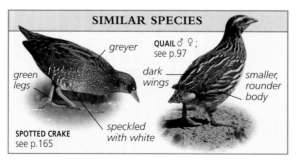

greyer

green legs

speckled with white

SPOTTED CRAKE
see p.165

QUAIL ♂ ♀;
see p.97

dark wings

smaller, rounder body

OCCURRENCE
Widespread but scarce or rare in France and C Europe; very rare in Ireland and W Scotland. Breeds in hayfields and wet grass with dense cover in spring and in late summer harvest (unable to survive in early-cut silage).

| Seen in the UK |
| J F M **A M J J A S** O N D |

| Length **27–30cm (10¹/₂–12in)** | Wingspan **46–53cm (18–21in)** | Weight **135–200g (5–7oz)** |
| Social **Family groups** | Lifespan **5–7 years** | Status **Vulnerable** |

Order **Gruiformes**	Family **Rallidae**	Species *Gallinula chloropus*

Moorhen 🔊 20

green legs

IN FLIGHT

rich brown back

red eye

slate-grey head

bright red bill with yellow tip

brown body

dull greenish yellow bill

JUVENILE

ADULT

bold white patch under tail

ADULT

slate-grey underside

diagonal white stripe

long green toes

glossy plumage

ADULT

A bird of the water's edge and nearby marshy ground, rather than open water, the Moorhen is widespread and surprisingly common in many areas as it can occupy anything from a wet ditch to a large lake. Small, loose groups move slowly, feeding on damp meadows, running to cover if disturbed, and even breaking into brief flight. Only rarely is a Moorhen seen way out on open water, looking a little uncomfortable (unlike the Coot) in such an exposed situation.

VOICE Loud, sudden, throaty or metallic notes from cover, *kurruk* or *kittik*, high *kik*, stuttering *kik-kikikikik-ik*.

NESTING Shallow bowl of leaves and stems, in vegetation from low reeds to high in trees but usually just above water, often in fallen or drooping branch; 5–11 eggs; 2 or 3 broods; April–August.

FEEDING Picks up seeds, fruit, shoots, roots, snails, insects, and occasionally eggs, from damp ground or shallows.

FLIGHT: low, fluttery, with long legs trailing; often scutters over water to nearest cover.

FIGHTING FRENZY
Moorhens fight furiously in the breeding season, kicking with their feet; females fight over males.

OCCURRENCE
Widespread except in Iceland and N Scandinavia; in summer only in N and E Europe. Breeds on small ponds with overgrown margins or overhanging branches; found in ditches, rivers, ponds, lakes, and reservoirs of all kinds. Feeds in small groups, on open, wet, grassy ground, even in hedges, usually near water.

Seen in the UK
J F M A M J J A S O N D

SIMILAR SPECIES

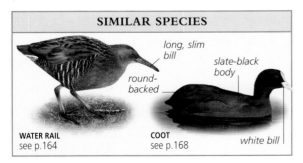

long, slim bill

round-backed

slate-black body

WATER RAIL
see p.164

COOT
see p.168

white bill

CLIMBER
Moorhens are surprisingly agile, climbing trees and dense hedgerow bushes.

Length **32–35cm (12½–14in)**	Wingspan **50–55cm (20–22in)**	Weight **250–420g (9–15oz)**
Social **Small flocks**	Lifespan **Up to 15 years**	Status **Secure**

Order **Gruiformes**	Family **Rallidae**	Species **Fulica atra**

Coot 🔊 21.I, 21.II

ADULT

blurred white face and throat

yellowish bill

JUVENILE

IN FLIGHT

rounded rump

slate-black body

intensely black head

red eye

white facial shield and bill

pale rear edge

ADULT

ADULT

large grey feet with lobed toes

These quarrelsome waterbirds are often found on wide open water on large reservoirs or flooded pits; they are rarely on the sea. They are also frequently seen feeding on nearby grassy banks. Feeding flocks are usually bigger and more coherent than the loose groups of Moorhens. Coots are also obviously bigger and more heavily built. Close views reveal their broadly lobed toes, more like a grebe's. While superficially duck-like, the Coot is really not very like any species of European duck.

VOICE Loud, explosive *kowk*, high, squeaky *teuwk*, *pik*, and similar notes; juvenile has loud whistling calls.

NESTING Large bowl of wet vegetation, in overhanging branches or reeds, or on mound of semi-floating debris at water's edge; 6–9 eggs; 1 or 2 broods; April–August.

FEEDING Dives easily to feed underwater, bouncing back up like a cork; eats grass, seeds, shoots, snails, tadpoles, and similar small aquatic creatures.

FLIGHT: usually low, quite quick but heavy, lacking agility; big feet trailing.

FAMILY GROUP
Coots often nest on town park lakes, where family groups are a familiar sight.

TERRITORIAL DISPLAY
Coots raise their wings and body feathers to impress and scare away rivals, and often resort to fights.

SIMILAR SPECIES

MOORHEN see p.167

slimmer

pointed tail with white beneath

slim, upturned bill

BLACK-NECKED GREBE winter, similar to juvenile; see p.110

OCCURRENCE
Widespread except in Iceland and N Scandinavia; in summer only in N and E Europe. Nests mostly on lakes and flooded pits, with marginal vegetation or overhanging branches. In winter, on bigger lakes and more open shores.

Seen in the UK
| J | F | M | A | M | J | J | A | S | O | N | D |

Length **36–38cm (14–15in)**	Wingspan **70–80cm (28–32in)**	Weight **600–900g (21–32oz)**
Social **Large winter flocks**	Lifespan **Up to 15 years**	Status **Secure**

CRANES AND BUSTARDS

CRANES ARE tall, upstanding, long-striding birds, heron-like but with smaller bills and thicker necks that broaden into the shoulders. They have dramatic courtship displays, with elegant "dancing" and trumpeting calls. They migrate to southern Europe and Africa each autumn, flying majestically in long lines or V-shaped flocks, and use traditional wintering sites and intermediate resting and feeding areas.

Bustards are threatened birds of dry, open landscapes. The Great Bustard is huge, the Little Bustard pheasant-sized and quick, more duck-like, in flight. They are unable to survive in intensively farmed countryside and, already much reduced, face further declines.

CHARISMATIC BIRD
The European Crane is well-known in the north, where flocks appear in spring before pairs disperse to nest, often making the air ring with their evocative calls.

GRUIDAE

OTIDIDAE

thick pale lower neck and chest

black neck with white "V" and broad white collar

LITTLE BUSTARD
p.171

heavily barred rufous upperparts

CRANE
p.170

GREAT BUSTARD
p.172

| Order **Gruiformes** | Family **Gruidae** | Species *Grus grus* |

Crane

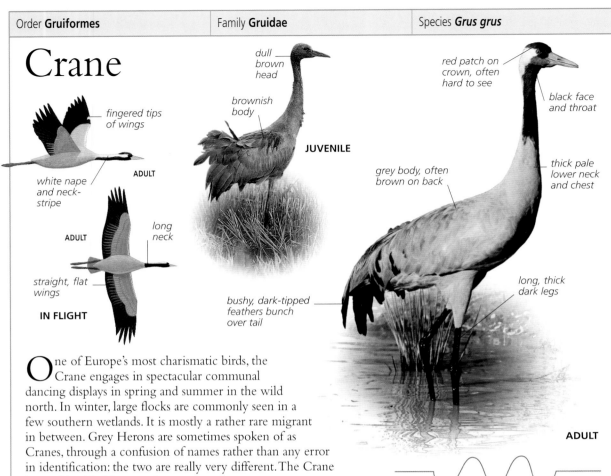

fingered tips of wings

white nape and neck-stripe

ADULT

ADULT

long neck

straight, flat wings

IN FLIGHT

dull brown head

brownish body

JUVENILE

red patch on crown, often hard to see

black face and throat

grey body, often brown on back

thick pale lower neck and chest

long, thick dark legs

bushy, dark-tipped feathers bunch over tail

ADULT

One of Europe's most charismatic birds, the Crane engages in spectacular communal dancing displays in spring and summer in the wild north. In winter, large flocks are commonly seen in a few southern wetlands. It is mostly a rather rare migrant in between. Grey Herons are sometimes spoken of as Cranes, through a confusion of names rather than any error in identification: the two are really very different. The Crane is considerably bigger and more dramatic than a heron.

VOICE Loud, deep, clanging *krro*; in spring, bugling notes as pairs display.

NESTING Big, rough mound of stalks and leaves on ground on which bird crouches, hard to see; 2 eggs; 1 brood; May–July.

FEEDING Strides majestically over ground, digging up roots, grain, and insect larvae; eats acorns in winter.

FLYING IN A GROUP
Crane flocks fly in lines, "V"s, and irregular packs.

FLIGHT: strong, direct, with head and legs outstretched, wings held straight and flat; shallow beats between short glides.

GROUP DISPLAYING
Large groups gather in spring to display, with graceful, rhythmic leaps and bows and loud trumpeting calls.

SIMILAR SPECIES

GREY HERON
see p.129

greyer outer wing

long white head plumes

smaller

neck curled back in flight

DEMOISELLE CRANE
escapee; see p.453

OCCURRENCE
Breeds in N Europe on remote bogs within forests or on wide, reedy marshes with little disturbance. Migrants on open ground near coasts. In winter, in rolling uplands, cork oak, and around large, boggy lakes in SW Europe.

| Seen in the UK |
| J F M A M J J A S O N D |

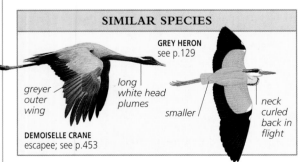

| Length **0.96–1.19m (3–4ft)** | Wingspan **1.8–2.22m (6–7¼ft)** | Weight **4.5–6kg (10–13lb)** |
| Social **Large winter flocks** | Lifespan **Up to 20 years** | Status **Vulnerable** |

| Order **Gruiformes** | Family **Otitidae** | Species ***Tetrax tetrax*** |

Little Bustard

fingered black wingtips

MALE

big white wing patches

IN FLIGHT

pheasant-like head and neck

barred back

spotted breast

FEMALE

small head

neck feathers inflated in display

mottled sandy upperparts

black neck with white "V" and broad white collar

white belly

long legs: the Little Bustard sinks into vegetation when alarmed, stands again when relaxed

MALE

Although it resembles the Great Bustard in shape and general appearance, the Little Bustard is smaller in size, and is able to fly low and fast, whirring almost like a big pigeon or partridge. It can be very difficult to see except in flight, occupying wide open plains but keeping to sufficient cover in which to hide itself expertly. Its size may be hard to judge and it may look larger than expected. The Little Bustard has declined in many areas in the face of agricultural intensification and is further threatened by irrigation schemes.

VOICE Male repeats short gruff note repeated every 10 seconds, in spring; otherwise very quiet except for noise from wings in flight. Female has low chuckling note.

NESTING Scrape on ground in thick cover; 3–5 eggs; 1 brood; April–June.

FEEDING Picks seeds, grain, shoots, buds, roots, and various insects from ground.

MALE DISPLAYING
In display, the male raises his head and neck feathers to show off the black and white pattern on the neck.

FLIGHT: fast, direct, partridge-like, with quick wingbeats and short glides with wings stiffly arched.

WINTER FLOCKS
Large groups gather outside the breeding season, flying fast over open, grassy plains.

SIMILAR SPECIES

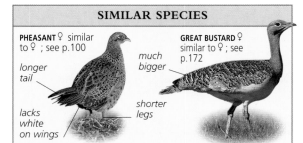

PHEASANT ♀ similar to ♀; see p.100

longer tail

lacks white on wings

GREAT BUSTARD ♀ similar to ♀; see p.172

much bigger

shorter legs

OCCURRENCE
Breeds in areas with open grass or cereals on rolling plains, often dry, stony places, in France (summer), Spain, Portugal, and Sardinia; local in Italy and Balkans. Rare vagrant outside usual breeding areas.

| Seen in the UK |
| J F M A M J J A S O N D |

| Length **40–45cm (16–18in)** | Wingspan **83–91cm (33–36in)** | Weight **600–900g (21–32oz)** |
| Social **Winter flocks** | Lifespan **Up to 10 years** | Status **Vulnerable** |

Order **Gruiformes**	Family **Otididae**	Species *Otis tarda*

Great Bustard

extensive white on wings

IN FLIGHT

MALE

short, broad tail

heavily barred rufous upperparts

grey head and neck

short, strong bill

rufous breast

black and rufous bars above

slim head and neck

FEMALE

MALE

This is one of the world's heaviest flying birds, the males being massive, heavy-bodied, thick-legged, strong-billed birds, and the females markedly smaller. They live in semi-natural steppe and remote areas of cereal cultivation, but agricultural intensification and irrigation threaten their future survival. Small groups are shy and easily disturbed, flying off powerfully with deep, slow wingbeats, revealing extensive areas of white. They are very rare outside their regular range, appearing at long intervals and quite erratically on open farmland in northwest Europe in winter or early spring.

VOICE Mostly silent.
NESTING Unlined scrape in soil; 2 or 3 eggs; 1 brood; April–June.
FEEDING Takes small rodents, reptiles, amphibians, and insects from ground.

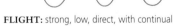

FLIGHT: strong, low, direct, with continual powerful wingbeats.

DISPLAY
A displaying male Great Bustard is a remarkable sight, turning his wings over to reveal large areas of white.

OCCURRENCE
Very local in Spain, Portugal, and E Europe; resident on open plains with dry grass or cereals in undisturbed areas, typically in areas with extensive views all round. Very rare vagrant elsewhere.

Seen in the UK
J F M A M J J A S O N D

SIMILAR SPECIES

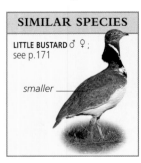

LITTLE BUSTARD ♂ ♀;
see p.171

smaller

TAKING OFF
The distinctive white underwings with black tips of the Great Bustard are clearly visible at take-off.

Length **90–105cm (35–41in)**	Wingspan **2.1–2.4m (7–7³/₄ft)**	Weight **8–16kg (18–35lb)**
Social **Small flocks**	Lifespan **15–20 years**	Status **Declining**

Families **Burhinidae, Recurvirostridae, Haematopodidae, Charadriidae, Scolopacidae, Glareolidae**

WADERS

CALLED SHOREBIRDS IN North America, and waders in Europe, some live far from any shore and several rarely wade. They are mostly long-legged but vary from short-billed to very long-billed, their beaks straight, curved down, or curved upwards. Some are among the world's longest-distance migrants. This large group includes the Oystercatcher, avocets, stilts, plovers, pratincoles, sandpipers, godwits, and curlews.

GROUPS

Plovers are short-billed birds: this group includes broad-winged lapwings and sharp-winged plovers, some "ringed" with black and white head and chest patterns. Pratincoles are plover-like but especially agile in flight. Long-legged, Avocets and stilts feed in shallow water. Small sandpipers can be abundant, flying in large flocks. Some feed on rocky shores, some on sand, others on mud or shallow water. Medium-sized sandpipers have longer legs and bills, are less gregarious, and have loud calls and striking

STUNNING FLOCKS
High-tide roosts bring waders together in dense packs, which make a spectacular sight when they take flight.

patterns in flight. Larger godwits have bright summer plumages, while curlews are much bigger with no clear differences according to age, sex, or season.

BURHINIDAE

long, dark-streaked, sandy brown body

RECURVIROSTRIDAE

black cap; upcurved bill

extremely long, dark pink legs

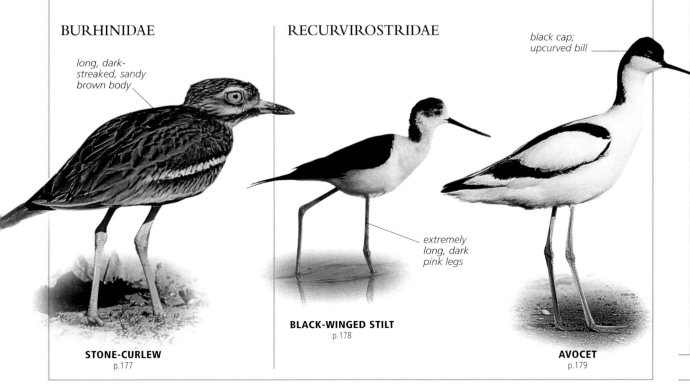

STONE-CURLEW
p.177

BLACK-WINGED STILT
p.178

AVOCET
p.179

Families **Burhinidae, Recurvirostridae, Haematopodidae, Charadriidae, Scolopacidae, Glareolidae**

HAEMATOPODIDAE

vivid orange-red bill

OYSTERCATCHER
p.180

CHARADRIIDAE

yellow, white, and black spangled parts

bold white band from forehead to side of chest

GOLDEN PLOVER
p.181

GREY PLOVER
p.182

SCOLOPACIDAE

streaked, Curlew-like pattern on body

long, evenly downcurved bill

WHIMBREL
p.188

CURLEW
p.189

pale grey back (can look dark on glistening mudflat)

bright ochre-buff on breast and head

dark back with even, pale buff scales

yellow legs

KNOT
p.193

RUFF
p.194

CURLEW SANDPIPER
p.195

TEMMINCK'S STINT
p.196

black cap extends into wispy crest

bright yellow eye ring

LITTLE RINGED PLOVER
p.184

orange legs brighter than Little Ringed Plover

RINGED PLOVER
p.185

broad white stripe over eye to nape

broken breast band

LAPWING
p.183

KENTISH PLOVER
p.186

DOTTEREL
p.187

long, fine-tipped, faintly upcurved bill

strongly patterned black and white head

very long legs

BLACK-TAILED GODWIT
p.190

BAR-TAILED GODWIT
p.191

TURNSTONE
p.192

long, tapered black bill, slightly curved

whitish and rufous edges to feathers

cream lines on back

jet black legs

SANDERLING
p.197

DUNLIN
p.198

PURPLE SANDPIPER
p.199

LITTLE STINT
p.200

Families **Burhinidae, Recurvirostridae, Haematopodidae, Charadriidae, Scolopacidae, Glareolidae**

SCOLOPACIDAE *continued*

bright red neck (less red on male)

RED-NECKED PHALAROPE
p.201

pearly grey back

GREY PHALAROPE
p.202

whitish crescent in front of closed wing

COMMON SANDPIPER
p.203

white-speckled dark grey-brown underparts

GREEN SANDPIPER
p.204

fine, long, red-based black bill

SPOTTED REDSHANK
p.205

long, grey-green legs

GREENSHANK
p.206

extremely long, greenish legs

MARSH SANDPIPER
p.207

cream-spotted brown back

WOOD SANDPIPER
p.208

dark brown; red legs

REDSHANK
p.209

striped crown with black centre

JACK SNIPE
p.210

dead-leaf pattern on upperparts

WOODCOCK
p.211

extremely long bill, angled down

SNIPE
p.212

GLAREOLIDAE

black line through eye extends around pale bib

COLLARED PRATINCOLE
p.213

176

Order **Charadriiformes**	Family **Burhinidae**	Species **Burhinus oedicnemus**

Stone-curlew

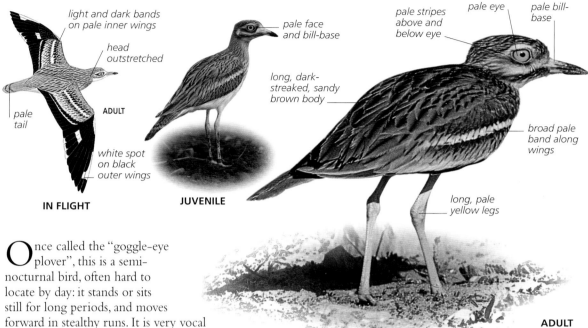

light and dark bands on pale inner wings

head outstretched

pale tail

ADULT

white spot on black outer wings

IN FLIGHT

pale face and bill-base

long, dark-streaked, sandy brown body

JUVENILE

pale stripes above and below eye

pale eye

pale bill-base

broad pale band along wings

long, pale yellow legs

ADULT

O nce called the "goggle-eye plover", this is a semi-nocturnal bird, often hard to locate by day: it stands or sits still for long periods, and moves forward in stealthy runs. It is very vocal in summer, and eerie, wild sounds are created as birds communicate over long distances. This bird does not cope well with modern development. Coastal populations, especially, have largely gone, and it is only liaison between conservationists and sympathetic farmers that has helped it to survive in some regions; it struggles on in disturbed heathland and dunes elsewhere.
VOICE Loud notes recall Curlew and Oystercatcher, but with wild, wailing quality at times; *kur-li, klip*, piping *keeee, krr-leee*, and variations.
NESTING Shallow scrape on ground lined with shells, stones, and rabbit droppings; 2 eggs; 1 or 2 broods; April–August.
FEEDING Tilts forwards, plover-like, to pick up beetles, worms, snails, frogs, lizards, and mice.

FLIGHT: usually low, fast, strong, with steady wingbeats and long glides.

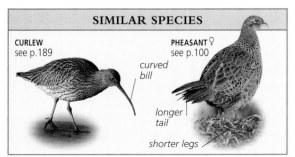

CAMOUFLAGED
Unless it moves, a sitting Stone-curlew is extremely difficult to see.

COURTING PAIR
The white patches on the wings and under the tail are revealed in courtship displays and confrontations.

SIMILAR SPECIES

CURLEW
see p.189

curved bill

PHEASANT ♀
see p.100

longer tail

shorter legs

OCCURRENCE
Summer visitor to S Great Britain, France, Spain, Portugal, and Mediterranean area; breeds on heaths, shingle, cereal and arable fields with light, stony soil and sparse spring crops. Reduced numbers in SW Europe in winter.

Seen in the UK
J F M **A M J J A S O** N D

Length **40–45cm (16–18in)**	Wingspan **77–85cm (30–34in)**	Weight **370–450g (13–16oz)**
Social **Autumn flocks**	Lifespan **Up to 10 years**	Status **Vulnerable**

| Order **Charadriiformes** | Family **Recurvirostridae** | Species *Himantopus himantopus* |

Black-winged Stilt

variable black or grey markings on head

needle-like bill

black upperparts glossed dark green

white neck

trailing, often crossed, legs

long white "V" on back from white tail

ADULT (WINTER)

pointed wings

IN FLIGHT

extremely long, dark pink legs

white underparts

white head

ADULT (WINTER)

pale edges to feathers on back

ADULT (SUMMER)

JUVENILE

In terms of leg length relative to size, the Black-winged Stilt represents the peak of development in the waders. The remarkable long legs enable the bird to wade out into deep waters; however, it picks its food from the water surface. This distinctive and elegant bird, one of Europe's most beautiful species, generally occurs in the Mediterranean region, with an extension northwards in France; it is typically associated with hot, open saltpans and coastal lagoons shimmering in the heat. It has some obvious similarities to its relative, the Avocet, but is essentially unique in Europe.

VOICE Noisy in summer, with strident, rasping *kyik kyik kyik* or *kreeek kreeek;* quiet in winter.

NESTING Shallow hollow in mud or sand, often on small islands in shallow water, lined with some grass or leaves; 3 or 4 eggs; 1 brood; April–June.

FEEDING Picks insects from wet mud, stems, and water surface, either tilting well forward or wading out into deeper water.

FLIGHT: strong, direct, quick, with long legs trailed (often crossed), pointed wings flicked in quite shallow beats; glides in wind.

MIGRANT FLOCKS
Before spreading out to pair and nest, flocks of Black-winged Stilts roost together in the shallows.

SIMILAR SPECIES

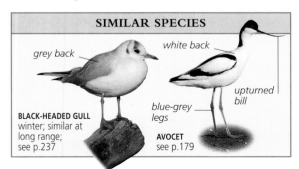

grey back

white back

upturned bill

BLACK-HEADED GULL
winter; similar at long range; see p.237

blue-grey legs

AVOCET
see p.179

OCCURRENCE
In Spain, Portugal, Mediterranean region, and W and N France; very rare vagrant farther north. Frequents shallow reedy pools, flooded fields and rice paddies, salt pans, and coastal lagoons, less commonly on sheltered estuaries.

| Seen in the UK |
| J F M A M J J A S O N D |

| Length **33–36cm (13–14in)** | Wingspan **70cm (28in)** | Weight **250–300g (9–11oz)** |
| Social **Family groups** | Lifespan **Up to 10 years** | Status **Secure** |

| Order **Charadriiformes** | Family **Recurvirostridae** | Species ***Recurvirostra avosetta*** |

Avocet

diagonal bar over inner wings

ADULT

squared black wingtips (larger on male)

IN FLIGHT

black cap

slim, bright white body

curved black band on each side of back

very fine, upcurved bill

leans forward while feeding

brown tips to feathers

JUVENILE

long blue-grey legs

ADULT

ADULT

ADULT

FLIGHT: quick, rather stiff; fast wingbeats; often in irregular packs.

If its special needs are met – essentially shallow slightly saline water and oozy mud, with drier islands – the Avocet may nest in quite large, loose, widely scattered colonies, not tight-packed like gulls or terns. However, it does form tightly packed flocks in winter. Flocks jostle shoulder-to-shoulder when feeding in a shallow tidal flow. With protection and management of habitats, Avocets have increased and spread in recent years.

VOICE Loud, fluty, somewhat Bee-eater-like (see p.272) *klute* or *kloop*.

NESTING Scrape on low islands or dry mud, bare or lined with grass and shell fragments; 3 or 4 eggs; 1 brood; April–July.

FEEDING Sweeps curved bill sideways through water to locate tiny shrimps and marine worms.

TIGHT, ELEGANT FLOCKS
In winter, Avocets form tight-packed, synchronized flocks, flying and feeding together in elegant groups.

SIMILAR SPECIES

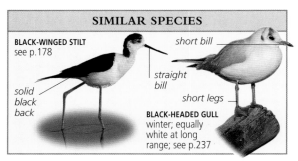

BLACK-WINGED STILT see p.178

solid black back

short bill

straight bill

short legs

BLACK-HEADED GULL winter; equally white at long range; see p.237

OCCURRENCE
Mostly found in S Baltic and North Sea coasts, Mediterranean area, and also SW Great Britain. Breeds on shallow, saline coastal lagoons and near muddy pools, at times on bare ground around ponds. In winter, on muddy estuaries.

Seen in the UK
J F M A M J J A S O N D

| Length **42–46cm (16½–18in)** | Wingspan **67–77cm (26–30in)** | Weight **250–400g (9–14oz)** |
| Social **Winter flocks** | Lifespan **10–15 years** | Status **Localized** |

Oystercatcher 22.I, 22.II

red eye

bulky, striking black and white body

vivid orange-red bill

IMMATURE

dark-tipped bill

long, broad white wingbar

white "V" on back

ADULT (SUMMER)

IN FLIGHT

white collar

dark-tipped bill

ADULT (SUMMER)

short, pale pink legs

ADULT (WINTER)

The dazzlingly patterned Oystercatcher is an extraordinarily distinctive bird in Europe, no other bird forming such tight-packed, often enormous, noisy flocks. Oystercatchers tend to "take over" and dominate whole estuaries with their clamorous presence. In some places, they have come into conflict with people for their supposed impact on commercial cockle fisheries. As cockles have declined on some estuaries, Oystercatchers have increasingly turned to farmland to feed.

VOICE Loud, strident *klip* or *kleep*, develops into penetrating *kleep-a-kleep, kleep-a-kleep*; shrill chorus from large flocks.

NESTING Shallow scrape in shingle or sand, often amongst rocks or grassy tussocks; 2 or 3 eggs; 1 brood; April–July.

FEEDING Probes for large marine worms and molluscs and prises shellfish from rocks and seaweed; also eats earthworms.

FLIGHT: fast, direct, on rapid wingbeats; flies into standing flocks, landing "on the run".

LARGE, NOISY FLOCKS
Oystercatchers feed in huge groups, and make the area ring with their ear-splitting chorus; they roost in tight flocks.

OCCURRENCE
Breeds on sandy, muddy, and rocky beaches, grassy islands, riverside grassland or shingle, and grassy fields along northern river valleys. Also found, at any time of year, on coasts. Rare migrant inland south of breeding areas.

Seen in the UK
J F M A M J J A S O N D

SIMILAR SPECIES

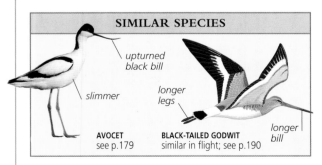

upturned black bill

slimmer

AVOCET
see p.179

longer legs

longer bill

BLACK-TAILED GODWIT
similar in flight; see p.190

Length **40–45cm (16–18in)**	Wingspan **80–85cm (32–34in)**	Weight **400–700g (14–25oz)**
Social **Large flocks**	Lifespan **Up to 15 years**	Status **Secure**

Order **Charadriiformes**	Family **Charadriidae**	Species *Pluvialis apricaria*

Golden Plover 🔊 25.I, 25.II

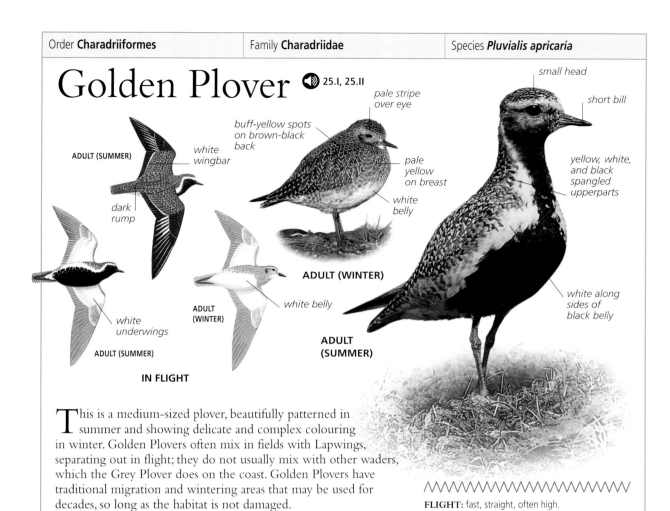

ADULT (SUMMER)

white wingbar

dark rump

pale stripe over eye

buff-yellow spots on brown-black back

pale yellow on breast

white belly

ADULT (WINTER)

small head

short bill

yellow, white, and black spangled upperparts

white along sides of black belly

white underwings

ADULT (SUMMER)

ADULT (WINTER)

white belly

ADULT (SUMMER)

IN FLIGHT

This is a medium-sized plover, beautifully patterned in summer and showing delicate and complex colouring in winter. Golden Plovers often mix in fields with Lapwings, separating out in flight; they do not usually mix with other waders, which the Grey Plover does on the coast. Golden Plovers have traditional migration and wintering areas that may be used for decades, so long as the habitat is not damaged.

FLIGHT: fast, straight, often high.

VOICE Plaintive, whistled *tleee*, higher *tlee, treeoleee*, and variants; *phee-oo, pheee-oo* in song-flight.

NESTING Shallow scrape, lined with scraps of lichen and heather, on ground in heather, grass, or bilberry, often in burned areas; 4 eggs; 1 brood; April–July.

FEEDING Takes variety of insects in summer, mostly earthworms in winter, often stolen by gulls.

SUBSPECIES

bolder black face and breast

white band

P. a. altifrons (N Europe)

HIGH-FLYING FLOCK
Flocks of Golden Plovers often fly high, stringing out into long lines or in irregular packs.

OCCURRENCE
Breeds in N Europe, on high moorland or northern tundra, both on limestone grassland and acid heath with patches of burnt heather or bilberry. Widespread in winter on low-lying arable fields and pastures, coastal salt marsh, sometimes on estuary mud.

Seen in the UK
J F M A M J J A S O N D

RESTING FLOCK
When feeding, flocks spread evenly over fields, tightening into packs if alarmed or roosting.

SIMILAR SPECIES

silvery grey above

GREY PLOVER
summer, similar to adult summer; see p.182

blacker below

Length **26–29cm (10–11½in)**	Wingspan **67–76cm (26–30in)**	Weight **140–250g (5–9oz)**
Social **Winter flocks**	Lifespan **Up to 10 years**	Status **Secure**

Order **Charadriiformes**	Family **Charadriidae**	Species *Pluvialis squatarola*

Grey Plover 🔊 24

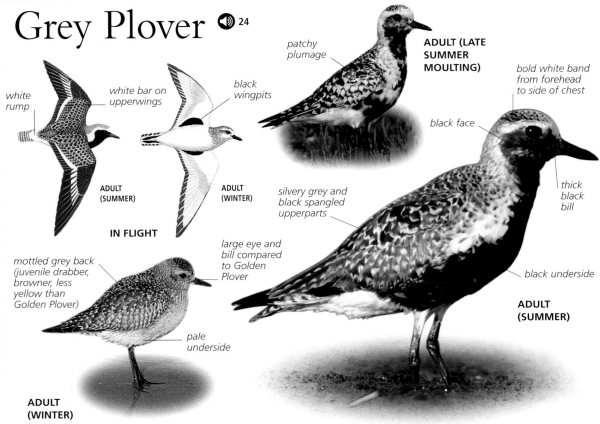

patchy plumage

ADULT (LATE SUMMER MOULTING)

white rump

white bar on upperwings

black wingpits

bold white band from forehead to side of chest

black face

silvery grey and black spangled upperparts

thick black bill

ADULT (SUMMER)

ADULT (WINTER)

IN FLIGHT

large eye and bill compared to Golden Plover

black underside

mottled grey back (juvenile drabber, browner, less yellow than Golden Plover)

pale underside

ADULT (WINTER)

ADULT (SUMMER)

This is primarily a coastal bird, scattered over mudflats when it is feeding and gathering in rather static flocks at high tide, unlike the large, mobile flocks of Golden Plovers, which prefer grassy fields or saltmarshes. It usually mingles with godwits, Curlews, and Redshanks. Grey Plovers are easy to identify, but some are yellow and may be confused with Golden Plovers. At long range, they can be mistaken for other medium-sized waders; close up they look pale, but far out on the mud they can look remarkably dark in winter plumage.

VOICE High, plaintive *twee-oo-wee!*; also loud, melancholy, fluted song. Unique triple-note call carries across shorelines.

NESTING Scrape on ground in short vegetation, usually on dry rises; 4 eggs; 1 brood; May–July.

FEEDING Pulls worms, molluscs, and crustaceans from mud, in winter; eats mainly insects in summer in Arctic tundra.

FLIGHT: quick, with deep wingbeats; sometimes quite active, twisting descent to roost.

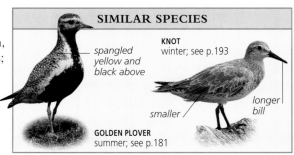

HIGH-TIDE ROOST
High tide forces dispersed feeding Grey Plovers to gather together in more compact flocks.

SIMILAR SPECIES

spangled yellow and black above

GOLDEN PLOVER
summer; see p.181

KNOT
winter; see p.193

smaller

longer bill

OCCURRENCE
Breeds on northern tundra. Found on large muddy estuaries, or on sandy or rocky shores, from autumn to spring. Flocks may roost on adjacent pasture, or shallow coastal lagoons. In winter on seashores. Rare bird inland.

Seen in the UK
J F M A M J J A S O N D

Length **27–30cm (10½–12in)**	Wingspan **71–83cm (28–33in)**	Weight **200–250g (7–9oz)**
Social **Winter flocks**	Lifespan **Up to 10 years**	Status **Secure†**

| Order **Charadriiformes** | Family **Charadriidae** | Species ***Vanellus vanellus*** |

Lapwing 🔊 26.I, 26.II

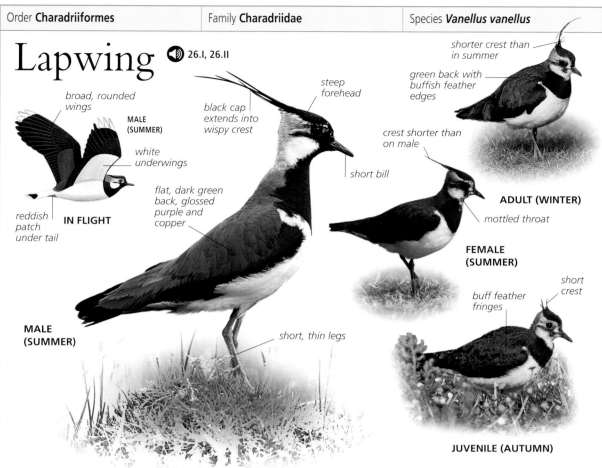

MALE (SUMMER)

broad, rounded wings

white underwings

reddish patch under tail

IN FLIGHT

black cap extends into wispy crest

steep forehead

short bill

flat, dark green back, glossed purple and copper

short, thin legs

shorter crest than in summer

green back with buffish feather edges

ADULT (WINTER)

crest shorter than on male

mottled throat

FEMALE (SUMMER)

buff feather fringes

short crest

JUVENILE (AUTUMN)

A familiar and much-loved part of the farmed countryside in Europe, the distinctive-looking Lapwing is sadly declining in most areas as farming systems change. It breeds in loose colonies scattered over suitable fields or moors, but gathers into flocks for the rest of the year, often mixed with Golden Plovers and Black-headed Gulls (see p.237).

VOICE Nasal, strained *weet* or *ee-wit*; wheezy variations on this theme; passionate song in spring, *whee-er-ee, a wheep-wheep!* accompanied by loud throbbing from wings.

NESTING Grass-lined shallow hollow on ground; 3 or 4 eggs; 1 brood; April–June.

FEEDING Often taps foot on ground to attract or reveal prey; tilts forwards to pick insects and spiders from ground, or pull earthworms from soil.

FLIGHT: unique flappy flight with steady beats of broad, round wings.

ROOSTING FLOCK
Flocks rest in tight groups; otherwise, they tend to be loosely scattered.

FLYING FLOCK
Flocks of Lapwings fly in lines, "V"s, or irregular masses, rising steadily as a group, often circling and returning.

OCCURRENCE
Breeds on wet moors, riverside pastures, upland fields, and farmland (decreasing), almost throughout Europe. In winter, moves south and west, feeds on arable fields, meadows, salt marsh and muddy reservoir edges; in estuaries in hard weather.

Seen in the UK
| J | F | M | A | M | J | J | A | S | O | N | D |

| Length **28–31cm (11–12in)** | Wingspan **70–76cm (28–30in)** | Weight **150–300g (5–11oz)** |
| Social **Winter flocks** | Lifespan **Up to 10 years** | Status **Secure†** |

| Order **Charadriiformes** | Family **Charadriidae** | Species *Charadrius dubius* |

Little Ringed Plover

plain wings (Ringed Plover has white wingbar)

small, banded head

ADULT

IN FLIGHT

indistinct pale area over eye

black bill

broken band

dull legs

JUVENILE

white line between brown crown and black forehead band

bright yellow eye-ring

sandy brown upperparts

long, tapered wingtips

stubby black bill

narrow black breast-band

dull pink legs

clean white underside

ADULT

Small and neat, precise in its movements as well as its appearance, this is a handsome little wader of freshwater shorelines and a variety of dry, rough, open spaces in the "waste ground" category. It is often at the waterside in spring and autumn, but as likely to breed on a patch of bulldozed rubble or coalmining waste. It tends to be irregular in occurrence, breeding for a few years and then moving on.

VOICE Short, abrupt, whistled *piu* or *p'ew*; song rolling, harsh *cree-cree-cree-cree* draws attention in flight.

NESTING Hollow in bare ground, usually hard to spot; 4 eggs; 1 brood; April–June.

FEEDING Stands upright, then runs forwards and tilts to pick insects and small aquatic invertebrates from ground.

FLIGHT: quick, low, direct with angled, pointed wings; song-flight rolling, bat-like.

SPRING DISPLAY
Noisy pairs display on the ground with drooped wings, and males also perform long, low song-flights over the territory.

OCCURRENCE
Widespread except in extreme north. Breeds in wide variety of natural, semi-natural, and derelict places, from sandy and shingly shores and gravel to flat, dry areas of waste ground, mining waste, and shingly riverbeds; scarce on sea coasts but occasional migrant on coastal lagoons.

Seen in the UK
J F **M A M J J A S O** N D

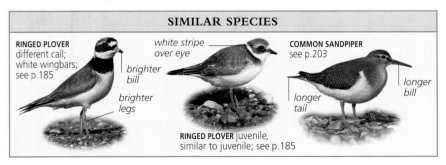

SIMILAR SPECIES

RINGED PLOVER different call; white wingbars; see p.185

brighter bill

brighter legs

white stripe over eye

RINGED PLOVER juvenile, similar to juvenile; see p.185

longer tail

COMMON SANDPIPER see p.203

longer bill

| Length **14–15cm (5¹/₂–6in)** | Wingspan **42–48cm (16¹/₂–19in)** | Weight **30–50g (1¹/₁₆–1³/₄ oz)** |
| Social **Winter flocks** | Lifespan **5–10 years** | Status **Secure†** |

Order **Charadriiformes**	Family **Charadriidae**	Species ***Charadrius hiaticula***

Ringed Plover 🔊 23

IN FLIGHT

- white tail sides
- bold white wingbar
- dark tail tip

ADULT (SUMMER)

- white stripe over eye
- short, black and orange bill
- black and white ringed head and throat
- pale sandy brown upperparts
- broad breast-band
- clean white underparts
- orange legs brighter than Little Ringed Plover

ADULT (SUMMER)

- weaker breast-band
- dull bill

ADULT (WINTER)

- white over eye
- dull head
- incomplete breast-band
- dull legs

JUVENILE

∧∧∧∧∧∧∧∧∧∧∧∧∧∧∧∧∧

FLIGHT: strong, fast, direct; shallow beats of angled wings; bat-like song-flight.

Along with the Dunlin, this species is one of the common "standards" by which other waders may be judged. There are several other "ringed" plovers with similarly patterned heads but none in Europe with such brightly coloured bill and legs. It is generally a coastal bird, although it moves inland, sometimes in places frequented by the Little Ringed Plover. In spring and autumn especially, substantial numbers may appear inland where conditions are right, sometimes pausing on migration for several days. It forms tightly packed flocks at high tide, often mixed with other waders. Usually, a large, tight flock will be more or less clearly separated, with each species bunched together within it.
VOICE Characteristic fluty whistle, a bright, mellow *too-li*; also sharp *queep*; repeated *too-wee-a too-wee-a* in song-flight. Rich double-note call unlike that of Little Ringed Plover.
NESTING Shallow scrape in sand or stones, lined with pebbles and grass stems; 4 eggs; 2 or 3 broods; April–August.
FEEDING Picks small insects and worms from ground.

"BROKEN WING LOOK"
If a predator threatens the nest or chicks, the parent plover feigns injury to lead it away.

SIMILAR SPECIES

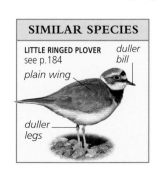

LITTLE RINGED PLOVER see p.184
- duller bill
- plain wing
- duller legs

OCCURRENCE
Breeds on sand and shingle beaches, near gravel pits inland. Found at any time of year mostly on broad beaches, including estuaries, of all kinds, but fewest on rocky shores. Widespread migrant inland and on coasts.

Seen in the UK
J F M A M J J A S O N D

Length **17–19cm (6¹/₂–7¹/₂in)**	Wingspan **48–57cm (19–22¹/₂in)**	Weight **55–75g (2–2⁵/₈oz)**
Social **Winter flocks**	Lifespan **5–10 years**	Status **Secure**

| Order **Charadriiformes** | Family **Charadriidae** | Species *Charadrius alexandrinus* |

Kentish Plover

swept-back, pointed wings

long white wingbar

MALE (SUMMER)

IN FLIGHT

earth brown upperparts

rufous-ginger cap (dull in winter)

black bar on forehead

JUVENILE

dark legs

white forehead

short black bill

black markings on sides of breast (browner in winter) never meet across chest

clean white underparts

dark legs

dusky eye patch

brown chest patch

clear white breast

dark legs

FEMALE

MALE (SUMMER)

Long gone from Kent except as an occasional visitor, the Kentish Plover is still quite widespread just across the English Channel, but is commonest around the Mediterranean. It prefers sandy places, such as the embankments and waste areas around salt pans and behind beaches, even around building sites near the shore. Rare migrants farther north are generally found within Ringed Plover flocks and their identification requires careful observation, especially when juveniles are about in late summer.

VOICE Short, sharp, whistled whip, whistled *bew-ip*; rolled trilling notes.

NESTING Shallow hollow in sand, lined with pebbles or shell fragments; 3 or 4 eggs; 2 broods; March–July.

FEEDING Takes small invertebrates such as flies and sandhoppers, from ground, tilting forward after short run in typical stop–start plover action.

FLIGHT: quick and dashing, on swept-back, pointed wings; glides in to land.

PALE BEACH PLOVER
A spring male has almost entirely white underparts, with small chest marks. Its short dashes on the beach give it a lively character.

SIMILAR SPECIES

RINGED PLOVER juvenile; see p.185

pale legs

LITTLE RINGED PLOVER juvenile; see p.184

no wing-stripe

pale legs

OCCURRENCE
Found mostly on sandy areas near shores, also beside freshwater lagoons and flooded areas of waste ground, on S North Sea and Channel coasts, W France, and Mediterranean area. Migrants rare on estuaries or inland waters.

| Seen in the UK |
| J F M A M J J A S O N D |

| Length **15–17cm (6–6¹/₂in)** | Wingspan **50cm (20in)** | Weight **40–60g (1⁷/₁₆–2¹/₈oz)** |
| Social **Small flocks** | Lifespan **10 years** | Status **Declining** |

Order **Charadriiformes**	Family **Charadriidae**	Species **Charadrius morinellus**

Dotterel

ADULT (SUMMER)

plain wings

IN FLIGHT

dull plumage (paler in winter)

MALE (SUMMER)

dark stripe through eye

black cap

broad white stripe over eye to nape

face pattern less sharp than female's

thin black and broad white bands around chest

duller underside than female's

black belly (white in winter)

pale "V" from over eyes to nape

black, buff, and apricot markings on upperparts

JUVENILE (AUTUMN)

white throat

brighter plumage than male's

rich rust-red underside with blackish belly

FEMALE (SUMMER)

One of the few European birds with reversed sexual roles, and females larger and brighter than males, the Dotterel is a mountain-top or tundra breeder that appears in small flocks at regular places in the lowlands, usually cereal fields, on migration. It is famously tame, almost recklessly so at times, and may be attracted to within a metre or two (3–7ft) by a whistled imitation of its call. Its future range may well diminish with the increasing effects of climate change.

VOICE Soft *pip pip* or sweet *wit-ee-wee*; rather silent outside breeding season.

NESTING Shallow scraped hollow on ground, usually under cover of low vegetation; 3 eggs; 1 brood; May–August.

FEEDING Eats flies, beetles, earthworms, spiders, and similar small terrestrial creatures, tilting forwards in typical plover fashion.

FLIGHT: quick, agile, with fast, deep wingbeats.

SIMILAR SPECIES

GOLDEN PLOVER
winter; see p.181

no white "V" over eye

no breast-band

BREEDING HABITAT
The Dotterel breeds in high, rolling, or flat-topped mountainous regions with low cover, or in tundra.

OCCURRENCE
Occupies wild northern tundra and mountainous areas with similar habitat south to Pyrenees, often with abundance of stones and scree. On migration, in lowland fields in traditional areas inland.

Seen in the UK
J F **M A M J J A S O** N D

Length **20–22cm (8–9in)**	Wingspan **57–64cm (22¹/₂–25in)**	Weight **90–145g (3¹/₄–5oz)**
Social **Small flocks**	Lifespan **5–10 years**	Status **Secure†**

| Order **Charadriiformes** | Family **Scolopacidae** | Species **Numenius phaeopus** |

Whimbrel

two dark bands along crown with narrow central line

bent bill

barred tail

long white "V" on back

deep-chested shape

plain, dark upperwings

streaked, Curlew-like pattern on body

dark breast

white belly

IN FLIGHT

quite short greyish legs

FLIGHT: fast, strong, quicker than Curlew; wingbeats quite quick and deep.

Superficially like the closely related Curlew, which is a more familiar bird all year round in Europe, the Whimbrel is a more northerly breeder and only a spring and autumn migrant elsewhere. This large wader would often be overlooked were it not for its call, although in fact its compact, dark, chunky form is really quite distinct from the lankier, paler Curlew. It is rather more squat and a little larger than the straight-billed godwits. As with most "streaky brown birds", a close view reveals an exquisite pattern of fine streaks, bars, and spots. In Europe, it is very much a ground or waterside bird, but in winter in Africa it often perches up on trees or even overhead cables.

VOICE Song loud, rich, rippling trill; in flight, loud, even, fast *pipipipipipipip* on one pitch.

NESTING Simple, shallow scrape on ground; 4 eggs; 1 brood; May–July.

FEEDING Takes insects, snails, earthworms, crabs, and marine worms.

MIGRANT WADER
For much of the summer, Whimbrels are on dry ground, but migrants may be seen wading at the edges of pools or along the sea shore.

SIMILAR SPECIES

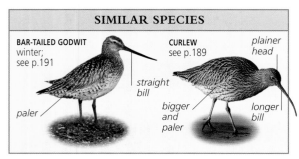

BAR-TAILED GODWIT
winter;
see p.191

straight bill

paler

CURLEW
see p.189

plainer head

bigger and paler

longer bill

OCCURRENCE

Breeds on open heaths and moors in far N and NW Europe; on migration, on many coasts. Flies over almost any open landscapes, especially moving north in spring, but prefers undisturbed estuaries and rarely lingers inland.

| Seen in the UK |
| J F M **A M J J A S O** N D |

| Length **40–46cm (16–18in)** | Wingspan **71–81cm (28–32in)** | Weight **270–450g (10–16oz)** |
| Social **Spring flocks** | Lifespan **10–15 years** | Status **Secure†** |

| Order **Charadriiformes** | Family **Scolopacidae** | Species *Numenius arquata* |

Curlew 🔊 32.I, 32.II

dark-tipped upperwings with pale inner half

broad "V" on rump

IN FLIGHT

often looks dark at distance

whiter belly

streaked brown above

spotted flanks

short greyish legs

head uniform or faintly banded along crown

gull-like shape

long, evenly downcurved bill

A breeding bird in much of Europe, the Curlew is also widespread and common on shorelines and around many inland waters. It is a big, gull-like wader of coasts and moors. Easily recognized both by its shape and its voice, in spring, it has one of the most beautiful of all European bird songs. Curlews at long range on mudflats or roosting on a sand spit tend to look large and in most circumstances rather dark, although close views, or bright sun, reveal a quite pale, sandy-brown colour.

VOICE Typical calls loud, full *whoy, haup, cur-li*, hoarse, throaty *cu-cu-cew*, longer, slow, repeated *cur-lew*; song begins slowly, accelerates into ecstatic, rich, bubbling trill.
NESTING Shallow hollow, lined with grass, on ground; 4 eggs; 1 brood; April–July.
FEEDING Probes and picks up worms, insects, crabs, starfish, and molluscs.

FLIGHT: strong, direct, gull-like, quite slow beats; often in lines or "V"s.

MIXED ROOST
Curlews stand tall beside godwits and other waders at high-tide roosts.

OPPORTUNIST
Curlews are able to use their long bills to feed on wave-washed rocks, as well as on mud.

OCCURRENCE
Breeds widely in N and W Europe, on riverside meadows, bogs in heaths, wet moors, and northern shores and islands. Winters on estuaries, especially larger, muddy ones, but also small creeks, salt marshes, and wet grassland.

Seen in the UK
J F M A M J J A S O N D

SIMILAR SPECIES

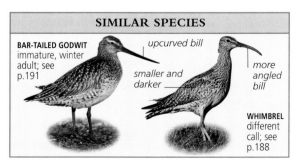

BAR-TAILED GODWIT immature, winter adult; see p.191

upcurved bill

smaller and darker

more angled bill

WHIMBREL different call; see p.188

| Length **50–60cm (20–23½ in)** | Wingspan **80–100cm (32–39in)** | Weight **575–950g (21–34oz)** |
| Social **Winter flocks** | Lifespan **10–20 years** | Status **Declining** |

| Order **Charadriiformes** | Family **Scolopacidae** | Species *Limosa limosa* |

Black-tailed Godwit

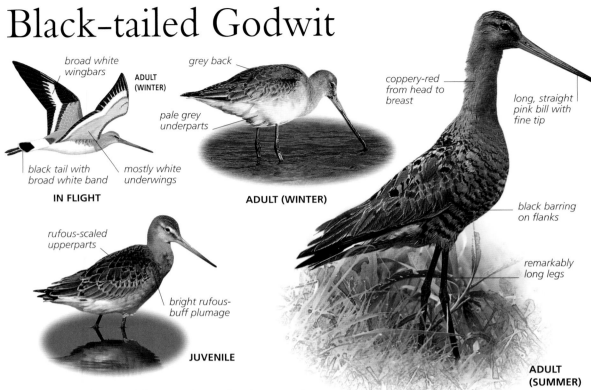

broad white wingbars

ADULT (WINTER)

grey back

pale grey underparts

black tail with broad white band

mostly white underwings

IN FLIGHT

ADULT (WINTER)

coppery-red from head to breast

long, straight pink bill with fine tip

black barring on flanks

remarkably long legs

rufous-scaled upperparts

bright rufous-buff plumage

JUVENILE

ADULT (SUMMER)

This is one of Europe's larger and more handsome waders, boldly patterned in flight (when it is unmistakable) and characterized by especially long legs. It usually stands with its body well forward, bill probing almost at its toes. The Black-tailed Godwit breeds in wet meadowland, where it is susceptible both to drainage and sudden spring floods. In winter, it resorts to relatively few estuaries, often rather narrow and enclosed with long, narrow areas of rich mud; these are occupied year after year. It is generally much less widespread than the Bar-tailed Godwit but may gather in hundreds in traditional wintering places. Spring flocks before migration look stunning in red plumage.

VOICE Noisy in spring with nasal *weeka-weeka-weeka*; quick *vi-vi-vi* in flight.

NESTING Shallow scrapes on ground in rich vegetation; 3 or 4 eggs; 1 brood; May–July.

FEEDING Probes deeply, often in water up to its belly, for worms, molluscs, and seeds.

FLIGHT: quick, direct flight with shallow, stiff wingbeats; head well outstretched, legs trail far beyond tail.

WINTER FLOCKS
From autumn to late winter, Black-tailed Godwits are found in flocks in quite small, sheltered, muddy estuaries.

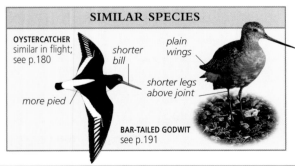

SIMILAR SPECIES

OYSTERCATCHER similar in flight; see p.180

shorter bill

more pied

plain wings

shorter legs above joint

BAR-TAILED GODWIT see p.191

OCCURRENCE
Breeds in N and W Europe, in wet meadowland and flooded pasture; otherwise, mostly coastal. Widely spread except in far N Scandinavia but everywhere localized, even in winter, when most are on traditional, muddy, narrow estuaries.

| Seen in the UK |
| J F M A M J J A S O N D |

| Length **36–44cm (14–17½in)** | Wingspan **62–70cm (24–28in)** | Weight **280–500g (10–18oz)** |
| Social **Winter flocks** | Lifespan **10–15 years** | Status **Vulnerable** |

| Order **Charadriiformes** | Family **Scolopacidae** | Species *Limosa lapponica* |

Bar-tailed Godwit

plain upperwings with darker tips

barred tail

IN FLIGHT

ADULT (WINTER)

streaked bright buff above

warm orange-buff on neck

JUVENILE

much smaller, less hunch-backed than Curlew

streaked grey-brown and buff (adult less chequered)

winter birds more boldly striped through eye than Black-tailed Godwit

long, fine-tipped, faintly upcurved bill

pinkish bill base

pale buff breast

shortish dark legs

deep coppery red underside (female paler)

MALE (SUMMER)

IMMATURE (1ST WINTER)

FLIGHT: quick, agile; legs not trailing much beyond tail; often acrobatic.

While Black-tailed Godwits breed in Europe, Bar-tailed Godwits breed in the far northern tundra, but they are otherwise much more widespread on shores of all kinds. They prefer extensive mudflats, groups scattering over them to probe for food, and are driven at high tide to large, mixed roosts where they tend to keep a little separate from the Curlews, Redshanks, and other species close by. Flocks flying to roost may arrive quite high up and dive down with much acrobatic twisting and rolling.

VOICE In flight, quick, yelping *kirruk kirruk*.

NESTING Small scrape on ground on drier patch in cold tundra; 4 eggs; 1 brood; May–July.

FEEDING Probes for large marine worms and molluscs.

HIGH-TIDE FLURRY
A rising tide pushes a group of godwits off a mud bank, to seek a safe roost on a nearby marsh.

SIMILAR SPECIES

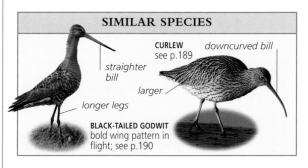

straighter bill

longer legs

CURLEW see p.189   *downcurved bill*

larger

BLACK-TAILED GODWIT bold wing pattern in flight; see p.190

OCCURRENCE
Arctic breeder on tundra; in Europe, mostly in scattered flocks on broad estuaries, but also seen in small numbers on smaller beaches and rocky shores, lingering until May and returning from July onwards.

| Seen in the UK |
| J F M A M J J A S O N D |

| Length **33–42cm (13–16½in)** | Wingspan **61–68cm (24–27in)** | Weight **280–450g (10–16oz)** |
| Social **Winter flocks** | Lifespan **10–15 years** | Status **Vulnerable** |

| Order **Charadriiformes** | Family **Scolopacidae** | Species *Arenaria interpres* |

Turnstone

strongly patterned black and white head

stout, tapered bill

white back

white patch and stripe on wings

black, white, and bright chestnut upperparts

bold black breast-band

ADULT (WINTER)

IN FLIGHT

white underside

irregular whitish marbling on dark head and neck

ADULT (SUMMER)

short, vivid orange legs

dark brown and black upperparts (more buff feather edges on juvenile)

dull brown-black breast

ADULT (WINTER)

While most waders like soft ground, chiefly mud or sand, the Turnstone is equally at home on rocks, although sandy beaches with a tangle of seaweed, shells, and small stones at the high-tide mark are ideal for it. It makes a good living searching through such debris, which is very rich in small invertebrates and regularly refreshed by high tides. Turnstones are typically noisy, active, and often quite tame.

VOICE Fast, hard, staccato calls, *tukatukatuk, teuk, tchik*.

NESTING Scantily lined scrape on ground close to shore in islands and on rocky coasts; 4 eggs; 1 brood; May–July.

FEEDING Stirs up and turns over weed, stones, shells, and beach debris to find invertebrates.

FLIGHT: fast, low, flickering.

TIGHT ROOST
High tide sees scores of Turnstones packed close together for an hour or two.

QUARRELSOME FEEDERS
Small groups of Turnstones bicker as they feed along the shore.

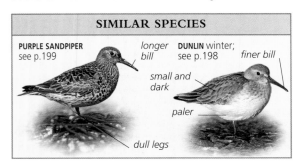

SIMILAR SPECIES

PURPLE SANDPIPER see p.199

longer bill

DUNLIN winter; see p.198

finer bill

small and dark

paler

dull legs

OCCURRENCE
Breeds on rocky coasts around Scandinavia. At other times, on sea coasts of all kinds, from open mud to rocks, but especially hard coasts and gravelly tidelines. Occasional migrants turn up inland but soon move on.

Seen in the UK
J F M A M J J A S O N D

| Length **21–24cm (8½–9½ in)** | Wingspan **44–49cm (17½–19½ in)** | Weight **80–110g (2⅞–3⅝ oz)** |
| Social **Flocks** | Lifespan **Up to 10 years** | Status **Secure** |

Knot

spangled chestnut, black, and buff on back

pale coppery red head

thin pale wingbar

pale coppery red underparts

pale stripe over eye

pale grey back (can look dark on glistening mudflat)

ADULT (WINTER)

IN FLIGHT

pale grey rump and tail

ADULT (SUMMER)

shortish, straight black bill (Grey Plover is short-billed, bigger, and rounder)

whitish belly

short grey legs

ADULT (WINTER)

lacy pattern of dark and light feather edges

apricot-tinged underparts

JUVENILE

Many waders flock together and some make dense packs when they roost at high tide, but few are as social at all times as the Knot. It forms enormous flocks, sometimes totalling hundreds of thousands. Such flocks flying over estuaries, moving to new feeding areas, or perhaps disturbed from a roost, are among the most dramatic of all bird spectacles. The rare solitary Knot is likely to be one of the occasional migrants that turn up near pools and reservoirs inland. In autumn, these may be juveniles and can be exceptionally tame, probably never having seen a human before in their short life. Knot flocks typically swarm over mudflats in slow, steady progession, heads down, feeding avidly.

VOICE Rather quiet; dull, short *nut*, occasionally bright, whistled note; no obvious flight note.

NESTING Shallow hollow on ground in cold tundra, usually near water; 3 or 4 eggs; 1 brood; May–July.

FEEDING Takes insects and plant material in summer, and molluscs and marine worms in winter.

FLIGHT: quick, strong; quite shallow wingbeats; flocks make coordinated movements.

ROOSTING
Knots and Dunlins stand shoulder to shoulder as they wait for the tide to recede.

SIMILAR SPECIES

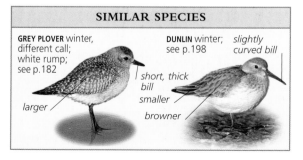

GREY PLOVER winter, different call; white rump; see p.182

larger

DUNLIN winter; see p.198

slightly curved bill

short, thick bill

smaller

browner

OCCURRENCE
Breeds in Arctic tundra. Found in W Europe from late summer to late spring; biggest numbers in winter in dense flocks on large muddy estuaries and in small numbers on wide variety of shorelines.

Seen in the UK
| J | F | M | A | M | J | J | A | S | O | N | D |

Length **23–27cm (9–10½in)**	Wingspan **47–54cm (18½–21½in)**	Weight **125–215g (4–8oz)**
Social **Large flocks**	Lifespan **Up to 10 years**	Status **Localized**

| Order **Charadriiformes** | Family **Scolopacidae** | Species *Philomachus pugnax* |

Ruff

thin white wingbar

white on sides of dark rump

JUVENILE (AUTUMN)

IN FLIGHT

plainer head than Redshank

boldly blotched back

legs may be red

smaller than male

FEMALE (SUMMER)

reddish legs

head often white

pale belly

MALE (WINTER)

short, faintly curved bill

bright ochre-buff on head and breast

male larger than female

white underside (Greenshank whiter below; Redshank browner on chest)

long, pale yellow-ochre legs

curly tufts of variable colours on crown

bright buff edges on dark brown feathers

broad feathery ruff of varying coloration

chequered back

MALE (SPRING)

JUVENILE (AUTUMN)

FLIGHT: rather slow, relaxed action with shallow, soft beats of rather long wings.

Male Ruffs in spring look extraordinary, and the females in summer are boldly blotched; in winter, they retain little individuality. Juveniles in autumn, which are most often seen in Europe, are much more consistent in appearance. They appear in mid-autumn on wet, muddy edges of lakes and reservoirs, looking quite sedate compared with smaller waders or even Redshanks, with a steady, plodding action that rarely gets close to a run.

VOICE Very quiet; occasionally low, gruff *wek*.

NESTING Grass-lined scrape, well hidden in deep vegetation at edge of marsh; 4 eggs; 1 brood; April–July.

FEEDING Probes in soft mud for worms, insects, insect larvae, and seeds.

DISPLAYING

Male Ruffs display in groups to females, with mock battles, their unusual breeding plumage creating a striking spectacle.

SIMILAR SPECIES

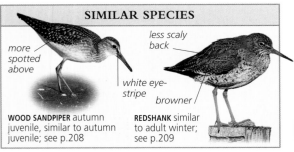

more spotted above

less scaly back

white eye-stripe

browner

WOOD SANDPIPER autumn juvenile, similar to autumn juvenile; see p.208

REDSHANK similar to adult winter; see p.209

OCCURRENCE

Breeds on wet meadowland; declining and local in NW Europe, more widespread in NE. At other times, on wet fields and marshes, muddy freshwater margins, most commonly autumn juveniles, some winter on western estuaries.

Seen in the UK
J F M A M J J A S O N D

| Length **20–32cm (8–12½in)** | Wingspan **46–58cm (18–23in)** | Weight **70–230g (2½–8oz)** |
| Social **Small flocks** | Lifespan **10–15 years** | Status **Secure†** |

| Order **Charadriiformes** | Family **Scolopacidae** | Species *Calidris ferruginea* |

Curlew Sandpiper

broad white wingbar

pale eye-ring and chin

ADULT (SUMMER)

long, slim, slightly downcurved bill

pale stripe over eye

dark back with even, pale buff scales

white rump (Dunlin's dark centred)

JUVENILE (AUTUMN)

IN FLIGHT

pale peachy buff breast

dark cap

white belly (never has black streaks or smudges like Dunlin)

long black legs

long wings

JUVENILE (AUTUMN)

no streaks below

JUVENILE (AUTUMN)

Very much linked with the Little Stint in birdwatchers' minds, as they are often found together, the Curlew Sandpiper tends to be scarcer in early spring in the Mediterranean and more erratic in western Europe in autumn. It follows the same early adult, later juvenile migration pattern as the Little Stint. It is noticeable in a group of Dunlins, being a little more elegant and elongated in its proportions.

VOICE Soft, trilling, rolled, rippled, *chirr-up* call draws attention in mixed flock.

NESTING Simple shallow scrape on ground; 4 eggs; 1 brood; May–July.

FEEDING Typically exploits longer legs and bill to wade deeply and probe into softer mud than Dunlin, in search of small worms.

FLIGHT: quick and direct with occasional erratic twists and tumbles.

WHITE RUMP
Only this bird and some much rarer sandpiper species have an unmarked white rump.

SPRING GEMS
Curlew Sandpiper migrants in spring, in coppery red breeding colour, are superbly colourful birds. They tend to wade more often and more deeply than Dunlins.

SIMILAR SPECIES

DUNLIN juvenile, similar to juvenile; see p.198

streaks on flanks and breast

shorter legs

KNOT summer; see p.193

larger and heavier

shorter legs

OCCURRENCE
Breeds in high Arctic region. Mostly in shallow fresh water and on muddy edges of coastal or inland lagoons. Chiefly adults in SE Europe in spring and in W Europe in late summer; followed by autumn juveniles.

| Seen in the UK |
| J F M **A M J J A S O** N D |

| Length **18–23cm (7–9in)** | Wingspan **38–41cm (15–16in)** | Weight **45–90g (1⅝–3¼oz)** |
| Social **Small groups** | Lifespan **Up to 10 years** | Status **Localized** |

| Order **Charadriiformes** | Family **Scolopacidae** | Species *Calidris temminckii* |

Temminck's Stint

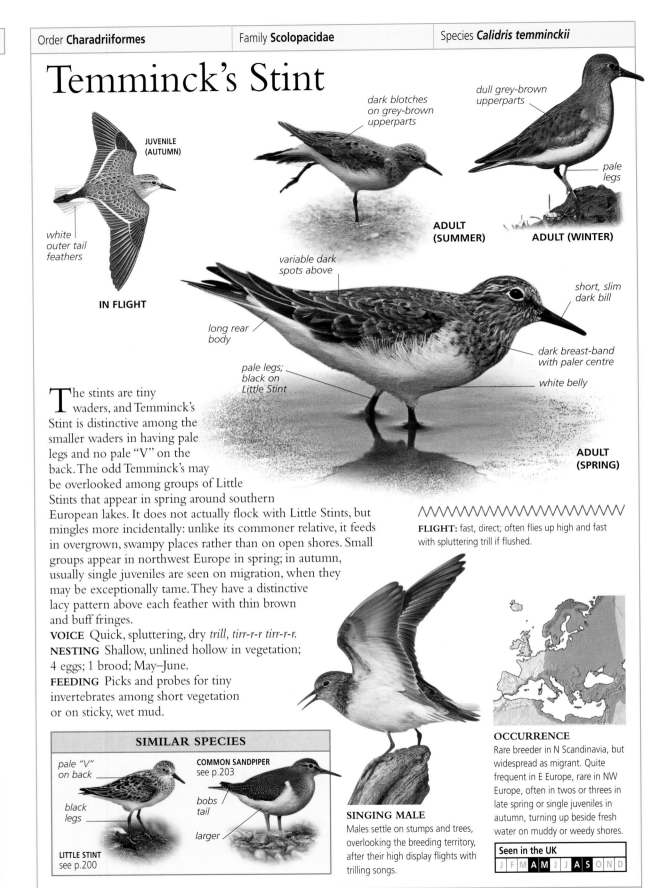

JUVENILE (AUTUMN)

dark blotches on grey-brown upperparts

dull grey-brown upperparts

pale legs

ADULT (SUMMER)

ADULT (WINTER)

white outer tail feathers

IN FLIGHT

variable dark spots above

short, slim dark bill

long rear body

pale legs; black on Little Stint

dark breast-band with paler centre

white belly

ADULT (SPRING)

The stints are tiny waders, and Temminck's Stint is distinctive among the smaller waders in having pale legs and no pale "V" on the back. The odd Temminck's may be overlooked among groups of Little Stints that appear in spring around southern European lakes. It does not actually flock with Little Stints, but mingles more incidentally: unlike its commoner relative, it feeds in overgrown, swampy places rather than on open shores. Small groups appear in northwest Europe in spring; in autumn, usually single juveniles are seen on migration, when they may be exceptionally tame. They have a distinctive lacy pattern above each feather with thin brown and buff fringes.

VOICE Quick, spluttering, dry *trill, tirr-r-r tirr-r-r.*
NESTING Shallow, unlined hollow in vegetation; 4 eggs; 1 brood; May–June.
FEEDING Picks and probes for tiny invertebrates among short vegetation or on sticky, wet mud.

FLIGHT: fast, direct; often flies up high and fast with spluttering trill if flushed.

SINGING MALE
Males settle on stumps and trees, overlooking the breeding territory, after their high display flights with trilling songs.

OCCURRENCE
Rare breeder in N Scandinavia, but widespread as migrant. Quite frequent in E Europe, rare in NW Europe, often in twos or threes in late spring or single juveniles in autumn, turning up beside fresh water on muddy or weedy shores.

Seen in the UK
J F M **A M** J J **A S** O N D

SIMILAR SPECIES

pale "V" on back

black legs

LITTLE STINT
see p.200

COMMON SANDPIPER
see p.203

bobs tail

larger

Length **13–15cm (5–6in)**	Wingspan **34–37cm (13½–14½in)**	Weight **20–40g (¹¹⁄₁₆–1⁷⁄₁₆ oz)**
Social **Solitary/Small flocks**	Lifespan **Up to 5 years**	Status **Secure†**

Order **Charadriiformes**	Family **Scolopacidae**	Species *Calidris alba*

Sanderling

back pure pearl-grey
(Dunlin drabber)

pale grey
head

**ADULT
(WINTER)**

light grey
back

broad white
wingbar

dark bend
of wings

IN FLIGHT

straight
black bill

jet black
legs

bright white
underparts

**ADULT
(WINTER)**

black spangling
on grey back

buff on
breast

JUVENILE (AUTUMN)

head and back marbled
buff, pale chestnut,
and black

pale chestnut breast
with dark marbling,
pure white below

ADULT (SUMMER)

The Sanderling stands out from the other birds belonging to the sandpiper family in appearance and general behaviour. In winter, it is the whitest, and it is particularly quick and nimble, darting along the waves as they move in and out, to snatch up tiny items rolled up by the surf. At high tide, Sanderlings and Dunlins often mix, the Sanderlings making a paler splash against the Dunlins in the packed roosting flock.

VOICE Sharp, hard, short *plit* or *twik* call, unlike Dunlin's thin *shrree*.

NESTING Scrape on ground part-filled with willow leaves by chance; 4 eggs; 1 brood; May–July.

FEEDING Snatches marine worms, crustaceans, molluscs, sandhoppers, and similar animal matter from beach.

FLIGHT: fast, low, often swirling round and back again; groups well coordinated.

FEEDING ALONG WAVES
Small groups of Sanderlings dash in and out along the edge of the waves on a sandy beach.

SIMILAR SPECIES

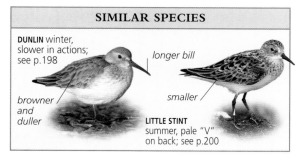

DUNLIN winter, slower in actions; see p.198

browner and duller

longer bill

smaller

LITTLE STINT summer, pale "V" on back; see p.200

OCCURRENCE
Breeds in northern tundra; otherwise, migrants in Europe from late summer to late spring. Wintering flocks typically on broad sandy beaches, but also found on shorelines of all kinds, sometimes inland in May.

Seen in the UK
J F M A M J J A S O N D

Length **20–21cm (8–8½in)**	Wingspan **36–39cm (14–15½in)**	Weight **50–60g (1¾–2⅛oz)**
Social **Winter flocks**	Lifespan **Up to 10 years**	Status **Secure**

Order **Charadriiformes**	Family **Scolopacidae**	Species *Calidris alpina*

Dunlin 🔊 27.I, 27.II

ADULT (WINTER)

dull grey-brown head and back

long, tapered black bill, slightly curved

thin white wingbar

dark centre to white-sided rump

JUVENILE

IN FLIGHT

dull grey-streaked breast

rich chestnut and black back

black and cream stripes on back

short black legs

black streaks and flank spots on underside (Curlew Sandpiper and Little Stint much cleaner)

large, squarish, black belly patch

fine dark streaks on whitish breast

JUVENILE

ADULT (SUMMER)

Widespread and covering a variety of wetland habitats, the Dunlin is the typical small wader of Europe, and is often used as the yardstick to assess other species. In spring, the streaked adults look rather sleek, while autumn juveniles have a certain brightness of colour and complexity of pattern. The Dunlin also has a distinctive call that allows it to be identified easily.

VOICE Thin, reedy, vibrant *shrree* or rasping *treerrr*; song-flight develops this into longer, trilled or pulsating "pea whistle".

NESTING Small, grass-lined, shallow scoop on ground or in grassy tussock; 4 eggs; 1 brood; May–July.

FEEDING Plods rather lethargically, on mud or drier shores, sometimes wading quite deeply, probing and picking up worms, insects, and molluscs.

REMARK Subspecies *C. a. alpina* (N Scandinavia) has long, curved bill and bright chestnut back in summer; *C. a. arctica* (Greenland) has short bill and dull body; *C. a. schinzii* (S Scandinavia and UK) has dull body.

FLIGHT: fast, dashing; flocks tight and well coordinated, often sweeping way out over sea and back again in spectacular manoeuvres.

WINTER ROOST

This group of Dunlins at high tide has been forced onto a small, exposed piece of rocky shore; when the tide recedes, they will disperse.

SIMILAR SPECIES

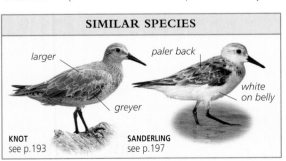

larger

greyer

KNOT
see p.193

paler back

white on belly

SANDERLING
see p.197

OCCURRENCE

In summer, breeds on wet moors, wet places on heaths, and northern isles, right up to the tundra, in far N and NW Europe. In all kinds of wet places from floods to wet fields but mostly on large estuaries.

Seen in the UK
J F M A M J J A S O N D

Length **16–20cm (6½–8in)**	Wingspan **35–40cm (14–16in)**	Weight **40–50g (1⁷/₁₆–1³/₄oz)**
Social **Flocks**	Lifespan **Up to 10 years**	Status **Common**

Order **Charadriiformes**	Family **Scolopacidae**	Species *Calidris maritima*

Purple Sandpiper

thin white wingbar

white sides to black rump

ADULT (WINTER)

IN FLIGHT

rufous on head

slightly curved, dark bill

whitish and rufous edges to feathers

dark streaks on breast

dull yellow legs

ADULT (SUMMER)

broad brownish streaks on cap

scaly pattern on wings

pale bill base

JUVENILE

drab grey-brown head and neck

dull yellow-based dark bill

dark back with scaly white feather edges

dark streaks on flanks

white belly

orange-yellow legs

ADULT (WINTER)

/\/\/\/\/\/\/\/\/\/\/\/\/\/\/\/\/\/\

FLIGHT: low, fast, darting flights from rock to rock.

Few waders are as tightly restricted to a particular habitat, or role, as the Purple Sandpiper, which is essentially a bird of the very edge of the surf, searching through wave-washed, seaweed-covered rocks for its food. Only rarely will one turn up inland. Unless it is with the more nervous Turnstones, it may well be absurdly tame. Like most waders, the Purple Sandpiper will not leave for its breeding grounds until mid-May and can return in July; hence it is present for most months of the year in western Europe, despite being a non-breeding visitor.
VOICE Simple, low, liquid *weet* or *weet-wit*.
NESTING Slight scrape on ground, on wide open tundra; 4 eggs; 1 brood; May–July.
FEEDING Variety of insects, spiders, and other invertebrates, chiefly periwinkles and similar molluscs in winter.

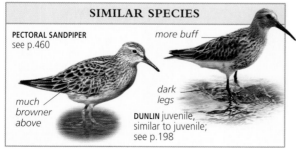

SIMILAR SPECIES

PECTORAL SANDPIPER see p.460

more buff

much browner above

dark legs

DUNLIN juvenile, similar to juvenile; see p.198

INCONSPICUOUS
A dark wader on dark, weedy rocks, the Purple Sandpiper is easily overlooked.

OCCURRENCE
Breeds in Iceland and Scandinavia, on tundra and mountains. Widespread in winter, preferring rocky shores, usually with plentiful seaweed, at times on bare rock and stony beaches, also piers, harbour walls, and other artificial sites.

Seen in the UK
J F M A M J J A S O N D

Length **20–22cm (8–9in)**	Wingspan **40–44cm (16–17½ in)**	Weight **60–75g (2⅛–2⅝oz)**	
Social **Small flocks**	Lifespan **Up to 10 years**	Status **Secure†**	199

Order **Charadriiformes**	Family **Scolopacidae**	Species *Calidris minuta*

Little Stint

strong cream "V" on back

JUVENILE (AUTUMN)

IN FLIGHT

most frequently seen plumage

cream lines on back

pale head with streaked cap

forked whitish line over eye

bright white underside; much cleaner than juvenile Dunlin

smaller than sparrow; much smaller than Dunlin or Starling

short black bill

black legs

pale breast with some flecks at sides

dull grey upperparts

black and cream mottling on rich brown upperparts

ADULT (WINTER)

ADULT (SUMMER)

JUVENILE (AUTUMN)

bright rufous body

FLIGHT: fast, twisting; often going well out over water and returning.

The smallest of the common waders, the Little Stint is a shoreline bird around the Mediterranean in spring, moving far to the north in summer. In autumn, adults move south early, to be followed by a larger wave of juveniles in western Europe, a pattern followed by several wader species. Little Stints are often mixed up with larger numbers of Dunlins and, sometimes, Curlew Sandpipers.

VOICE Hard, dry, sharp *tip* or *trip*, sometimes *ti-ti-trip*.
NESTING Small, shallow scrape on ground, close to water; 4 eggs; 1 brood; May–July.
FEEDING Scampers about at water's edge, finding tiny animal matter; does not often wade deeply.

MIGRANT JUVENILE
Most autumn migrants are clean, bright, well-marked juveniles. They are often remarkably tame.

SIMILAR SPECIES

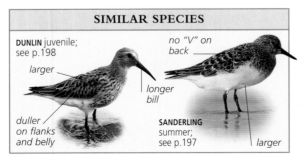

DUNLIN juvenile; see p.198

larger

longer bill

duller on flanks and belly

no "V" on back

SANDERLING summer; see p.197

larger

OCCURRENCE
Breeds on tundra. On migration, appears beside all kinds of muddy pools and lagoons, less so on sea coast. Adults found mostly in SE Europe in spring; majority in W Europe being small parties of juveniles in autumn.

Seen in the UK
J F M **A M** J **J A S O N** D

Length **12–14cm (4¾–5½in)**	Wingspan **34–37cm (13½–14½in)**	Weight **20–40g (¹¹⁄₁₆–1⁷⁄₁₆oz)**
Social **Small flocks**	Lifespan **Up to 10 years**	Status **Secure†**

| Order **Charadriiformes** | Family **Scolopacidae** | Species ***Phalaropus lobatus*** |

Red-necked Phalarope

blackish wings with strong white stripe

FEMALE (SUMMER)

looks very dark

IN FLIGHT

dark face

white throat

bright red neck (less red on male)

long buff stripes on dark grey back

fine black bill (never shows yellow of Grey Phalarope)

FEMALE (SUMMER)

black cap

black mask

striped blackish back

all-black bill

JUVENILE

black eye patch

grey above

needle-fine bill

ADULT (WINTER)

A tiny, delicate wader, the Red-necked Phalarope spends much of its time at sea, swimming with foreparts held up, and tail and wingtips upswept. This is a common breeder in the far north and winters in large numbers in the Middle East, but is a rare bird in most of Europe. In much of western Europe, it is an occasional autumn migrant, usually in juvenile plumage, and is much less frequent inland than the Grey Phalarope. Careful observation is required to be certain of identification in non-breeding plumages.
VOICE Sharp *twik* and quick, twittering notes.
NESTING Small, round hollow in grass tussocks in wet marshes; 4 eggs; 1 brood; April–July.
FEEDING Feeds at water's edge on insects, or picks insects from water surface, often spinning like a top.

FLIGHT: fast, low, darting flight, with fluttering effect, on broad-based wings.

SIMILAR SPECIES

GREY PHALAROPE winter, similar to adult winter; see p.202

thicker bill

paler back

MARSHLAND NESTER
In summer, the shallows of reedy lakes or stony pools on northern islands are the best places to look for the Red-necked Phalarope (male pictured).

OCCURRENCE
Breeds on northern pools and wet marshes in extreme N and NW Europe. Winters at sea. Rare migrants in spring and autumn, mostly juveniles, on coastal lagoons; much less often storm-blown inland than Grey Phalarope.

| Seen in the UK |
| J F M **A M J J A S O** N D |

| Length **17–19cm (6½–7½ in)** | Wingspan **30–34cm (12–13½ in)** | Weight **25–50g (⅞–1¾ oz)** |
| Social **Winter flocks** | Lifespan **Up to 10 years** | Status **Secure†** |

| Order **Charadriiformes** | Family **Scolopacidae** | Species *Phalaropus fulicarius* |

Grey Phalarope

broad white wingbar

FEMALE (SUMMER)

IN FLIGHT

black face

thick, dark-tipped yellow bill

orange-red underside

FEMALE (SUMMER)

MALE (SUMMER)

black eye patch

yellow-based dark bill

pearly grey back

white cheeks

ADULT (WINTER)

white underparts

Like other phalaropes, this species has "reversed roles": females are brighter than males, and the males incubate eggs and rear the chicks. A more northerly breeding bird than other phalaropes, it is, however, the most common along European coasts in autumn, sometimes turning up inland after autumn gales. Its frequent swimming is distinctive but inland it is sometimes seen wading on muddy shorelines like other waders. At sea, it is easily overlooked but sometimes gathers in small, swimming groups which fly off low and fast if disturbed by a ship.

VOICE High *prip* or *whit*.
NESTING Small, grassy hollow in northern tundra; 4 eggs; 1 brood; June–July.
FEEDING Picks invertebrates from mud and surface of water, often while swimming.

FLIGHT: slightly fluttery, erratic, low flight over waves with shallow wingbeats.

SWIMMING JUVENILE
The Grey Phalarope swims on the open sea and may be brought close inshore by autumn gales, but is usually able to cope with rough seas.

SIMILAR SPECIES

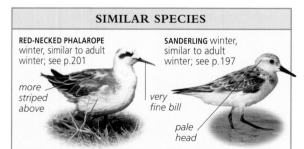

RED-NECKED PHALAROPE winter, similar to adult winter; see p.201

more striped above

SANDERLING winter, similar to adult winter; see p.197

very fine bill

pale head

OCCURRENCE
Rare breeder in Iceland. Otherwise lives at sea, sometimes off headlands on migration in storms; a few may be blown onto all kinds of shores and inland pools by gales in autumn, but always rare, especially in breeding plumage.

Seen in the UK
J F M **A M** J J **A S O N** D

Length **20–22cm (8–9in)**	Wingspan **37–40cm (14½–16in)**	Weight **50–75g (1¾–2⅝oz)**
Social **Small flocks**	Lifespan **Up to 10 years**	Status **Secure†**

Order **Charadriiformes**	Family **Scolopacidae**	Species **Actitis hypoleucos**

Common Sandpiper 🔊 28

bold white wingbar

stiff, bowed wings

dark, mottled back (plainer in winter)

mid-brown above

pale flecking along feather edges

ADULT (SUMMER)

dark tail with white sides

IN FLIGHT

long tail

greenish or dull ochre legs

ADULT (SUMMER)

pale-based, dark-tipped bill

greyish breast, paler in centre

white crescent in front of closed wing

JUVENILE (AUTUMN)

A common wader, the Common Sandpiper is much more widespread than its northern counterparts, the Green and Wood Sandpipers. A few may even be seen in winter, although the great majority go to Africa for the winter months. Typically a freshwater bird, it may also sometimes be seen on rocky sea shores. It is usually found in small numbers, of rarely more than ten or so together, strung out along a shore rather than in tight groups; they usually hold their heads low and swing their tails up and down in a constant swaying bob.

VOICE Loud, ringing, sharp *tew-tew-tew* or *tyew-yu-yu*; many summer calls include fast, trilling *teu-i teu-i teu-i*, *chip, tidledi tidledi tidledi*.

NESTING Small, grass-lined hollow on ground, often on grassy banks; 4 eggs; 1 brood; April–July.

FEEDING Skips and saunters along waterside, snatching insects and also some worms and molluscs.

FLIGHT: highly characteristic, low over water, with stiff, flickering beats of bowed wings.

BATHING
All waders bathe regularly, even in cold weather, to help keep their plumage in tiptop condition.

SIMILAR SPECIES

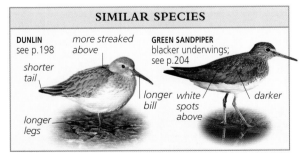

DUNLIN
see p.198

more streaked above

shorter tail

longer legs

GREEN SANDPIPER
blacker underwings;
see p.204

longer bill

white spots above

darker

OCCURRENCE
Breeds on rocky streams and lake sides with shingle and grassy banks locally throughout Europe. On migration, in all kinds of waterside habitats, from reservoirs and streams to muddy estuaries and even rocky foreshores.

Seen in the UK
J F **M A M J J A S O** N D

Length **19–21cm (7½–8½in)**	Wingspan **32–35cm (12½–14in)**	Weight **40–60g (1⁷/₁₆–2⅛oz)**
Social **Small flocks**	Lifespan **Up to 10 years**	Status **Secure**

| Order **Charadriiformes** | Family **Scolopacidae** | Species ***Tringa nebularia*** |

Greenshank 🔊 31

plain upperwings

greyish
upperparts

pale head
and neck

slightly
upturned
bill

longer necked,
less squat
than Redshank
or Ruff

white
wedge
on back

**ADULT
(WINTER)**

IN FLIGHT

clear white
underparts

long, grey-
green legs

**ADULT
(WINTER)**

pale, scaly edges
on upperparts

JUVENILE

blackish spots
above

streaked
breast

ADULT (SUMMER)

FLIGHT: fast, strong, with regular wingbeats; looks long and tapered.

One of the most beautiful of waders, despite a lack of strong colour or pattern, the Greenshank is an elegant, delicate-looking bird. It is, nevertheless, noticeably bigger than a Redshank, being part way to a godwit in size. It is easily located by its loud, ringing calls that echo around estuaries and inland pools. In summer, this is a bird of wild and remote places, shy and difficult to observe.

VOICE Main call distinctive, loud, ringing, far-carrying even-pitch *tew-tew-tew*, without accelerating "bounce" or hysterical quality of Redshank.

NESTING Scrape on ground, often near logs, stones, or posts, in grass or heather; 4 eggs; 1 brood; May–July.

FEEDING Probes while wading in shallow water, often very active, running and chasing fish; eats worms, insects, and crustaceans.

ELEGANT STANCE
Its long bill and long legs help to give the Greenshank a particularly delicate, elegant appearance.

SIMILAR SPECIES

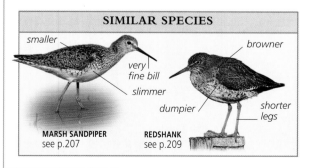

smaller

very
fine bill

slimmer

MARSH SANDPIPER
see p.207

browner

dumpier

shorter
legs

REDSHANK
see p.209

OCCURRENCE
Breeds on moorland near northern pools in NW Europe. On migration near water, including reservoirs well inland, but not often on exposed mudflats of larger estuaries. Winters in more sheltered salt-marsh creeks.

Seen in the UK
| J | F | M | A | M | J | J | A | S | O | N | D |

| Length **30–35cm (12–14in)** | Wingspan **53–60cm (21–23½in)** | Weight **140–270g (5–10oz)** |
| Social **Small flocks** | Lifespan **Up to 10 years** | Status **Secure** |

Order **Charadriiformes**	Family **Scolopacidae**	Species *Tringa stagnatilis*

Marsh Sandpiper

long, white "V" on back

dark wings

JUVENILE (AUTUMN)

narrow, angled wing shape

IN FLIGHT

very pale head and neck

pale line over eye

greyish upperparts with buff spots

straight, fine dark bill

extremely long greenish legs

white underparts

JUVENILE (AUTUMN)

grey-brown upperparts with dark spots

spotted flanks

streaked chest

ADULT (SUMMER)

/\/\/\/\/\/\/\/\/\/\/\/\/\/\/\/\/\/\/\

FLIGHT: quick, direct with quite fast wingbeats; legs trail beyond tail.

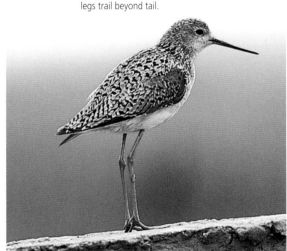

Its long legs and very fine, straight bill make this a particularly delicate and elegant wader, almost a stilt among the sandpipers. It is markedly smaller than a Redshank but needs to be carefully distinguished, when seen on its own, from a Greenshank, also a rather refined-looking bird. It typically stalks daintily around the edge of freshwater muddy pools. Marsh Sandpipers are generally rare in western Europe, although they may be seen regularly in a few areas of southeast Europe.

VOICE Quick, sharp *kyew* or high *kyu kyu kyu*.

NESTING Scantily lined scrape in grassy bog and marsh, or on open boggy clearing in northern forest; 4 eggs; 1 brood; May–July.

FEEDING Picks small insects and crustaceans from mud or water surface.

PALE LOOKS

A browner bird in summer, with black spots above, the Marsh Sandpiper ususally looks pale and rather colourless.

OCCURRENCE

Breeds in N and extreme E Europe, in forest clearings. Migrates through eastern Mediterranean, rare farther west in late spring or autumn. Mostly in freshwater marshes and lagoons.

Seen in the UK
J F M A M J J A S O N D

SIMILAR SPECIES

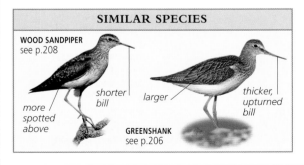

WOOD SANDPIPER see p.208

more spotted above

shorter bill

larger

thicker, upturned bill

GREENSHANK see p.206

Length **22–25cm (9–10in)**	Wingspan **50cm (20in)**	Weight **80–90g (2⅞–3¼oz)**
Social **Small flocks**	Lifespan **Up to 10 years**	Status **Secure†**

| Order **Charadriiformes** | Family **Scolopacidae** | Species *Tringa glareola* |

Wood Sandpiper

small pale spots on back

cream-spotted brown back (more boldly marked than Green Sandpiper, Redshank, or Ruff)

pale stripe extends well behind eye (short on Green Sandpiper)

ADULT (SUMMER)

no white on upperwings

white rump

pale underwings

narrow bars on tail

JUVENILE (AUTUMN)

IN FLIGHT

white underside

long, yellow-ochre legs

streaked breast

straight, dark-tipped bill

JUVENILE (AUTUMN)

With a typically *Tringa* sandpiper form and bobbing action, the Wood Sandpiper is, however, a more elegant and longer-legged bird than the Green Sandpiper. It is noticeably less thickset than the larger Redshank and Greenshank. It is basically a freshwater bird, not seen on open sea shores, often found in weedy pools or paddling about on floating vegetation. Many Wood Sandpipers pass through eastern and southern Europe in spring, when they are scarce migrants in western Europe. In autumn, they are still relatively uncommon but more regular and predictable in western Europe, especially in August. Ones and twos then appear on sheltered muddy shores of reservoirs or on lagoons near the coast, feeding in a rather nervous, jumpy manner, easily disturbed and ready to fly off at some height.

VOICE Distinctive quick, sharp *chiff-iff-iff-iff*.
NESTING Small, leaf-lined scrape on ground, occasionally old nest in tree; 4 eggs; 1 brood; May–July.
FEEDING Steps delicately over vegetation, picking up insects and small aquatic invertebrates.

FLIGHT: strong, quick, light, with flicking wingbeats; often rises high if disturbed.

SHALLOW WATER WADER
The Wood Sandpiper feeds beside muddy pools or on shallow floods, flying off quickly and noisily if disturbed.

SIMILAR SPECIES

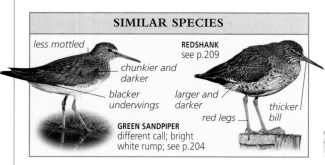

less mottled

REDSHANK see p.209

chunkier and darker

blacker underwings

larger and darker

red legs

thicker bill

GREEN SANDPIPER different call; bright white rump; see p.204

OCCURRENCE
Summer visitor, breeding in N and NE Europe. Migrants widespread in south and west, most on muddy pools, weedy fringes of shallow lagoons, salt pans, and often near coast, but not on estuarine mud.

| Seen in the UK |
| J F M **A M J J A S O** N D |

| Length **19–21cm (7½–8½in)** | Wingspan **36–40cm (14–16in)** | Weight **50–90g (1¾–3¼oz)** |
| Social **Small flocks** | Lifespan **Up to 10 years** | Status **Declining** |

| Order **Charadriiformes** | Family **Scolopacidae** | Species *Tringa totanus* |

Redshank 🔊 30

broad white band on upperwings

white underwings

white rump

ADULT

barred tail

IN FLIGHT

dark brown head and upperparts

pale eye-ring

straight, red-based bill

whitish belly with black spots

bright red legs

ADULT (SUMMER)

plain brown above

no spots below

ADULT (WINTER)

lacy buff feather edges

yellowish orange legs

JUVENILE

Its noisy behaviour makes the widespread Redshank one of the most obvious shoreline birds. It roosts in tight flocks at high tide, looking noticeably dark brown compared with paler godwits and Knots. It is declining fast in areas where farmland is drained or agriculture intensified, and has also been affected by the loss of salt-marsh habitats. Nevertheless, it remains frequent on many coasts.

VOICE Loud, ringing calls, "bouncing" *tyew-yu-yu, teu, teu-hu*, sharp annoyed *tewk, tewk;* song *tu-yoo tu-yoo tu-yoo* commonly heard on coasts.

NESTING Simple, sparsely lined hollow on ground, often with grass intertwined above it, forming canopy; 4 eggs; 1 brood; April–July.

FEEDING Probes and picks from mud, taking insects, earthworms, marine worms, crustaceans, and molluscs.

FLIGHT: fast, direct, gliding to ground; raises wings as it settles.

DENSE ROOSTS
Flocks of Redshanks are pushed tightly together by the rising tide. They tend to remain separate from other waders.

SIMILAR SPECIES

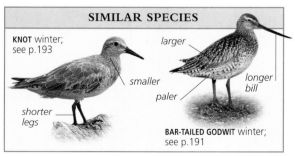

KNOT winter; see p.193

smaller

shorter legs

larger

paler

longer bill

BAR-TAILED GODWIT winter; see p.191

OCCURRENCE
Breeds on salt marshes, wet pastures, near freshwater pools, and on wet upland moors in N and E Europe. Otherwise, in wet places, on fresh water and salt coasts; mostly on estuaries but likely in almost any small creek or marsh.

Seen in the UK
J F M A M J J A S O N D

| Length **27–29cm (10½–11½in)** | Wingspan **45–52cm (18–20½in)** | Weight **85–155g (3–5oz)** |
| Social **Winter flocks** | Lifespan **Up to 10 years** | Status **Declining** |

Order **Charadriiformes**	Family **Scolopacidae**	Species *Lymnocryptes minimus*

Jack Snipe

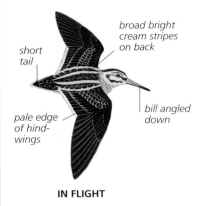

short tail

broad bright cream stripes on back

pale edge of hind-wings

bill angled down

IN FLIGHT

looks very dark on upperside

striped crown with black centre

bill shorter than that of Snipe

striped, green-glossed back

streaked flanks

short greenish legs

FLIGHT: quite slow, almost flitting compared with Snipe's; wings angled back.

While Snipe are generally skulking but often feed in the open, Jack Snipe almost never do, keeping to the depths of deep vegetation in very wet places. These handsome little birds are generally seen as they fly up, practically only when they are almost trodden on, and even then they go just a short distance before dropping down again. Close views on the ground are mostly restricted to spells of very cold weather when they are forced into unexpected places or stand out on ice. Wintering birds regularly appear at traditional places year after year, even in very small, marshy spots near pools or at the upper edge of estuarine salt marshes. In favoured spots, groups of ten or twenty Jack Snipe may feed in loose flocks, flying up singly.

VOICE Usually quiet; muffled "galloping" *og-ogok og-ogok* sound in display flight.

NESTING Hollow in dry hummock of grass or moss in bog; 4 eggs; 1 brood; May–July.

FEEDING Walks forward with bouncy action, probing for insect larvae, worms, and seeds.

WELL HIDDEN
The Jack Snipe feeds in dense vegetation in wet places and is very difficult to see on the ground.

SIMILAR SPECIES

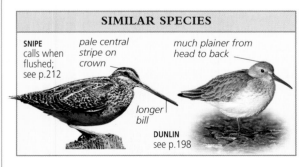

SNIPE calls when flushed; see p.212

pale central stripe on crown

longer bill

much plainer from head to back

DUNLIN see p.198

OCCURRENCE
Breeds in northern bogs; more southerly in winter. Outside the breeding season, in very wet grass, rushy places with standing water and mud, edges of reedbeds, and upper edges of weedy salt marshes, in deep cover.

Seen in the UK
J F M A M J J A S O N D

Length **17–19cm (6½–7½in)**	Wingspan **30–36cm (12–14in)**	Weight **35–70g (1¼–2½oz)**
Social **Small flocks**	Lifespan **5–10 years**	Status **Vulnerable†**

| Order **Charadriiformes** | Family **Scolopacidae** | Species **Scolopax rusticola** |

Woodcock

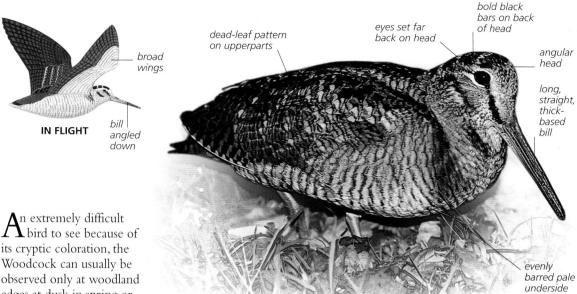

broad wings

IN FLIGHT

bill angled down

dead-leaf pattern on upperparts

eyes set far back on head

bold black bars on back of head

angular head

long, straight, thick-based bill

evenly barred pale underside

An extremely difficult bird to see because of its cryptic coloration, the Woodcock can usually be observed only at woodland edges at dusk in spring or summer while "roding". This is a mysterious territorial or courtship display at just over treetop height, involving fast quivering of bowed wings with regular grunts and whistles. At other times, it remains determinedly out of sight in thick vegetation on the woodland floor, or feeding in wet ditches or bogs at night. Only rarely, usually in severe weather, is it seen on the ground. If disturbed, it gets up with a clatter and flies off quite low and fast, sometimes turning back in a wide arc.

VOICE In display, diagnostic sharp, high whistle and deep throaty grunt, *tsi-wip grr grrr, tsi-wip grr grrr*.

NESTING Slight hollow in dead leaves, under brambles, or other cover in woods; 4 eggs; 1 brood; March–August.

FEEDING Probes for worms, beetles, and seeds in rich leaf mould, muddy ditches, and streamsides.

FLIGHT: quite quick and direct; flies up with loud wing noise and dashes away in zigzag.

EVENING FLIGHT
In summer, Woodcocks fly over regular circuits above woodland areas at dusk.

DIFFICULT TO SPOT
A Woodcock on its nest, or resting on the ground, is exceedingly difficult to see even at very close range.

SIMILAR SPECIES

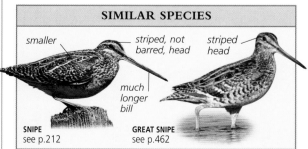

smaller

striped, not barred, head

striped head

much longer bill

SNIPE
see p.212

GREAT SNIPE
see p.462

OCCURRENCE
Widespread except in Iceland and most of Spain and Portugal; many move west and south in winter. Breeds in woodland of all kinds with soft, damp earth, bogs, and ditches nearby; frequents similar areas in winter in small numbers.

Seen in the UK
J F M A M J J A S O N D

| Length **33–38cm (13–15 in)** | Wingspan **55–65cm (22–26in)** | Weight **250–420g (9–15oz)** |
| Social **Family groups** | Lifespan **Up to 10 years** | Status **Vulnerable** |

Order **Charadriiformes**	Family **Scolopacidae**	Species **Gallinago gallinago**

Snipe 🔊 33.I, 33.II

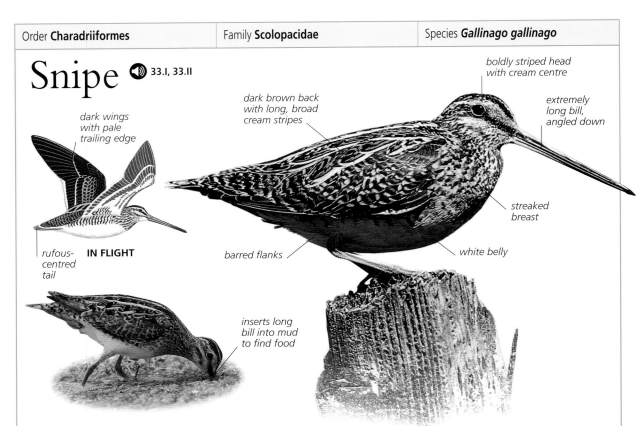

dark brown back with long, broad cream stripes

boldly striped head with cream centre

extremely long bill, angled down

dark wings with pale trailing edge

streaked breast

rufous-centred tail

IN FLIGHT

barred flanks

white belly

inserts long bill into mud to find food

The Snipe needs floods and oozy, watery mud, which allow its extraordinarily long, thin bill to be inserted into the ground to detect and grasp worms; it cannot survive for very long without soft ground. With the increasing drying out or tidying up of the modern landscape, with water constrained into firm channels, the Snipe and its remarkable spring displays have disappeared from vast areas of its former range. It is still seen at the edge of marshes, or occasionally flushed from almost underfoot amongst wet rushes. While displaying, it has a high, steeply undulating flight and dives with its tail fanned out.

VOICE Sharp, short, rasping *scaap!*; in spring, bright, rhythmic, musical *chip-per, chip-per, chip-per* from perch; also short, wavering, throbbing "bleat" from tail feathers in switchback display flight. It almost always calls loudly in flight.

NESTING Grass-lined shallow scrape in dense vegetation; 4 eggs; 1 or 2 broods; April–July.

FEEDING Probes deeply in soft mud for worms.

FLIGHT: quick, rolling from side to side with flicked beats of angled-back wings; sudden, fast escape flight; settles with sudden flurry of wings.

RESTING
This medium-sized wader may sit quietly for long spells beside a tussock of rushes or grass and is less active than most other waders.

SIMILAR SPECIES

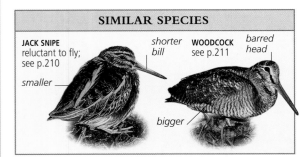

JACK SNIPE reluctant to fly; see p.210

shorter bill

smaller

WOODCOCK see p.211

barred head

bigger

OCCURRENCE
Prefers wet marshes and boggy heaths at all times, breeding through NW and N Europe. Outside breeding season, in all kinds of freshwater marshes with shallow water and soft mud, moving to coasts in freezing conditions.

Seen in the UK
J F M A M J J A S O N D

Length **25–28cm (10–11in)**	Wingspan **37–43cm (14½–17in)**	Weight **80–120g (2⅞–4oz)**
Social **Small flocks**	Lifespan **5–10 years**	Status **Secure†**

Collared Pratincole

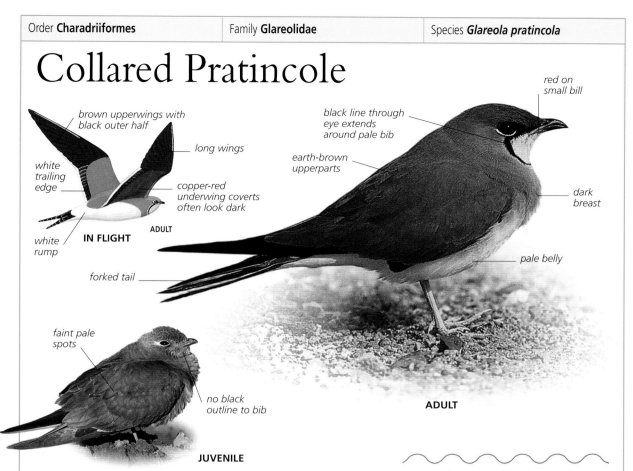

brown upperwings with black outer half

long wings

white trailing edge

copper-red underwing coverts often look dark

ADULT

IN FLIGHT

white rump

forked tail

red on small bill

black line through eye extends around pale bib

earth-brown upperparts

dark breast

pale belly

ADULT

faint pale spots

no black outline to bib

JUVENILE

A lovely, specialized wader with an aerial feeding technique that has helped it to evolve a swallow-like form, the Collared Pratincole is basically a Mediterranean bird which occasionally strays farther north. It is a long, tapered bird, but when standing hunched up with its feathers fluffed out, it can look dumpy, almost round except for its protruding wingtips and tail. In the air, however, it has the skill and manoeuvrability of a Black Tern (see p.231). The Collared Pratincole often feeds in small parties.

VOICE Sharp, far-carrying, tern-like *kit, kitik,* rhythmic *kirri-tik-kit-ik.*

NESTING Shallow scrape in dry mud on ground; loose colonies; 2 or 3 eggs; 1 brood; April–July.

FEEDING Catches insects in bill while flying.

FLIGHT: elegant, swooping action; changing pace and direction, with erratic twists and turns.

RESTING BETWEEN MEALS
The Collared Pratincole typically sits or stands on the ground for long spells between bouts of feeding.

OCCURRENCE
In summer, in S Spain, Portugal, France, Italy, and Balkans, rare vagrant elsewhere. In extensive areas of flat, dry mud and damp pasture, bare ground, marshes and deltas drained for farmland, and salt pans.

Seen in the UK
J F M A M J J A S O N D

SIMILAR SPECIES

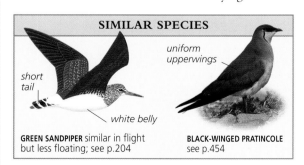

short tail

white belly

GREEN SANDPIPER similar in flight but less floating; see p.204

uniform upperwings

BLACK-WINGED PRATINCOLE see p.454

Length **24–28cm (9½–11in)**	Wingspan **60–70cm (23½–28in)**	Weight **50–80g (1¾–2⅞oz)**
Social **Flocks**	Lifespan **Up to 5 years**	Status **Endangered**

Family **Alcidae**

AUKS

Exclusively seabirds, auks come to land only to breed on cliff ledges or in burrows, in noisy colonies, and spend the winter at sea. They swim and dive expertly, using their wings underwater, but in flight their small wings whirr rapidly to keep them airborne. They are vulnerable to pollution, often forming the bulk of the victims of oil spills around Atlantic coasts. Some Puffin colonies have disappeared as tunnelling birds have eroded all the available soil; others have suffered from declining fish stocks.

PENGUIN-LIKE
Auks look like northern equivalents of the southern hemisphere penguins.

ALCIDAE

deep, triangular bill, patterned blue-grey, orange, yellow, and red

smoky black body

PUFFIN
p.215

BLACK GUILLEMOT
p.216

thick, flattened bill

white underparts, rounded against black throat

RAZORBILL
p.217

tiny bill

LITTLE AUK
p.218

GUILLEMOT
p.219

| Order **Charadriiformes** | Family **Alcidae** | Species *Fratercula arctica* |

Puffin

plain black wings

disc-like, grey-white sides of face

dark eye

deep, triangular bill, patterned blue-grey, orange, yellow, and red

black upperparts and neck

ADULT (SUMMER)

ADULT (SUMMER)

IN FLIGHT

white underside

/\/\/\/\/\/\/\/\/\/\/\/\/\/\/\

FLIGHT: fast, direct, low; wingbeats quick, whirring; may fly higher over sea or when circling colony in flocks.

Puffins are more localized than Guillemots, requiring more earth in which to burrow or cavities in rocks in summer. In winter, they are far out in the Atlantic. Summer birds are often seen flying by from headlands in the north and west. However, winter ones, which lack the distinctive bill ornamentation seen in the breeding season, are generally rare close inshore. Occasionally, Puffins are blown far inland during autumn gales, and turn up in unexpected places.

VOICE At nest, loud, cooing growl, *aaarr, karr-oo-arr.*

NESTING Occupies ready-made burrow, digs burrow into soft earth, or finds cavity between boulders; 1 egg; 1 brood; May–June.

FEEDING Dives from water surface to catch fish and squid.

grooves on bill increase with age

vivid orange legs

ADULT (SUMMER)

smaller, duller bill (loses colourful sheath)

dusky grey face (as on juvenile)

ADULT (WINTER)

RUNNING THE GAUNTLET
Adults bringing fish back to their chicks are often harried by gulls.

OCCURRENCE
Breeds on coasts and islands from Iceland south to NW France, in cliff cavities, scree, or on grassy slopes. Widespread migrant offshore, but scarce in winter when most are far out in Atlantic. Very rare inland after storms.

| Seen in the UK |
| J F **M A M J J A S O N** D |

SIMILAR SPECIES

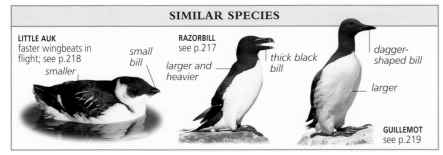

LITTLE AUK
faster wingbeats in flight; see p.218
smaller
small bill

RAZORBILL
see p.217
larger and heavier
thick black bill

dagger-shaped bill
larger

GUILLEMOT
see p.219

| Length **26–29cm (10–11½in)** | Wingspan **47–63cm (18½–25in)** | Weight **310–500g (11–18oz)** |
| Social **Small flocks** | Lifespan **10–20 years** | Status **Vulnerable** |

| Order **Charadriiformes** | Family **Alcidae** | Species **Cepphus grylle** |

Black Guillemot

black bars on wing patch, otherwise same plumage as winter adult

dark around eye

neat black head

small, sharp, dagger-like bill

oval patch on each wing, above and below

JUVENILE

ADULT (SUMMER)

IN FLIGHT

smoky black body

bold white wing patch

bright red legs

ADULT (SUMMER)

Not nearly so social as other auks, the Black Guillemot prefers small islets and rocky headlands around northern coasts and archipelagos where pairs swim about close to the shore all year round. These expert swimmers and divers often penetrate far into deep inlets in quiet, calm waters. In summer, their unique plumage makes them easily identifiable. They are distinctive in winter as well, when they retain the clean white wing patches, but have a mottled white back and a dusky cap and eye patch. The winter bird may be confused with similar-looking species when encountered unexpectedly but the pale head and pointed bill help to separate it from ducks and grebes.

VOICE Shrill, high whistle extends into fast trill; quick, thin *sip-sip-sip* notes occasionally heard.

NESTING Crevice or cavity between boulders, or hole in harbour wall; 1 egg; 1 brood; May–June.

FEEDING Dives underwater to catch small fish and crustaceans.

FLIGHT: low, quick, direct, with fast, whirring wingbeats.

AWKWARD ON LAND
Black Guillemots usually sit horizontally, less upright than Guillemots, and are not as agile on land as Puffins.

OCCURRENCE
Breeds on coasts in N Europe, usually around rocky islets with boulders and cavities in rocks. Usually resident, only rare elsewhere in winter. Strictly marine, extremely rare inland.

| Seen in the UK |
| J F M A M J J A S O N D |

SIMILAR SPECIES

no white wing patches

PUFFIN
see p.215

stumpy tail

thick bill

blacker above

GUILLEMOT winter;
see p.219

black wings

SLAVONIAN GREBE winter; see p.109

| Length **30–32cm (12–12½in)** | Wingspan **52–58cm (20½–23in)** | Weight **340–450g (12–16oz)** |
| Social **Family groups** | Lifespan **Up to 10 years** | Status **Declining** |

| Order **Charadriiformes** | Family **Alcidae** | Species *Alca torda* |

Razorbill

black cap

pointed tail

ADULT (SUMMER)

broad white sides to dark rump

IN FLIGHT

ADULT (WINTER)

white throat and breast

horizontal white line in front of eye

white line on bill

black head (juvenile has dusky cheeks)

thick, flattened bill

black upperparts

white underside; pointed against black throat (rounded on Guillemot)

ADULT (SUMMER)

Not usually so numerous as the Guillemot, the Razorbill is nevertheless a frequent constituent of northwest European seabird colonies, often less conspicuous because of its preference for cavities rather than open ledges. In summer, Razorbills tend to come into sheltered bays and estuaries more than Guillemots. They are usually best separated by structure, the often cocked, pointed tail being a useful feature compared with the short, square tail of a Guillemot; the head and bill shape are the best clues at close range.

VOICE Prolonged, tremulous growls and various grunting sounds at colony, deep *urrr*.

NESTING On sheltered ledge or cavity between boulders; 1 egg; 1 brood; May–June.

FEEDING Dives, often very deep, from surface to catch fish using its wings underwater.

FLIGHT: fast, low, direct with quick, almost whirring wingbeats; swoops upwards to land on cliff.

STRONG FLIERS
Razorbills fly strongly despite their small wings, with constant wingbeats unlike superficially similar shearwaters.

OCCURRENCE
Breeds on rocky coasts from Iceland south to NW France, usually on cliffs with cavities or boulder scree. Widespread in winter but scarce inshore. Very rare inland even after gales.

Seen in the UK
| J | F | M | A | M | J | J | A | S | O | N | D |

SIMILAR SPECIES

smaller head

sharp, dagger-like bill

GUILLEMOT see p. 219

BRUNNICH'S GUILLEMOT see p. 466

stubby, triangular bill

PUFFIN see p. 215

colourful, triangular bill

| Length **37–39cm (14½–15½in)** | Wingspan **63–67cm (25–26in)** | Weight **590–730g (21–26oz)** |
| Social **Small flocks** | Lifespan **10–20 years** | Status **Secure** |

Order **Charadriiformes**	Family **Alcidae**	Species **Alle alle**

Little Auk

white sides of neck curve up behind cheeks

black face and cap

short, stumpy black bill

slender wings

WINTER

white trailing edge

IN FLIGHT

white streaks on shoulders

WINTER

black of back extends as point on to sides of chest

/\/\/\/\/\/\/\/\/\/\/\/\/\/\/\/\

FLIGHT: fast, low; wings longish, slightly swept back, wader-like but blur of fast, whirring beats.

all-black head and breast

SUMMER

The smallest auk, and the most northerly, the Little Auk is rare in most of Europe and best known as a late autumn visitor to the North Sea. In some years, large numbers appear; in others it is scarce. In "good" years, autumn gales may sweep a few score well inland. Breeding colonies are often huge, with vast numbers of birds flying around overhead, often very high up beside towering cliffs. Fit, alert birds swim with head raised and tail cocked, while tired or sick birds are hunched, with drooped wings. They are vulnerable to predators such as gulls and skuas, or even crows, when they are exhausted and driven inland. The Little Auk opens its wings slightly as it dives for food, and returns to the surface like a cork, swimming buoyantly on the roughest seas, but tired birds on inland waters are often inactive.

VOICE Shrill, twittering, or chattering notes and trills; silent at sea.

NESTING Burrow high above shore; 1 egg; 1 brood; June.

FEEDING Dives for fish, plankton, and crustaceans.

STORM-BLOWN MIGRANTS
In late autumn, some Little Auks are driven close inshore by gales. They may be seen unexpectedly over rocky shores or wide, sandy beaches.

SIMILAR SPECIES

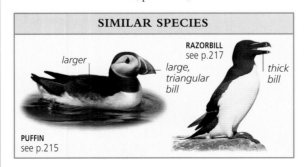

larger

PUFFIN
see p.215

large, triangular bill

RAZORBILL
see p.217

thick bill

OCCURRENCE
Breeds in Arctic on islands. Mostly rare late autumn or winter visitor to NW Europe, sometimes briefly numerous in North Sea after northerly gales. Rare storm-blown birds appear inland.

Seen in the UK
J F M A M J J A S O N D

Length **17–19cm (6½–7½in)**	Wingspan **40–48cm (16–19in)**	Weight **140–170g (5–6oz)**
Social **Small flocks**	Lifespan **Up to 10 years**	Status **Secure†**

Order **Charadriiformes**	Family **Alcidae**	Species *Uria aalge*

Guillemot

sharp, dagger-like bill

long neck

white underparts, rounded against black throat

upright posture

ADULT (SUMMER)

dark brown to black above

swims low

ADULT (SUMMER)

black line through eye on white face (juvenile similar)

dark cap

ADULT (WINTER)

short, square tail

dark rump with narrow white sides

white trailing edge

ADULT (SUMMER)

IN FLIGHT

Guillemots, along with Kittiwakes, usually dominate seabird colonies which densely pack ledges of sea cliffs in summer. Guillemots swim offshore in large "rafts" under the cliffs. Off headlands, they are often seen flying by, low and fast. In winter, however, they are seen inshore only during or after gales. In the south of their range, they are quite brown and easily distinguished from Razorbills; northern birds are blacker and more difficult to identify.

VOICE At colony, loud, whirring, growling chorus, *arrrr-rr-rr*; juveniles make surprisingly loud, musical whistle at sea.

NESTING On bare ledge on sheer cliff; 1 egg; 1 brood; May–June.

FEEDING Dives from surface to catch fish deep underwater.

FLIGHT: low, fast, direct, little manoeuvrability; fast, almost whirring wingbeats; sweeps up to land on ledge.

LARGE FLOCKS
Large flocks of Guillemots swim on the sea under the breeding colonies.

LARGE COLONIES
Guillemots crowd steep coastal cliffs in the breeding season; colonies are often mixed with Razorbills.

OCCURRENCE
Breeds on rocky coasts of Iceland, Scandinavia, Great Britain, Ireland, NW France, Spain, and Portugal, chiefly on sheer cliffs, also on flat-topped stacks. Scarce inshore in winter, even after gales.

Seen in the UK
J F M A M J J A S O N D

SIMILAR SPECIES

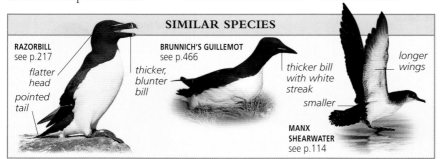

RAZORBILL
see p.217

flatter head

pointed tail

BRUNNICH'S GUILLEMOT
see p.466

thicker, blunter bill

thicker bill with white streak

smaller

longer wings

MANX SHEARWATER
see p.114

Length **38–54cm (15–21½in)**	Wingspan **64–73cm (25–29in)**	Weight **850–1,130g (30–40oz)**
Social **Small flocks**	Lifespan **10–20 years**	Status **Secure**

SKUAS, TERNS, AND GULLS

LARIDAE *continued*

pure white head
from January to
early autumn

IMMATURE

brown
blotched
body

HERRING GULL
p.242

yellow legs

blackish
bill

IMMATURE

YELLOW LEGGED GULL
p.243

long, white
wingtips

barring on oatmeal-
brown body

IMMATURE
(1ST WINTER)

ICELAND GULL
p.244

short, white
wingtips

pink and
black bill

IMMATURE
(1ST WINTER)

GLAUCOUS GULL
p.245

whitish head

black back

IMMATURE
(1ST WINTER)

GREAT BLACK-BACKED GULL
p.246

| Order **Charadriiformes** | Family **Stercorariidae** | Species ***Stercorarius pomarinus*** |

Pomarine Skua

IN FLIGHT

blunt tail — **ADULT (DARK FORM)**

spoon-like tail — white flash — **ADULT (PALE FORM)**

broad bars under tail — brown body — blunt, twisted tail

JUVENILE (AUTUMN; DARK FORM)

thick bill with pink-brown base — black cap — brown breast-band — brown back — white underside — grey legs

ADULT (SUMMER; PALE FORM)

Pomarine Skuas breed in the far north and so are seldom seen in Europe in summer, but in spring small groups of adults pass by southern and western headlands in a short, concentrated migration; in autumn, larger numbers can be seen around west European coasts over a period of many weeks. These migrants, however, tend to be widely scattered and often far offshore, and it usually needs a good onshore wind to bring some within easy range of a birdwatcher ashore.

VOICE Usually silent away from breeding sites.

NESTING Shallow scrape on open ground in Arctic tundra; 2 eggs; 1 brood; June.

FEEDING Eats lemmings and seabirds in summer; otherwise fish, stolen from other birds, and offal.

FLIGHT: direct flight steady, strong, straight; wingbeats smooth and powerful; piracy involves fast, active chase.

AUTUMN MIGRANT
Tired migrants after gales may rest on beaches and forage for food like gulls.

OCCURRENCE
Breeds in extreme NE Europe. On passage mostly found in North Sea and Atlantic. Occurs in variable numbers: usually scarce; at times concentrated movements in spring; occasional larger, more prolonged influxes in late autumn to North Sea; very few in winter.

| Seen in the UK |
| J F M **A M** J J **A S O N** D |

SIMILAR SPECIES

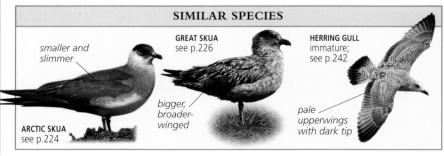

smaller and slimmer — **ARCTIC SKUA** see p.224

GREAT SKUA see p.226 — bigger, broader-winged

HERRING GULL immature; see p.242 — pale upperwings with dark tip

| Length **46–51cm (18–20in)** | Wingspan **1.13–1.25m (3¾–4ft)** | Weight **550–900g (20–32oz)** |
| Social **Small flocks** | Lifespan **10–15 years** | Status **Secure†** |

| Order **Charadriiformes** | Family **Stercorariidae** | Species **Stercorarius parasiticus** |

Arctic Skua

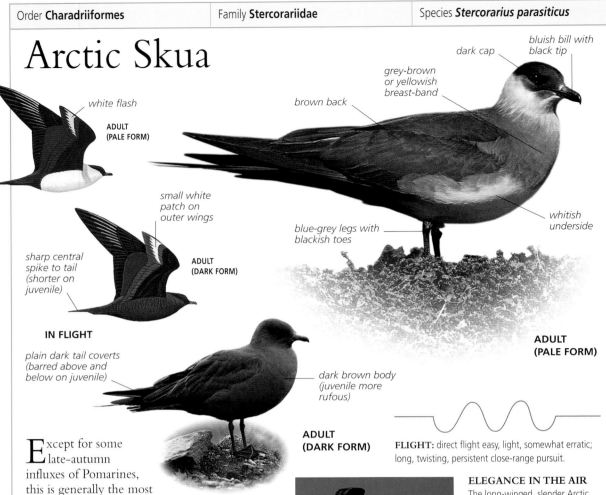

white flash

ADULT (PALE FORM)

small white patch on outer wings

ADULT (DARK FORM)

sharp central spike to tail (shorter on juvenile)

IN FLIGHT

plain dark tail coverts (barred above and below on juvenile)

dark cap

bluish bill with black tip

grey-brown or yellowish breast-band

brown back

blue-grey legs with blackish toes

whitish underside

ADULT (PALE FORM)

dark brown body (juvenile more rufous)

ADULT (DARK FORM)

FLIGHT: direct flight easy, light, somewhat erratic; long, twisting, persistent close-range pursuit.

Except for some late-autumn influxes of Pomarines, this is generally the most common skua in Europe. Learning the variety in this species will help identification of the rarer skuas. On its breeding grounds, it is a magnificent, dynamic bird, with fast, swooping, high display flights; it also attacks human intruders with great courage. At sea, it is a pirate, chasing other seabirds in order to make them disgorge fish. Its swift, dogged, and acrobatic pursuit of terns and small gulls, often in pairs, is always exciting to watch.

VOICE In summer, loud, nasal, wailing *ahh-yeow, eee-air, ka-wow* etc; silent at sea.
NESTING Hollow on ground in moss or heather; 2 eggs; 1 brood; May–June.
FEEDING Robs terns and gulls of fish; also catches fish, small birds, and voles, and eats some berries and insects.

ELEGANCE IN THE AIR
The long-winged, slender Arctic Skua, with its central tail spike, is one of the most beautifully shaped seabirds in flight.

OCCURRENCE
Breeds on northern moors and islands from Scotland north into Arctic. Spring and autumn migrants appear off most European coasts, especially North Sea and Atlantic; usually most common skua in early autumn. Rare inland after storms.

| Seen in the UK |
| J F **M A M J J A S O N** D |

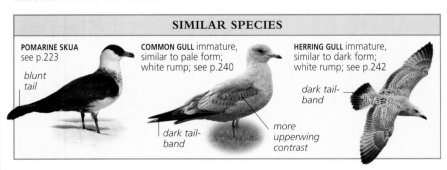

SIMILAR SPECIES

POMARINE SKUA see p.223

blunt tail

COMMON GULL immature, similar to pale form; white rump; see p.240

dark tail-band

HERRING GULL immature, similar to dark form; white rump; see p.242

dark tail-band

more upperwing contrast

| Length **37–44cm (14½–17½in)** | Wingspan **0.97–1.15m (3¼–3¾ft)** | Weight **380–600g (13–21oz)** |
| Social **Small flocks** | Lifespan **10–15 years** | Status **Secure†** |

| Order **Charadriiformes** | Family **Stercorariidae** | Species ***Stercorarius longicaudus*** |

Long-tailed Skua

grey-brown upperwings with very thin pale flash

black cap

narrow dark wings

thick-necked shape

ADULT (SUMMER)

dark trailing edge

thick neck, flat belly

grey-brown above

ADULT (SUMMER)

IN FLIGHT

white breast

brown body

very long, whip-like central tail spike

dark belly

broad pale bars under tail

ADULT (SUMMER)

JUVENILE (DARK FORM)

This is usually the rarest of the smaller skuas, with occasional large numbers moving north in a few days in spring off western headlands and small numbers over longer periods in autumn, especially in the North Sea. In its breeding areas in the far north, it is very bold and may even perch on people's heads. It flies low and easily, rather tern-like, and rarely chases other seabirds. Like some other birds that feed mostly on lemmings in summer, its numbers vary from year to year according to the lemming population. Unlike the Arctic Skua, it does not have a dark form when adult, but juveniles are variable and have paler forms with whiter head, cold grey body, and whiter belly.

VOICE Wailing gull-like squeal and high alarm notes in summer; silent at sea.

NESTING Hollow on ground in tundra or on high mountains; 2 eggs; 1 brood; June.

FEEDING Eats mostly lemmings, voles, and small birds in summer; at sea, feeds on offal and fish, mostly self-caught.

FLIGHT: direct flight light, erratic, often lifting briefly before dropping to sea.

AGRESSIVE ADULT
Breeding birds fly around intruders, calling loudly, their flexible tail spikes very obvious.

SIMILAR SPECIES

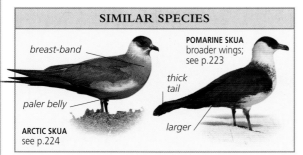

breast-band

POMARINE SKUA broader wings; see p.223

thick tail

paler belly

ARCTIC SKUA see p.224

larger

OCCURRENCE
Breeds in N and W Norway and extreme N Sweden. Migrates through North Sea, around Western Isles of Scotland and off W Spain and Portugal in brief spring movement of adults in flocks and more prolonged autumn passage.

| Seen in the UK |
| J F **M A M J J A S O** N D |

| Length **35–41cm (14–16in)** | Wingspan **1.05–1.12m (3½–3¾ft)** | Weight **250–450g (9–16oz)** |
| Social **Small flocks** | Lifespan **Up to 10 years** | Status **Secure†** |

| Order **Charadriiformes** | Family **Stercorariidae** | Species *Stercorarius skua* |

Great Skua

long, broad wings taper to point

bold white patch on outer wings

ADULT

belly always dark

IN FLIGHT

dark cap

pale brown to cream streaks on neck and mantle

streaked dark brown upperparts

stout, hooked dark bill

uniformly dark underparts (juveniles often blacker)

ADULT

thick blackish legs

ADULT

The largest, heaviest, boldest, and most predatory skua, the Great Skua is literally hair-raising near the nest as it zooms in at head-height at intruders. It has increased greatly in recent years, to the detriment of some other seabird species. In most of western Europe it is a migrant (to and from Africa) in spring and autumn, best seen from headlands in periods of strong onshore winds. Usually it is less numerous than the Arctic Skua in such circumstances. It accompanies gulls and Gannets in flocks around trawlers, and at coastal freshwater lakes in its breeding areas in the north.
VOICE Barking *uk-uk-uk*, deep *tuk-tuk*; silent at sea.
NESTING Simple hollow on ground on moorland; 2 eggs; 1 brood, May–June.
FEEDING Steals fish from other seabirds up to size of Gannet; kills many birds up to size of Kittiwake; eats much offal, carrion, and eggs.

FLIGHT: low, direct, heavy, with slow wingbeats; chase fast, rather brief.

FLASHING WING PATCHES
Great Skuas display on their breeding grounds, showing off their bold white wing patches.

OCCURRENCE
Breeds from Scotland northwards, on islands and remote moors and hills. Widespread off W European coasts and out at sea in spring and autumn; sometimes brought closer inland by gales and often passing longer headlands in any weather. Rare in winter.

| Seen in the UK |
| J F **M A M J J A S O N** D |

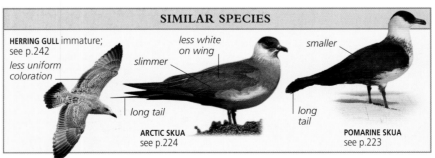

SIMILAR SPECIES

HERRING GULL immature; see p.242
less uniform coloration

slimmer

less white on wing

long tail

ARCTIC SKUA see p.224

smaller

long tail

POMARINE SKUA see p.223

| Length **50–58cm (20–23in)** | Wingspan **1.25–1.4m (4–4½ft)** | Weight **1.2–2kg (2¾–4½lb)** |
| Social **Small flocks** | Lifespan **10–20 years** | Status **Secure** |

| Order **Charadriiformes** | Family **Sternidae** | Species ***Sterna albifrons*** |

Little Tern

blackish streak at wingtips

short, white forked tail

ADULT (SUMMER)

IN FLIGHT

streaky crown

dark chevrons on back

blackish bill

JUVENILE

black stripe through eye

white forehead

black cap

black nape

pale grey back

sharp yellow bill with tiny dark tip

white forehead

pure white underside

ADULT (SUMMER)

orange to yellow legs

ADULT (SUMMER)

Small, quick, nervous, and now rare, the Little Tern is a lively coastal bird; it is rare inland but seen along most coasts. Its pale colours and small size are usually obvious at first glance, especially in flight. At its nesting colonies, it is noisy and aggressive to intruders but easily disturbed – most colonies are on popular beaches and now succeed only if specially protected. Climate change, causing a rise in the sea level, also threatens this bird, which often nests right at the edge of the sea and risks losing eggs and chicks to high tides.

VOICE Sharp, high, rapid chattering *kirri-kirri-kirri* and *kitititit*.

NESTING Shallow scoop on sand or shingle beach; 2 or 3 eggs; 1 brood; May–June.

FEEDING Plunges for fish after a brief, whirring hover, fast but light with quick "smack" into water, often near beach.

FLIGHT: quick, flickering wingbeats; hovers briefly with very quick whirring beats.

TINY TERNS
Little Terns are smaller and whiter than Common Terns which do not have white foreheads all year round.

OCCURRENCE
Breeds on narrow sand and shingle beaches, very locally south from Baltic, thriving only where protected; also inland in S Spain and Portugal and E Europe. Mostly coastal migrant in spring and autumn, rare inland.

| Seen in the UK |
| J F **M A M J J A S O** N D |

SIMILAR SPECIES

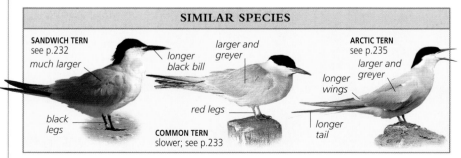

SANDWICH TERN see p.232 much larger

longer black bill

black legs

larger and greyer

red legs

COMMON TERN slower; see p.233

ARCTIC TERN see p.235

larger and greyer

longer wings

longer tail

| Length **22–24cm (9–9½in)** | Wingspan **48–55cm (19–22in)** | Weight **50–65g (1¾–2⅜oz)** |
| Social **Small flocks** | Lifespan **Up to 10 years** | Status **Declining** |

227

Order **Charadriiformes**	Family **Sternidae**	Species *Sterna nilotica*

Gull-billed Tern

thin dark band on outer wings

white head with black eye patch

thick bill

round black cap

stout black bill

ADULT (SUMMER)

pale grey tail

ADULT (WINTER)

pale grey back

pale grey bloom on wing feathers wears off to reveal black

IN FLIGHT

ADULT (SUMMER)

white underside

black legs

One of the more localized terns of Europe, this is unlike the black-capped, pale grey and white "sea terns" in structure and behaviour but similar in general appearance. It is a bird of freshwater marshes and coastal lagoons although it does migrate over the sea. In winter, in Africa, Gull-billed Terns feed over the open plains with huge numbers of animals; in Europe, they also feed over fields where livestock disturb insects, which they snatch in the air like giant swallows. They require careful separation from Sandwich Terns outside their usual range, but in reality are generally relatively easy to identify. Despite having a characteristically grey tail, they usually look very pale, especially in winter.

VOICE Nasal, deep *gur-wik*, laughing notes and rattling call.

NESTING Grass-lined small hollow on sand or mud near water; 3 eggs; 1 brood, May–June.

FEEDING Takes most food while flying, dipping to snatch insects from ground or in air; eats some small birds, rodents, and frogs.

FLIGHT: direct flight easy, languid, slightly more gull-like than smaller terns.

FLIGHT PATTERN
In flight, the Gull-billed Tern shows long, tapered wings with dusky trailing edges towards sharp tips.

OCCURRENCE
Breeds and feeds around lagoons, rice paddies, marshes, wet fields, and high grassland, mostly in S and E Europe, very locally in North Sea area. Generally only very rare migrant outside S Europe, usually on or near coast.

Seen in the UK
J F M A M J J A S O N D

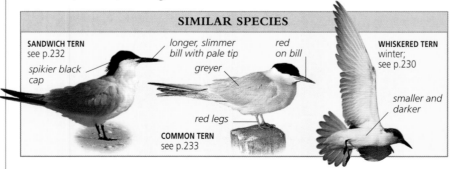

SIMILAR SPECIES

SANDWICH TERN see p.232
spikier black cap

longer, slimmer bill with pale tip
greyer
red legs
COMMON TERN see p.233

red on bill

WHISKERED TERN winter; see p.230
smaller and darker

Length **35–42cm (14–16½in)**	Wingspan **76–86cm (30–34in)**	Weight **200–250g (7–9oz)**
Social **Flocks**	Lifespan **Up to 10 years**	Status **Endangered†**

| Order **Charadriiformes** | Family **Sternidae** | Species ***Sterna caspia*** |

Caspian Tern

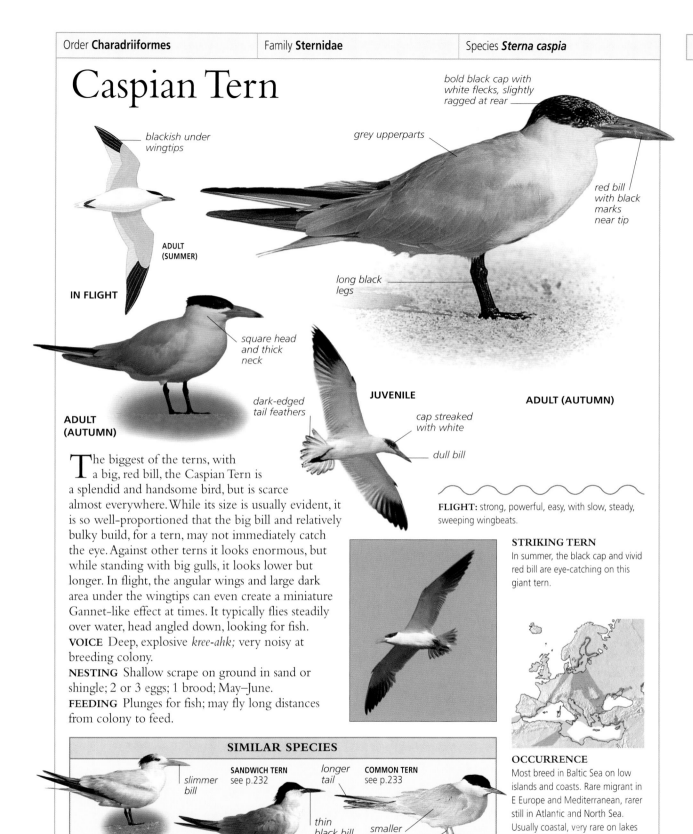

bold black cap with white flecks, slightly ragged at rear

grey upperparts

red bill with black marks near tip

long black legs

blackish under wingtips

ADULT (SUMMER)

IN FLIGHT

square head and thick neck

dark-edged tail feathers

ADULT (AUTUMN)

JUVENILE

cap streaked with white

dull bill

ADULT (AUTUMN)

The biggest of the terns, with a big, red bill, the Caspian Tern is a splendid and handsome bird, but is scarce almost everywhere. While its size is usually evident, it is so well-proportioned that the big bill and relatively bulky build, for a tern, may not immediately catch the eye. Against other terns it looks enormous, but while standing with big gulls, it looks lower but longer. In flight, the angular wings and large dark area under the wingtips can even create a miniature Gannet-like effect at times. It typically flies steadily over water, head angled down, looking for fish.
VOICE Deep, explosive *kree-ahk;* very noisy at breeding colony.
NESTING Shallow scrape on ground in sand or shingle; 2 or 3 eggs; 1 brood; May–June.
FEEDING Plunges for fish; may fly long distances from colony to feed.

FLIGHT: strong, powerful, easy, with slow, steady, sweeping wingbeats.

STRIKING TERN
In summer, the black cap and vivid red bill are eye-catching on this giant tern.

OCCURRENCE
Most breed in Baltic Sea on low islands and coasts. Rare migrant in E Europe and Mediterranean, rarer still in Atlantic and North Sea. Usually coastal, very rare on lakes and reservoirs inland.

Seen in the UK
J F M A M J J A S O N D

SIMILAR SPECIES

slimmer bill

SANDWICH TERN see p.232

longer tail

COMMON TERN see p.233

thin black bill

smaller

ROYAL TERN paler under wingtips in flight; see p.465

smaller and slimmer

short red legs

| Length **48–55cm (19–22in)** | Wingspan **0.96–1.11m (3–3¾ft)** | Weight **200–250g (7–9oz)** |
| Social **Flocks** | Lifespan **Up to 10 years** | Status **Endangered†** |

| Order **Charadriiformes** | Family **Sternidae** | Species *Chlidonias hybridus* |

Whiskered Tern

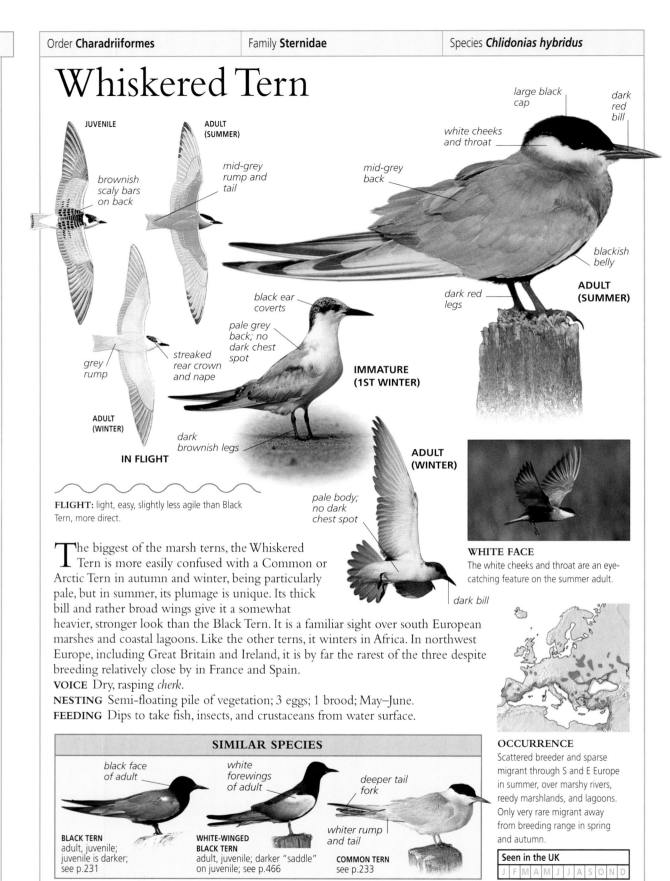

JUVENILE

brownish scaly bars on back

ADULT (SUMMER)

mid-grey rump and tail

grey rump

streaked rear crown and nape

ADULT (WINTER)

IN FLIGHT

FLIGHT: light, easy, slightly less agile than Black Tern, more direct.

large black cap

dark red bill

white cheeks and throat

mid-grey back

blackish belly

dark red legs

ADULT (SUMMER)

black ear coverts

pale grey back; no dark chest spot

IMMATURE (1ST WINTER)

dark brownish legs

pale body; no dark chest spot

ADULT (WINTER)

dark bill

WHITE FACE
The white cheeks and throat are an eye-catching feature on the summer adult.

The biggest of the marsh terns, the Whiskered Tern is more easily confused with a Common or Arctic Tern in autumn and winter, being particularly pale, but in summer, its plumage is unique. Its thick bill and rather broad wings give it a somewhat heavier, stronger look than the Black Tern. It is a familiar sight over south European marshes and coastal lagoons. Like the other terns, it winters in Africa. In northwest Europe, including Great Britain and Ireland, it is by far the rarest of the three despite breeding relatively close by in France and Spain.

VOICE Dry, rasping *cherk*.
NESTING Semi-floating pile of vegetation; 3 eggs; 1 brood; May–June.
FEEDING Dips to take fish, insects, and crustaceans from water surface.

OCCURRENCE
Scattered breeder and sparse migrant through S and E Europe in summer, over marshy rivers, reedy marshlands, and lagoons. Only very rare migrant away from breeding range in spring and autumn.

| Seen in the UK |
| J F M A M J J A S O N D |

SIMILAR SPECIES

black face of adult

white forewings of adult

deeper tail fork

whiter rump and tail

BLACK TERN
adult, juvenile; juvenile is darker; see p.231

WHITE-WINGED BLACK TERN
adult, juvenile; darker "saddle" on juvenile; see p.466

COMMON TERN
see p.233

| Length **24–28cm (9¾–11in)** | Wingspan **57–63cm (22½–25in)** | Weight **70–80g (2½–2⅞oz)** |
| Social **Flocks** | Lifespan **Up to 10 years** | Status **Declining** |

Order **Charadriiformes**	Family **Sternidae**	Species *Chlidonias niger*

Black Tern

black head

dark smoky grey
upperparts

blackish
bill

blackish legs

white
under
tail

ADULT (SUMMER)

grey
rump
and tail

sharp
wings

**ADULT
(SUMMER)**

IN FLIGHT

white
forehead

browner body
than adult's

dark
forewings

grey rump
and tail

dark chest
spot

dark
chest spot

black cap extends
as rounded lobe
onto ear coverts

JUVENILE

ADULT (WINTER)

Marsh terns (*Chlidonias* spp.) are small and delicate, dipping to the surface of water to feed rather than plunging like the sea terns (*Sterna* spp.). Of the three species, the Black Tern is the most widespread and the most uniformly dark-coloured in summer. In much of west Europe, migrating Black Terns move through in large groups, unexpectedly and quickly in spring, but small numbers are much more predictable over longer periods in autumn. Larger flocks in autumn may contain the rarer White-winged Black Tern, and so are always worth close scrutiny.

VOICE Short, low, squeaky calls, *kik, kik-keek*.
NESTING Nest of stems and waterweed in marsh; 3 eggs; 1 brood; May–June.
FEEDING Dips to take insects, small fish, crustaceans, and amphibians from water.

FLIGHT: light, buoyant, easy turns and dips to water; direct flight quite straight, rhythmic.

TYPICAL PERCH
Black Terns often settle on posts and buoys in lakes and reservoirs between bouts of feeding.

OCCURRENCE
Mostly breeds in E Europe. Widespread migrant through Europe, especially common in autumn in W, over marshes, lagoons, salt pans, and reservoirs; local in W and S Europe in summer. Occasional big flocks inland, but erratic.

Seen in the UK
J F M **A M** J J **A S O** N D

SIMILAR SPECIES

WHISKERED TERN summer, similar to adult summer; see p.230

red bill

white cheeks

bigger

WHITE-WINGED BLACK TERN summer, similar to adult summer; dark underwings; see p.466

strikingly pied

dark underwings

LITTLE GULL similar in flight; see p.238

Length **22–24cm (9–9½in)**	Wingspan **63–68cm (25–27in)**	Weight **50–75g (1¾–2⅝oz)**
Social **Flocks**	Lifespan **Up to 10 years**	Status **Declining**

Order **Charadriiformes**	Family **Sternidae**	Species ***Sterna sandvicensis***

Sandwich Tern

dark corners to tail

indistinct barring above

JUVENILE

white forehead after July

ADULT (WINTER)

ADULT (WINTER)

dark streaks on outer wings

long, slim black bill with pale yellowish tip

black cap, spiky at rear

very pale silver-grey back

white underside

ADULT (SUMMER)

white body (grey on Common and Arctic Terns)

short white tail with shallow fork

very pale silver-grey wings, blacker towards tip

black legs

IN FLIGHT

ADULT (SUMMER)

A large, active, noisy bird, the Sandwich Tern has a spiky crest, a long, sharp bill, and long, angular wings which are often held away from the body and slightly drooped. It seems almost to swagger, much more so than the smaller Common and Arctic Terns. It is equally distinctive in flight, looking very white, which helps to emphasize its size. The Sandwich Tern plunges for fish from a good height, with a loud "smack" as it enters the water. It is easily disturbed at the nesting colony and prone to desert, even after several good breeding seasons.

VOICE Distinctive, loud, harsh, rhythmic *kerr-ink* or *kear-ik* call attracts attention.
NESTING Shallow scoop in sand or shingle; 1 or 2 eggs; 1 brood; May–June.
FEEDING Catches fish, especially sandeels, in dive from air.

FLIGHT: strong, direct; wings long and angular, tail short; regular shallow wingbeats.

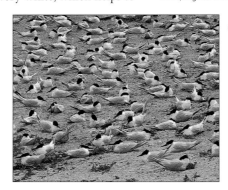

BUSY COLONY
Sandwich Tern colonies on sand dunes are large, containing hundreds of nests.

OCCURRENCE
Widespread but local breeding bird north to Baltic. Prefers sandy coasts, shallow coastal lagoons, and offshore islands. Rare migrant inland but quite widely seen on all kinds of coasts.

Seen in the UK
J F **M A M J J A S O** N D

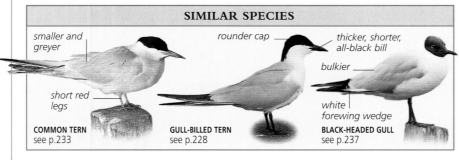

SIMILAR SPECIES

smaller and greyer

short red legs

COMMON TERN
see p.233

rounder cap

thicker, shorter, all-black bill

GULL-BILLED TERN
see p.228

bulkier

white forewing wedge

BLACK-HEADED GULL
see p.237

Length **36–41cm (14–16in)**	Wingspan **95–105cm (37–43in)**	Weight **210–260g (7–9oz)**
Social **Flocks**	Lifespan **Up to 10 years**	Status **Declining**

Order **Charadriiformes**	Family **Sternidae**	Species ***Sterna hirundo***

Common Tern 🔊 37.I, 37.II

black cap

grey upperparts

black-tipped, bright red bill

outer primaries darker than inner ones, with darker "notch" effect

thick dark band on underwings

ADULT (SUMMER)

translucent patch on underwings

long bill

pale short primaries contrast with darker longer ones (all the same shade on Arctic Tern)

long neck

IN FLIGHT

grey below

red legs (longer than Arctic Tern)

ADULT (SUMMER)

upperwing has blackish front, pale centre, darker grey band close to trailing edge

dark nape

pale base to bill

slight washed-ginger barring on upperparts

dark shoulder

white forehead

dark shoulder

pale grey rump

long legs

(underwing same as adult)

JUVENILE

dark streaks on outer wings

ADULT (WINTER)

The most likely tern to be seen inland over most of Europe, the Common Tern is very much a bird of the coast in most of its range. It is a typical black-capped, pale-bodied tern, rather grey, with a red bill and legs. It usually plunge-dives for prey, which it may quickly swallow or carry off in its bill for either its mate or chicks back at the nest. The Common Tern often mixes with Arctic and Sandwich Terns.

VOICE Grating, thin, falling *kreee-yair*, sharp *kik kik*, ringing *keeer*, rapid *kirrikirrikirrik*.

NESTING Scrape in sand or dry earth on ground; 2–4 eggs; 1 brood; May–June.

FEEDING Plunges from air for fish and aquatic invertebrates; picks some insects and fish from water surface in flight.

FLIGHT: steady, relaxed, shallow, spring-like wingbeats; soars high above colony.

DIVING FOR FISH
The Common Tern is a classic plunge-diver, hovering before diving headlong for fish.

OCCURRENCE
Widespread, breeding inland in C and E Europe, mostly in coasts in W Europe, but also locally on gravel pits and shingly rivers. Migrant almost everywhere on coasts; moderately common inland on freshwater areas.

Seen in the UK
J F **M A M J J A S O** N D

SIMILAR SPECIES

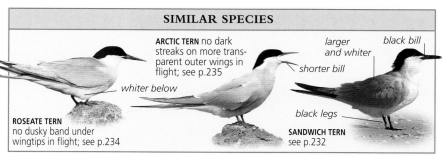

ARCTIC TERN no dark streaks on more transparent outer wings in flight; see p.235

whiter below

ROSEATE TERN no dusky band under wingtips in flight; see p.234

larger and whiter

black bill

shorter bill

black legs

SANDWICH TERN see p.232

Length **31–35cm (12–14in)**	Wingspan **82–95cm (32–37in)**	Weight **90–150g (3¼–5oz)**
Social **Flocks**	Lifespan **Up to 10 years**	Status **Secure**

233

| Order **Charadriiformes** | Family **Sternidae** | Species *Sterna dougallii* |

Roseate Tern

ADULT (LATE SUMMER)

blackish streaks on wingtips

IN FLIGHT

long white tail streamers

black-brown bars on back

dusky forehead

blackish legs (reddish on Common and Arctic terns)

JUVENILE

pale underwings with no dark trailing edge

ADULT

black bill (red base more extensive in late summer)

smooth black cap

very pale grey upperparts

longish red legs

white underside, flushed pink (grey on Common and Arctic terns)

ADULT (SPRING)

FLIGHT: quite quick like large Little Tern; stocky body, shortish wings, and very long tail distinctive.

With worldwide populations going down, this is a rare bird in Europe, having suffered prolonged declines, which are as much to do with problems in West Africa where it spends the winter as in Europe. It forms a similar trio with Common and Arctic Terns but is rather easier to tell from them, with some features more reminiscent of the whiter, larger Sandwich Tern. Like the Arctic Tern, but unlike the Common Tern, it is unlikely to be seen in northwest Europe in full winter plumage.

VOICE Harsh croaking note and musical, quick *chu-vik*, unlike other terns.

NESTING Grassy nest often in tall vegetation or under shelter of tussock; 1 or 2 eggs; 1 brood; May–June.

FEEDING Plunges for fish, especially sandeels and sprats after fast, winnowing hover.

ELEGANT DISPLAY
Roseate Terns are at their most graceful when they are displaying during their spring courtship.

OCCURRENCE
Scattered very locally in Great Britain, Ireland, and NW France, breeding in small numbers on vegetated islands. Scarce or rare migrant off headlands or at mouth of estuaries; extremely rare inland.

| Seen in the UK |
| J F M **A M J J A S O** N D |

SIMILAR SPECIES

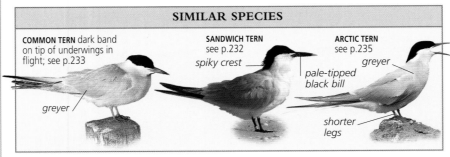

COMMON TERN dark band on tip of underwings in flight; see p.233

greyer

SANDWICH TERN see p.232
spiky crest

ARCTIC TERN see p.235
greyer

pale-tipped black bill

shorter legs

| Length **33–38cm (13–15in)** | Wingspan **75–80cm (30–32in)** | Weight **95–130g (3⅜–5oz)** |
| Social **Small flocks** | Lifespan **Up to 10 years** | Status **Endangered** |

Order **Charadriiformes**	Family **Sternidae**	Species ***Sterna paradisaea***

Arctic Tern

thin line on outer edge of wings

round black cap (with whiter forehead in winter)

short red bill (blacker in winter)

very pointed, tapered wingtips

ADULT

very pale outer wings; no dark notch

thin dark line on translucent flight feathers

white tail with long outer streamers

ADULT (SUMMER)

looks neckless, short-billed

closed wingtip, all same shade of grey

grey back

grey underside

IN FLIGHT

short red legs

ADULT (SUMMER)

dark "V"s and crescent markings on back

black bill

upperwing has blackish front, grey centre, and white rear

white rump

short legs

JUVENILE

underwing same as adult

A more northerly bird and more strictly maritime than the Common Tern, the Arctic Tern forms the slightly more elegant half of one of the really difficult species pairings in Europe. Good views are usually needed to separate it from the Common Tern. Arctic Terns breed as far north as any bird, and also winter as far south as almost any other: they are often credited with enjoying more hours of daylight than any other bird on earth.

VOICE Grating, sharp *kee-yaah*, rising *pee-pee-pee, kik, kreerr*.
NESTING Scrape in sand or shingle, or hollow in rock; 2 eggs; 1 brood; May–June.
FEEDING Plunges for fish, often pausing at intervals before final dive; takes some insects from water surface.

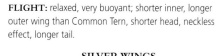

FLIGHT: relaxed, very buoyant; shorter inner, longer outer wing than Common Tern, shorter head, neckless effect, longer tail.

SILVER WINGS
The uniformly pale silver-grey upperwing of the adult shows well here as it feeds its chick.

OCCURRENCE
Breeds in far north, south to Great Britain, mostly on offshore islands, also on sandy or gravelly beaches. Migrant around North Sea and Atlantic coasts, usually scarce inland, occasional flocks appearing briefly in spring.

Seen in the UK
J F **M A M J J A S O** N D

SIMILAR SPECIES

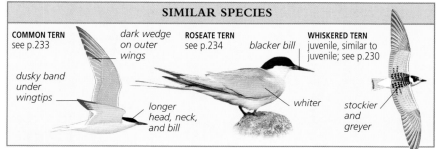

COMMON TERN see p.233

dark wedge on outer wings

dusky band under wingtips

longer head, neck, and bill

ROSEATE TERN see p.234

blacker bill

whiter

WHISKERED TERN juvenile, similar to juvenile; see p.230

stockier and greyer

Length **32–35cm (12½–14in)**	Wingspan **80–95cm (32–37in)**	Weight **80–110g (2⅞–4oz)**
Social **Flocks**	Lifespan **Up to 10 years**	Status **Secure**

Order **Charadriiformes** | Family **Laridae** | Species *Larus tridactyla*

Kittiwake 🔊 36

black triangle on wingtips without white spots

pale outer wings

ADULT

black collar

JUVENILE

black zigzag on wings

IN FLIGHT

dark eye

white head and breast

pale green-yellow bill

blue-grey back

short blackish legs

ADULT (SUMMER)

collar turns grey

black zigzag mark across upperwings (like Little Gull)

IMMATURE (1ST WINTER)

dull and dingy by summer

IMMATURE (1ST SUMMER)

dark ear-spot

grey back of head

all-black wingtips

ADULT (WINTER)

FLIGHT: easy, elegant; in wind, bounds in series of steep, arcing banks over waves on angled wings.

One of the most maritime of the gulls, the Kittiwake mostly comes to land only to breed, but some also visit freshwater pools near the coast to drink and bathe. Flocks of non-breeding immatures may loaf about on beaches or flat rocky platforms during the summer and a few, in winter, stay around harbours. Most feed well out at sea and, in winter, lead a tough life, enduring the gales and rain of mid-ocean for months. In summer, they nest on sheer cliffs with the tiniest of ledges and make the coast ring to their distinctive calls. They are often in large colonies close to Guillemots, Razorbills, and Puffins (see pp.219, 217, and 215).

VOICE Ringing, nasal, rhythmic *kiti-a-wake!* often repeated in summer, also high, thin mewing note.

NESTING Nest of weed on tiny ledge on sheer cliff or seaside building; 2 or 3 eggs; 1 brood, May–June.

FEEDING Takes mostly fish from surface or in shallow dive; eats offal from trawlers.

EYE-CATCHING CHICK
The sharp black collar and wing markings are obvious on juveniles.

OCCURRENCE
Breeds on sheer northern and western coastal cliffs often in mixed seabird colonies. Widespread at sea in winter but scarcer on coasts; common off headlands and rare but regularly inland on migration.

| Seen in the UK |
| J F M A M J J A S O N D |

SIMILAR SPECIES

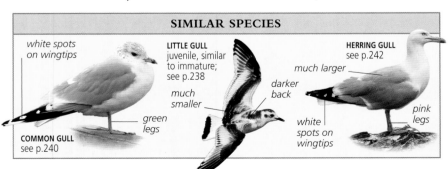

white spots on wingtips

green legs

COMMON GULL
see p.240

LITTLE GULL
juvenile, similar to immature; see p.238

much smaller

darker back

white spots on wingtips

HERRING GULL
see p.242

much larger

pink legs

Length **38–40cm (15–16in)** | Wingspan **0.95–1.1m (3–3½ft)** | Weight **300–500g (11–18oz)**

Social **Large flocks** | Lifespan **Up to 10 years** | Status **Secure**

| Order **Charadriiformes** | Family **Laridae** | Species ***Chroicocephalus ridibundus*** |

Black-headed Gull 🔊 34.I, 34.II

brown on neck and back

JUVENILE

white front of outer wings

black on trailing edge

white outer edge of dark grey underwings

ADULT (WINTER)

very pale grey back

IN FLIGHT

dark brown hood

deep red bill

black-tipped bill

neck and back become grey

brown band on wings

orange-buff legs

IMMATURE (1ST WINTER)

long, dark wing point

white head with dark ear spot

vivid red bill

deep red legs

white front of wings

dark rear edge

vivid red legs

ADULT (WINTER)

ADULT (SUMMER)

dark tail tip

IMMATURE (1ST WINTER)

Common and familiar, this is a small, agile, very white-looking gull and is never truly black-headed: it is one of the "hooded" gulls with a dark brown head when breeding but a pale head with a dark ear spot in other plumages. Its dark underwing gives a flickering effect in flight. It has always been a frequent bird inland, by no means confined to the sea or the coast. Numbers have increased somewhat with extra reservoirs and flooded pits providing safe roosts and refuse tips offering an abundance of food.

VOICE Loud, squealing, laughing, and chattering calls, *kwarr, kee-arr, kwuk, kuk-kuk*.
NESTING Pile of stems on ground in vegetation, on marsh; 2 or 3 eggs; 1 brood; May–June.
FEEDING Takes worms, seeds, fish, and insects, from ground and water; catches insects in flight.

FLIGHT: light, buoyant, very agile; glides a lot; steady beats of pointed wings.

OCCURRENCE
Breeds from coastal marshes to upland pools, widespread but local. Often numerous and widespread at other times, from coasts to farmland, reservoirs, refuse tips, and along rivers through towns and cities; in summer, even high on hills.

Seen in the UK
J F M A M J J A S O N D

SIMILAR SPECIES

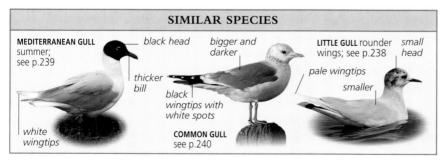

MEDITERRANEAN GULL summer; see p.239

black head

thicker bill

white wingtips

COMMON GULL see p.240

black wingtips with white spots

bigger and darker

LITTLE GULL rounder wings; see p.238

pale wingtips

small head

smaller

| Length **34–37cm (13½–14½in)** | Wingspan **1–1.1m (3¼–3½ft)** | Weight **225–350g (8–13oz)** |
| Social **Large flocks** | Lifespan **10–15 years** | Status **Secure** |

| Order **Charadriiformes** | Family **Laridae** | Species *Hydrocoloeus minutus* |

Little Gull

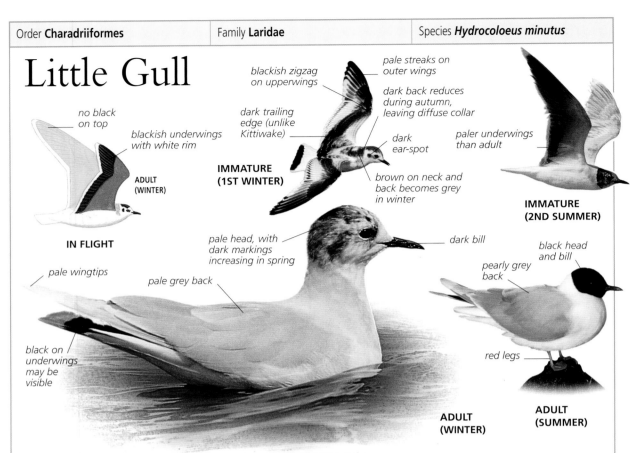

no black on top

blackish underwings with white rim

ADULT (WINTER)

IN FLIGHT

blackish zigzag on upperwings

dark trailing edge (unlike Kittiwake)

IMMATURE (1ST WINTER)

pale streaks on outer wings

dark back reduces during autumn, leaving diffuse collar

dark ear-spot

brown on neck and back becomes grey in winter

paler underwings than adult

IMMATURE (2ND SUMMER)

pale head, with dark markings increasing in spring

dark bill

pale wingtips

pale grey back

black head and bill

pearly grey back

black on underwings may be visible

red legs

ADULT (WINTER)

ADULT (SUMMER)

Short-legged, small-billed, delicate and elegant, this gull is reminiscent of the marsh terns, feeding like a Black Tern over open water. It tends to appear over lakes and reservoirs in small groups in spring and autumn, also like the terns, but immatures may linger for weeks in the summer. It combines the typical "hooded" gull sequence of plumages with a strongly contrasted immature pattern rather like that of the Kittiwake. In most of Europe, it is much less abundant than other gulls, with which it often associates.

VOICE Low, rapid tern-like calls, *kek-kek-kek, akar akar akar.*

NESTING Grassy nest on ground or in dense marsh vegetation; 3 eggs; 1 brood; May–June.

FEEDING Mostly picks up insects, aquatic invertebrates, and fish from surface of water in dipping flight.

FLIGHT: light, buoyant, erratic; shallow, quick flicks of wings, frequent turns.

PALE UPPERWINGS
Adults have no trace of black on the upperside of the wings.

OCCURRENCE
Mostly breeds in E Europe, on wet grassy marshes and floods; at other times, on coastal lagoons (around coasts and over reservoirs on migration). Winters west to Ireland and frequent migrant on W. European coasts, but mostly scarce and somewhat erratic inland.

Seen in the UK
| J | F | M | A | M | J | J | A | S | O | N | D |

SIMILAR SPECIES

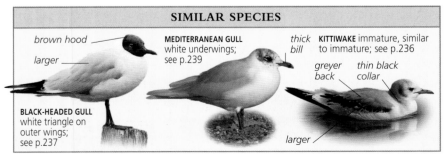

brown hood

larger

BLACK-HEADED GULL white triangle on outer wings; see p.237

MEDITERRANEAN GULL white underwings; see p.239

thick bill

KITTIWAKE immature, similar to immature; see p.236

greyer back

thin black collar

larger

| Length **25–27cm (10–10½in)** | Wingspan **70–77cm (28–30in)** | Weight **90–150g (3¼–5oz)** |
| Social **Small flocks** | Lifespan **5–10 years** | Status **Declining** |

| Order **Charadriiformes** | Family **Laridae** | Species *Larus melanocephalus* |

Mediterranean Gull

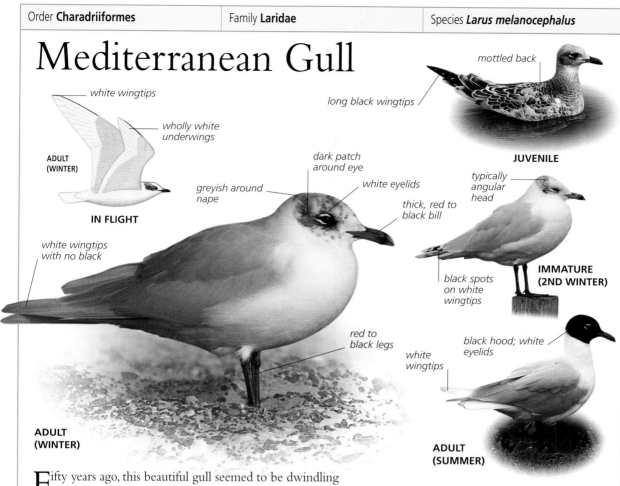

white wingtips

wholly white underwings

ADULT (WINTER)

IN FLIGHT

white wingtips with no black

greyish around nape

dark patch around eye

white eyelids

thick, red to black bill

red to black legs

ADULT (WINTER)

mottled back

long black wingtips

JUVENILE

typically angular head

black spots on white wingtips

IMMATURE (2ND WINTER)

white wingtips

black hood; white eyelids

ADULT (SUMMER)

Fifty years ago, this beautiful gull seemed to be dwindling towards eventual oblivion, but recently its numbers have staged a remarkable recovery; it has spread, albeit patchily, to areas of western Europe far beyond its previous range. Along the North Sea and English Channel coasts, it has become a regular nonbreeding visitor, and now nests in a number of Black-headed Gull colonies. It is a tricky bird to find among large numbers of more common gulls in some plumages, but breeding plumage adults are highly distinctive.

VOICE Nasal, rising and falling *eeu-err eeu-err*.
NESTING Grass-lined nest on sand, shingle, or in marsh; 3 eggs; 1 brood; May–June.
FEEDING Forages for fish, aquatic invertebrates, worms, and offal on beaches, and in tips, fields, and sewage outflows.

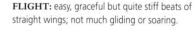

FLIGHT: easy, graceful but quite stiff beats of straight wings; not much gliding or soaring.

IMMATURE
The immature (1st winter) has a dark mask, a pale grey panel on the upperwing, and black wingtips and tail tip. Common Gull immatures are darker on the back.

OCCURRENCE
Breeds on shallow lagoons and coastal marshes, scattered and rare in W Europe, more common in SE. In winter, on estuaries, beaches, lakes, harbours, and at times at tips but rare far inland, mostly in E Europe, increasing in NW Europe, especially English Channel region.

Seen in the UK
J F M A M J J A S O N D

SIMILAR SPECIES

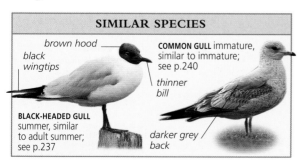

brown hood

black wingtips

BLACK-HEADED GULL summer, similar to adult summer; see p.237

COMMON GULL immature, similar to immature; see p.240

thinner bill

darker grey back

| Length **36–38cm (14–15in)** | Wingspan **0.98–1.05m (3¼–3½ft)** | Weight **200–350g (7–13oz)** |
| Social **Flocks** | Lifespan **10–15 years** | Status **Secure** |

Order **Charadriiformes**	Family **Laridae**	Species **Larus canus**

Common Gull

large white spot on black wingtips

ADULT (WINTER)

grey back

dull yellow-green bill

grey-brown markings on head

black eye (pale on Herring Gull)

mid-grey back

IMMATURE (1ST WINTER)

brown wings, fading to buff, with dark brown tips

bold white crescent between grey back and black wingtips

black band on white tail

IN FLIGHT

white spots on wingtips

IMMATURE (1ST WINTER)

green to yellow-green legs

mid-grey back

buff-grey bill with black tip

dark eye

long, slim shape

no red on small, slim bill

ADULT (WINTER)

wings fade much paler

pale buff-pink legs

white head

green legs

IMMATURE (1ST SUMMER)

ADULT (SUMMER)

FLIGHT: fluent, easy, relaxed; few glides; little or no soaring.

Rather like the Herring Gull in its general pattern, the Common Gull is not as common in many areas and even in winter, when it is more widespread, it has a curiously local distribution. In England, for example, it is abundant on fields in some counties but quite scarce in others nearby. It becomes adult in three years, taking longer than the smaller gulls but a year or two less than the larger species. The plumage changes that occur with age and season are easily seen but, as with other gulls, male and female are alike.

VOICE Loud, high, nasal, squealing *kee-ee-ya, kee-ar-ar-ar-ar*, short *gagagaga*.
NESTING Pad of grass on ground or low stump; 2 or 3 eggs; 1 brood; May–June.
FEEDING Takes worms, insects, fish, and molluscs from ground or water.

OCCURRENCE
Widespread but local, breeding on coasts and moors in N and NW Europe. In winter, on farmland, especially grassy pastures, all kinds of coasts, reservoirs, some on tips, but generally more unevenly distributed than Black-headed Gull and less universally common.

Seen in the UK
J F M A M J J A S O N D

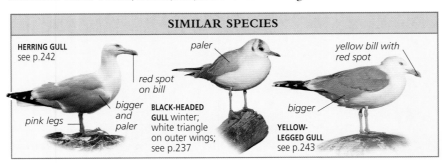

SIMILAR SPECIES

HERRING GULL see p.242

paler

yellow bill with red spot

red spot on bill

bigger and paler

BLACK-HEADED GULL winter; white triangle on outer wings; see p.237

bigger

YELLOW-LEGGED GULL see p.243

pink legs

Length **38–44cm (15–17½ in)**	Wingspan **1.05–1.25m (3½–4ft)**	Weight **300–500g (11–18oz)**
Social **Flocks**	Lifespan **Up to 10 years**	Status **Declining**

| Order **Charadriiformes** | Family **Laridae** | Species **Larus fuscus** |

Lesser Black-backed Gull

black wingtips with one white primary spot

dark grey band on underwings

ADULT (WINTER)

IN FLIGHT

white head

yellow bill with red spot

ADULT (SUMMER)

bright yellow legs

densely streaked grey-brown head

slaty grey back

dull yellow legs

ADULT (WINTER)

flight feathers all dark

mottled, dark brown body

black bill

IMMATURE (1ST YEAR)

long wingtips

back turns dark grey

IMMATURE (2ND YEAR)

In summer, this is a remarkably handsome gull, immaculate in slate-grey and pure white with vivid yellow legs and bill. It has the same basic pattern as other "white-headed" gulls, with black and white wingtips. The black areas have extra pigment that strengthens them, while the white spots are weak and disappear as the feathers become old and worn. The Lesser Black-backed Gull used to be a summer visitor to west Europe, but has established large wintering populations inland. Nevertheless, it remains a strong migrant and can often be seen in spring and autumn, high overhead, flying over land.

VOICE Deep, throaty, wailing calls, various barks, yelps, *kyow, kyow-yow-yow, ga-ga-ga*.

NESTING Pile of grass on ground; 2 or 3 eggs; 1 brood; May.

FEEDING Takes fish, worms, molluscs, and edible refuse; feeds on seabirds in summer.

REMARK Subspecies *L. f. graellsii* (NW Europe) is palest above; *L. f. fuscus* (Scandinavia) is smaller, blacker, white-headed all year, long-winged, and more marine.

FLIGHT: majestic, gliding, soaring; direct flight powerful with regular beats.

HANDSOME ADULT
In spring, this is one of the most immaculate of European gulls, with a vivid bill and leg colours.

SUBSPECIES

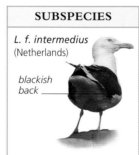

L. f. intermedius (Netherlands)

blackish back

OCCURRENCE
Breeds on cliffs, islands, moorland, and rooftops in N and NW Europe. In winter, at tips and reservoirs, on beaches, and often on farmland; most migrate south to Africa but many remain in W Europe. Parties often fly high over land in spring.

Seen in the UK
J F M A M J J A S O N D

SIMILAR SPECIES

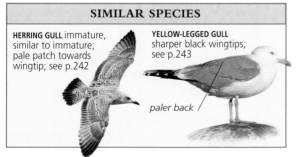

HERRING GULL immature, similar to immature; pale patch towards wingtip; see p.242

YELLOW-LEGGED GULL sharper black wingtips; see p.243

paler back

| Length **52–67cm (20½–26in)** | Wingspan **1.28–1.48m (4¼–4¾ft)** | Weight **650–1,000g (23–36oz)** |
| Social **Large flocks** | Lifespan **10–15 years** | Status **Secure** |

| Order **Charadriiformes** | Family **Laridae** | Species ***Larus argentatus*** |

Herring Gull  35

yellow bill with red spot

pure white head from January to early autumn

pale grey back

pale pink legs

white spots on black wingtips

pale area behind bend of wing in flight

brown blotched body

IMMATURE

ADULT (WINTER)

ADULT (SUMMER)

IN FLIGHT

grey-brown-streaked head (white on Yellow-legged Gull)

ADULT (WINTER)

ADULT (SUMMER)

Often considered a nuisance in towns, where it breeds on rooftops and begins calling very loudly, early on summer mornings, the Herring Gull has actually declined over much of Europe. It is mainly a bird of sea cliffs in summer, but roams over all kinds of shorelines and far inland, feeding on tips and roosting on large reservoirs. Flocks returning to evening roosts in long lines or "V"s look dramatic. In winter, groups typically forage around outflows from pipes and sewers, around small harbours, or out on the mudflats at low tide.

VOICE Loud, squealing notes, yelps, barks, *kyow, kee-yow-yow-you, ga-ga-ga, kuk-kuk*.

NESTING Grass-lined nest on ground, cliff ledge, or building; 2 or 3 eggs; 1 brood; May.

FEEDING Takes fish, molluscs, insects, offal, and scraps of all kinds from ground or water.

FLIGHT: steady, powerful, with continual easy wingbeats; masterly soaring, gliding.

IMMATURE
It takes about four years for the clear grey back and whiter underside to gradually appear.

SUBSPECIES

L. a. argentatus (Scandinavia) winter

larger, darker grey

less black on wingtips

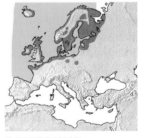

OCCURRENCE
Breeds widely in NW Europe on cliffs, islands, and rooftops. Wide-spread in winter on beaches, reservoirs, often abundant at refuse tips, frequent on adjacent farmland, and likely to fly over almost anywhere from time to time.

Seen in the UK
| J | F | M | A | M | J | J | A | S | O | N | D |

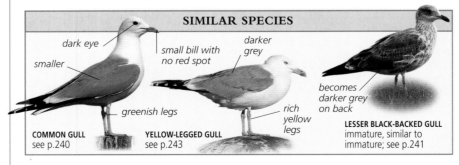

SIMILAR SPECIES

dark eye

smaller

greenish legs

COMMON GULL see p.240

small bill with no red spot

rich yellow legs

YELLOW-LEGGED GULL see p.243

darker grey

becomes darker grey on back

LESSER BLACK-BACKED GULL immature, similar to immature; see p.241

| Length **55–67cm (22–26in)** | Wingspan **1.3–1.6m (4¼–5¼ft)** | Weight **750–1,250g (27–45oz)** |
| Social **Flocks** | Lifespan **10–20 years** | Status **Secure** |

| Order **Charadriiformes** | Family **Laridae** | Species *Larus (cachinnans) michahellis* |

Yellow-legged Gull

sharply defined wingtips with extensive black and white

white head, smudged grey only in late summer

mid-grey back

vivid yellow bill with large red spot

long black wingtips, white spots wear off in summer

blackish bill

pale to deep yellow legs

ADULT

IN FLIGHT

IMMATURE (1ST WINTER)

ADULT

Until recently, this was considered a race of the Herring Gull. It is essentially the Mediterranean replacement of the more northerly Herring Gull, with darker races on the Atlantic islands (Canaries, Azores, and Madeira) and other birds, perhaps of a different species again (the Caspian Gull, *L. cachinnans*), to the east. In Asia, the situation is more complex, with intermediate forms between Herring and Lesser Black-backed Gulls. Mediterranean Yellow-legged Gulls are big, handsome birds, with a close relationship to the typical seaside Herring Gull of northwest Europe. They now breed side-by-side in a few places, without hybridizing.

VOICE Deeper than Herring Gull's, more like Lesser Black-backed Gull's.
NESTING Pile of grass on ground, cliff ledge, or building; 2 or 3 eggs; 1 brood; May.
FEEDING Takes aquatic invertebrates, molluscs, fish, and offal, from water or ground.

FLIGHT: strong, easy, elegant, with powerful, shallow wingbeats.

L. cachinnans

Caspian Gull
A rare visitor to W Europe.

pear-shaped head

more white in wingtip

long bill

long legs

BOLD PATTERN
The adult Yellow-legged Gull has a striking contrast between the extensive black wingtip and the rest of the underwing.

OCCURRENCE
In summer, breeds in S Europe, chiefly on rocky islands and off-shore stacks but often scavenges around docks and towns. In late summer/autumn, moves north; frequent in Low Countries and SE England, on tips and beaches.

Seen in the UK
J F M A M J J A S O N D

SIMILAR SPECIES

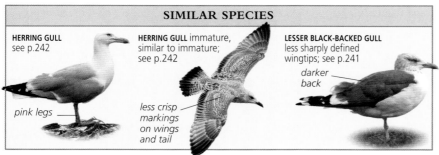

HERRING GULL see p.242

pink legs

HERRING GULL immature, similar to immature; see p.242

less crisp markings on wings and tail

LESSER BLACK-BACKED GULL less sharply defined wingtips; see p.241

darker back

Length **55–65cm (22–26in)**	Wingspan **1.3–1.5m (4¼–5ft)**	Weight **750–1200g (27–43oz)**	
Social **Flocks**	Lifespan **Up to 10 years**	Status **Secure†**	243

Order **Charadriiformes**	Family **Laridae**	Species *Larus glaucoides*

Iceland Gull

wingtips fade to buff-ivory

barring on oatmeal-brown body

round head

blackish bill with dull pale base

long wings

clouded buff-brown and grey from head to breast

short, pale yellow bill with red spot

pale grey back

IMMATURE (1ST WINTER)

rounded, dumpy body

IMMATURE (1ST WINTER)

long wings, pot-belly

ADULT (WINTER)

wingtips extend well beyond tail

short legs

ADULT (WINTER)

IN FLIGHT

white wingtips

It is unusual to find two species so closely matched in plumage colour and pattern as Iceland and Glaucous Gulls. The Iceland Gull is nearly always the scarcer of the two, but appears inland as well as around coastal harbours (and well out at sea) in ones and twos – visitors from Arctic Greenland. It is a handsome bird, especially in summer plumage; like Glaucous Gulls, the oatmeal-coloured immatures with ivory wingtips are striking. To separate the two species, details of shape and structure are more important than plumage.

VOICE Shrill squealing notes and barking calls like Herring Gull.
NESTING Small grassy nest on cliff ledges or ground; 2 or 3 eggs; 1 brood; June.
FEEDING Fish, molluscs, crustaceans, rubbish, and offal, from water, fields, and rubbish tips.

FLIGHT: steady, easy, heavy-bellied; wings rather straight-out, taper to point.

IMMATURE
In their first and second years, Iceland Gulls fade almost to white by summer, and are difficult to age with certainty.

OCCURRENCE
Breeds in Greenland. In winter, common in Iceland, much scarcer in Great Britain and Ireland, often following fishing vessels and seen around harbours; generally rare in mainland Europe. Usually in flocks of more common gulls at tips, reservoirs, and beaches.

Seen in the UK
J F M A M J J A S O N D

SIMILAR SPECIES

GLAUCOUS GULL see p.245

GLAUCOUS GULL immature, similar to immature; see p.245

dark-tipped pink bill

blackish wingtip

shorter wings

thicker bill

larger

HERRING GULL immature, similar to immature; see p.242

dark wingtips

Length **52–60cm (20½–23½in)**	Wingspan **1.3–1.45m (4¼–4¾ft)**	Weight **750–1000g (27–36oz)**
Social **Flocks**	Lifespan **Up to 10 years**	Status **Secure†**

Order **Charadriiformes**	Family **Laridae**	Species *Larus hyperboreus*

Glaucous Gull

buff wingtips

pure white wingtips

mottled brown plumage

JUVENILE (1ST WINTER)

cloudy grey-brown from head to chest

pale grey upperparts

red spot on large yellow bill

underparts darker than back

short white wingtips extend just beyond tail

long wings, heavy, flat-bellied body

JUVENILE (1ST WINTER)

ADULT (WINTER)

IN FLIGHT

barred, mottled, oatmeal-brown plumage

pale pink bill with sharp black tip

pale pink legs

ADULT (WINTER)

ivory-buff wingtips

pale tail

IMMATURE (1ST WINTER, FADED)

FLIGHT: majestic, often gliding and soaring; strong, deep wingbeats in rather sluggish direct flight.

This is essentially a winter bird in Europe (although it does breed in Iceland and Spitsbergen), hanging on into early spring in northwest Europe while the snow lingers farther north. This fiercely predatory gull follows fishing fleets and is found around northern harbours, but also joins inland gull flocks, feeding on refuse tips and roosting on reservoirs. Finding "white-winged" gulls (Glaucous and Iceland) in winter flocks is an interesting challenge: distinguishing between the two can be difficult.

VOICE Much like Herring Gull, wailing and yapping notes.
NESTING Pad of grass and stems on cliff ledges or ground; 2 or 3 eggs; 1 brood; May–June.
FEEDING Takes fish, invertebrates, and all kinds of offal and rubbish; more predatory in summer.

ELEGANT SUMMER PLUMAGE
In its immaculate grey and white summer plumage, the Glaucous Gull is a handsome bird.

SIMILAR SPECIES

ICELAND GULL see p.244

smaller bill

blackish wingtip

dark tail-band and wingtips

slightly smaller

HERRING GULL immature, similar to immature; see p.242

OCCURRENCE

Breeds locally in Iceland. Scarce in winter on beaches, around harbours, tips, and reservoirs in NW Europe, usually among flocks of more common gulls, and often in groups of gulls around trawlers far out at sea.

Seen in the UK
J F **M A** M J J A S O **N D**

Length **62–70cm (24–28in)**	Wingspan **1.42–1.62m (4¾–5¼ft)**	Weight **1–2kg (2¼–4½lb)**
Social **Flocks**	Lifespan **10–20 years**	Status **Secure**

| Order **Charadriiformes** | Family **Laridae** | Species *Larus marinus* |

Great Black-backed Gull

large white patch on outer wings

broad wings

faint markings on white head

dark flight feathers

ADULT (SUMMER)

black back darker than Lesser Black-backed Gull

ADULT (WINTER)

IN FLIGHT

huge size obvious

ADULT (WINTER)

large yellow bill with red spot

white underside

pale greyish, whitish, or pink legs

whitish head

black bill

IMMATURE (1ST WINTER)

pale head

chequered back

IMMATURE (2ND WINTER)

white head

ADULT (SUMMER)

FLIGHT: strong but heavy, with slow, deep, sweeping wingbeats.

This is the world's largest gull, heavily built, big-billed, and fiercely predatory. The size of the bill is a good guide to its identity even in immature plumages. It is generally less abundant than the Herring Gull, although it does form flocks of hundreds in areas where it is common, even in summer when such gatherings follow trawlers off northern Scotland. In winter, it usually forms only a small proportion of the gull flocks on reservoirs inland but is widespread on many coasts.

VOICE Deep, barking notes, hoarse *yowk*, gruff *ow-ow-ow*.
NESTING Shallow grass- or weed-lined scrape on cliff ledge or pinnacle; 3 eggs; 1 brood; May–June.
FEEDING Bold and predatory in summer, eating seabirds and voles; eats fish, crustaceans, offal, and rubbish from sea, beaches, and tips.

POWERFUL PRESENCE
Very big and strongly contrasted in pattern, Great Black-backed Gulls always dominate other gulls.

SIMILAR SPECIES

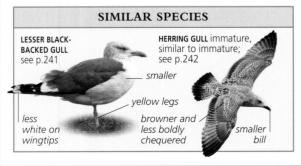

LESSER BLACK-BACKED GULL see p.241

smaller

yellow legs

less white on wingtips

HERRING GULL immature, similar to immature; see p.242

browner and less boldly chequered

smaller bill

OCCURRENCE
Widespread in NW Europe, on rocky coasts, breeding sparsely on rock pinnacles and offshore stacks, often in flocks around coastal pools. In winter, on beaches, harbours, tips, and reservoirs, increasingly inland in W Europe.

| Seen in the UK |
| J F M A M J J A S O N D |

| Length **64–78cm (25–31in)** | Wingspan **1.5–1.7m (5–5½ft)** | Weight **1–2.1kg (2¼–4¾lb)** |
| Social **Flocks** | Lifespan **10–20 years** | Status **Secure** |

PIGEONS AND DOVES

This is a rather artificial distinction in a large family found worldwide. In Europe, some species commonly called "doves" (Stock Dove and Rock Dove) are more like the "pigeons", being rather larger and heavier, and shorter-tailed, than the more delicate species that are usually called doves.

They are all round-bodied, soft-plumaged birds whose plumage seems to come away easily in an "explosion" of feathers in any collision, perhaps helping them to escape predators. They have short, usually red, legs and small bills with the nostrils in a fleshy bump at the base (the "cere"). Bill and leg colours can be bright and obvious but are of limited value in identification, which more often rests on differences in wing and tail patterns. There is little variation in appearance between sexes and seasons, and juveniles look much like their parents, generally being a little duller or marked with paler feather edges.

Pigeons and doves drink by sucking water up in a continuous draught, unlike other birds that have to raise their heads to tip water back into their throats. They have loud, simple vocalizations with little obvious differentiation between calls and songs, which are good identification clues, but, with the exception of the Collared Dove, do not call in flight. Their wings make loud clapping sounds, either in display or in a sudden take-off when disturbed, serving the purpose of alarm calls.

Nests are flimsy affairs and eggs are always pure, unmarked white. Shells may be found on the ground where they are dropped, far from the nest, by the parent birds after hatching. Breeding seasons are long and nesting is timed to coincide with a local abundance of food.

COLUMBIDAE

glossed purple and green neck

ROCK DOVE
p.248

glossed green neck

STOCK DOVE
p.249

bold white patch on each side of neck

WOODPIGEON
p.250

pale, grey-brown body

COLLARED DOVE
p.251

dark spots on bright brown back

TURTLE DOVE
p.252

CUCKOOS

CUCULIDAE

Europe has two species but there are many cuckoos elsewhere; they are brood parasites, laying their eggs in the nests of other species which then unwittingly rear their young. Hence cuckoos are not found in family groups.

Cuckoos have short, curved bills, small heads, rather long, broad tails, and wide-based, tapered wings that give a curious appearance in flight, beating below body level.

grey bars on white underside

CUCKOO
p.253

247

| Order **Columbiformes** | Family **Columbidae** | Species ***Columba livia*** |

Rock Dove

IN FLIGHT
- big white patch on back
- dark tail band
- white underwings

FERAL/TOWN PIGEON
- larger white spot

- tiny white spot on bill
- glossed purple and green neck
- pale grey back
- two broad, long black bars across wings
- dark underside

FLIGHT: fast, swooping; direct flight striking, wings swept back, quick, deep beats; swoops and glides around cliffs with wings in steep "V".

Ancestor of the domestic pigeon, the Rock Dove as a truly wild bird has long been "diluted" in most areas by domestic pigeons reverting to the wild state: the "feral" or town pigeons. Only in parts of northwest Europe do "pure" Rock Doves with immaculate plumage patterns persist: in most areas, different plumage patterns and colours are found even in groups living entirely wild. The true Rock Dove is a bird of cliffs, inland or – more particularly in Europe – by the sea, from which it moves to fields to feed each day.

VOICE Deep, rolling, moaning coo, *oo-ooh-oorr, oo-roo-coo*.
NESTING Loose, untidy, sparse nest on ledge, in cavity; 2 eggs; 3 broods; all year.
FEEDING Forages on fields and rough grassland, picking up seeds, buds, berries, and small invertebrates.

DASHING FLIGHT
Rock Dove flocks fly with dash and agility, their white underwings flashing conspicuously.

CLIFF-EDGE BREEDER
Ledges and cavities form ideal sheltered nest sites for wild Rock Doves; this nesting bird has a "pure" plumage pattern with an unspotted back.

OCCURRENCE
Breeds on coasts of Great Britain and Ireland all year round; also on cliffs in S Europe. Feral birds widespread, from coastal cliffs to city centres, often feeding on arable land.

| Seen in the UK |
| J F M A M J J A S O N D |

SIMILAR SPECIES

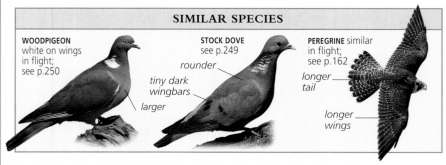

WOODPIGEON white on wings in flight; see p.250
- larger

STOCK DOVE see p.249
- rounder
- tiny dark wingbars

PEREGRINE similar in flight; see p.162
- longer tail
- longer wings

| Length **31–35cm (12–14in)** | Wingspan **63–70cm (25–28in)** | Weight **250–350g (9–13oz)** |
| Social **Flocks** | Lifespan **Up to 10 years** | Status **Secure** |

Order **Columbiformes**	Family **Columbidae**	Species *Columba oenas*

Stock Dove 🔊 38

glossed green neck

blue-grey body

deep wine-pink breast

bright pink-red legs

IN FLIGHT

black wingtips and trailing edge

pale midwings

grey underwings

two short dark bars on wings

dark band on tail

small, neat, round head

A handsome bird of farms and parkland, the Stock Dove also finds suitable nesting habitat in quarries and crags in remote upland areas. It is slightly smaller than a Woodpigeon, rounder and blunter-winged than a "racing pigeon" or the town pigeon of city streets. It is easily overlooked, but its song is distinctive and display flights frequently draw attention to it. It regularly mixes with more numerous Woodpigeons, Jackdaws (see p. 299), and Rooks (see p. 300) when feeding in fields, and roosts with them in woodland.

VOICE Deep, rhythmic, booming coo, repeated several times with increased emphasis, *ooo-woo ooo-woo*.

NESTING Tree hole, ledge, or cavity in building, quarry, or cliff; 2 eggs; 2 or 3 broods; all year.

FEEDING Forages widely on ground looking for seeds, grain, buds, shoots, roots, leaves, and berries; does not visit gardens.

FLIGHT: fast, powerful; deep wingbeats; display flight gliding on steeply raised wings, rocking from side to side.

SOCIAL BIRD
Evening gatherings can be large, but are often out-numbered by Woodpigeons in woodland roosts.

GROUND FEEDER
Seeds are washed up at the edges of freshwater floods, and Stock Doves often gather to exploit such abundance of food.

OCCURRENCE
Widespread except in far N Europe: mostly summer visitor in E Europe, resident in W. In wide variety of places from lowland farms and parkland with many old trees to higher moors with cliffs and quarries.

Seen in the UK
J F M A M J J A S O N D

SIMILAR SPECIES

ROCK DOVE see p.248 — bigger wingbars

longer neck

WOODPIGEON juvenile; see p.250

white on wings

larger

long tail

FERAL/TOWN PIGEON more pointed wings; often white underwings; see p.248

Length **32–34cm (12½–13½in)**	Wingspan **63–69cm (25–27in)**	Weight **290–330g (11–12oz)**
Social **Flocks**	Lifespan **Up to 10 years**	Status **Secure**

Order **Columbiformes**	Family **Columbidae**	Species *Columba palumbus*

Woodpigeon 🔊 39

large white patch across midwings

rump paler than back

ADULT

IN FLIGHT

white on wings

no white on neck

duller, less clean look

JUVENILE

bold white patch on each side of neck

grey back

deep pink breast

dull red legs

dark band on tail

ADULT

A large, handsome, boldly marked pigeon, the Woodpigeon is wild and shy in country areas where it is regularly shot, but visits gardens and becomes surprisingly tame in town parks where it is left undisturbed. It sometimes forms huge flocks, becoming a dramatic part of the country scene in winter. In summer, its dreamy, cooing song is one of the typical sounds of woodland areas. Single birds in flight can easily be mistaken for birds of prey, but any reasonable view should remove such confusion.
VOICE Distinctive, husky, muffled, multi-syllable cooing, oft-repeated, *coo-coo-cu, cu-coo, cook*; commonly heard loud wing clatter.
NESTING Thin platform of twigs in tree or bush; 2 eggs; 1 or 2 broods; April–September (often autumn).
FEEDING Eats buds, leaves, berries, and fruit in trees; also takes food from ground; feeds at bird-tables where undisturbed.

FLIGHT: quick, direct, with even, strong, deep wingbeats; takes off with powerful clatter when surprised; flies in large flocks.

FORAGING ON GROUND
The Woodpigeon often forages for food on the ground, sometimes in flocks. It takes seeds, grain, and shoots.

OCCURRENCE
Widespread except in Iceland and N Scandinavia; summer visitor only in N and E Europe. Breeds in variety of woodland and farmland with trees, town parks, and big gardens. In winter, flocks feed on open ground but farmland remains most important.

Seen in the UK
J F M A M J J A S O N D

SIMILAR SPECIES

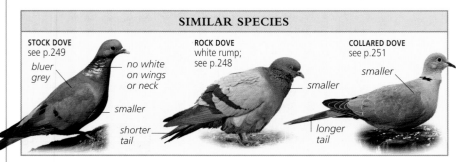

STOCK DOVE
see p.249

bluer grey

smaller

shorter tail

ROCK DOVE
white rump; see p.248

no white on wings or neck

smaller

COLLARED DOVE
see p.251

smaller

smaller

longer tail

Length **40–42cm (16–16½in)**	Wingspan **75–80cm (30–32in)**	Weight **480–550g (17–20oz)**
Social **Large flocks**	Lifespan **Up to 10 years**	Status **Secure**

Order **Columbiformes**	Family **Columbidae**	Species *Streptopelia decaocto*

Collared Dove 🔊 40

ADULT

grey area on upperwings

dark wingtips

small head

plain tail with whitish tip above, black base and broad white tip beneath

IN FLIGHT

dark eyes

thin black collar on back of neck

pale, grey-brown body

subtle pink head and breast

red legs

no collar

sandy buff body

JUVENILE

ADULT

A remarkable natural phenomena of the 20th century is the spread of the Collared Dove from extreme southeast Europe to the entire continent. It is now a common sight in farms and suburbs in Europe, its rather monotonous triple coo a familar sound on summer mornings. It relies heavily on food provided incidentally by people, and on food put out for smaller birds. Its display flights are seen over any built-up area with scattered trees in gardens or parks.

VOICE Loud, frequently-repeated three-syllable coo, *cu-cooo-cuk*, or *cooo-coo-coo* characteristic, unusually for a pigeon also a flight call, a nasal *gwurrrr*.

NESTING Small platform of twigs and rubbish; 2 eggs; 2 or 3 broods, sometimes more; all year.

FEEDING Picks up grain, seeds, buds, and shoots from ground in parks, fields, paddocks, and gardens; often at bird-tables for seeds.

FLIGHT: quick, direct, often long, straight flight, or rising to glide down in wide arcs on flat wings angled back, bowed; whistle from wings in short flights.

CONIFERS PREFERRED
Collared Doves typically roost and nest in tall, dense coniferous trees, often in parks or gardens.

OCCURRENCE
In most of Europe except S Spain, Portugal, and N Scandinavia. Breeds in pines and warm wooded areas in S Europe. Widespread in gardens, parks, and around farm buildings in W Europe after massive expansion of range.

Seen in the UK
J F M A M J J A S O N D

SIMILAR SPECIES

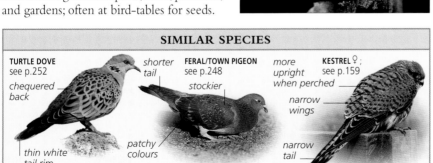

TURTLE DOVE see p.252

chequered back

shorter tail

thin white tail rim

FERAL/TOWN PIGEON see p.248

stockier

patchy colours

more upright when perched

KESTREL ♀; see p.159

narrow wings

narrow tail

Length **31–33cm (12–13in)**	Wingspan **47–55cm (18½–22in)**	Weight **150–220g (5–8oz)**
Social **Small flocks**	Lifespan **Up to 10 years**	Status **Secure†**

Order **Columbiformes**	Family **Columbidae**	Species *Streptopelia turtur*

Turtle Dove  41

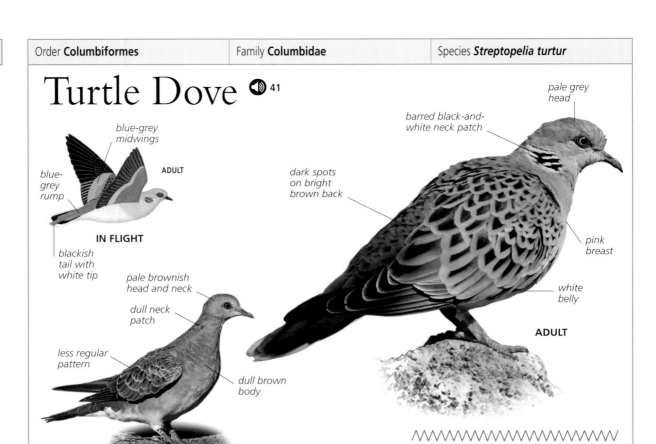

blue-grey midwings

blue-grey rump

ADULT

IN FLIGHT

blackish tail with white tip

pale brownish head and neck

dull neck patch

less regular pattern

dull brown body

JUVENILE

pale grey head

barred black-and-white neck patch

dark spots on bright brown back

pink breast

white belly

ADULT

FLIGHT: light, buoyant, springy, often rolling sideways; wings angled back, flicking downward beats; also glides on flat wings in display flight.

Unlike the Collared Dove, the Turtle Dove remains a country bird, inhabiting well-wooded areas and farmland with large hedges. Flocks form on stubble fields in late summer, eager for spilled grain. The Turtle Dove is also seen along coasts in spring, migrating in small groups as it returns north from Africa. It is in serious decline as agricultural intensification takes hold over most of Europe, and its purring song, which used to be a frequent feature of high summer, is now heard less often.

VOICE Deep, purring, pleasant crooning *rooorrrr rooorrrr*.

NESTING Small platform of thin twigs in hedge or tree; 2 eggs; 2 or 3 broods; May–July.

FEEDING On ground, takes seeds and shoots of arable weeds.

HEDGEROW SPECIALIST
The loss of tall, dense hedgerows has caused a widespread decline in Turtle Doves.

SIMILAR SPECIES

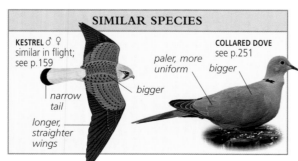

KESTREL ♂ ♀ similar in flight; see p.159

narrow tail

longer, straighter wings

bigger

COLLARED DOVE see p.251

paler, more uniform

bigger

OCCURRENCE
Summer visitor to most of Europe except Iceland, Ireland, and Scandinavia, declining in numbers. In wooded farmland, broadleaved woods with sunny clearings, and thick, old hedgerows.

Seen in the UK

| J | F | M | **A** | **M** | **J** | **J** | **A** | **S** | O | N | D |

Length **26–28cm (10–11in)**	Wingspan **47–53cm (18½–21in)**	Weight **130–180g (4–6oz)**
Social **Small flocks**	Lifespan **Up to 10 years**	Status **Declining**

Order **Cuculiformes**	Family **Cuculidae**	Species ***Cuculus canorus***

Cuckoo

long wings

dark, white-spotted tail

dark wingtips

ADULT

IN FLIGHT

pale spot on nape

JUVENILE

wedge-shaped tail

head held up

short, thick, curved bill with yellow base

yellow eye

pale band under wings

medium grey head and chest

ADULT

medium grey upperparts

grey bars on white underside

ADULT

brown back with dark bars

broad rufous and black bars on tail

JUVENILE

A quite large, long-winged, long-tailed grey bird, the Cuckoo in spring is familiar to everyone by its call, but not its appearance. In late summer, it is the barred, brown young cuckoo that is more often seen, calling loudly to be fed by almost any bird that passes by: its call and huge, orange gape prove irresistible. Cuckoos typically perch low down in or on the edge of trees, but are often easier to see singing or calling from telephone wires; they also fly about from tree to tree, giving frequent short flight views.

VOICE Familiar loud, bright *cuc-coo*, sometimes *cuc-cuc-coo*, much more staccato than Collared Dove song, also rough, laughing *wha-wha-wha*; loud, rich, throaty bubble from female and loud, thin, hissing *ssee-ssee-ssee* from juvenile begging food.

NESTING Lays eggs in other birds' nests; 1–25 (usually 9) eggs per female, 1 per nest; May–June.

FEEDING Drops to ground to pick up large, hairy caterpillars; also eats small insects.

FLIGHT: low, direct, heavy-looking, with head held up, wings below horizontal; quick, deep downward beats; often swoops up to perch.

GIANT CHICK
Cuckoo chicks are often reared by small foster parents such as Reed Warblers.

OCCURRENCE
Summer visitor to all Europe except Iceland, migrating back south early in autumn. On bushy moorland and heaths, in woods and well-wooded farmland, around reedbeds, and wherever small birds nest.

Seen in the UK
J F M **A M J J A S** O N D

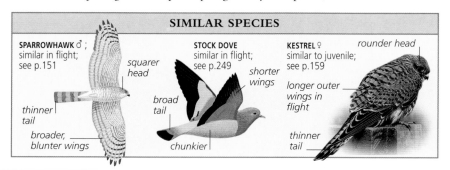

SIMILAR SPECIES

SPARROWHAWK ♂; similar in flight; see p.151

squarer head

thinner tail

broader, blunter wings

STOCK DOVE similar in flight; see p.249

broad tail

chunkier

shorter wings

KESTREL ♀ similar to juvenile; see p.159

rounder head

longer outer wings in flight

thinner tail

Length **32–34cm (12½–13½in)**	Wingspan **55–65cm (22–26in)**	Weight **105–130g (4–5oz)**
Social **Solitary**	Lifespan **Up to 10 years**	Status **Secure**

OWLS

Owls ARE highly specialized birds with exellent eyesight and hearing. They see well in low light levels, although not in complete darkness, and many can pinpoint prey accurately by sound alone. Their ears are slightly asymmetrical in size, shape, or position. As an owl twists and bobs its head it is using this adaptation to get a "fix" on its prey. Owls often attract the attention of smaller birds (and also sometimes Jays and Magpies) if discovered by day, and mixed parties "mob" the roosting owl with loud calls, sometimes diving at its head.

BREEDING STRATEGIES

Some owls have fixed territories and rear small broods each year. Others are nomads, breeding wherever there is plentiful food. They lay many eggs but only in years with an adequate food supply do many chicks survive. The Tawny and

FORWARD SWING
Owls can swing their whole body forward, to grip with their feet, whether while landing or striking prey.

Tengmalm's are typical woodland owls; Scops and Little Owls live in more open surroundings. The Eagle Owl is a massive and powerful predator, as is the Snowy Owl of far northern tundras. The two "eared" owls are very alike in plumage pattern and overall shape, and both hunt over open ground. The Short-eared Owl lives on heaths and marshes and hunts even in full daylight, while the Long-eared inhabits woods and thickets and hunts only after dark.

TYTONIDAE

rounded, heart-shaped facial disc

BARN OWL
p.256

STRIGIDAE

broad pale grey "V" between eyes

short white eyebrows

SCOPS OWL
p.257

PYGMY OWL
p.258

cream-buff spots on liver-brown back

LITTLE OWL
p.259

STRIGIDAE *continued*

black eyes

large, deep orange eye, circled with black

ear tufts

ear tufts

dark surround to bright orange-red eye

pale, streaked underside

TAWNY OWL
p.261

LONG-EARED OWL
p.262

blackish ring around large, cold yellow eye

high brow

EAGLE OWL
p.260

SHORT-EARED OWL
p.263

TENGMALM'S OWL
p.264

Family **Caprimulgidae**

NIGHTJARS

CAPRIMULGIDAE

SUMMER MIGRANTS to Europe, taking advantage of an abundance of moths, the nightjars spend the day motionless and emerge at dusk to feed. These remarkably agile, light-weight fliers have large wings and tail. Their strange songs help to identify them. Nightjars occupy open heathland or lightly wooded places with wide clearings, often flying around bush tops and trees to catch moths in flight. They may approach people at dusk, as if curious about intruders in their territory.

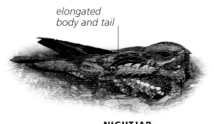

elongated body and tail

NIGHTJAR
p.265

Order **Strigiformes**	Family **Tytonidae**	Species **Tyto alba**

Barn Owl 🔊 43

rounded, heart-shaped facial disc

black eyes

white underwings

short tail

big head

pale buff upperside

thin dark bars on outer wings

grey and black spots

white underside

IN FLIGHT

This medium-sized owl is sometimes about by day in winter and often hunts well before dark in summer if it has a family to feed. At such times, it allows excellent views, frequently at roadsides where verges may provide almost the only remaining rough grassland over which it can hunt for voles. It also frequents marshy areas such as rough grassland beside reedbeds or along embankments. There is little chance of misidentification, although other owls appearing "white" in car headlights at night can be taken for it.

VOICE Hissing, snoring calls from nest, nasal *hi-wit*, rolling, shrill shriek and high squeal of alarm.

NESTING Big hole in tree, stack of hay bales, or building; 4–7 eggs; 1 brood; May–June.

FEEDING Hunts from perch or in low flight; catches voles, mice, rats, and occasionally birds.

FLIGHT: light, agile, spring-like action with quite quick, deep wingbeats; hovers and plunges head-first into long grass with wings pulled back.

NIGHT HUNTER
Barn Owls are usually nocturnal but they may be seen before sunset if they have young to feed.

SUBSPECIES

T. a. guttata
(C and E Europe)

deep orange-buff underside

OCCURRENCE
Widespread but sparse in SE, C, and W Europe; absent from Iceland, Scandinavia, and NE Europe. Breeds and hunts in open areas, from farmland to marshes with reedbeds and moors, and young plantations.

Seen in the UK
J F M A M J J A S O N D

SIMILAR SPECIES

SHORT-EARED OWL
see p.263

yellow eyes

dark wing patches

much larger

whiter above

SNOWY OWL
see p.468

TAWNY OWL
see p.261

browner

streaked below

Length **33–39cm (13–15½in)**	Wingspan **85–93cm (34–37in)**	Weight **290–460g (10–16oz)**
Social **Family groups**	Lifespan **5–10 years**	Status **Declining**

Order **Strigiformes**	Family **Strigidae**	Species *Otus scops*

Scops Owl

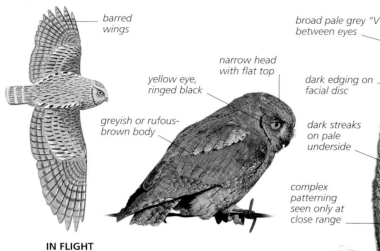

barred wings

narrow head with flat top

yellow eye, ringed black

greyish or rufous-brown body

angular corners of head sometimes raised as ear tufts

broad pale grey "V" between eyes

dark edging on facial disc

dark streaks on pale underside

complex patterning seen only at close range

IN FLIGHT

FLIGHT: short, quick flights, only faintly undulating, with bursts of fast wingbeats.

In Mediterranean villages and woodlands in summer, the dawn and dusk calling of the Scops Owl is commonplace. Tracking one down by carefully approaching the sound is sometimes difficult but usually just a matter of time; with patience it may be possible to see one really well if it perches close to a streetlight. Daytime views are much more difficult to get: it is practically impossible to track one down to a visible roost. Little Owls are often found in nearby areas, so identification requires some care: Little Owls are resident but in most areas Scops Owls are found in summer; Scops Owls are more often seen on roofs and church towers in the middle of villages, while Little Owls tend to be on the edges of villages or around farm outbuildings and isolated barns; Little Owls look more rounded or squat, with a broad, flatter head, while a Scops Owl is more tapered towards the wingtips and tail. They are only very rare vagrants (usually in spring) north of their regular breeding range.

VOICE Distinctive at dusk: single, fluty, indrawn musical whistle, *pew* or *tyuh* repeated unvaryingly every 2–3 seconds.

NESTING Cavity in tree, wall, or building; 4 or 5 eggs; 1 brood; April–June.

FEEDING Mostly drops down from perch to take large insects.

OCCURRENCE
In summer, found widely in S Europe and north to C France and Alps. In small towns, parks, and wooded areas, often around older buildings and churchyards, but also in mixed woodland. Most migrate in winter, only some staying in S Europe.

Seen in the UK
J F M A M J J A S O N D

SIMILAR SPECIES

flatter head

stockier

LITTLE OWL
see p.259

TENGMALM'S OWL different habitat; see p.264

rounder head

black eye

much larger

TAWNY OWL
see p.261

Length **19–21cm (7½–8½in)**	Wingspan **47–54cm (18½–21½in)**	Weight **150g (5oz)**
Social **Family groups**	Lifespan **Up to 10 years**	Status **Declining†**

| Order **Strigiformes** | Family **Strigidae** | Species *Glaucidium passerinum* |

Pygmy Owl

broad, round wings

brown back with fine white specks

large head

stocky, narrow tail

often raises or twists tail

IN FLIGHT

crown darker than in adults

JUVENILE

heavily barred sides of breast and flanks

short white eyebrows

small yellow eyes

largely white belly with dark streaks

narrow barred tail

ADULT

FLIGHT: fast, bounding like a woodpecker, with deep undulations.

Europe's smallest owl, the Pygmy Owl is active at dusk and dawn rather than in deep darkness, and can sometimes be seen perched on top of a tall, prominent coniferous tree. It may also be attracted to imitations of its call. It hunts small and medium-sized birds, taking some that are larger than itself in a bold, fearless chase.

VOICE A thin, thrush-like *seeee*; in autumn, a series of thin, forced, squeaky whistles, usually 5–10, rising in pitch; in spring, repetition of a simple piping note, similar to Bullfinch's call, 7–8 notes per 10 seconds.

NESTING Uses a natural or an old woodpecker hole in a tree; 4–7 eggs; 1 brood; April–June.

FEEDING Hunts small mammals (mostly mice and voles) and woodland birds, taking them by surprise on perch or on the ground.

SIMILAR SPECIES

wider, whiter facial disc

TENGMALM'S OWL white spots along shoulder, larger eyes; see p.264

JAUNTY TAIL
A typically rounded owl, the Pygmy Owl nevertheless has a medium-length tail that is often slightly raised at an angle to its back, giving it an alert, perky look.

OCCURRENCE
Prefers coniferous (especially spruce) or mixed forest in Alpine areas and far north of Europe; rarely moves out of breeding areas.

| Seen in the UK |
| J F M A M J J A S O N D |

| Length **15–19cm (6–7½in)** | Wingspan **32–39cm (12½ –15½in)** | Weight **50–65g (1¾–2⅜oz)** |
| Social **Family groups** | Lifespan **5–10 years** | Status **Secure** |

Order **Strigiformes**	Family **Strigidae**	Species *Athene noctua*

Little Owl

shortish, round wings, barred brown and cream

dark crown with small white spots

broad head

flattish white eyebrows

large, pale yellow eye, circled with black

cream-buff spots on liver-brown back

complex wavy streaks on pale underside

IN FLIGHT

Widespread in Europe, and long established after introduction in Great Britain, the Little Owl is small, chunky, flat-headed, and short-tailed. It can appear very round when perched out in the open by day, but may stretch upwards to look more elongated when alarmed. It hunts at dusk but sometimes perches quite openly in daylight, often attracting the noisy attention of small birds. Its undulating flight may briefly recall a woodpecker or large thrush. In much of Europe, it lives in ruins and old barns with tiled roofs, or on hillsides with boulders and stones scattered over the slopes.

VOICE Loud, musical, plaintive calls, rising *keeeooo*, sharper *werro!*, short *kip kip kip*.

NESTING In long, narrow hole in tree, bank, or building; 2–5 eggs; 1 brood; May–July.

FEEDING Mostly takes small rodents and large insects from ground; also picks small birds and earthworms from ground.

FLIGHT: distinctive bounding, undulating action with bursts of quick wingbeats between downward swoops; sweeps upwards to perch.

SQUAT SILHOUETTE
A rounded, short-tailed, thin-legged shape on a post or branch at dusk is likely to be a Little Owl.

OCCURRENCE
Widespread resident through Europe north to Great Britain and Baltic. In great variety of terrain, on open rocky slopes and islands, farmland and parkland with old trees and outbuildings, and even semi-desert areas with tumbled rocks and cliffs.

Seen in the UK
J F M A M J J A S O N D

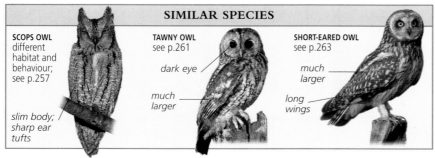

SIMILAR SPECIES

SCOPS OWL different habitat and behaviour; see p.257

slim body; sharp ear tufts

TAWNY OWL see p.261

dark eye

much larger

SHORT-EARED OWL see p.263

much larger

long wings

Length **21–23cm (8½–9in)**	Wingspan **50–56cm (20–22in)**	Weight **140–200g (5–7oz)**
Social **Family groups**	Lifespan **Up to 10 years**	Status **Declining**

Order **Strigiformes**	Family **Strigidae**	Species **Bubo bubo**

Eagle Owl

dark brown wings with pale patch

black marbling and streaking on back

IN FLIGHT

large ear tufts, usually angled in shallow "V"

boldly streaked, pale underside

pale "V" on upper face

large, deep orange eye, circled with black

pale area around bill

A massive bird and one of Europe's most powerful predators, the Eagle Owl even kills other owls and birds of prey quite frequently in order to claim exclusive use of territory. Mostly a bird of cliffs and crags, it may be found relatively close to villages and farms, although many prefer wilder places remote from human habitation. It calls early in the year and is not at all easy to track down in summer: careful scrutiny of likely ledges, cavities, and trees on occupied cliffs is often to no avail. Its huge eyes give it really exceptional vision in poor light and hunting is almost entirely nocturnal.

VOICE Song deep, booming hoot, *oo-hu*, carries 2–4km (1–2½ miles); also loud barking alarm notes, *kvek*, *kwa*, or *kwa-kwa-kwa*.

NESTING Unlined cavity in tree or deep, sheltered cliff ledge; 2 or 3 eggs; 1 brood; April–May.

FEEDING Eats wide range of food from small rodents to much bigger prey, including birds such as crows, pigeons, and thrushes, and small mammals such as rats, hares, and squirrels.

REMARK Subspecies *B. b. ascalaphus* (Middle East) is paler, buff below, and has yellower eyes.

FLIGHT: direct, heavy but strong; wings slightly arched in frequent glides; head looks very big.

ELUSIVE OWL
Good camouflage colours render immobile Eagle Owls extremely difficult to spot.

OCCURRENCE
Widespread in mainland Europe but everywhere sparse, mostly in forested mountain areas with gorges, cliffs, and deep ravines with deep caves or large ledges, and on big, old trees. Not found outside usual breeding areas.

Seen in the UK
J F M A M J J A S O N D

SIMILAR SPECIES

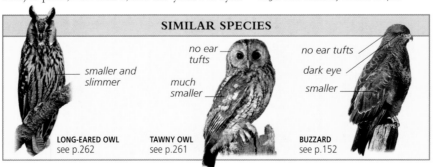

smaller and slimmer

no ear tufts

much smaller

no ear tufts

dark eye

smaller

LONG-EARED OWL
see p.262

TAWNY OWL
see p.261

BUZZARD
see p.152

Length **59–73cm (23–29in)**	Wingspan **1.38–1.7m (4½–5½ft)**	Weight **1.5–3kg (3¼–6½lb)**
Social **Solitary**	Lifespan **10–20 years**	Status **Vulnerable**

Order **Strigiformes**	Family **Strigidae**	Species **Strix aluco**

Tawny Owl 🔊 42.I, 42.II

large, round head

obvious facial disc

large black eye

pale spots and bars on wings

brown back with diagonal row of white spots on each side

short wings and tail

ADULT

IN FLIGHT

ADULT

red-brown or grey-brown body (downy juvenile pale grey)

pale, streaked underside

ADULT

FLIGHT: strong, quite swift, but heavy, undulating; deep, regular flappy wingbeats, short glides.

This is the owl that hoots after dark, but its vocal repertoire is quite wide and a loud, yapping *ke-wik* is heard more regularly through the year than the beautiful, wavering hoot. It is often seen merely as a large, big-headed silhouette, or a vague shape flying from a roadside pole at night. Sometimes small birds mob it by day, giving away its presence, or it can be found in trees or ivy above splashes of white droppings. It can then be watched quite closely with care and proves to be an impressive and extremely beautiful bird.

VOICE Loud, excited yapping notes, variations on nasal *ke-wick!* or *keeyip*; long, musical, breathy, quavering hoot, *hoo hoo-hoooo hoo-ho-ho*.

NESTING Hole in tree or building or in old stick nest of crow or Magpie; 2–5 eggs; 1 brood; April–June.

FEEDING Drops down to take voles, mice, rats, frogs, beetles, and earthworms from ground; catches many small birds as they are roosting or incubating eggs at night.

NIGHT HUNTER
Tawny Owls begin to call at dusk but only start to hunt when it is quite dark.

OCCURRENCE
Resident almost throughout Europe except Iceland, Ireland, and N Scandinavia. In all kinds of woodland and wooded areas such as farmland with tall hedges and trees, and large gardens with conifers or evergreen broadleaves.

Seen in the UK
| J | F | M | A | M | J | J | A | S | O | N | D |

SIMILAR SPECIES

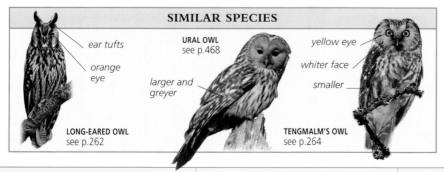

ear tufts

orange eye

LONG-EARED OWL
see p.262

URAL OWL
see p.468

larger and greyer

yellow eye

whiter face

smaller

TENGMALM'S OWL
see p.264

Length **37–39cm (14½–15½in)**	Wingspan **94–104cm (37–41in)**	Weight **330–590g (12–21oz)**
Social **Family groups**	Lifespan **Up to 10 years**	Status **Secure**

Order **Strigiformes**	Family **Strigidae**	Species *Asio otus*

Long-eared Owl

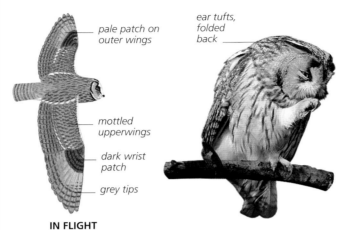

pale patch on outer wings

ear tufts, folded back

dark surround to bright orange-red eye

long ear tufts, raised

mottled upperwings

dark wrist patch

grey tips

IN FLIGHT

closely streaked underside

FLIGHT: looks like Short-eared Owl but less often seen by day; slightly shorter-winged, less buoyant.

A large, handsome bird, the Long-eared Owl can be found in winter in communal roosts of a handful to twenty or more birds. These are often found in tall pines, but also resort to dense thickets of willow and hawthorn, where they can be extremely hard to see; they are sometimes given away by splashes of droppings and regurgitated pellets on the ground. Unless disturbed, they rarely move before it is almost dark: only occasionally may one be found hunting in better light, inviting confusion with the Short-eared Owl.

VOICE Song deep, moaning, short hoot, *oo oo oo* or *uh uh*; juvenile begs for food with high, sharp, "squeaky-hinge" *eee-ip*.
NESTING Old nest of crow, squirrel drey, or under thick growth of bracken or brambles; 3–5 eggs; 1 brood; March–June.
FEEDING Hunts from perch or in flight, catching small rodents and roosting birds.

WINTER ROOST
A typical winter roost is in a thorn or willow thicket, or a pine tree, where several owls may sit close together, relaxed, or stretched tall when alarmed.

OCCURRENCE
Widespread except in Iceland and N Scandinavia; summer only in NE Europe. Mostly breeds in coniferous woodland and shelter belts near moors, heaths, and marshes; roosts in thickets, old, tall hedgerows, belts of willows near marshes, and similar low, thick cover near open ground.

Seen in the UK
J	F	M	A	M	J	J	A	S	O	N	D

SIMILAR SPECIES

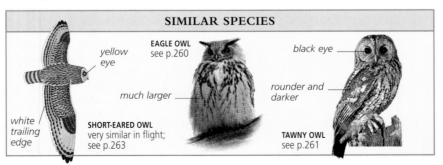

yellow eye

EAGLE OWL see p.260

black eye

much larger

rounder and darker

white trailing edge

SHORT-EARED OWL very similar in flight; see p.263

TAWNY OWL see p.261

Length **35–37cm (14–14½in)**	Wingspan **84–95cm (33–37in)**	Weight **210–330g (7–12oz)**
Social **Roosts in small flocks**	Lifespan **10–15 years**	Status **Secure**

Short-eared Owl

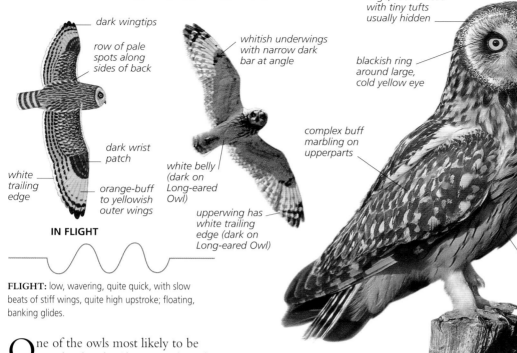

- large, round head with tiny tufts usually hidden
- blackish ring around large, cold yellow eye
- complex buff marbling on upperparts
- buff-white underside with fine dark streaks
- whitish underwings with narrow dark bar at angle
- white belly (dark on Long-eared Owl)
- upperwing has white trailing edge (dark on Long-eared Owl)

- dark wingtips
- row of pale spots along sides of back
- dark wrist patch
- white trailing edge
- orange-buff to yellowish outer wings

IN FLIGHT

FLIGHT: low, wavering, quite quick, with slow beats of stiff wings, quite high upstroke; floating, banking glides.

One of the owls most likely to be seen by day, the Short-eared Owl hunts like a harrier, flying low over open ground, often in good light well before dusk. Its numbers and distribution reflect the fluctuating numbers of voles. It may appear for a year or two in suburban areas where fields are neglected in advance of development, but is more frequent on coastal marshes and, in summer, over upland moors and young conifer plantations. It is easily confused with the Long-eared Owl in flight but Long-eareds are more strictly nocturnal.

VOICE Nasal bark, *kee-aw*, or hoarse, whip-like *ke-ow*; song deep, soft, quick booming hoot, *boo-boo-boo-boo* in display flight.
NESTING Unlined scrape on ground; 4–8 eggs; 1 or 2 broods; April–July.
FEEDING Hunts by flying slowly over ground or watching from perch; eats small rodents and some birds.

FLOATING HUNTER
Its long wings and light weight make the Short-eared Owl remarkably buoyant when hunting.

OCCURRENCE
Widespread through Europe, mostly in N but erratically in S, largely linked with variations in prey populations. In all kinds of rough grassland, marshes, heaths, upland moors, and plantations; many temporary habitats occupied opportunistically when large numbers of voles present.

Seen in the UK
J F M A M J J A S O N D

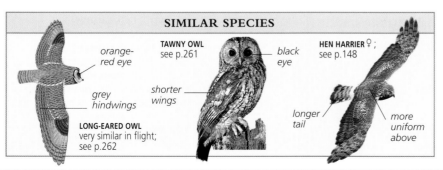

SIMILAR SPECIES

- **LONG-EARED OWL** very similar in flight; see p.262 — orange-red eye, grey hindwings
- **TAWNY OWL** see p.261 — black eye, shorter wings
- **HEN HARRIER** ♀; see p.148 — longer tail, more uniform above

| Length **34–42cm (13½–16½in)** | Wingspan **90–105cm (35–41in)** | Weight **260–350g (9–13oz)** |
| Social **Small flocks** | Lifespan **10–15 years** | Status **Vulnerable†** |

263

SWIFTS

No BIRDS ARE MORE specialized than the swifts. They have tiny bills but large, broad mouths, in which they catch flying insects. Their tapered bodies and long, stiff, scythe-shaped wings are aerodynamically ideal for sustained flight. They have a good turn of speed when displaying, but generally feed at a much slower pace, gliding efficiently and turning dextrously to catch their small prey.

Their feet are minute, with all four toes pointing forwards, just enough to cling to a rough surface but making it impossible for them to perch on a wire or twig: indeed, they are almost never seen settled, as they dive into the cavities where they nest with no discernible pause.

Swifts have become extremely reliant on buildings for nesting and few now nest in holes in cliffs or trees. However, modern buildings are useless to swifts, which concentrate on old housing and structures such as church towers. They must face an uncertain future.

Identification can be difficult, but good views reveal differences in shape and pattern between most species. All are social birds, often flying in fast-moving, closely packed groups during the breeding season. They arrive late in spring and leave in late summer for Africa. There, they feed in company with several similar species of African-nesting swifts.

APODIDAE

SCREAMING PARTY

A group of Swifts dashes by at rooftop height, with a chorus of loud, piercing calls. This seems to have a strong social significance within the birds of a breeding colony.

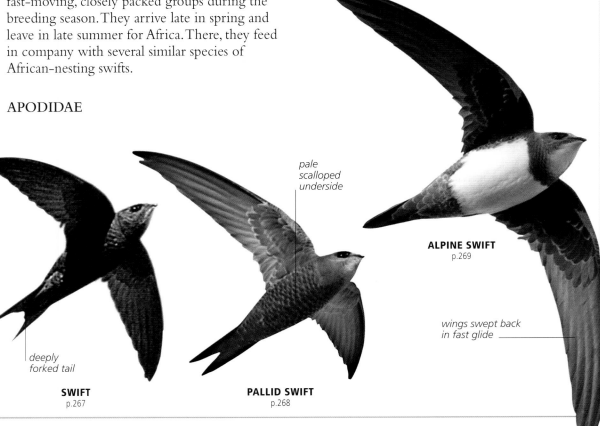

pale scalloped underside

ALPINE SWIFT
p.269

wings swept back in fast glide

deeply forked tail

SWIFT
p.267

PALLID SWIFT
p.268

Order **Apodiformes**	Family **Apodidae**	Species *Apus apus*

Swift 🔊 45

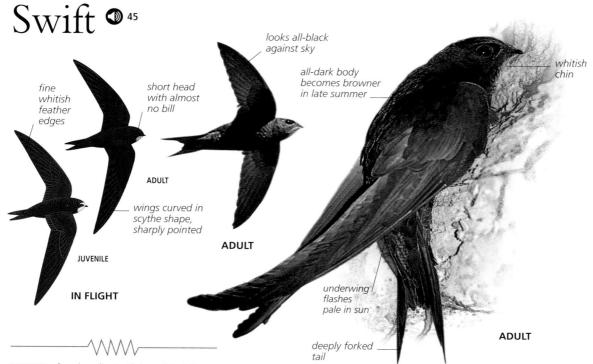

looks all-black against sky

all-dark body becomes browner in late summer

whitish chin

fine whitish feather edges

short head with almost no bill

ADULT

wings curved in scythe shape, sharply pointed

ADULT

JUVENILE

IN FLIGHT

underwing flashes pale in sun

deeply forked tail

ADULT

FLIGHT: often slow, direct with long glides between flurries of deep wingbeats, wings very rigid; also faster dashing flights in groups with flickering wingbeats.

No other bird is more aerial than the Swift. Immature Swifts may spend three years aloft before returning to breed: like seabirds, they come to land only to breed. Swifts appear in Europe late in spring and depart for Africa early in autumn. Flocks fly very high or at rooftop level; they are seldom seen perching on wires or clinging to walls or roofs like martins or swallows. Their scythe-like wings and loud, screaming calls make identification simple, but Pallid Swifts in southern Europe complicate the matter.

VOICE Loud, screeching, shrill screams from flocks, *shrreeee, sirrr*.

NESTING Feather-lined cavity in building, more rarely in cliff; 2 or 3 eggs; 1 brood; May–June.

FEEDING Entirely aerial, taking flying insects in bill.

HIDDEN NESTER
Swifts nest deep inside holes, mostly in older buildings, rarely now in cliffs.

SCREAMING PARTIES
Feeding Swifts fly much more slowly than may be thought, but noisy, chasing groups are genuinely fast.

OCCURRENCE
Widespread in summer except in Iceland; arrives late and leaves early. Feeds over any kind of open area, including towns, villages, and some larger S European cities, but needs old civic buildings and housing for nesting, usually excluded from new developments.

Seen in the UK
J F M **A M J J A S** O N D

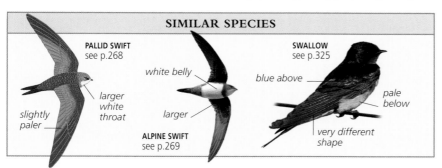

SIMILAR SPECIES

PALLID SWIFT see p.268

slightly paler

larger white throat

white belly

larger

ALPINE SWIFT see p.269

SWALLOW see p.325

blue above

pale below

very different shape

Length **16–17cm (6½in)**	Wingspan **42–48cm (16½–19in)**	Weight **36–50g (1¼–1¾oz)**
Social **Flocks**	Lifespan **Up to 10 years**	Status **Secure**

| Order **Apodiformes** | Family **Apodidae** | Species ***Apus pallidus*** |

Pallid Swift

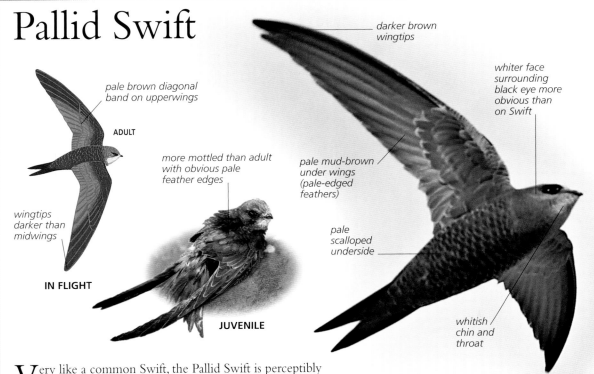

darker brown wingtips

pale brown diagonal band on upperwings

ADULT

whiter face surrounding black eye more obvious than on Swift

more mottled than adult with obvious pale feather edges

pale mud-brown under wings (pale-edged feathers)

wingtips darker than midwings

pale scalloped underside

IN FLIGHT

JUVENILE

whitish chin and throat

V ery like a common Swift, the Pallid Swift is perceptibly paler and more mud- or clay-brown when seen well; it is also broader-winged, shorter-tailed, and broader-headed. As it is so often seen flying against the sky, a good view may not be easy to achieve, but if it flies low against a darker background, plumage differences should be visible. In many southern European towns, both species are common, allowing opportunities both for comparison and confusion. Pallid Swifts have sometimes been seen well north of their usual range very late in the summer or autumn when swifts have normally long since migrated to Africa, but this alone is not sufficient to identify late swifts such as Pallid: very occasionally, a common Swift will linger as well. Such isolated wanderers outside the normal range need careful observation for positive identification.
VOICE Scream like common Swift's but usually falls in pitch, more disyllabic, lower in pitch, but hard to separate with certainty.
NESTING Unlined cavity in roof space, old building, or wall; 2 or 3 eggs; 1 brood; May–June.
FEEDING Like Swift, takes food entirely in air, catching flying insects and drifting spiders in its mouth.

ADULT

FLIGHT: fast, typical swift-type flight on stiff, scythe-like wings, with slightly slower wingbeats; longer glides than Swift on average.

OCCURRENCE
Breeds in S Europe, mostly near coasts; more inland in S Spain and Italy, in older areas of towns and villages. Migrates to Africa in winter, but only very rare vagrant north of breeding range. Feeds over all kinds of open countryside.

| Seen in the UK |
| J F M A M J J A S O N D |

SIMILAR SPECIES

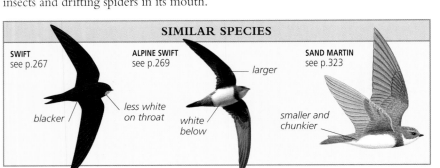

SWIFT
see p.267

ALPINE SWIFT
see p.269

larger

SAND MARTIN
see p.323

blacker

less white on throat

white below

smaller and chunkier

| Length **16–18cm (6½–7in)** | Wingspan **39–46cm (15½–18in)** | Weight **50g (1¾oz)** |
| Social **Flocks** | Lifespan **10–15 years** | Status **Secure†** |

Order **Apodiformes**	Family **Apodidae**	Species *Apus melba*

Alpine Swift

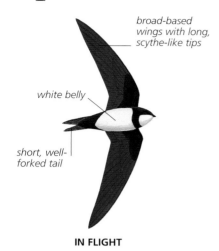

broad-based wings with long, scythe-like tips

white belly

short, well-forked tail

IN FLIGHT

FLIGHT: strong, powerful, sweeping flight; easy, graceful beats of stiff, scythe-like wings.

The Alpine Swift is a uniquely large and powerful swift in Europe. It likes cliffs in mountainous areas but is not always found at high altitude, breeding in many lower, deeply incised ravines and in tall buildings in towns and old villages. It often mixes with other swift species and Crag Martins and is frequently seen in the air with Choughs (see pp.294–5). It is not always easy to judge size, so silhouette views are not necessarily adequate to identify possible Alpine Swifts out of their usual range: the possibility of partly white Swifts should also be borne in mind.
VOICE Loud chorus of Greenfinch-like trills, rising and falling and changing in speed and pitch, *titititititi-ti-ti-ti-ti-ti ti ti.*
NESTING Shallow cup of grass and stems in cavity in building or cliff; 2 or 3 eggs; 1 brood; April–June.
FEEDING Exclusively aerial, catching insects in air in open mouth.

wings swept back in fast glide

white chin and throat, difficult to see

white underside

dark under tail

dark brown breast-band

STRONG FLIER
In silhouette, an Alpine Swift may momentarily recall a Hobby.

OCCURRENCE
Summer visitor and breeder in S Europe north to Alps. Over all kinds of open country, especially hill towns, cliffs, and gorges, from sea level to high hills, between April and September. Rarely strays north in spring.

Seen in the UK
J F M A M J J A S O N D

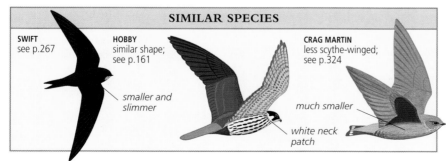

SIMILAR SPECIES

SWIFT
see p.267

HOBBY
similar shape;
see p.161

smaller and slimmer

CRAG MARTIN
less scythe-winged;
see p.324

much smaller

white neck patch

Length **20–23cm (8–9in)**	Wingspan **51–58cm (20–23in)**	Weight **75–100g (2⅝–3⅝oz)**
Social **Flocks**	Lifespan **10–15 years**	Status **Secure†**

KINGFISHERS, BEE-EATERS, ROLLERS, AND HOOPOE

THESE SPECIES AND the woodpeckers fall between the swifts and nightjars and the Passerines (perching or song birds), and are often termed "near passerines". They are grouped here for convenience rather than for any more meaningful relationship.

KINGFISHERS
The "original" kingfisher from which the family takes its name, the European species is a fish-eater (others are dry land birds that catch insects). It is often first detected by its piercing call or a splash as it dives for a fish.

BEE-EATERS
True to their name, bee-eaters do eat bees and wasps, wiping away their stings against a perch, but they also catch many other insects in their acrobatic, swooping and gliding flight. They breed socially, nesting in burrows in earth banks.

ROLLERS
Rather crow-like in character, rollers are much more colourful, especially in mid-summer when paler feather edges have worn away to reveal the rich colours beneath.

HOOPOE
Striking and boldly patterned in a photograph, the Hoopoe is often surprisingly inconspicuous

FISHING
The European Kingfisher plungedives into water to catch fish. A transparent membrane protects the eye during the dive.

as it shuffles on the ground in the dappled light and shade beneath a hedge or in a sunny orchard. Only when it flies does it suddenly catch the eye.

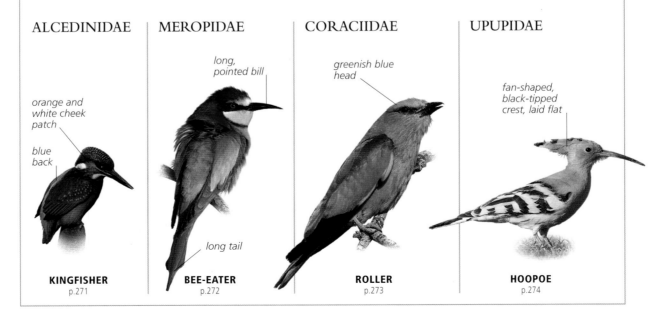

ALCEDINIDAE
orange and white cheek patch

blue back

KINGFISHER
p.271

MEROPIDAE
long, pointed bill

long tail

BEE-EATER
p.272

CORACIIDAE
greenish blue head

ROLLER
p.273

UPUPIDAE
fan-shaped, black-tipped crest, laid flat

HOOPOE
p.274

Order **Coraciiformes**	Family **Alcedinidae**	Species **Alcedo atthis**

Kingfisher

vivid blue streak

MALE

IN FLIGHT

barred blue cap

orange and white cheek patch

black bill

electric blue upperside (slightly duller on juvenile)

white chin

black bill with red base

rusty orange underside

MALE

tiny red legs (blackish on juvenile)

MALE

FEMALE

Most people seeing one for the first time think the Kingfisher is surprisingly small: about Starling-sized or a little less. It is also, despite its bright colours, much more unobtrusive than may be expected: in the dappled shade of flickering foliage above rippling water, it can be extremely hard to see. Often it is the sharp call that gives it away, followed by a glimpse of a bright blue back flying off over water, but sometimes it allows really close, clear views. It is particularly vulnerable to hard winter weather and populations consequently fluctuate greatly from year to year. Occasionally a Kingfisher will take advantage of a garden pond full of goldfish, but visits are usually brief as Kingfishers are shy birds and quick to fly off if disturbed.

VOICE Quite loud, sharp, high *kit-cheeee* or *cheee*; also high, fast trill in spring.
NESTING Deep tubular tunnel, lined with fish bones, in soft earth cliff over water; 5–7 eggs; 2 broods; May–July.
FEEDING Catches fish, small aquatic invertebrates, and amphibians in dive from perch or mid-air hover.

FLIGHT: low, direct, fast; poor manoeuvrability; quick, almost whirring wingbeats.

DRAMATIC DIVER
The Kingfisher makes a dramatic dive usually from a low perch, but sometimes also from a brief hover.

OCCURRENCE
In most of Europe, N to S Scotland, Baltic, and extreme S Scandinavia. Summer visitor only in north and east of range, from which birds move southwest in winter. Along rivers and canals, on marshes, flooded pits, and coastal areas including salt-marsh creeks, especially in winter.

Seen in the UK
J F M A M J J A S O N D

Length **16–17cm (6½in)**	Wingspan **24–26cm (9½–10in)**	Weight **35–40g (1¼–1⁷⁄₁₆oz)**
Social **Pairs**	Lifespan **5–10 years**	Status **Declining**

Order **Coraciiformes**	Family **Meropidae**	Species ***Merops apiaster***

Bee-eater

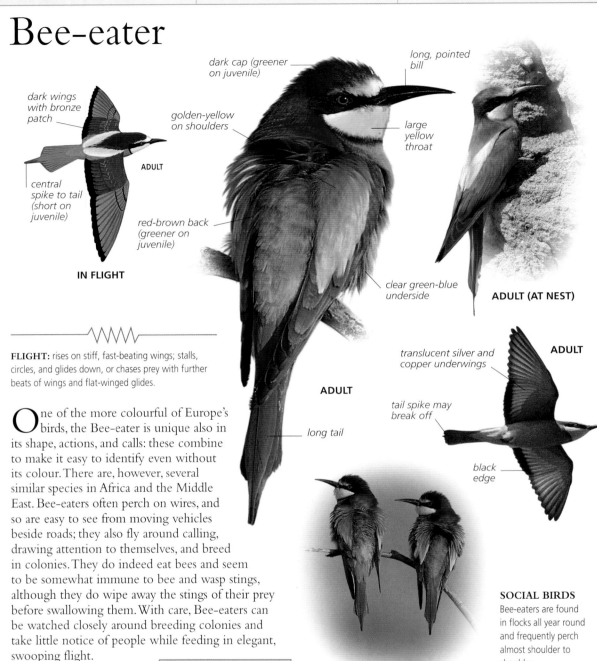

dark wings with bronze patch

dark cap (greener on juvenile)

golden-yellow on shoulders

long, pointed bill

large yellow throat

ADULT

central spike to tail (short on juvenile)

red-brown back (greener on juvenile)

IN FLIGHT

clear green-blue underside

ADULT (AT NEST)

translucent silver and copper underwings

ADULT

tail spike may break off

black edge

ADULT

long tail

FLIGHT: rises on stiff, fast-beating wings; stalls, circles, and glides down, or chases prey with further beats of wings and flat-winged glides.

One of the more colourful of Europe's birds, the Bee-eater is unique also in its shape, actions, and calls: these combine to make it easy to identify even without its colour. There are, however, several similar species in Africa and the Middle East. Bee-eaters often perch on wires, and so are easy to see from moving vehicles beside roads; they also fly around calling, drawing attention to themselves, and breed in colonies. They do indeed eat bees and seem to be somewhat immune to bee and wasp stings, although they do wipe away the stings of their prey before swallowing them. With care, Bee-eaters can be watched closely around breeding colonies and take little notice of people while feeding in elegant, swooping flight.

VOICE Distinctive, far-carrying, deep, quite liquid notes, *prroop prroop*.
NESTING Burrows in sandy banks or even flatter ground; 4–7 eggs; 1 brood; May–June.
FEEDING Catches insects in flight, in prolonged, fast swoops and slow, gliding flights or sallies from perch.

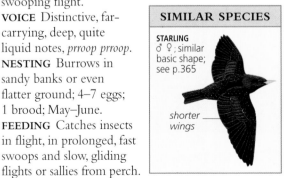

SOCIAL BIRDS
Bee-eaters are found in flocks all year round and frequently perch almost shoulder to shoulder.

SIMILAR SPECIES

STARLING
♂ ♀; similar basic shape; see p.365

shorter wings

OCCURRENCE
Breeds in S and E Europe, north to C France and Alps in summer. Rare migrant farther north in spring or autumn. Usually in warm, often sandy areas with orchards, bushy areas, open grassland, and low earth cliffs.

Seen in the UK
J F M A M J J A S O N D

Length **27–29cm (10½–11½in)**	Wingspan **36–40cm (14–16in)**	Weight **50–70g (1¾–2½oz)**
Social **Flocks**	Lifespan **5–10 years**	Status **Declining**

Order **Coraciiformes**	Family **Coraciidae**	Species **Coracias garrulus**

Roller

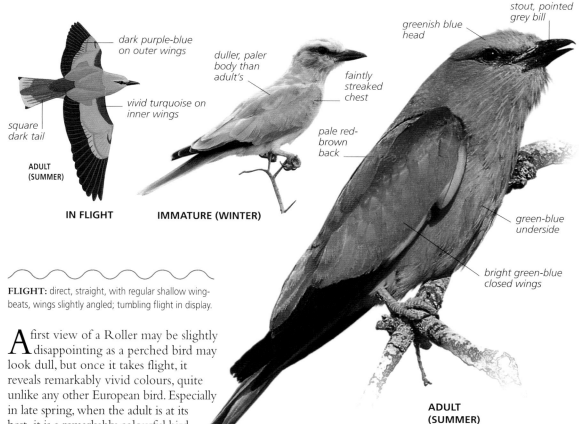

dark purple-blue
on outer wings

vivid turquoise on
inner wings

square
dark tail

**ADULT
(SUMMER)**

IN FLIGHT

duller, paler
body than
adult's

faintly
streaked
chest

pale red-
brown
back

IMMATURE (WINTER)

greenish blue
head

stout, pointed
grey bill

green-blue
underside

bright green-blue
closed wings

**ADULT
(SUMMER)**

FLIGHT: direct, straight, with regular shallow wing-beats, wings slightly angled; tumbling flight in display.

A first view of a Roller may be slightly disappointing as a perched bird may look dull, but once it takes flight, it reveals remarkably vivid colours, quite unlike any other European bird. Especially in late spring, when the adult is at its best, it is a remarkably colourful bird, as the dull, pale feather edges of winter wear away. Rollers are most characteristic of southeast Europe, less common in southwest Europe, and rare vagrants north of the breeding range. They often perch on wires or telephone poles, and so are usually quite easy to detect; they may also perch on the sides of trees, tucked in the shade under the canopy. In Africa, Rollers concentrate more around herds of animals or at fires, where insects are forced to fly and reveal themselves. They are very rare in summer in Europe north of their usual range.

VOICE Hard, crow-like *rak*, *rak-aaak*, or *rack-ak-ak*.
NESTING Hole in tree, wall, or building, or old crow nest in tree; 4–7 eggs; 1 brood; May–June.
FEEDING Eats large insects and small rodents, usually caught on ground after flurried drop from perch.

HOLE NESTER
A big, rotten cavity in an old tree is a typical nest site for this bold, strong bird.

OCCURRENCE
Breeds in S Europe, north to Baltic States in E Europe; present from May to August. In variety of open countryside with orchards, woods, bushes, and rough grassland, often perched on overhead wires or tops of isolated bushes.

Seen in the UK
J F M A M J J A S O N D

Length **30cm (12in)**	Wingspan **52–57cm (20½–22½in)**	Weight **120–190g (4–7oz)**
Social **Small flocks**	Lifespan **Up to 10 years**	Status **Declining†**

Order **Coraciiformes**	Family **Upupidae**	Species ***Upupa epops***

Hoopoe 🔊 46

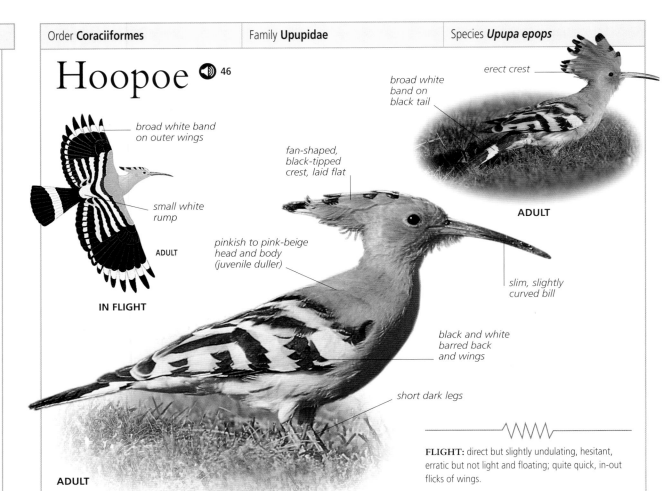

broad white band on outer wings

small white rump

ADULT

IN FLIGHT

broad white band on black tail

erect crest

ADULT

fan-shaped, black-tipped crest, laid flat

pinkish to pink-beige head and body (juvenile duller)

slim, slightly curved bill

black and white barred back and wings

short dark legs

ADULT

FLIGHT: direct but slightly undulating, hesitant, erratic but not light and floating; quite quick, in-out flicks of wings.

Unobtrusive in the dappled shade of a tree in a sunny Mediterranean grove, the Hoopoe bursts to vivid life as it takes flight, revealing a dazzling pattern of black and white. When it settles, it may raise its unique fan-shaped crest, which may also be fanned in flight. It calls from a rooftop or tree, with crest raised, but otherwise remains quiet and quite hard to spot. Its colours are quite subdued, not vivid, but the pattern is remarkable and quite unlike any other European bird: identification is simple.

VOICE Soft, low, quite quick, hollow *poop-poop-poop*, often repeated; hoarse *scheer*.

NESTING Hole in tree or wall; 5–8 eggs; 1 brood; April–July.

FEEDING Walks on ground, probing and picking with bill, taking grubs, insects, and worms.

BLUR OF WHITE
As it alights at the nest, the Hoopoe spreads its wings in a blur of black and white.

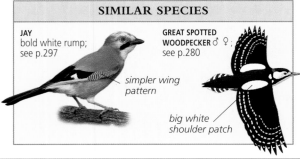

SIMILAR SPECIES

JAY
bold white rump;
see p.297

GREAT SPOTTED WOODPECKER ♂ ♀;
see p.280

simpler wing pattern

big white shoulder patch

OCCURRENCE
Widespread breeder and scarce migrant outside breeding areas north to Baltic; seen in summer only, except in S Spain, S Portugal, and Balearics. In open woodland, parks, gardens, old villages and farmsteads, and orchards.

Seen in the UK
J	F	M	A	M	J	J	A	S	O	N	D
		M	A	M		J	A	S			

Length **26–28cm (10–11in)**	Wingspan **44–48cm (17½–19in)**	Weight **60–75g (2⅛–2⅝oz)**
Social **Family groups**	Lifespan **5–10 years**	Status **Secure**

WOODPECKERS AND WRYNECK

MOST WOODPECKERS are tied to tree habitats but accept a wide variety of species, age, and size of tree: the Great Spotted Woodpecker may feed in willow thickets in winter, but needs bigger branches in which to nest. Others feed on the ground: the Green Woodpecker, an ant-eater, feeds on grassland much more than in trees. Others, however, are more exacting and require large amounts of dead wood, and struggle to survive in modern forests with intensive management.

There are two main groups in Europe: the green woodpeckers and the pied, or spotted, woodpeckers. Green and Grey-headed Wood-peckers are large, rather plainly patterned, with loud, laughing calls. Spotted woodpeckers are

NUT-FEEDER
Great Spotted Woodpeckers wedge large nuts and seeds into bark for easier feeding, hammering them open with their bills.

boldly barred and spotted with black and white and have varying amounts of red. They have short, sharp calls but also frequently "drum" in spring, hammering their bills hard against a resonant branch in a short, rapid drum-roll.

WRYNECK

The Wryneck is brown, barred, and streaked, and unlike other woodpeckers in its posture (it does not so often use its tail as a prop), although its calls and general behaviour indicate a close relationship. An ant-eating specialist, it often feeds on the ground. It is a migrant.

CAMOUFLAGE
Cryptic coloration makes the Wryneck difficult to see amongst branches, but it is a beautiful bird if seen well.

PICIDAE

grey-brown upperparts with complex patterning

greenish yellow rump

small red forehead patch

WRYNECK
p.277

GREEN WOODPECKER
p.278

GREY-HEADED WOODPECKER
p.279

PICIDAE *continued*

white shoulder

large vivid red patch under tail

GREAT SPOTTED WOODPECKER
p.280

finely streaked yellow-buff underside

MIDDLE SPOTTED WOODPECKER
p.281

black cheek patch

barred back

LESSER SPOTTED WOODPECKER
p.282

yellow cap

THREE-TOED WOODPECKER
p.283

dagger-like pale bill with dark tip

all black

BLACK WOODPECKER
p.284

| Order **Piciformes** | Family **Picidae** | Species *Jynx torquilla* |

Wryneck

barred wings

pale grey crown

black-brown central back stripe

short, slim, pointed bill

long, dark eye-stripe

grey-brown upperparts with complex patterning

finely barred, bright buff throat

fine dark bars on pale grey tail

broad, round-tipped tail

IN FLIGHT

pale spots on wings

frequently on the ground

A somewhat aberrant woodpecker, the Wryneck can appear more like a big warbler or small, slim thrush at times, depending on the circumstances. It moves about on the ground, flits up into trees or bushes, or slips through foliage, but also clambers around on thick branches and trunks of trees, although it usually perches across branches rather than upright like the more common woodpeckers. At moderate range it is rather dull and inconspicuous, but close views reveal both an intricate pattern and clean, bright golden-buff colours. Once located, the Wryneck may often be watched really closely for long periods, especially on migration. Occasionally it may then turn up in unexpected places such as parks and gardens.
VOICE Quick, repeated, nasal notes, *kwee-kee-kee-kee-kee-kee-kee*, lower than Kestrel's or Lesser Spotted Woodpecker's.
NESTING Existing hole in tree or wall; 7–10 eggs; 1, occasionally 2, broods; May–June.
FEEDING Often on ground, eating ants and ant larvae; various other insects, spiders, woodlice, and some berries.

FLIGHT: usually short flights, slightly undulating, with flurries of quick beats.

CRYPTIC COLOURS
The mottled pattern of a Wryneck gives excellent camouflage against the bark of a tree and it can be very difficult to spot.

OCCURRENCE
Widespread but scarce in summer, except in Iceland, Ireland, UK, and N Scandinavia. Breeds in farmed countryside with trees, copses, and more extensive pine or mixed forest; migrants often near coast. Scarce migrant in UK, chiefly in autumn.

| Seen in the UK |
| J F **M A M J J A S O N** D |

SIMILAR SPECIES

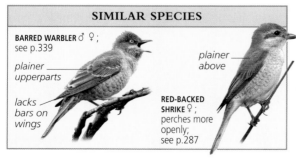

BARRED WARBLER ♂ ♀; see p.339

plainer upperparts

lacks bars on wings

plainer above

RED-BACKED SHRIKE ♀; perches more openly; see p.287

| Length **16–17cm (6½in)** | Wingspan **25–27cm (10–10½in)** | Weight **30–45g (1¹/₁₆–1⁵/₈oz)** |
| Social **Solitary** | Lifespan **5–10 years** | Status **Declining** |

Order **Piciformes**	Family **Picidae**	Species *Picus viridis*

Green Woodpecker 🔊 47

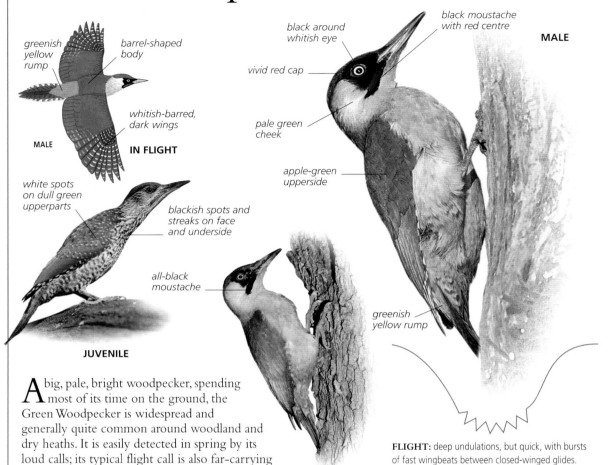

MALE

greenish yellow rump

barrel-shaped body

whitish-barred, dark wings

MALE

IN FLIGHT

black around whitish eye

black moustache with red centre

vivid red cap

pale green cheek

apple-green upperside

greenish yellow rump

white spots on dull green upperparts

blackish spots and streaks on face and underside

all-black moustache

JUVENILE

FEMALE

FLIGHT: deep undulations, but quick, with bursts of fast wingbeats between closed-winged glides.

A big, pale, bright woodpecker, spending most of its time on the ground, the Green Woodpecker is widespread and generally quite common around woodland and dry heaths. It is easily detected in spring by its loud calls; its typical flight call is also far-carrying and distinctive. It excavates its own nest hole but has a less powerful bill than the pied woodpeckers, feeding far less on insect larvae in timber or under bark, and it rarely drums. It is a typical woodpecker in flight, having a deeply undulating action and a final upward swoop to a perch.

VOICE Loud, shrill, bouncing *keu-keu-keuk*; song ringing, slightly descending, laughing *kleu-kleu-kleu-keu-keu*.

NESTING Large hole in tree, 6.5cm (2½in) in diameter; 5–7 eggs; 1 brood; May–July.

FEEDING Eats ants and ant eggs and larvae, using long, sticky tongue to probe nests.

SUBSPECIES

greyer head

black around eye

P. v. sharpei ♂ (Spain, Portugal)

SIMILAR SPECIES

grey crown and red forehead

GOLDEN ORIOLE ♀ similar to ♂ ♀; see p.286

lacks head pattern

darker wings

GREY-HEADED WOODPECKER ♂ ♀; see p.279

OCCURRENCE
Widespread resident except in Iceland, Ireland, and most of N Scandinavia. In or around broadleaved and mixed woodland and heath-like places with bushes and clumps of trees. Regularly feeds on large lawns and other open grassy areas with ants.

Seen in the UK
J F M A M J J A S O N D

Length **30–33cm (12–13in)**	Wingspan **40–42cm (16–16½in)**	Weight **180–220g (6–8oz)**
Social **Solitary**	Lifespan **5–10 years**	Status **Declining**

Order **Piciformes**	Family **Picidae**	Species *Picus canus*

Grey-headed Woodpecker

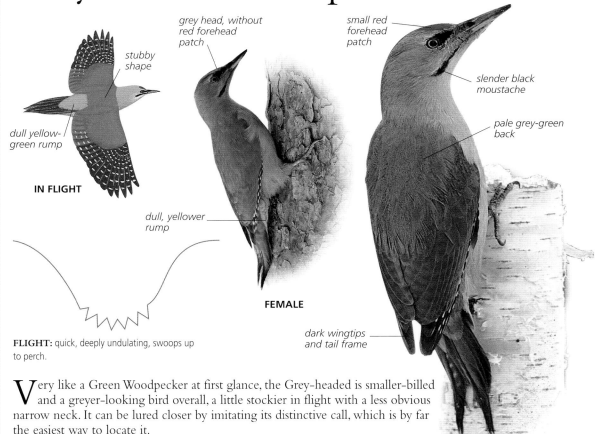

grey head, without red forehead patch

stubby shape

dull yellow-green rump

IN FLIGHT

small red forehead patch

slender black moustache

pale grey-green back

dull, yellower rump

FEMALE

dark wingtips and tail frame

MALE

FLIGHT: quick, deeply undulating, swoops up to perch.

Very like a Green Woodpecker at first glance, the Grey-headed is smaller-billed and a greyer-looking bird overall, a little stockier in flight with a less obvious narrow neck. It can be lured closer by imitating its distinctive call, which is by far the easiest way to locate it.

VOICE Series of loud notes, less ringing than the Green Woodpecker, becoming both slower and lower in pitch, characteristically 'fading out'.

NESTING Excavates hole in tree, such as aspen, beech, or oak, about 5.5cm (2¼ in) across; 7–9 eggs; 1 brood; April–July.

FEEDING Eats insects, especially ants taken from ground, but more varied diet than the Green Woodpecker, including sap from holes in bark.

SIMILAR SPECIES

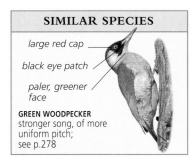

large red cap

black eye patch

paler, greener face

GREEN WOODPECKER
stronger song, of more uniform pitch;
see p.278

TYPICAL WOODPECKER
Like other woodpeckers, the Grey-headed uses its stiff, spread tail as a support when perching upright on a branch or stump, but it can hang beneath a branch, clinging on with its strong feet and claws.

OCCURRENCE
Widespread resident in forests, parks, and riverside trees, also in higher upland mixed forest, in central and eastern Europe.

Seen in the UK
J F M A M J J A S O N D

Length **27–30cm (11–12in)**	Wingspan **38–40cm (15–16in)**	Weight **125–165g (4–6oz)**
Social **Family groups**	Lifespan **5–10 years**	Status **Declining**

| Order **Piciformes** | Family **Picidae** | Species **Dendrocopos major** |

Great Spotted Woodpecker 🔊 48.I, 48.II

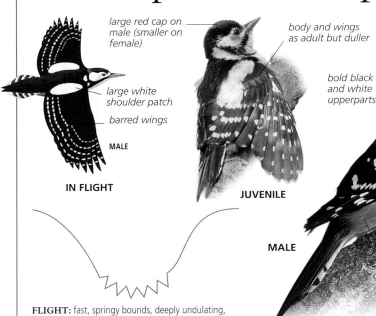

large red cap on male (smaller on female)

large white shoulder patch

barred wings

MALE

IN FLIGHT

body and wings as adult but duller

JUVENILE

red patch on back of head

bold black and white upperparts

bright buff underside

MALE

large vivid red patch under tail

no red on nape

FLIGHT: fast, springy bounds, deeply undulating, with short bursts of whirring wingbeats.

The common woodpecker in most areas, the Great Spotted announces itself in spring by a loud drumming: a rapid, abrupt "drum roll" made by hammering the bill against a resonant branch. It uses its stiff tail as a prop, so that it can grip a branch while resting upright against it. Its grip is secure enough to enable it to hang beneath a branch without the help of its tail for a time. Careful observation is required to be sure of woodpecker identification where several woodpecker species overlap.

VOICE Loud, hard, explosive *tchik!*, less often fast, chattering rattle of alarm; loud, fast, very short drum.

NESTING Digs hole, diameter 5–6cm (2–2¼in), in trunk or branch of tree; 4–7 eggs; 1 brood; April–June.

FEEDING Finds insects and larvae beneath bark, digging them out with strong bill; also takes seeds and berries; visits gardens for nuts, seeds, cheese, and fat.

NUT FEEDER
The woodpecker uses its stiff tail to help secure a good grip on a basket of peanuts.

FEMALE

OCCURRENCE
Inhabits mature woods and even scrub, locally visiting gardens, all over Europe except for Iceland, Ireland, and extreme N Scandinavia. Some migrants from N Europe move south and west in winter, turning up on open islands at times.

| Seen in the UK |
| J F M A M J J A S O N D |

SIMILAR SPECIES

SYRIAN WOODPECKER ♂ ♀; see p.470

different neck pattern

paler under tail

round head

small bill

paler under tail

MIDDLE SPOTTED WOODPECKER ♂ ♀; see p.281

no shoulder patch

much smaller

LESSER SPOTTED WOODPECKER ♂ ♀; see p.282

| Length **22–23cm (9in)** | Wingspan **34–39cm (13½–15½in)** | Weight **70–90g (2½–3¼oz)** |
| Social **Solitary** | Lifespan **5–10 years** | Status **Secure** |

Order **Piciformes**	Family **Picidae**	Species **Dendrocopos medius**

Middle Spotted Woodpecker

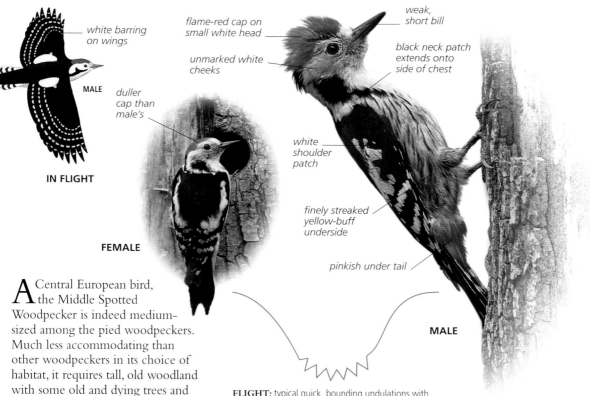

white barring on wings

MALE

IN FLIGHT

flame-red cap on small white head

unmarked white cheeks

duller cap than male's

FEMALE

weak, short bill

black neck patch extends onto side of chest

white shoulder patch

finely streaked yellow-buff underside

pinkish under tail

MALE

FLIGHT: typical quick, bounding undulations with bursts of fast wingbeats.

A Central European bird, the Middle Spotted Woodpecker is indeed medium-sized among the pied woodpeckers. Much less accommodating than other woodpeckers in its choice of habitat, it requires tall, old woodland with some old and dying trees and some small clearings, with a selection of decaying branches in which it can nest. In many regions with intensively managed woodland and young plantations, it is unable to survive. It is relatively quiet and fond of staying high in trees, so it is easy to overlook. It calls more often in spring in preference to drumming.
VOICE Song quite slow repetition of nasal *kvek-kvek-kvek-kvek*, infrequent weak *kik*, quick, rhythmic *kuk-uk kuk-uk- kuk-uk*; does not drum.
NESTING Excavates hole in rotten branch, 4cm (1½in) diameter; 4–7 eggs; 1 brood; May–June.
FEEDING Finds insects, larvae, and sap in high branches, often in dead or dying wood.

SECRETIVE NESTER
Until the young begin to call, this woodpecker, like other wood-peckers, is quiet around the nest.

OCCURRENCE
Breeds locally in N Spain, France, and east to E Europe and Balkans. Mostly in older woodland with some dead and decaying trees and usually not able to survive in over-managed woods and young or uniform plantations.

Seen in the UK
J F M A M J J A S O N D

SIMILAR SPECIES

GREAT SPOTTED WOODPECKER ♂ ♀; see p.280

slightly different head pattern

bolder red under tail

bigger bill

barred above

LESSER SPOTTED WOODPECKER ♂ ♀; see p.282

different head pattern

lacks white shoulder patch

SYRIAN WOODPECKER ♂ ♀; see p.470

Length **19–22cm (7½–9in)**	Wingspan **35cm (14in)**	Weight **60–75g (2⅛–2⅝oz)**
Social **Solitary**	Lifespan **5–10 years**	Status **Secure**

Order **Piciformes**	Family **Picidae**	Species **Dendrocopos minor**

Lesser Spotted Woodpecker

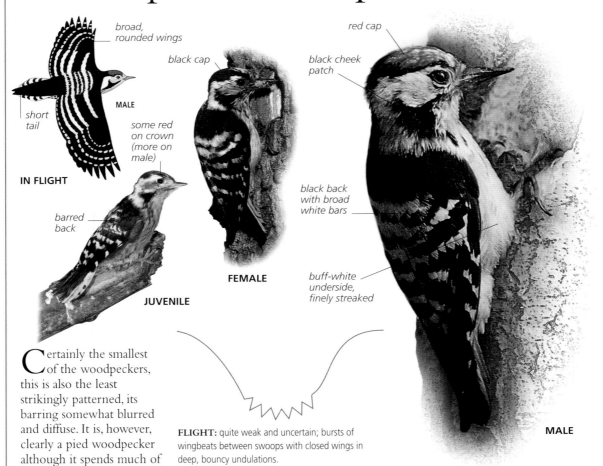

broad, rounded wings

black cap

red cap

black cheek patch

MALE

short tail

IN FLIGHT

some red on crown (more on male)

black back with broad white bars

barred back

FEMALE

buff-white underside, finely streaked

JUVENILE

MALE

FLIGHT: quite weak and uncertain; bursts of wingbeats between swoops with closed wings in deep, bouncy undulations.

Certainly the smallest of the woodpeckers, this is also the least strikingly patterned, its barring somewhat blurred and diffuse. It is, however, clearly a pied woodpecker although it spends much of its time in the higher, more slender branches of trees, unlike the others. It prefers limes, elms, and other trees with very upright twigs, and uses its tail as a prop like most other woodpeckers, clinging more or less upright to its perch. Because of its size and generally quiet demeanour, it is easy to overlook, but in most areas is genuinely rather scarce.

VOICE Sharp, weak *tchik*, nasal, peevish *pee-pee-pee-pee-pee-pee* especially in spring; weak drum.

NESTING Hole in tree, 3cm (1¼in) in diameter; 4–6 eggs; 1 brood; May–June.

FEEDING Chips out insects and their larvae from beneath loose or rotten bark; also takes insects from thick, woody plant stems close to ground.

OCCURRENCE
In most of Europe except Iceland, Ireland, N UK, and much of Spain and Portugal. Widespread in woodland, copses, orchards, and tall hedges with old or diseased trees. Resident, except for local movements which take it into gardens and parks.

Seen in the UK
J F M A M J J A S O N D

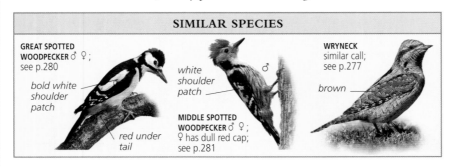

SIMILAR SPECIES

GREAT SPOTTED WOODPECKER ♂ ♀; see p.280

bold white shoulder patch

red under tail

white shoulder patch

MIDDLE SPOTTED WOODPECKER ♂ ♀; ♀ has dull red cap; see p.281

WRYNECK similar call; see p.277

brown

Length **14–15cm (5½–6in)**	Wingspan **25–27cm (10–10½in)**	Weight **18–22g (⅝–¹³/₁₆oz)**
Social **Solitary**	Lifespan **5–10 years**	Status **Secure**

Three-toed Woodpecker

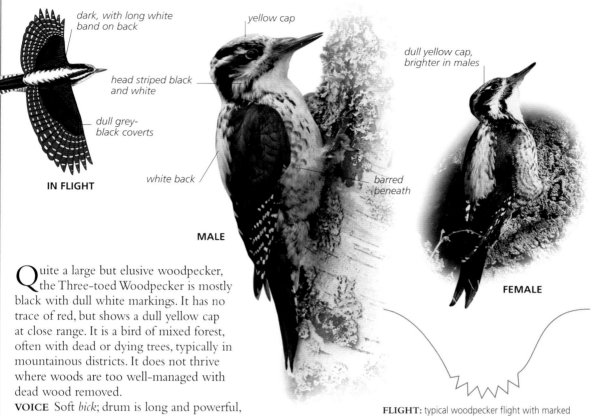

dark, with long white band on back

head striped black and white

dull grey-black coverts

IN FLIGHT

yellow cap

white back

barred beneath

MALE

dull yellow cap, brighter in males

FEMALE

Quite a large but elusive woodpecker, the Three-toed Woodpecker is mostly black with dull white markings. It has no trace of red, but shows a dull yellow cap at close range. It is a bird of mixed forest, often with dead or dying trees, typically in mountainous districts. It does not thrive where woods are too well-managed with dead wood removed.

VOICE Soft *bick*; drum is long and powerful, accelerating slightly.

NESTING Excavates a cavity about 5cm (2in) wide in trunk or larger branch of dead or dying tree; 3–5 eggs; 1 brood; April–June.

FEEDING Drills holes to take sap from bark and strips bark from dead trees to reach insect larvae beneath, especially of wood-boring beetles.

FLIGHT: typical woodpecker flight with marked undulations and short bursts of quick, deep wingbeats.

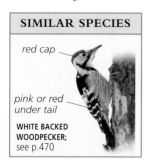

SIMILAR SPECIES

red cap

pink or red under tail

WHITE BACKED WOODPECKER; see p.470

NEST CAVITY
In typical woodpecker style, Three-toed Woodpecker excavates a nest hole in a pine tree and can most easily be located there in summer as it returns to feed its young.

OCCURRENCE
Scarce resident woodpecker of mixed or coniferous forest, especially mature spruce, with dead trees intermixed, in north and, very locally, central Europe.

Seen in the UK
J F M A M J J A S O N D

Length **22–24cm (9–9½in)**	Wingspan **34–38cm (13½–15in)**	Weight **65–75g (2⅜–2⅝oz)**
Social **Family groups**	Lifespan **5–10 years**	Status **Secure**

Order **Piciformes**	Family **Picidae**	Species ***Dryocopus martius***

Black Woodpecker

rounded wings with fingered tips

slim neck

MALE

IN FLIGHT

red only on back of head

FEMALE

dagger-like pale bill with dark tip

bold red cap

white eye

glossy black plumage

tail used as prop while perching upright

MALE

Easily the largest of the woodpeckers, but not correspondingly any easier to see, the Black Woodpecker is common in woodland areas with big, mature beech or pine trees. In winter, it wanders quite freely through big gardens and suburban parks. In much of Europe it is associated with mountainous areas, but it is common in the lowlands of northwest Europe. It can usually be detected by its loud calls, or its bursts of "machine-gun" drumming, but is usually shy and not easy to approach.

VOICE Loud, high, plaintive, long *pyuuu*; loud, rolling, far-carrying *krri-krri-krri-krri-krri*; loud Green Woodpecker-like laugh, louder, more irregular. Long, loud drumming.

NESTING Large oval hole, 9 x 12cm (3½ x 5in) in diameter, in big tree; 4–6 eggs; 1 brood; April–June.

FEEDING Digs insect larvae from tree branches and trunks and fallen timber; eats ants on ground.

FLIGHT: direct, strong, not undulating; head up, wings beat mostly below body level; swoops up to perch with quick flurry.

HEAVY FLIGHT
Although large and heavy, the shape is typical thin-necked, square-winged woodpecker in flight.

SIMILAR SPECIES

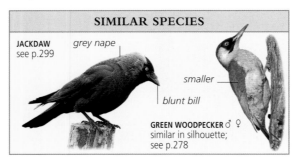

JACKDAW see p.299

grey nape

smaller

blunt bill

GREEN WOODPECKER ♂ ♀ similar in silhouette; see p.278

OCCURRENCE
Breeds widely from N Spain east through France, north to Scandinavia; absent from most of Italy, UK, and Iceland. Associated with big trees in mature woods or clumps within patchy forest. Wanders more widely in winter.

Seen in the UK
| J | F | M | A | M | J | J | A | S | O | N | D |

Length **40–46cm (16–18in)**	Wingspan **67–73cm (26–29in)**	Weight **250–370g (9–13oz)**
Social **Solitary**	Lifespan **Up to 10 years**	Status **Secure**

ORIOLES

Several species look vividly coloured in books; some may disppoint a little in real life. The male Golden Oriole, however, is always a wonderful sight if seen well, a vivid buttercup yellow and black. Strangely, it is remarkably elusive.

The song is loud and obvious, always an easy clue to the presence of an oriole, but seeing it is still difficult most of the time. Orioles live in dense foliage, typically in poplars or oak woodland, and even their bright colours are hard to spot in the dappled light and shade of a wind-swept leafy canopy.

Male and female usually differ but old females become almost as bright as males. In winter, orioles migrate to Africa, where they come into contact with several other similar species. In Europe, nothing else is similar except for a poorly seen Green Woodpecker in flight, which recalls the female oriole, and no other bird is remotely so yellow and black as the adult male.

ORIOLIDAE

vivid yellow and intense black plumage

GOLDEN ORIOLE
p.286

SHRIKES

Thrush-like in shape and general form, shrikes have stout, sharp, hooked bills and strong feet, and they are every bit as predatory as the small falcons. They drop to the ground onto prey, from a perch, or catch insects and birds in flight: a shrike will pursue and catch a bird almost as big as itself.

Shrikes are migratory, the Great Grey Shrike moving to western Europe in winter, others going to Africa. Most species are suited to warmer parts of southern and eastern Europe, where there are abundant large insects. Intensive farming in many areas has reduced their numbers and the Red-backed Shrike has only recently been lost as a breeding bird in parts of its original range.

Some species have obvious sexual differences in plumage, others are more or less the same. Identification is likely to pose problems only with migrant juveniles in autumn. Then precise details of bill, head, wing, and tail patterns are necessary to confirm more general impressions of size and shape.

LANIIDAE

rufous-brown back

thick black bill

cold grey upperparts

rufous cap

RED-BACKED SHRIKE
p.287

LESSER GREY SHRIKE
p.288

GREAT GREY SHRIKE
p.289

WOODCHAT SHRIKE
p.290

| Order **Passeriformes** | Family **Oriolidae** | Species *Oriolus oriolus* |

Golden Oriole 🔊 83

bold yellow rump

quite pointed wingtips

FEMALE

MALE

black wings

IN FLIGHT

vivid yellow and intense black plumage

bright pink-red bill

no black mask

long, slim body

greener plumage than male's

FEMALE

pale yellow rear flanks

MALE

FLIGHT: thrush-like, direct, slightly undulating, swift from tree to tree.

Despite its bright colour and loud, unique song, the Golden Oriole mostly remains hidden in dense foliage. Occasional brief glimpses may be followed by a longer view as it flies to the next belt of woodland, looking like a large, fast-flying yellow thrush. Females look more like Green Woodpeckers in flight but lack the yellow rump; they look quite different when perched. Orioles prefer leafy belts of oak, chestnut, and especially poplars.

VOICE Hoarse, strained, Jay-like or cat-like *meeaik*, fast *gigigi*; song far-carrying, loud, short phrase, very full-throated and fluty with yodelling quality, such as *wee-dl-eyo, wee-weo-we-weedl*, or *weeeoo*.

NESTING Shallow nest of grass and strips of bark, slung beneath horizontal fork in branch, high in tree; 3 or 4 eggs; 1 brood; May–June.

FEEDING Takes caterpillars and other invertebrates from foliage within dense tree canopy; also feeds on berries in late summer and autumn.

STUNNING COLOURS
One of Europe's most striking birds, good views of an adult male reveal a bird of brilliant yellow and intense black.

SIMILAR SPECIES

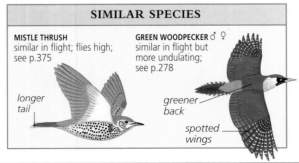

MISTLE THRUSH
similar in flight; flies high; see p.375

longer tail

GREEN WOODPECKER ♂ ♀
similar in flight but more undulating; see p.278

greener back

spotted wings

OCCURRENCE
Breeds in extreme E England, more commonly in Europe north to S Finland. From April to September, in open or dense woodland, poplar plantations, riverside forest, wooded parks, and similar places with airy, leafy trees.

| Seen in the UK |
| J F M **A M J J A S** O N D |

| Length **22–25cm (9–10½in)** | Wingspan **35cm (14in)** | Weight **55g (2oz)** |
| Social **Solitary** | Lifespan **Up to 5 years** | Status **Secure** |

Order **Passeriformes**	Family **Laniidae**	Species *Lanius collurio*

Red-backed Shrike

grey rump

MALE

FEMALE

rufous tail

IN FLIGHT

/\/\/\/\/\/\/\/\/\/\/\/\/\/\/\/\

FLIGHT: jerky, bounding; bursts of wingbeats, tail sometimes waved.

dark patch behind eye

warm brown back

slight scaly barring on greyish buff underside

FEMALE

pale powder-blue head

black mask

rufous-brown back

shell-pink underside

black tail with white sides

MALE

Sadly diminished in much of its range, and gone from the UK as a breeding bird, the Red-backed Shrike is still moderately common in places where traditional farming leaves plenty of hedges, bushes, and rough grassland with an abundance of large insects. It perches prominently, looking for prey, which it catches in a sudden flurry on the ground. Large items are brought back up to a perch, sometimes to be impaled on a thorn for easy manipulation or for storage.
VOICE Harsh *hek*, harder *chek*; song low, rambling, some bright warbling and mimicry.
NESTING Untidy nest of grass, moss, feathers, and refuse in bush; 5 or 6 eggs; 1 brood; May–June.
FEEDING Watches from perch, and drops to ground to catch beetles and other large insects; catches some insects in flight; also feeds on small lizards and small rodents.

HIDDEN FEMALE
While males often perch on bush tops, the females, which are drabber, tend to be inconspicuous when breeding, perching low down on hedges or bushes.

OCCURRENCE
Breeds in mainland Europe, except in N Scandinavia and S Spain; now rare migrant in UK (where it was breeding species earlier). In farmland with hedges, thorn bushes, and bushy slopes, from April to October, when some migrants linger near coasts.

Seen in the UK
J F **M A M J** J **A S O** N D

SIMILAR SPECIES

PENDULINE TIT ♂ ♀ ; acrobatic in foliage; see p.305

tiny

NIGHTINGALE similar to ♀ ; see p.379

rufous tail

plainer

LINNET ♂ ♀ ; more social; see p.419

tiny

Length **16–18cm (6½–7in)**	Wingspan **24–27cm (9½–10½in)**	Weight **25–30g (⅞–1¹/₁₆oz)**
Social **Solitary**	Lifespan **3–5 years**	Status **Declining†**

Order **Passeriformes**	Family **Laniidae**	Species *Lanius minor*

Lesser Grey Shrike

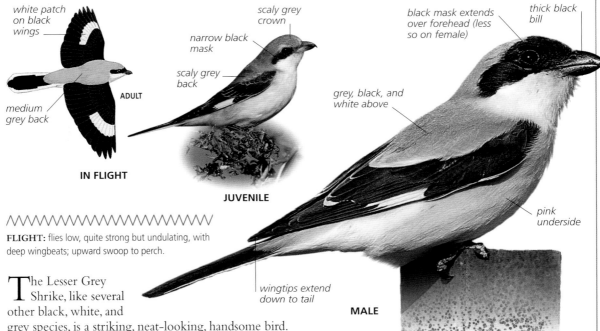

white patch on black wings

medium grey back

ADULT

IN FLIGHT

scaly grey crown

narrow black mask

scaly grey back

JUVENILE

black mask extends over forehead (less so on female)

thick black bill

grey, black, and white above

pink underside

wingtips extend down to tail

MALE

/\/\/\/\/\/\/\/\/\/\/\/\/\/\/\

FLIGHT: flies low, quite strong but undulating, with deep wingbeats; upward swoop to perch.

The Lesser Grey Shrike, like several other black, white, and grey species, is a striking, neat-looking, handsome bird. Usually quite obvious, it behaves like other shrikes in perching on open perches much of the time (but, like them, can be frustratingly elusive on occasions). It frequently flies out to chase prey, or drops to the ground with a flurry of white-barred wings. It is the southeastern counterpart of the Great Grey Shrike, preferring warm, dry summer climates and migrating to Africa in winter. In parts of eastern Europe, migrants follow long lines of trees beside roads in areas that are otherwise open, treeless croplands. Shrikes have strong claws to catch and hold prey and use their hooked bills to kill and tear up small birds and voles. Food may be impaled on thorns to make it easier to deal with.

VOICE Short, hard *tchek tchek*; brief, bright, screaming note serves as song often given in hovering flight.

NESTING Untidy structure of grass and twigs high in bush or tree; 5–7 eggs; 1 brood; May–June.

FEEDING Watches from high perch such as overhead wire and drops onto lizards, beetles, and small birds.

SPRING MALE
The male Lesser Grey Shrike looks handsome with his bold mask and pink underside.

SIMILAR SPECIES

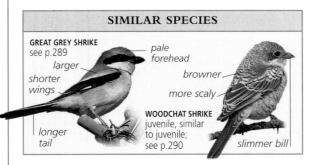

GREAT GREY SHRIKE
see p.289

larger

shorter wings

pale forehead

longer tail

WOODCHAT SHRIKE
juvenile, similar to juvenile; see p.290

browner

more scaly

slimmer bill

OCCURRENCE
Breeds in extreme S France, Italy, and more widely in Balkans and E Europe. Seen from March to September, chiefly in open places with scattered trees, bushes, orchards, and avenues, and in woodland edges.

Seen in the UK
J F M A M J J A S O N D

Length **19–21cm (7½–8½in)**	Wingspan **30cm (12in)**	Weight **30g (1¹⁄₁₆oz)**
Social **Solitary**	Lifespan **3–5 years**	Status **Declining†**

Great Grey Shrike

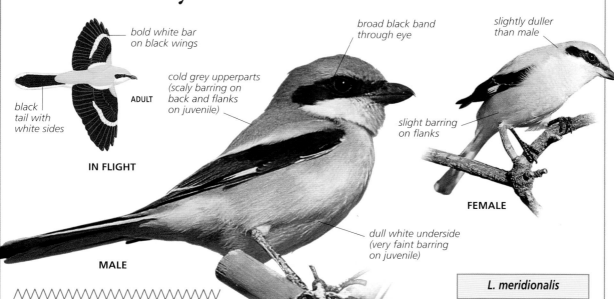

bold white bar
on black wings

black
tail with
white sides

ADULT

IN FLIGHT

broad black band
through eye

cold grey upperparts
(scaly barring on
back and flanks
on juvenile)

slightly duller
than male

slight barring
on flanks

FEMALE

MALE

dull white underside
(very faint barring
on juvenile)

FLIGHT: bounding, undulating; flurries of
wingbeats between glides; upward swoop to perch.

This is the largest European shrike, boldly
patterned in clean grey, white, and black. It can be
easy to find or surprisingly elusive. Like other shrikes,
it is fond of prominent perches and can be seen at a
great distance as a white dot on a bush top. It balances
by swaying its tail and leaning forward, sometimes
lurching at an odd angle, before diving to the ground
in a flurry of white-barred wings to catch its prey.
VOICE Dry trill and various short, hard notes; song
short, simple, squeaky notes.
NESTING Untidy grassy nest in thick bush; 5–7 eggs;
1 brood; May–July.
FEEDING Watches from perch, swoops onto small
rodents, small birds, big insects, and lizards.
REMARK Subspecies *L. e. algeriensis* (N Africa) is
darker above and grey below; *L. e. pallidirostris* (Asia,
vagrant in W Europe) has pale bill, mask only behind
eye, and more white on wings.

L. meridionalis

Southern Grey Shrike
Formerly considered a
subspecies of *L. excubitor*,
now a species in its own right;
S France, Spain, Portugal,
N Africa.

darker
above

greyer
or
pinker
below

BALANCING ACT
The long tail acts as a balance for
a Great Grey Shrike perched on a
thin branch.

OCCURRENCE
Breeds widely but very locally,
in Scandinavia, across N Europe
to France, Spain, and Portugal;
N European birds winter west
to UK, south to N Italy. Breeds
in birch woods, wooded bogs;
in dry, hotter, scrubby areas in
S Europe; in winter, on bushy,
heathy, or boggy ground.

Seen in the UK
| J | F | M | A | M | J | J | A | S | O | N | D |

SIMILAR SPECIES

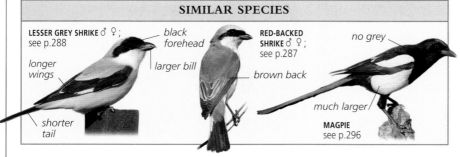

LESSER GREY SHRIKE ♂ ♀;
see p.288

longer
wings

black
forehead

larger bill

shorter
tail

**RED-BACKED
SHRIKE** ♂ ♀;
see p.287

brown back

no grey

much larger

MAGPIE
see p.296

| Length | **22–26cm (9–10in)** | Wingspan | **30cm (12in)** | Weight | **30–40g (1¹/₁₆–1⁷/₁₆oz)** |
| Social | **Solitary** | Lifespan | **3–5 years** | Status | **Declining** |

Order **Passeriformes**	Family **Laniidae**	Species *Lanius senator*

Woodchat Shrike

bold white
shoulder patch

white
rump

MALE

IN FLIGHT

grey-brown
upperparts

white scapulars
with thin dark
crescents (brown
on Red-backed
Shrike)

pale underside,
barred grey

JUVENILE

rufous cap

black and
white back

white below

black tail with
white at base

MALE

pale area
around eye

white "V"
on back

FEMALE

Like other shrikes, Woodchat Shrikes can be remarkably obvious, perched on overhead wires, trees, or bush tops, or frustratingly difficult to find as they skulk in thick cover. They are strikingly patterned and easy to identify. In parts of southern Europe, they are common and many can be seen, for example, on a long journey by road. They feed on insects and small birds, watching for these from a perch and chasing or diving on them from above.
VOICE Short, chattering, hard notes in alarm; song loud, quick, varied jumble of squeaks and warblers.
NESTING Untidy nest of grass and stems in low bush; 5 or 6 eggs; 1 brood; April–July.
FEEDING Takes big insects from ground or in air, after watching from open perch; also catches small birds, rodents, and lizards.

FLIGHT: strong, quick, low, undulating; swoops up to perch.

BOLD PERCH
A feeding Woodchat Shrike keeps a keen eye out for large insects and other potential prey. It may chase and capture smaller birds.

OCCURRENCE
Breeding bird in Spain, Italy, S France, east to Balkans, and locally in C Europe. Seen from April to October, in bushy areas, open countryside, orchards, overgrown old gardens, and other generally bushy places.

Seen in the UK
J F M A M J J A S O N D

SIMILAR SPECIES

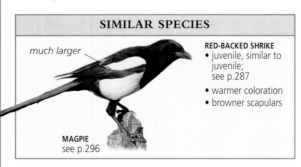

much larger

RED-BACKED SHRIKE
• juvenile, similar to juvenile;
  see p.287
• warmer coloration
• browner scapulars

MAGPIE
see p.296

Length **17–19cm (6½–7½in)**	Wingspan **25–30cm (10–12in)**	Weight **25–35g (⅞–1¼oz)**
Social **Solitary**	Lifespan **3–5 years**	Status **Vulnerable**

CROWS

THERE IS CONSIDERABLE VARIATION in the crow family within Europe. The Jay is colourful and strongly-patterned. Like most crows, it is heavily persecuted and very shy, but where unmolested it becomes much bolder. The Siberian Jay has quite different colours: it is a bird of far northern forests.

Magpies are striking, long-tailed, pied birds. They are generally disliked because of their liking for small birds' eggs and chicks. They are, nevertheless, fascinating and handsome.

The Alpine Chough is a high-altitude species, coming lower in winter, and often mixing with the Chough, itself a social bird where common. Choughs are also found on coastal cliffs in north-west Europe.

The world's largest crow is the Raven. The all-black Carrion Crow is widely spread but replaced in parts of Europe by the grey-and-black Hooded Crow.

MAGPIE
A highly social bird, the Magpie has increased in suburban areas, where ornamental shrubberies and parks provide ideal habitat. It is immediately distinctive.

CORVIDAE

bright orange-buff area under tail

SIBERIAN JAY
p.293

slightly curved, tapered, bright red bill

CHOUGH
p.294

short, pale yellow bill

ALPINE CHOUGH
p.295

long blackish tail, glossed purple and green

MAGPIE
p.296

blue panel on wings

JAY
p.297

bold white spots all over dark brown body

NUTCRACKER
p.298

CORVIDAE *continued*

paler grey nape and cheeks

JACKDAW
p.299

peaked crown

ROOK
p.300

tight body feathering

CARRION CROW
p.301

pale grey body

HOODED CROW
p.302

very thick bill

RAVEN
p.303

Order **Passeriformes**	Family **Corvidae**	Species ***Perisoreus infaustus***

Siberian Jay

rust orange wingbars

bright tail patch

IN FLIGHT

head looks dark

bright buff underside

WINTER

bright orange-buff area under tail

bright sides to tail

pale spot above bill

dark hood

short, thick bill

pale buffish beneath

rufous wing patch

ADULT

FLIGHT: smooth, fast, silent, acrobatic in tight spaces.

A small jay, with a marked resemblance to the common Jay, especially in the expression of its face and bill, the Siberian Jay is a bird of the far north, often seen around camps and near villages in forest areas. It is usually bold and tame in the presence of people and eager to feed on scraps. It is, nevertheless, a rather elusive bird in many parts of its range until a family party appears nearby and seems to inspect human intruders, with quiet, soft, clucking calls. Only when an adult feeds its young can the two be separated easily.

VOICE Mostly silent, but occasional weak mewing or feeble screaming calls, especially from small groups.

NESTING Nest a loose platform of twigs and lichens close to trunk of conifer; 3–4 eggs; 1 brood; May–July.

FEEDING Eats almost anything of suitable size, from insects and small mammals to birds' eggs, seeds, berries, and scraps from human camps. It hides food in crevices in bark and under piles of pine needles for winter use.

CAMP FOLLOWER
Often tame or oblivious of human presence, the Siberian Jay frequently forages around woodland camp sites in the summer, taking whatever edible scraps it can find.

SIMILAR SPECIES

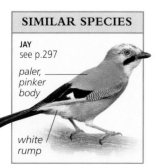

JAY
see p.297

paler, pinker body

white rump

OCCURRENCE
N Scandinavia; a resident in deep pine and spruce forest, especially older trees with an abundance of lichen and some undergrowth. Almost unknown outside its breeding range.

Seen in the UK
J F M A M J J A S O N D

Length **30cm (12in)**	Wingspan **40–45cm (16–18in)**	Weight **80–100g (2⅞–3⅝oz)**
Social **Small flocks**	Lifespan **Up to 10 years**	Status **Secure**

Order **Passeriformes**	Family **Corvidae**	Species **Pyrrhocorax pyrrhocorax**

Chough

ADULT

long, square, fingered wings (greyer flight feathers below)

IN FLIGHT

slightly curved, tapered, bright red bill

orange-red bill, paler than adult's

glossy black body

JUVENILE

square tail

red legs

ADULT

Superficially like a Jackdaw (see p.299), the Chough is blacker, glossier, and altogether more showy. It is also more acrobatic in the air. It is usually found in small groups; where Choughs are common, much bigger flocks form, often mixed with Alpine Choughs. These big flocks may roam widely over mountain pastures or through green valleys below when peaks are in snow. In northwest Europe, Choughs are more coastal and always scarcer than in the mountains.

VOICE Loud calls distinctive, with explosive, ringing, piercing quality, *pee-yaa* or *chia*, some shorter *chuk* and *kwarr* sounds.

NESTING Nest of sticks lined with wool and hair, in sea cave and cavity in cliff or old ruin; 3–5 eggs; 1 brood; May–July.

FEEDING Eats ants from old pastures, insects dug up from beneath soil, and lichen on rocks, prised up with bill.

FLIGHT: exuberant, bouncy, aerobatic; soars well and often; deeply undulating at times; dives into caves or to cliff ledge with wings angled back.

AGILE BIRD
The aerobatic Chough makes long, steep, rapid descents and fast, upward swoops, in pairs or small flocks in concert.

OCCURRENCE
Breeds very locally on coasts of Ireland, W Britain, N and W France, and uplands of Spain, Portugal, S France, Italy, and Balkans. Frequents gorges, crags, high altitude pastures and coastal cliffs, quarries, and grassland; on beaches in winter.

Seen in the UK
J F M A M J J A S O N D

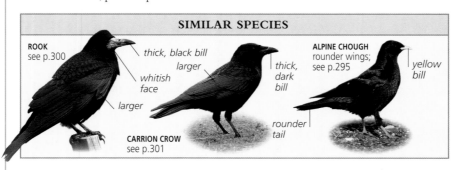

SIMILAR SPECIES

ROOK see p.300 — thick, black bill — larger — whitish face — larger

CARRION CROW see p.301 — thick, dark bill

ALPINE CHOUGH rounder wings; see p.295 — yellow bill — rounder tail

Length **37–41cm (14½–16in)**	Wingspan **68–80cm (27–32in)**	Weight **280–360g (10–13oz)**
Social **Flocks**	Lifespan **5–10 years**	Status **Vulnerable**

| Order **Passeriformes** | Family **Corvidae** | Species *Pyrrhocorax graculus* |

Alpine Chough

long, fingered, slightly rounded wings

two-toned underwing

neater, smoother shape than Chough

glossy black body

short, pale yellow bill

IN FLIGHT

rounded tip to long tail

red legs

A true high altitude specialist, the Alpine Chough nevertheless occasionally descends to valley fields, sometimes with Choughs; it is found in some rugged coastal areas as well, especially in winter. Its calls while it is circling about mountain peaks are distinctive. The Alpine Chough's wings are less rectangular and less straight-edged than the Chough's: the two choughs can be distinguished by their shape at distances which make fine details, such as bill colour, impossible to see.

VOICE Strange, rippling or sizzling *zirrrr* or hissy *chirrish*, penetrating *zeee-up*.

NESTING Bulky nest of stems in cliff cavity; 3–5 eggs; 1 brood; May–July.

FEEDING Forages on grassland; eats insects, other invertebrates, berries, seeds, and scraps from around ski lifts.

FLIGHT: superb, elegant, light flight, soaring and circling on spread wings, often in swirling flocks; frequent aerobatics.

CIRCLING FLOCKS
This bird is often seen in swirling flocks, circling and crossing against a backdrop of mountain peaks.

FORAGING
The Alpine Chough typically forages on alpine grassland, often in large, active flocks.

SIMILAR SPECIES

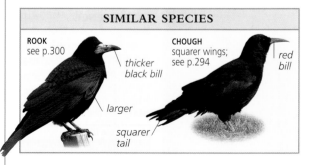

ROOK see p.300

thicker black bill

larger

CHOUGH squarer wings; see p.294

red bill

squarer tail

OCCURRENCE
Breeding bird and resident all year in Pyrenees, Alps, Italy, and Balkans. From coasts in winter to highest peaks, on cliffs and alpine pasture. Often found around ski resorts and lifts.

| Seen in the UK |
| J F M A M J J A S O N D |

| Length **36–39cm (14–15½in)** | Wingspan **65–74cm (26–29in)** | Weight **250–350g (9–13oz)** |
| Social **Flocks** | Lifespan **5–10 years** | Status **Secure†** |

| Order **Passeriformes** | Family **Corvidae** | Species *Nucifraga caryocatactes* |

Nutcracker

bold white spots all over dark brown body

dark brown cap (streaked on juvenile)

thick, dagger-like bill

thickset body

dark blue-black wings

ADULT

unspotted, brownish black wings (spotted on juvenile)

white-tipped black tail

IN FLIGHT

white under tail

ADULT

This pale-spotted, crow-like bird often perches conspicuously on top of tall conifers; it is unmistakable in a good view. Generally easy to find in its restricted range, the Nutcracker sometimes appears far outside this usual breeding distribution. It is subject to occasional population booms which, if they combine with a local food shortage, trigger large-scale movements, or irruptions, as birds are forced to fly far and wide in an effort to survive. Many do not survive these long journeys and the population quickly subsides. These invaders are nearly all of the slender-billed eastern race from Russia, although a few northern thick-billed birds are involved. These birds, far from their usual home, are often surprisingly tame.

VOICE Occasional long, drawn out, hard rattle in spring and summer; otherwise silent.

NESTING Nest of twigs lined with grass and moss, near trunk of tree, usually spruce; 3 or 4 eggs; 1 brood; May–July.

FEEDING Eats some large insects but mainly seeds of hazel, pine, and spruce, constantly collected and stored in summer and re-found with great accuracy (even under snow) in winter.

FLIGHT: rather Jay-like on broad wings but short-tailed, strong, direct, with upward swoop to treetop perch; drops steeply from perch.

UNIQUE BIRD
The Nutcracker is unique-looking, but novice birdwatchers may at times mistake Starlings and young Mistle Thrushes for the rarer, larger bird.

SIMILAR SPECIES	**SUBSPECIES**
STARLING winter; see p.365 *much smaller* 	*N. c. macrothyncus* (NE Europe, Asia); more white on tail; slimmer bill 

OCCURRENCE
Breeds in S Scandinavia, east from Baltic and in mountain areas of C and E Europe. In forests with spruce, hazel, and pine. Resident except when seed crops fail: birds then move south and west to find food, rarely in mass emigrations.

Seen in the UK
J F M A M J J A S O N D

| Length **32–35cm (12½–14in)** | Wingspan **49–53cm (19½– 21in)** | Weight **120–170g (4–6oz)** |
| Social **Small flocks** | Lifespan **Up to 5 years** | Status **Secure†** |

Order **Passeriformes**	Family **Corvidae**	Species **Corvus monedula**

Jackdaw 🔊 86

paler grey nape and cheeks

jet black cap

grey-black body

whitish eye

short, thick bill

dark grey underwings

IN FLIGHT

FLIGHT: straight, easy, fluent with constant floppy wing action, looser, slower than pigeon's; agile, aerobatic in wind, soaring well, usually in parties.

A small crow, the Jackdaw is an expert flier, and flocks frequently soar and glide over woods or around cliffs and quarries. They combine aerobatics with much calling, which makes them easy to identify, but they can be overlooked when feeding in mixed flocks on fields with Rooks. Even in flight, in the flurry of a large flock, their presence may not be immediately obvious although the size difference is marked. Jackdaws are often quite tame and frequently perch on buildings, large and small. Small parties often fly off, high and straight, with sudden bursts of speed.

VOICE Noisy; many calls based on short, sharp, yapping sound, *kyak* or *tjak!*, with slightly squeaky, bright quality, some longer calls like *chee-ar*.

NESTING Pile of sticks lined with animal dung, mud, roots, moss, and hair, in hole in tree or building, or old chimney; 4–6 eggs; 1 brood; April–July.

FEEDING Forages on ground, taking earthworms, seeds, and scraps; also takes caterpillars from foliage, and berries.

WOODLAND CROW
Jackdaws like cliffs and old buildings but are equally at home in treetops in mature woodland.

SIMILAR SPECIES

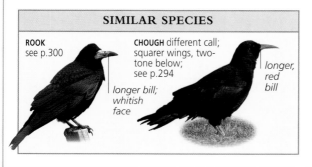

ROOK see p.300

longer bill; whitish face

CHOUGH different call; squarer wings, two-tone below; see p.294

longer, red bill

OCCURRENCE
Breeds in most of Europe except Iceland and N Scandinavia. In NE Europe only in summer, but resident elsewhere. In towns, parks, woods, farmland with scattered woodland, and gardens around old houses with open chimneys.

Seen in the UK
| J | F | M | A | M | J | J | A | S | O | N | D |

Length **33–34cm (13–13½in)**	Wingspan **67–74cm (26–29in)**	Weight **220–270g (8–10oz)**	
Social **Flocks**	Lifespan **5–10 years**	Status **Secure†**	

| Order **Passeriformes** | Family **Corvidae** | Species *Corvus frugilegus* |

Rook 🔊 83

ADULT   **IN FLIGHT**

wings more pointed than Carrion Crow's

peaked crown

glossy black body

bill tapers to point

bare white skin around bill base

face dark at first

thin bill

soars well

narrow, rounded tail recalls Raven (Carrion Crow's is squarer)

rounded tail

loose, ragged thigh feathers

ADULT

JUVENILE

Where common, the Rook is one of the familiar birds of farmland and village, part of the traditional rural scene. It is a big crow, and a very social one, often mixed with Jackdaws, Stock Doves (see p. 249) and Black-headed Gulls (see p. 237). It nests in colonies in treetops, its big, stick nests always easy to see except at the height of summer when foliage conceals them. Occasionally, a pair or two will nest in more isolated situations, inviting confusion with Carrion Crows; the two species can sometimes be quite difficult to tell apart.

VOICE Loud, raucous, relaxed cawing, *caaar*, *grah-gra-gra*, variety of higher, strangled or metallic notes especially around colony.

NESTING Big nest of sticks in tree, lined with grass, moss, and leaves; 3–6 eggs; 1 brood; March–June.

FEEDING Eats worms, beetle larvae, seeds, grain, and roots from ground, especially ploughed fields or stubble, usually in flocks; also forages along roadsides for insects and large road casualties.

FLIGHT: direct, steady, evenly flapping; aerobatic around colony, twisting and diving; soars well.

AT COLONY
Typically, scores of Rook nests may be seen close together near tops of trees in a visually obvious colony.

OCCURRENCE
Absent from Iceland, Scandinavia, and Mediterranean area, but resident across W and C Europe and summer visitor to NE Europe. Typically in farmland with scattered trees, parks, large gardens, and villages with spinneys of tall trees.

Seen in the UK
| J | F | M | A | M | J | J | A | S | O | N | D |

SIMILAR SPECIES

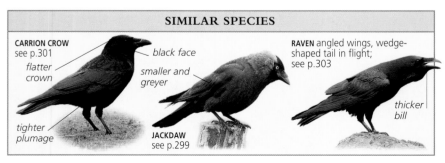

CARRION CROW see p.301

black face

flatter crown

smaller and greyer

RAVEN angled wings, wedge-shaped tail in flight; see p.303

thicker bill

tighter plumage

JACKDAW see p.299

| Length **44–46cm (17½–18in)** | Wingspan **81–99cm (32–39in)** | Weight **460–520g (17–19oz)** |
| Social **Flocks** | Lifespan **5–10 years** | Status **Secure** |

Order **Passeriformes**	Family **Corvidae**	Species ***Corvus corone***

Carrion Crow 🔊 88

broad, flat-topped head

glossy black body

thick, arched bill

tight body feathering

square wingtips

soars rarely

square tail

IN FLIGHT

The all-black Carrion Crow is easily confused with the larger Raven and especially the Rook, which may be almost inseparable at long range. Carrion Crows tend to be more solitary than Rooks, but will often roost in scores, even hundreds, together and occasionally feed in flocks. More usefully, crows nest solitarily while Rooks breed in colonies or, at least, small clusters of nests. Carrion Crows are often wary in open countryside but can become bold in gardens and town parks if not persecuted.

VOICE Loud, harsh, grating *caw, krra krra krra*, metallic *konk, korr*, and variants.
NESTING Big stick nest, shallower than Rook's, in tree or bush; 4 or 6 eggs; 1 brood; March–July.
FEEDING On ground, takes all kinds of invertebrates, eggs, grain, and various scraps; often in pairs, sometimes big flocks on fields spread with manure.

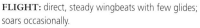

FLIGHT: direct, steady wingbeats with few glides; soars occasionally.

JAUNTY AIR
A bold, upright stance and confident, long-striding walk are characteristic of the Carrion Crow.

OCCURRENCE
Breeds in Great Britain and east to Denmark and C Europe; Hooded breeds in Ireland, Scotland, Isle of Man, and N and E Europe; winter visitor to North Sea coasts. In all kinds of open areas from upland moors to farmland and suburbs.

Seen in the UK
| J | F | M | A | M | J | J | A | S | O | N | D |

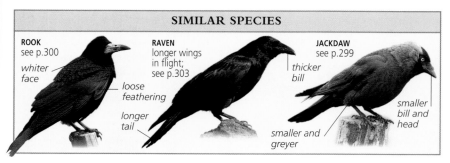

SIMILAR SPECIES

ROOK see p.300
whiter face
loose feathering
longer tail

RAVEN longer wings in flight; see p.303

JACKDAW see p.299
thicker bill
smaller bill and head
smaller and greyer

Length **44–51cm (17½–20in)**	Wingspan **93–104cm (37–41in)**	Weight **540–600g (19–21oz)**
Social **Occasional flocks**	Lifespan **5–10 years**	Status **Secure**

Order **Passeriformes**	Family **Corvidae**	Species ***Corvus cornix***

Hooded Crow

heavy black bill

black hood

flattish head similar to Carrion Crow

IN FLIGHT

bold grey and black pattern

pale grey body

black tail and wings

ADULT AT NEST

Hooded Crows are familiar birds of open landscapes and also suburban areas, such as refuse tips, in much of Europe, replacing the Carrion Crow with just a very narrow line of hybridization where they meet in places such as northern Scotland. They have recently been treated as separate species. Hooded Crows have much the same seemingly intelligent character as Carrion Crows, being wild and shy where (as is often the case) they are persecuted, but bold, opportunistic, and approachable around towns and villages.

VOICE Loud, harsh, grating *caw, krra krra krra*, metallic *konk, korr*, and variants.

NESTING Big stick nest, shallower than Rook's, in tree or bush; 4 or 6 eggs; 1 brood; March–July.

FEEDING On ground, takes all kinds of invertebrates, eggs, grain, and various scraps; often in pairs, sometimes big flocks on fields spread with manure.

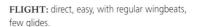

FLIGHT: direct, easy, with regular wingbeats, few glides.

OPPORTUNISTIC FEEDER
Hooded Crows eat almost anything edible, including rabbits and birds killed by road traffic, refuse from tips, and many large insects.

SIMILAR SPECIES

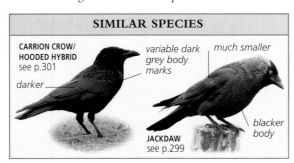

CARRION CROW/ HOODED HYBRID see p.301

darker

variable dark grey body marks

much smaller

JACKDAW see p.299

blacker body

OCCURRENCE
Breeds in open countryside, E Europe from Italy and Scandinavia eastwards, Scotland, Ireland, Isle of Man; more widespread but local in winter in NW Europe.

Seen in the UK
J F M A M J J A S O N D

Length **44–51cm (18–20in)**	Wingspan **93–104cm (37–41in)**	Weight **540–600g (19–21oz)**
Social **Small flocks**	Lifespan **5–10 years**	Status **Secure**

Order **Passeriformes**	Family **Corvidae**	Species **Corvus corax**

Raven 🔊 89

soars brilliantly

long, angular, fingered wings

wingbeats often noisy

all-black plumage

large head

very thick, long, arched bill

loose throat feathers can be expanded as beard

IN FLIGHT

wedge or diamond-shaped tail

long tail

The world's largest crow, the Raven is a heavier, longer-winged, and longer-tailed bird than the Carrion Crow. Its heavier, longer bill is evident at close range. At long range, it is best identified by its shape, manner of flight, calls, and often the situation that it is in. Ravens are often found over the wildest and craggiest country, even over high, barren peaks, although they also frequent softer, wooded or farmed land where Carrion Crows are equally likely. Where they are unmolested, they may be seen flying over coastal towns, and rarely they even nest on tall buildings.

VOICE Important clue: loud, abrupt, echoing *crronk crronk crronk* or *prruk prruk*, metallic *tonk*; various clicking, rattling or quiet musical notes, sometimes in rambling subsong audible at close range.

NESTING Huge nest of thick sticks, wool, grass, and heather, used for many years, under overhang on cliff, or in tall tree; 4–6 eggs; 1 brood; February–May.

FEEDING Eats almost anything; catches small mammals and birds, eats meat from dead sheep and road-killed rabbits; forages for scraps on shore or at refuse tips; eats invertebrates and grain.

FLIGHT: bold, strong, acrobatic, wings often angled; unique roll onto back and back again; soars brilliantly.

LARGE HEAD
The feathers of the crown, chin, and throat can be raised to exaggerate the size of a Raven's head.

SIMILAR SPECIES

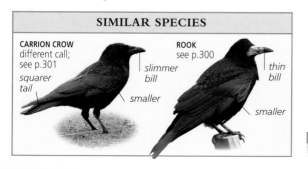

CARRION CROW different call; see p.301
squarer tail

ROOK see p.300
slimmer bill
smaller

thin bill
smaller

OCCURRENCE
In most of Europe except lowland Britain, France, Low Countries, and east to Denmark; year-round resident in large forests, mountain regions, open moorland, and hills with crags and isolated trees.

Seen in the UK
| J | F | M | A | M | J | J | A | S | O | N | D |

Length **54–67cm (21½–26in)**	Wingspan **1.2–1.5m (4–5ft)**	Weight **0.8–1.5kg (1¾–3⅓lb)**
Social **Pairs/Small flocks**	Lifespan **10–15 years**	Status **Secure†**

TITS AND ALLIES

I N THIS GROUP, there are the "true" tits, in the genus *Parus*, together with several "imposters": the Bearded Tit (really a parrotbill), the Long-tailed Tit, and the Penduline Tit. All are more or less social, the Penduline Tit least so; most are woodland birds, but Bearded and Penduline Tits live in or around reedbeds.

The true tits are rather small or very small birds (the Coal Tit is one of Europe's tiniest species), either green, blue, yellow, and white or mixtures of dull grey-brown, buff, white, and black. Several species come to garden feeders and are familiar favourites (especially the Blue and Great Tits, but even Crested Tits visit feeders where they are common).

These are primarily woodland species, timing their breeding to the sudden appearance of vast quantities of caterpillars on tree foliage (something that climate change is throwing out

of synchronization, causing severe problems in places). Garden breeders do less well, but survive the winter better with artificial feeding. In winter, mixed groups of tits forage through woods, hedgerows, and gardens, typically mostly Blue and Great Tits with Coal, Marsh, and Willow Tits in ones and twos around the fringe. They take advantage of greater numbers, being better able to find food and also to spot approaching predators since many pairs of eyes are better than one.

SUSPENDED NEST
The nest of a Penduline Tit is a masterpiece of construction hung from a slender, swaying twig.

REMIZIDAE

broad black mask

PENDULINE TIT
p.305

PARIDAE

bright blue cap, surrounded by white

BLUE TIT
p.306

bright yellow underparts

GREAT TIT
p.307

TIMALIIDAE

big black "moustache"

BEARDED TIT
p.312

AEGITHALIDAE

pointed crest, mottled black and white

CRESTED TIT
p.308

white cheeks

COAL TIT
p.309

big, dull black cap

WILLOW TIT
p.310

pale cheeks; slim neck

MARSH TIT
p.311

long, slim black tail with white sides

LONG-TAILED TIT
p.313

Penduline Tit

deep red-brown band across wings

MALE

IN FLIGHT

narrower black mask

FEMALE

plain head gradually develops black

uniform, plain, pale brown back

plain dark tail

JUVENILE

pale grey head

sharp, narrow, triangular bill

red-brown back

broad black mask

buffish underside

MALE

FLIGHT: quick, erratic, bounding undulations with bursts of wingbeats.

A small, neat, well-patterned bird, the Penduline Tit can be hard to spot (even though easily heard) in tall riverside treetops; in winter, it is often in lower bushes in and around reedbeds and may be easier to find. It is usually close to water, although sometimes several fields away in lines of trees along little more than a ditch or beside a damp meadow. It is common in southeast Europe, but spreading in the west, with increasing appearances in the UK.

VOICE Distinctive high, far-carrying, pure whistle, *psieeee*, longer than similar Reed Bunting note (see p.436); song simple mix of trills and calls.

NESTING Remarkable hanging nest of plant down and cobwebs with tubular entrance high on side, dangling from slim twig; 6–8 eggs; 1 brood; May–June.

FEEDING Eats small insects and reed seeds, in acrobatic tit-like manner.

BOTTLE NEST
The remarkable flask-shaped nest of the Penduline Tit has a short entrance tube high up on one side.

OCCURRENCE
In summer, widespread north to Baltic; in winter, in Mediterranean area, rare in UK. Breeds in and around wetlands with tall trees, especially willows and poplars, and in reeds or reedmace; in winter, mostly in reedbeds.

Seen in the UK
J F M A M J J A S O N D

SIMILAR SPECIES

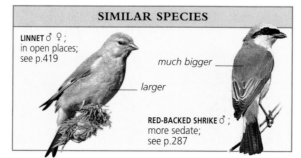

LINNET ♂ ♀; in open places; see p.419

larger

much bigger

RED-BACKED SHRIKE ♂; more sedate; see p.287

Length **10–11cm (4–4¼in)**	Wingspan **20cm (8in)**	Weight **8–10g (⁵⁄₁₆ –³⁄₈oz)**
Social **Small flocks**	Lifespan **3–5 years**	Status **Secure†**

Order **Passeriformes**	Family **Paridae**	Species ***Cyanistes caeruleus***

Blue Tit  78

white bars on blue wings

bright blue cap, surrounded by white

dark line through eye

white cheeks

MALE

IN FLIGHT

narrow black chin

pale yellow underside with thin, dark central streak

blue tail (brightest in spring male)

MALE

greenish cap

dull yellow underside

JUVENILE

slightly less blue than male

FEMALE

Common, noisy, colourful, and tame, the Blue Tit is a favourite garden bird, coming to feeders of all kinds. It also nests in boxes in larger gardens, but is generally not very successful there as broods of young require prodigious numbers of caterpillars to thrive. Only large woodland areas provide sufficient supplies. Even there, in recent years, Blue Tits have had reduced breeding success as the chicks hatch when food supplies have declined, caterpillars having appeared earlier due to climate change.

VOICE Thin, quick, *tsee-tsee-tsee*, harder *tsee-see-sit*, scolding *churrrrr*; song trilled, slurred *tsee-tsee-tsee-tsisisisisisi*.

NESTING Small, mossy cup, lined with hair and feathers, in hole in tree or wall or nest box; 7–16 eggs; 1 brood; April–May.

FEEDING Takes many seeds, nuts, insects, and spiders; visits garden feeders and bird-tables frequently in large numbers.

FLIGHT: fast, undulating over long distance; bursts of whirring wingbeats; sudden stop on perch.

AT NUT BASKET
The Blue Tit uses its quick actions and surefootedness to dash in to a peanut basket and hang, often upside down, to reach its food.

OCCURRENCE
Present almost throughout Europe except in N Scandinavia and Iceland, all year. In woods of all kinds, parks, gardens, and bushy places. In winter, quite often in reedbeds and even wandering around edges of salt marshes.

Seen in the UK
J F M A M J J A S O N D

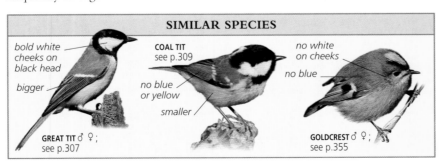

SIMILAR SPECIES

bold white cheeks on black head

bigger

GREAT TIT ♂ ♀; see p.307

COAL TIT see p.309

no blue or yellow

smaller

no white on cheeks

no blue

GOLDCREST ♂ ♀; see p.355

Length **11.5cm (4½ in)**	Wingspan **17–20cm (6½–8in)**	Weight **9–12g ($^{11}/_{32}$–$^{7}/_{16}$ oz)**
Social **Loose flocks**	Lifespan **2–3 years**	Status **Secure**

| Order **Passeriformes** | Family **Paridae** | Species **Parus major** |

Great Tit 🔊 76

pale wingbar

MALE

blue-grey wings

IN FLIGHT

shiny black head

white cheek patch

green back

bright yellow underparts

grey tail with white sides

MALE

dull greenish black head

yellower cheeks

JUVENILE

thick black stripe down underside

MALE

VVVVVVVVVVVVVVVVVVVV

FLIGHT: strong, undulating, with abrupt bursts of whirring wingbeats.

narrower black stripe down underside

FEMALE

A big, colourful, bold, and aggressive tit, the Great Tit is well known almost everywhere and is one of the most familiar of garden birds; it is also common in woodland and on bushy hillsides. It feeds more on the ground than the smaller tits, lacking their extreme lightness and agility, but is still an acrobatic bird, moving more energetically and erratically than the woodland warblers. In spring, it has a simple but remarkably fine and appealing song.

VOICE Extremely varied, often confusing; calls include ringing *chink* or *pink*, piping *tui tui tui*, nasal *churrr*; song variation on repeated two-syllable phrase, sharp, metallic, strident, musical, or grating, with varying emphasis, *tea-cher tea-cher tea-cher* or *seetoo seetoo seetoo*.

NESTING Cup of moss, leaves, and grass in natural hole, woodpecker hole, or nest box; 5–11 eggs; 1 brood; April–May.

FEEDING Eats insects, seeds, berries, and nuts, especially tree seeds in autumn and winter, many from ground; commonly visits bird-tables and feeders.

SIMILAR SPECIES

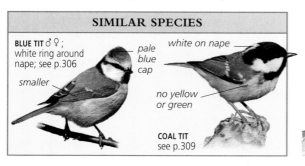

BLUE TIT ♂♀; white ring around nape; see p.306

smaller

pale blue cap

white on nape

no yellow or green

COAL TIT see p.309

OCCURRENCE

Breeds and all-year round resident almost throughout Europe except in Iceland. All-year in wide variety of mixed woodland, parks, and gardens; in S Europe, also on warm, scrubby hillsides.

Seen in the UK
J F M A M J J A S O N D

| Length **14cm (5½in)** | Wingspan **22–25cm (9–10in)** | Weight **16–21g (⁹⁄₁₆–¾oz)** |
| Social **Loose flocks** | Lifespan **2–3 years** | Status **Secure** |

Order **Passeriformes**	Family **Paridae**	Species *Lophophanes cristatus*

Crested Tit

brown tail

brown wings

warm brown back

IN FLIGHT

pointed crest, mottled black-and white

white face with black edge to cheek

plain wings

black bib

buff underside

FLIGHT: weak, flitting, quite quick, with bursts or flurries of wingbeats.

There are titmice with crests elsewhere in the world but this one is unique in Europe and thus easy to identify even in silhouette. It is a pine forest specialist, in the UK restricted to northern Scotland and more numerous in ancient pine forest than in newer plantations. It seems to require some variation, with clearings and forest edge, as well as dead or dying tree stumps in which it can nest. It is easily located by its distinctive stuttering call.

Like other titmice, the Crested Tit is almost oblivious to the presence of people and can be watched feeding from very close range.

VOICE Quick, low, rather soft trill or stutter, *b'd-rrrr-rup*, also usual thin, high tit *zit* or *zee*.

NESTING Soft cup in hole in decaying tree stump; 5–7 eggs; 1 brood; April–June.

FEEDING Eats small insects and spiders; in winter, takes seeds, many from stores made in spring; visits feeders in wooded gardens.

PINE SPECIALIST
Although found in mixed woods in Europe, most Crested Tits prefer old pine forest, with dead trees and stumps to excavate a nest-hole in.

OCCURRENCE
Breeds in N Scotland and most of mainland Europe except extreme N Scandinavia, Italy, and most of Balkans. Present all year round mostly in conifer forest, and locally in deciduous woods in mainland Europe.

Seen in the UK
J F M A M J J A S O N D

SIMILAR SPECIES

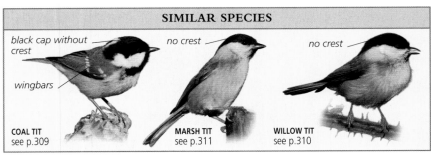

black cap without crest

wingbars

COAL TIT
see p.309

no crest

MARSH TIT
see p.311

no crest

WILLOW TIT
see p.310

Length **11.5cm (4½in)**	Wingspan **17–20cm (6½–8in)**	Weight **10–13g (³⁄₈–⁷⁄₁₆oz)**
Social **Loose flocks**	Lifespan **2–3 years**	Status **Secure**

Order **Passeriformes**	Family **Paridae**	Species ***Periparus ater***

Coal Tit  77

white nape patch

black head

greyish back

big black bib

white cheeks

bright buff underside

ADULT

dark wings with two white bars

ADULT

IN FLIGHT

white nape

yellower cheeks

JUVENILE

One of Europe's smallest birds, the Coal Tit is everywhere associated with conifers, even isolated pines within a wood of deciduous trees. It is also a garden bird in many places. In autumn and winter, it regularly joins other tits in large, shapeless, roaming flocks that wander through woods and gardens in search of food. Woods often seem birdless until such a flock comes by, when suddenly there is too much to see at once. Coal Tits typically exploit their minute weight by searching the thinnest twigs.

VOICE Call high, sweet, sad *tseu* or *tsoooo*, thin *tseee*, bright *psueet*; song bright, quick, repetitive, high *wi-choo wi-choo wi-choo* or *sweetu sweetu sweetu*.
NESTING Hair-lined small cup of moss and leaves in hole in stump, tree, wall, or ground, or in small-holed nest box; 7–11 eggs; 1 brood; April–June.
FEEDING Finds tiny insects, spiders, and their eggs in foliage; eats many seeds and nuts; visits feeders frequently, often dashing off with food to eat nearby; hides much food in tufts of pine needles.
REMARK Subspecies *P. a. ledouci* (N Africa) has yellow cheeks and underside.

/\/\/\/\/\/\/\/\/\/\/\/\/\/\/\/\/\

FLIGHT: weak, flitting, with whirring wingbeats, with sudden "stop" on perch.

TINY BUT FEARLESS
Coal Tits take very little notice of people and may forage through shrubberies almost within arm's length if one keeps still and quiet.

OCCURRENCE
Breeds through all of Europe except Iceland and extreme N Scandinavia. Lives all year round in mixed but mainly coniferous woodland, wooded parks, and gardens close to conifer woods.

Seen in the UK
J F M A M J J A S O N D

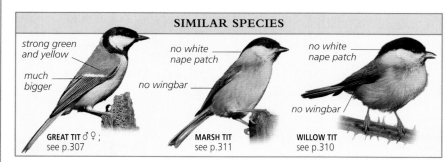

SIMILAR SPECIES

strong green and yellow

much bigger

GREAT TIT ♂♀; see p.307

no white nape patch

no wingbar

MARSH TIT see p.311

no white nape patch

no wingbar

WILLOW TIT see p.310

Length **11.5cm (4½in)**	Wingspan **17–21cm (6½–8½in)**	Weight **8–10g (5/16–3/8oz)**
Social **Loose flocks**	Lifespan **2–3 years**	Status **Secure**

Order **Passeriformes**	Family **Paridae**	Species **Poecile montana**

Willow Tit 🔊 79

small, rounded, plain brown wings

IN FLIGHT

big, dull black cap

appears big-headed and bull-necked

pale cheeks extending well back

dull brown back

black chin larger than on Marsh Tit

pale panel on wings

dull grey-buff underside

rich orange-buff flanks

∧∧∧∧∧∧∧∧∧∧∧∧∧∧∧∧∧∧∧

FLIGHT: low, quick, whirred wingbeats like other small tits.

Much like a Marsh Tit, the Willow Tit is more often found in places with few mature trees but plenty of lower bushes, such as old hedgerows and extensive willow carr over peat bogs. It is, however, also found in woods and the two often overlap. It often visits gardens but, like the Coal Tit, tends to take a morsel of food and fly away with it to eat elsewhere, so its visits are short. The Willow Tit appears big-headed and bull-necked, and uses its distinctive low, harsh, buzzy calls frequently, helping to separate it from the Marsh Tit. It requires patience and experience to learn how to separate the two on sight.
VOICE Thin *zi zi* combined with distinctive deep, nasal, scolding, buzzing *chair chair chair* is an important clue; song rarely varied warble, more commonly full, piping *tyoo tyoo tyoo*.
NESTING Excavates own hole in rotten stump; 6–9 eggs; 1 brood; April–June.
FEEDING Agile and acrobatic but generally quite sluggish, taking insects, seeds, and berries; comes to feeders in gardens for nuts and sunflower seeds.

PEANUT RAIDER
The Willow Tit is quite a frequent visitor to hanging feeders in gardens. The characteristic pale wing panel is visible in this picture.

OCCURRENCE
Breeds in N and E Europe, west to UK and E France, and south to C Italy and Balkans, present year round. Occupies coniferous and birch forest, mixed woods, thickets and hedgerows, and often visits gardens.

Seen in the UK
J F M A M J J A S O N D

SIMILAR SPECIES

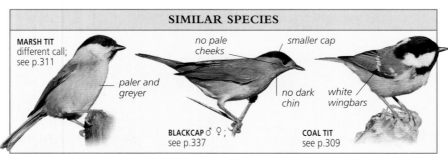

MARSH TIT different call; see p.311

paler and greyer

no pale cheeks

smaller cap

no dark chin

BLACKCAP ♂ ♀; see p.337

white wingbars

COAL TIT see p.309

Length **11.5cm (4½in)**	Wingspan **17–18cm (6½–7in)**	Weight **9–11g (¹¹⁄₃₂–³⁄₈oz)**
Social **Loose flocks**	Lifespan **2–3 years**	Status **Declining**

| Order **Passeriformes** | Family **Paridae** | Species **Poecile palustris** |

Marsh Tit 🔊 80

big, shiny black cap extending to back of neck

small, rounded grey-brown wings

neat, plain grey-brown upperparts

no pale panel on wings

small black chin

pale cheeks; slim neck

pale grey-buff underside

slim plain tail

IN FLIGHT

Marsh and Willow Tits are remarkably alike and present a real identification challenge even to experienced birdwatchers: their calls are helpful. It is essential to know the most frequently used ones to tell the two apart: a distinct *pit-chew* call is the best clue to a Marsh Tit, quite unlike any call used by the Willow. Both birds visit gardens but much less persistently than the Blue Tits and Great Tits. Marsh Tits like the vicinity of mature deciduous trees, especially beeches and oaks, although they often forage at a low level among thick undergrowth. They join mixed winter tit flocks, but usually only one or two hang around at the edges. Like most tits, Marsh Tits use existing holes in which to nest, while Willow Tits excavate their own.

VOICE Loud calls, often *titi-zee-zee-zee, tchair,* most distinctive a unique loud, bright, *pit-chew* diagnostic; song infrequent, quick, rippling *schip-schip-schip-schip*.

NESTING Grass and moss cup in existing hole in tree or wall, not often in nest box; 6–8 eggs; 1 brood; April–June.

FEEDING Mostly feeds on insects and spiders in summer; takes seeds, berries, and nits in autumn and winter, often from ground.

⋀⋀⋀⋀⋀⋀⋀⋀⋀⋀⋀⋀⋀⋀⋀⋀⋀

FLIGHT: low, weak, flitting, with bursts of wingbeats, similar to close relatives.

NEAT APPEARANCE
A glossy cap and uniform wings give the Marsh Tit a neat look, and are good identification clues to distinguish it from the Willow Tit.

OCCURRENCE
Breeds in S UK, S Scandinavia, and east across middle Europe, south to Italy and Balkans. In deciduous woodland and parkland with tall trees, sometimes visiting gardens to use feeders especially in winter.

| Seen in the UK |
| J F M A M J J A S O N D |

SIMILAR SPECIES

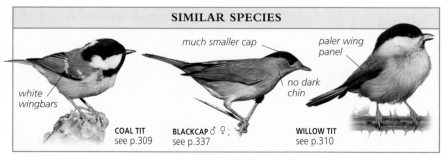

white wingbars

COAL TIT see p.309

much smaller cap

BLACKCAP ♂ ♀; see p.337

paler wing panel

no dark chin

WILLOW TIT see p.310

| Length **11.5cm (4½in)** | Wingspan **18–19cm (7–7½in)** | Weight **10–12g (³⁄₈–⁷⁄₁₆oz)** |
| Social **Loose flocks** | Lifespan **2–3 years** | Status **Secure** |

Order **Passeriformes**	Family **Timaliidae**	Species **Panurus biarmicus**

Bearded Tit

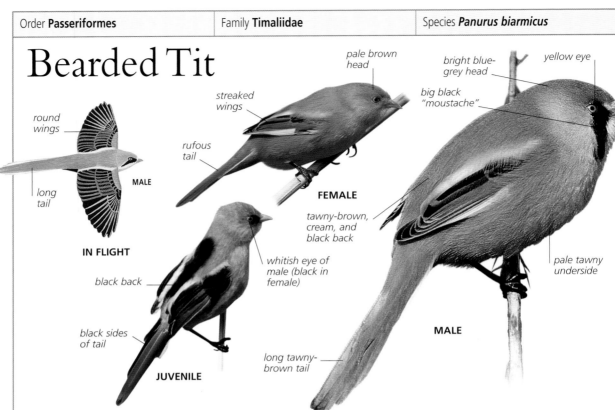

round wings

long tail

MALE

IN FLIGHT

streaked wings

rufous tail

pale brown head

FEMALE

tawny-brown, cream, and black back

black back

whitish eye of male (black in female)

black sides of tail

JUVENILE

long tawny-brown tail

bright blue-grey head

yellow eye

big black "moustache"

MALE

pale tawny underside

More closely related to the babblers and parrotbills of Asia than the tits, the Bearded Tit is one of Europe's most restricted birds in terms of habitat. It is entirely dependent on extensive reedbeds, although it will briefly occupy reedmace or tall, wet grass in winter when population pressure forces some to leave reedbeds to look for new sites. It can be hard to see, especially on windy days, but since it calls frequently it can usually be located quite easily. If nothing else, a glimpse of a tawny, long-tailed shape flitting across a gap in the reeds is usually forthcoming.

VOICE Variations on loud, metallic, "pinging" *psching*, *pink*, or *ping*, sometimes in loud, ringing chorus, often sporadic; quiet at times.

NESTING Deep cup of leaves, stems, and reed flower heads in reed litter under standing reeds in water; 5–7 eggs; 2 or 3 broods; April–August.

FEEDING Takes caterpillars from reed stems and reed seeds from flower heads and leaf litter or mud below.

FLIGHT: low, fast, over reeds; whirring, trailing long tail.

REEDBED INHABITANT
Patience and calm weather are needed for a view like this of a Bearded Tit in its reedbed habitat, but they are sometimes surprisingly tame.

SIMILAR SPECIES

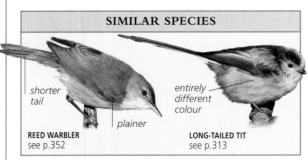

shorter tail

plainer

REED WARBLER
see p.352

entirely different colour

LONG-TAILED TIT
see p.313

OCCURRENCE
Extremely local, breeding in reedbeds in E UK, N and S France, Low Countries, Baltic area, E Spain, S Portugal, Italy, and SE Europe. Resident in winter, occasionally spilling into other wetland habitats.

Seen in the UK
J F M A M J J A S O N D

Length **12.5cm (5in)**	Wingspan **16–18cm (6½–7in)**	Weight **12–18g (7/16–5/8oz)**
Social **Loose flocks**	Lifespan **2–3 years**	Status **Secure†**

Order **Passeriformes**	Family **Aegithalidae**	Species **Aegithalos caudatus**

Long-tailed Tit 🔊 81

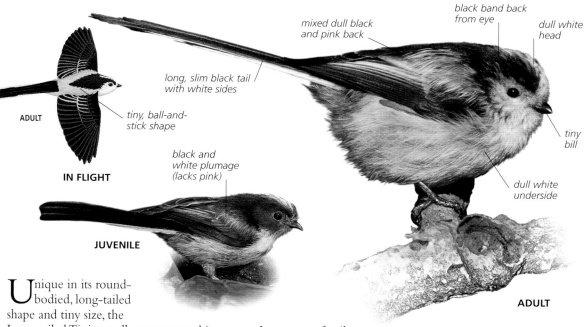

mixed dull black and pink back

black band back from eye

dull white head

long, slim black tail with white sides

ADULT

tiny, ball-and-stick shape

IN FLIGHT

tiny bill

black and white plumage (lacks pink)

dull white underside

JUVENILE

ADULT

Unique in its round-bodied, long-tailed shape and tiny size, the Long-tailed Tit is usually encountered in groups. In summer, family parties move noisily through bush tops or undergrowth; in winter sometimes much larger parties tend to string out as they feed, often crossing a gap between trees in a thin, erratic stream. Their high calls resemble those of other tits and Goldcrests (see p.355), but have a particularly shapeless, needle-like character, intermixed with low, abrupt notes that are immediately distinctive.

VOICE High, thin, colourless *seee seee seee* without emphasis or fuller quality of Goldcrest; short, abrup low *trrp*, longer *zerrrp* or *tsirrup!*.

NESTING Unique rounded and domed nest with side entrance into soft, springy, elastic ball of lichen, moss, cobweb, and feathers, in low, thorny bush; 8–12 eggs; 1 brood; April–June.

FEEDING Groups move about, often in single file, picking tiny spiders and insects from twigs and foliage; eat a few seeds; increasingly visit feeders in gardens.

FLIGHT: weak, quick; whirring wingbeats; often in groups, single file, flitting from bush to bush, tail bouncing along behind.

AMAZING NEST
The Long-tailed Tit's globular or bottle-shaped nest, coated with lichen, is a remarkable structure. It expands as the chicks grow.

OCCURRENCE
Breeding bird in all of Europe except Iceland, present all year. Occupies mixed or deciduous woods with bushy undergrowth, scrub, and tall old hedgerows; increasingly visits gardens.

Seen in the UK

J	F	M	A	M	J	J	A	S	O	N	D

LONG AND SLIM
A Long-tailed Tit looks quite slender in summer; it is in cold winter conditions that it looks like a round ball with a tail attached.

Length **14cm (5½in)**	Wingspan **16–19cm (6½–7½in)**	Weight **7–9g (¼–11⁄32oz)**
Social **Flocks**	Lifespan **2–3 years**	Status **Secure**

LARKS

Almost exclusively ground birds, except when singing, larks are quite stocky but long-bodied birds, bulkier than pipits or wagtails, but less chunky than most finches. They have thick, triangular bills – between the insect-eating bill of a pipit and the seed-eating bill of a finch, reflecting their varied diet. They fly strongly, with quite long, often rather angular wings.

Larks have faintly short legs but long toes and claws, especially the claws on their hind toes, which seem to be an adaptation for walking through grassy vegetation. They move easily on the ground in a quick walk or run, and feed on open ground: the Skylark, for example, will rarely feed in the shadow of a hedge, preferring to keep well out in the open where it has a good all-round view and from where it can fly in any direction.

Being birds of open spaces, larks have no perches from which to sing: they do sing from the ground or a post, but are at their best when singing in a special high song-flight. The Skylark sings for minutes on end in a simple rising hover, while the Woodlark flies in wide circles as it sings.

STRONG FLIER
Skylarks have large wings: these give them a slightly thrush-like look in flight but their more angular wings have a straighter rear edge.

Plumages of males and females, winter and summer, are usually much alike, although some juveniles are sufficiently different to look unusual. Some species are very difficult to identify: calls and songs are helpful but awkward plumage points such as underwing colour (on Crested and Thekla Larks) can be hard to see and structural differences (such as wingtip length on Short-toed and Lesser Short-toed Larks) can be frustratingly difficult to be sure of.

ALAUDIDAE

large black neck patch

CALANDRA LARK
p.315

thick, finch-like bill

SHORT-TOED LARK
p.316

sharp, pointed crest, raised

CRESTED LARK
p.317

fan-shaped crest, raised

THEKLA LARK
p.318

white stripes over eye

WOODLARK
p.319

white edge to tail

buff breast

SKYLARK
p.320

pale yellow face and throat (juvenile lacks yellow)

SHORELARK
p.321

Calandra Lark

streaked cap

boldly marked head

large, triangular, pale bill with dark ridge

dark cheeks, white-edged below

blackish underwings with white edge

dark upperwings with broad white edge

closely streaked back

large black neck patch

white underparts with fine streaks on chest

IN FLIGHT

FLIGHT: low, heavy, shallow but sometimes quick wingbeats; song-flight high, drifting, with unusually slow wingbeats.

This large, hefty lark of Mediterranean regions is characteristic of open plains: either dry steppe grassland or cultivated areas with vast expanses of corn. Calandra Larks may also congregate in marshy areas, especially in salty areas in depressions or near the sea; non-breeding flocks sometimes number scores or hundreds. Like most larks, they sing in flight, drifting around high up with slow, stiff wingbeats. Resident in southwest Europe, more migratory in southeast Europe, Calandras are extremely rare vagrants farther north outside their breeding range.

VOICE Dry, sizzly or trilling *schrreeup*; song in high flight prolonged, rich, varied like Skylark's but slower.

NESTING Grass cup on ground, in vegetation; 4–7 eggs; 2 broods; April–June.

FEEDING Searches ground for seeds, shoots, and insects.

VARIABLE PATCH
The black neck patch is obvious when the lark raises its head, but is hidden when it is hunched down.

OCCURRENCE
Resident in Spain, Portugal, S France, Italy, and locally in Balkans; extremely rare outside its usual breeding range. In farmland and open, dry, stony grassland in lowlands, sometimes in flocks in saline depressions with shrubby growth.

| Seen in the UK |
| J F M A M J J A S O N D |

SIMILAR SPECIES

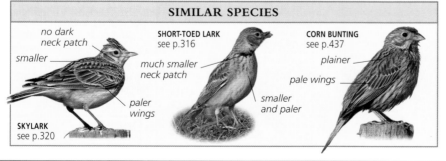

no dark neck patch

smaller

SHORT-TOED LARK
see p.316

much smaller neck patch

CORN BUNTING
see p.437

plainer

pale wings

paler wings

smaller and paler

SKYLARK
see p.320

| Length **17–20cm (6½–8in)** | Wingspan **35–40cm (14–16in)** | Weight **45–50g (1⅝–1¾oz)** |
| Social **Flocks** | Lifespan **Up to 5 years** | Status **Declining†** |

| Order **Passeriformes** | Family **Alaudidae** | Species *Galeridae theklae* |

Thekla Lark

fan-shaped crest, raised

straight bill

dark stripe below eye

pale around eye

closely streaked, grey-brown back

sharp, thick dark streaks on whitish breast (softer on Crested Lark)

grey underwings

plain upperwings

rufous rump more contrasted than that of Crested Lark

IN FLIGHT

crest may be held flat

blackish tail with pink-brown centre and rusty sides

white belly

Like the Crested Lark, the Thekla Lark has a quite obvious pointed crest, although somewhat blunter and more fan-shaped than the Crested's sharp spike. Theklas are found less in cereal fields than Cresteds, although they do occupy stony slopes with small corn fields separated by bushes and hedges. More usually, they occupy orchards and clearings in open woodland, or rough, open areas of stony grassland and barren rocky slopes. Plumage and call differences from other larks (especially the Crested Lark) are very subtle, and its identification is often based on a combination of factors.

VOICE Full-throated, musical call, *tu-tewi, tew-tewi-loo*, variable number of notes; song varied, liquid, rich warble in flight, much like Crested Lark's.

NESTING Shallow hollow on ground, in grass or other vegetation; 3–5 eggs; 2 broods; April–June.

FEEDING Picks shoots, seeds, and insects from ground.

FLIGHT: series of quick flaps between short glides; high, soaring song-flight.

CAMOUFLAGE
Thekla and Crested Larks vary only slightly in colour, both often matching the general colour of the local rocks and soil.

OCCURRENCE
Breeds in Spain, Portugal, and very locally in S France. In dry, cultivated areas with trees, rocky, grassy hillsides, and mountain slopes, either open and treeless, or bushy slopes with scattered boulders and taller trees. Strictly resident.

| Seen in the UK |
| J F M A M J J A S O N D |

SIMILAR SPECIES

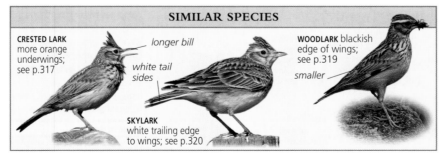

CRESTED LARK more orange underwings; see p.317

longer bill

white tail sides

SKYLARK white trailing edge to wings; see p.320

WOODLARK blackish edge of wings; see p.319

smaller

| Length **15–17cm (6–6½in)** | Wingspan **30–35cm (12–14in)** | Weight **30g (1¹⁄₁₆oz)** |
| Social **Small flocks** | Lifespan **Up to 5 years** | Status **Vulnerable** |

Order **Passeriformes**	Family **Alaudidae**	Species ***Lullula arborea***

Woodlark 🔊 51

short crest

buff-black-buff wing patch

ADULT

very short tail, white at corners

IN FLIGHT

rich tan and black streaks on cap

long whitish stripe over eye to back of neck

black-streaked, bright buff-brown back (juvenile spotted paler)

black and white wing patch

ADULT

dark-edged, rufous cheeks

dark streaks on whitish chest

whitish belly

ADULT

FLIGHT: distinctively floppy, on rounded wings, tail very short, in series of deep undulations.

One of the smallest and prettiest of the larks, the Woodlark is principally a bird of open woodland, woodland clearings, sandy heaths, and felled or replanted conifer plantations on sandy soils. In early spring, males sing from trees or in a wandering, circling song-flight, producing a highly distinctive song. In winter, small flocks wander widely over cultivated ground and under thinly scattered trees. When feeding Woodlarks are approached, they may fly off at some distance, or crouch and rely on camouflage to avoid detection, not flying up until the last moment.
VOICE Call varies on three-syllable pattern, first low and quiet *t'loo-i* or *ti-loooi*; song rich, slow, fluty diminuendos, *tlootlootloo*, *twee twee twee twee*, *dyoo dyoo dyoo dyoo*, *dlui dlui dlui*, in high, circling flight.
NESTING Hair- and grass-lined nest on ground near bush; 3 or 4 eggs; 2 broods; April–June.
FEEDING Picks up insects and small seeds from ground, often on bare, sandy patches.

GROUND FEEDER
The Woodlark spends most of its time feeding or standing on the ground, or on low logs and stumps.

OCCURRENCE
Widespread up to S Great Britain and S Scandinavia; in summer, only in north and east of range, breeding in open woodland, on bushy heaths, and especially in felled woodland such as extensive conifer plantations with areas of bare, sandy ground and short grass. On fields in winter.

Seen in the UK
J F M A M J J A S O N D

SIMILAR SPECIES

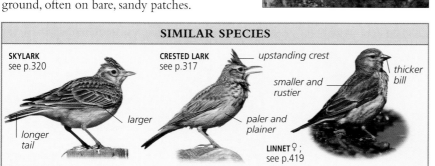

SKYLARK see p.320

longer tail

larger

CRESTED LARK see p.317

upstanding crest

paler and plainer

smaller and rustier

thicker bill

LINNET ♀; see p.419

Length **15cm (6in)**	Wingspan **27–30cm (10½–12in)**	Weight **24–36g (⁷/₈–1⁵/₁₆ oz)**
Social **Winter flocks**	Lifespan **Up to 5 years**	Status **Vulnerable**

Order **Passeriformes**	Family **Alaudidae**	Species **Alauda arvensis**

Skylark 🔊 49.I, 49.II

often flat, blunt, brown and black streaked crest

whitish over eye

pale-centred cheeks

grey underwings

dark under tail

dark stripes on back

closely streaked, pale to warm tan-brown above

buff breast

white belly

long hind claw

wide white sides to blackish tail

whitish band along trailing edge

IN FLIGHT

neatly defined gorget

FLIGHT: variable, wings usually stiff, straight rear edge, angled front; erratic bursts of wingbeats and swooping glides with wings closed.

Widespread but declining in the face of intensive agriculture, the Skylark is the classic lark of European farmland as well as heaths and upland grass or heather moors. It is a rather large lark, between a sparrow and a thrush in size, and has a distinctive appearance in flight with its quite angular, straight-edged wings and short tail. In hard weather, large flocks may pass over by day heading for milder refuges, looking very like flocks of Redwings (see p.375). When feeding, flocks tend to move more loosely, looking uncoordinated in comparison with most finch and bunting flocks.

VOICE Calls chirruping *shrrup, trrup*, higher *seee*; song from perch or in high, soaring flight, fast, rich, continuous outpouring, at distance sounding thinner and high-pitched.

NESTING Grassy cup on ground, in crop or grass; 3–5 eggs; 2 or 3 broods; April–July.

FEEDING Forages on ground in grass or on bare earth, eating seeds, shoots, grain, and insects.

SONG-FLIGHT
The Skylark rises vertically in song-flight with constant flickering; it has a rising hover and final steep plunge.

OCCURRENCE
Widespread except in Iceland, breeding on open moorland, heaths, cultivated areas in lowlands, especially cereals, and extensive pastures. In winter, widespread on arable land, with birds from N and E Europe moving south and west in sizeable flocks.

Seen in the UK
J F M A M J J A S O N D

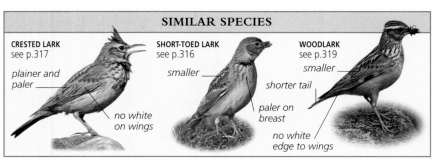

SIMILAR SPECIES

CRESTED LARK see p.317

plainer and paler

no white on wings

SHORT-TOED LARK see p.316

smaller

shorter tail

paler on breast

WOODLARK see p.319

smaller

no white edge to wings

Length **18–19cm (7–7½in)**	Wingspan **30–36cm (12–14in)**	Weight **33–45g (1³⁄₁₆–1⅝oz)**
Social **Flocks**	Lifespan **Up to 5 years**	Status **Vulnerable**

| Order **Passeriformes** | Family **Alaudidae** | Species *Eremophila alpestris* |

Shorelark

pale centre to dark tail — plain wings

ADULT (WINTER)
IN FLIGHT

tiny "horns" — primrose-yellow face

ADULT (SUMMER)

black pattern on head duller than in summer

pale yellow face and throat (juvenile lacks yellow)

mid-brown upperparts

broad black upper chest-band

variable brown lower chest-band

white underparts

ADULT (WINTER)

An odd distribution sees the Shorelark as a breeding bird in upland Scandinavia and mountains of southeast Europe and North Africa; in between, principally around the North and Baltic Seas, it is a winter bird. As such, it prefers sandy shores and beaches with strandlines and quiet little wet and marshy spots where the receding tide leaves little pools and patches of shrubby vegetation. It may mix with Snow Buntings in such places. Before leaving in spring, groups of Shorelarks may develop full summer colours as the dull feather edges of winter wear away, creating a pattern quite unlike other European larks. Breeding birds in the Balkans are greyer, less brown, on the back than northern birds.

VOICE Pipit-like, thin *tseeeep* or louder *seep-seep*; prolonged repetition of quiet warbling song from perch or in flight.

NESTING Hair-lined grass cup, on ground; 4 eggs; 1 or 2 broods; May–July.

FEEDING Creeps about unobtrusively on ground, taking seeds, insects, crustaceans, and tiny molluscs.

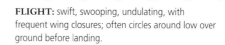

FLIGHT: swift, swooping, undulating, with frequent wing closures; often circles around low over ground before landing.

HANDSOME LARK
Inconspicuous as it feeds on the ground, the Shorelark is nevertheless an attractive bird close up.

SIMILAR SPECIES

plain head pattern

ROCK PIPIT see p.408

smaller and slimmer

streaked chest

SKYLARK see p.320

longer legs

ROCK PIPIT see p.408 — SKYLARK see p.320

OCCURRENCE
Breeds in mountainous areas of Scandinavia. In winter, quite scarce and local around Baltic and North Seas. Mostly coastal, on beaches and marshes especially just around high-tide mark, less often on nearby open arable land.

Seen in the UK
| J | F | M | A | M | J | J | A | S | O | N | D |

| Length **14–17cm (5½–6½in)** | Wingspan **30–35cm (12–14in)** | Weight **35–45g (1¼–1⅝oz)** |
| Social **Small flocks** | Lifespan **Up to 5 years** | Status **Secure†** |

MARTINS AND SWALLOWS

Collectively known as "hirundines", these are highly aerial birds, feeding almost entirely by catching flying insects while on the wing. They have tiny bills but wide mouths. Their feet are very small, but strong enough to give a good grasp on a wire or twig: they perch frequently, unlike swifts.

MARTINS

Martins are stockier than most swallows and lack the very long outer tail feathers that are so prominent on their close relatives. They have rather broad-based wings that taper to a point and deeply forked tails on a barrel-shaped body. House Martins make obvious mud-pellet nests on buildings while Sand Martins tunnel into sand cliffs in sizeable colonies.

SWALLOWS

The most elegant of the group, the swallows tend to feed lower down than the martins, and on larger prey, with a more fluent, swooping flight. All have elongated outer tail feathers, which are longest on the oldest and fittest males. Red-rumped Swallows make mud-pellet nests like House Martins, but with an entrance "porch", while Swallows nest inside small buildings in a more concealed position.

SUN-BATHING
House Martins take advantage of a rooftop in warm autumn sunlight. Their white rumps are fluffed out and obvious.

In late summer and autumn, flocks of swallows and martins gather together, often using reed-beds as roosts, before migrating to Africa. House Martins seem to remain at great heights while in Africa and are little observed. Swallows from Europe occupy different parts of southern Africa during their stay there.

HIRUNDINIDAE

all-brown upperparts

SAND MARTIN
p.323

whitish collar below dark cheek

CRAG MARTIN
p.324

deep rust-red chin

SWALLOW
p.325

white rump

HOUSE MARTIN
p.326

pale throat

RED-RUMPED SWALLOW
p.327

Order **Passeriformes**	Family **Hirundinidae**	Species **_Riparia riparia_**

Sand Martin

brown upperwings

often perches on earth bank or at entrance to nest-hole

all-brown upperparts

brown breast-band

upright posture while perching on twigs or wires

white underparts

brown wings (juvenile has pale feather edges on wings)

IN FLIGHT

ADULT

ADULT

ADULT

The hirundines – swallows and martins – are all small, aerial birds but the Sand Martin is the smallest, with the weakest, most fluttering flight. This is belied by the fact that it is the earliest to arrive in Europe each spring, often reaching the UK in early March. Conditions at this time are still very taxing for a bird that relies on flying insects for food. It is then that Sand Martins are restricted almost entirely to lakes and reservoirs where early insects are most reliable. They soon concentrate on their traditional colonies, but are also quick to exploit new possibilities, even small roadside cuttings and sand quarries which may only be suitable for a year or two. Artificial embankments specially made for them are successful.

VOICE Low, dry, slightly rasping or chattering _chrrrrp_; song rambling, chattering, weak twitter.

NESTING Bores long hole into earth or soft sandstone; 4 or 5 eggs; 2 broods; April–July.

FEEDING Aerial; catches insects in flight, often over water; sometimes feeds on bare ground.

∿∿∿∿ ∿∿∿∿ ∿∿∿

FLIGHT: weak, fluttery, with fast in-out wing flicks, wings angled well back; faster when flocks going to roost in reedbeds, or if predator nearby.

COLONIES
Sand Martin colonies are easy to see in earth banks and sand quarries, but restricted to a few localities.

OCCURRENCE
Breeds in earth cliffs, sandy river banks, and gravel pits throughout Europe except Iceland. Widespread in river valleys, typically near water, and most often over water in early spring, but also in moorland areas with eroded earth cliffs.

Seen in the UK
J F M **A M J J A S O** N D

SIMILAR SPECIES

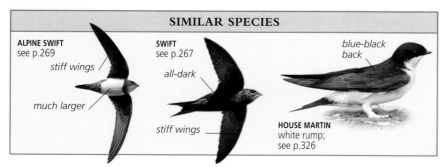

ALPINE SWIFT
see p.269
stiff wings
much larger

SWIFT
see p.267
all-dark
stiff wings

blue-black back

HOUSE MARTIN
white rump;
see p.326

Length **12cm (4¾ in)**	Wingspan **26–29cm (10–11½ in)**	Weight **13–14g (⁷⁄₁₆–½ oz)**
Social **Small flocks**	Lifespan **Up to 5 years**	Status **Declining**

| Order **Passeriformes** | Family **Hirundinidae** | Species ***Ptynoprogne rupestris*** |

Crag Martin

whitish collar below dark cheek

dull brown back and rump with grey tinge

finely streaked, brownish grey chin

brownish grey underside, whitest on throat

brown tail with oblong white spots near tip

blackish wedge on underwings

dusky rear underbody

IN FLIGHT

Of the European martins, this is the largest and the most uniform in colour, and the finest flier. It soars and floats confidently near cliffs, using the upcurrents expertly, often sweeping backwards and forwards across the cliff face like the end of a pendulum, neatly tilting over and turning at the end of each traverse. It is often accompanied by smaller, less accomplished House Martins during these flights.
VOICE Short, high, metallic clicking notes, dry *tshirr*; fast, twittering song.
NESTING Mud nest under overhang of cliff or building or in cave; 4 or 5 eggs; 1 or 2 broods; April–June.
FEEDING Aerial, catching flying insects and drifting spiders in air.

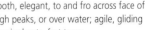

FLIGHT: smooth, elegant, to and fro across face of cliff, around high peaks, or over water; agile, gliding with few quick wingbeats, fast turns.

TAIL SPOTS
The white spots on the tail of the Crag Martin are diagnostic but sometimes difficult to see.

MUD GATHERER
Close views are often possible as Crag Martins almost ignore people as they gather at a puddle to collect mud for their nests.

OCCURRENCE
All year in S Europe, breeding in Spain, Portugal, Alps, Italy, and Balkans. Typically in mountainous areas or lowlands with gorges and broad, stony riverbeds; also in older parts of towns around Mediterranean, using buildings as cliffs; often at great altitude around peaks.

Seen in the UK
J F M A M J J A S O N D

SIMILAR SPECIES

SAND MARTIN
see p.323

stiff wings

much smaller

whiter below

HOUSE MARTIN
white rump; see p.326

much bigger

ALPINE SWIFT
see p.269

whiter below

Length **14–15cm (5½–6in)**	Wingspan **32cm (12½in)**	Weight **20–25g (¹¹/₁₆–⁷/₈oz)**
Social **Small flocks**	Lifespan **Up to 5 years**	Status **Secure**

Swallow 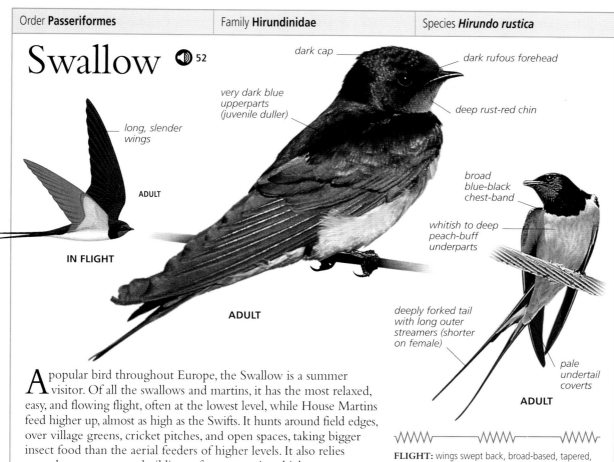 🔊 52

long, slender wings

IN FLIGHT

ADULT

dark cap

very dark blue upperparts (juvenile duller)

dark rufous forehead

deep rust-red chin

ADULT

broad blue-black chest-band

whitish to deep peach-buff underparts

deeply forked tail with long outer streamers (shorter on female)

pale undertail coverts

ADULT

A popular bird throughout Europe, the Swallow is a summer visitor. Of all the swallows and martins, it has the most relaxed, easy, and flowing flight, often at the lowest level, while House Martins feed higher up, almost as high as the Swifts. It hunts around field edges, over village greens, cricket pitches, and open spaces, taking bigger insect food than the aerial feeders of higher levels. It also relies nowadays on access to buildings of some sort in which to nest.
VOICE Calls distinctively liquid but Goldfinch-like *swit-swit-swit*, nasal *vit-vit-vit*, *tsee-tsee*; song quick, chirruping, twittering warble with slurred trills.
NESTING Open-topped cup of mud and straw, on beam or ledge in outbuilding, shed, or barn; 4–6 eggs; 2 or 3 broods; April–August.
FEEDING Flies low, swerving to catch flying insects in its mouth, mostly large flies.

FLIGHT: wings swept back, broad-based, tapered, flicked in shallow backward wingbeats; fluent and graceful with much swerving, rolling from side to side.

AGILE FLIGHT
Swallows dive from the nest to dash out through an open door or window.

AUTUMN FLOCKS
Before migrating in autumn, Swallows and House Martins gather in substantial, twittering flocks on overhead wires.

OCCURRENCE
In summer, throughout Europe except Iceland. Often near water, especially in spring and autumn, feeding over grassy or cultivated river valleys, open space, or rich farmland with hedgerows; nests in and around farms and villages but not often in suburbia.

Seen in the UK
J F **M A M J J A S O** N D

SIMILAR SPECIES

RED-RUMPED SWALLOW
pale rump; see p.327

all-dark

pale chin

black under tail

SWIFT
never on open perch; see p.267

thin wings

much smaller

HOUSE MARTIN
white rump; see p.326

Length **17–19cm (6½–7½ in)**	Wingspan **32–35cm (12½–14in)**	Weight **16–25g (⁹⁄₁₆–⁷⁄₈ oz)**
Social **Migrant flocks**	Lifespan **Up to 5 years**	Status **Declining**

| Order **Passeriformes** | Family **Hirundinidae** | Species *Delichon urbicum* |

House Martin

brown-black wings

ADULT

dark forked tail with no streamers

dark underwings

nest exposed on outside wall

ADULTS (AT NEST)

white rump (darker on juvenile)

IN FLIGHT

blue-black cap

blue-black back

white throat

ADULT

white underside

white-feathered legs

As its name suggests, this is the hirundine most closely associated with buildings, although apart from using houses to nest in, it really has no need for people at all. It is not a garden bird, but an aerial one, feeding over the housetops, coming down only to pick up mud with which to fashion its distinctive nest. In many parts of south Europe, it still breeds in remote places, nesting on cliffs high in the mountains.

VOICE Hard, quick, chirping *prrit* or *chrrit, tchirrip*; song twittering improvisation of similar notes.

NESTING Enclosed mud nest with entrance at top, fixed under eaves or overhang; 4 or 5 eggs; 2 or 3 broods; April–September.

FEEDING Catches flying insects and drifting spiders, high in air, in its mouth.

FLIGHT: stiff, circling, with flurries of flicked wingbeats and long glides; less fluent than Swallow.

AUTUMN FLOCK
House Martin flocks gather on wires in autumn before migrating south to Africa.

OCCURRENCE
In summer, in all of Europe except Iceland, typically abundant in south over towns, villages, open areas, mountain gorges, reservoirs, and reed beds. In N and W Europe, typical breeding bird of modern suburbs as well as older farmsteads and villages, but now rare on natural cliffs.

Seen in the UK
| J | F | **M** | **A** | **M** | **J** | **J** | **A** | **S** | **O** | N | D |

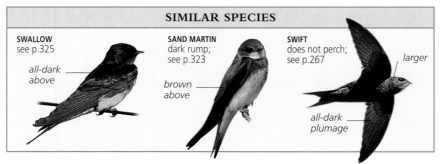

SIMILAR SPECIES

SWALLOW
see p.325

all-dark above

SAND MARTIN
dark rump; see p.323

brown above

SWIFT
does not perch; see p.267

larger

all-dark plumage

| Length **12cm (4¾in)** | Wingspan **26–29cm (10–11½in)** | Weight **15–21g (⁹/₁₆– ³/₄oz)** |
| Social **Flocks** | Lifespan **Up to 5 years** | Status **Secure** |

| Order **Passeriformes** | Family **Hirundinidae** | Species *Hirundo daurica* |

Red-rumped Swallow

rufous collar

dull orange-buff rump

ADULT

ADULT

thick tail streamers

blackish tail looks stuck-on

ADULT

IN FLIGHT

rufous sides of head and nape patch

dark blue cap

dark blue back (duller on juvenile)

blackish tail without white spots

black under tail

pale throat

faintly streaked, pale rufous-cream underside

ADULT

FLIGHT: quite stiff, wings held rather straight out; shallow flaps, long, circling, flat glides.

Once picked out from a flock of Swallows, the Red-rumped Swallow stands out because of its shape and actions as much as its pattern. It looks stiffer, straighter-winged, and slightly less fluent and relaxed in flight, enough to make it distinct to a practised eye. It is a bird of south Europe but appears with increasing frequency (if still unpredictably) farther north in spring and late autumn. In summer, it prefers areas with cliffs and gorges, inland or on the coast, nesting in caves or beneath natural overhangs as much as on buildings.

VOICE Quite distinct thin *queek* or *tsek*, sharper *keeer*; song lower, harsher than Swallow's.

NESTING Semi-spherical mud nest with entrance tube, under overhang, in cave or under eaves; 3–5 eggs; 2 or 3 broods; April–June.

FEEDING Takes insects in air, catching them in its mouth like other members of its family.

RUFOUS COLLAR
The rufous patch between the dark cap and back is easy to see on a perched bird, less so in flight.

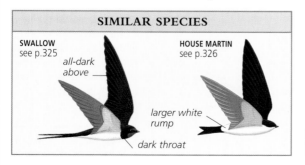

SIMILAR SPECIES

SWALLOW see p.325

all-dark above

HOUSE MARTIN see p.326

larger white rump

dark throat

OCCURRENCE
Bird of S Europe, especially Spain, Portugal, and Balkans; rare migrant farther north in late spring or autumn. Often in mountain areas with cliffs, also around coastal cliffs, gorges, and older towns and villages in summer.

| Seen in the UK |
| J F M A M J J A S O N D |

| Length **14–19cm (5½–7½in)** | Wingspan **30–35cm (12–14in)** | Weight **20g (11/16oz)** |
| Social **Small flocks** | Lifespan **Up to 5 years** | Status **Secure** |

WARBLERS AND ALLIES

Mostly smaller than the thrushes and chats, most warblers fall into several neat groups, best recognized by their generic names.

Locustella warblers have grasshopper- or cricket-like songs; they are hard to see and identify, and have narrow heads, wings with curved outer edges, and long undertail coverts beneath a rounded tail.

Acrocephalus warblers are mostly reedbed birds, with spiky bills, flattish heads, rather long tails, and strong feet that give a grip on upright stems. Their songs are rich and hurried, often with repetitive patterns; calls are churring.

Hippolais warblers are green or pale brown, with spike-like bills. Wing length is a vital clue to their identity. They have short undertail coverts and square tails, and hurried, rambling songs.

Sylvia warblers are small, lively, perky birds with short bills, often peaked heads, and slim, sometimes

DIMORPHISM
A few species, such as the Blackcap (male pictured), have different male and female plumages.

cocked tails; some have brightly coloured eye-rings. Their calls are short and hard ("tak") but their songs are often beautifully rich.

Phylloscopus warblers are mostly green and yellowish; they are delicate, slipping easily through foliage, have sweet "hooeet" calls, and distinctive songs.

SYLVIIDAE

dark reddish brown upperparts

CETTI'S WARBLER
p.332

wing and tail with yellow-green feather edges

BONELLI'S WARBLER
p.333

long, wide-yellow stripe over eye

WOOD WARBLER
p.334

greenish, with dark legs

CHIFFCHAFF
p.335

yellow-green,
with pale legs

WILLOW WARBLER
p.336

stocky build, black
or brown cap

BLACKCAP
p.337

pale buff-brown
upperparts

GARDEN WARBLER
p.338

dark around bright
yellow eye

BARRED WARBLER
p.339

whitish
underside,
washed pale
pink

LESSER WHITETHROAT
p.340

bright chestnut
panel on wings

WHITETHROAT
p.341

pale spots on
dark rust-
brown throat

DARTFORD WARBLER
p.342

pinkish red
from chin
to breast

SUBALPINE WARBLER
p.343

SYLVIIDAE *continued*

red eye ring

SARDINIAN WARBLER
p.344

finely streaked crown and cheeks

GRASSHOPPER WARBLER
p.345

curved outer edge to closed wings

SAVI'S WARBLER
p.346

orange-pink bill with dark ridge

ICTERINE WARBLER
p.347

pale yellow between eye and bill (no dark line)

MELODIOUS WARBLER
p.348

bold white wedge-shaped stripe over eye

MOUSTACHED WARBLER
p.349

soft greyish streaks
on tawny back

SEDGE WARBLER
p.350

yellowish white
underside

MARSH WARBLER
p.351

long tail
with pale
undertail
coverts

REED WARBLER
p.352

GREAT REED WARBLER
p.353

broad, slightly
rounded, dark
brown tail

short tail
dipped black
and white

broad-white
"V"-shaped bar
on wings

bronze-yellow
sides of neck

FAN-TAILED WARBLER
p.354

GOLDCREST
p.355

FIRECREST
p.356

| Order **Passeriformes** | Family **Sylviidae** | Species *Phylloscopus sibilatrix* |

Wood Warbler

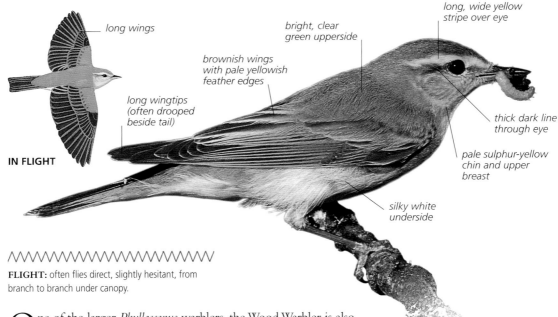

long wings

long wingtips (often drooped beside tail)

IN FLIGHT

bright, clear green upperside

brownish wings with pale yellowish feather edges

long, wide yellow stripe over eye

thick dark line through eye

pale sulphur-yellow chin and upper breast

silky white underside

FLIGHT: often flies direct, slightly hesitant, from branch to branch under canopy.

One of the larger *Phylloscopus* warblers, the Wood Warbler is also the brightest, with areas of pure lemon yellow and clear green. It is restricted to high woodland with open space beneath the trees and far less generally distributed than the Willow Warbler or the Chiffchaff. It is also curiously rare away from its nesting woods, not usually seen near the coast during migration. It is best located by its characteristic song in early summer, becoming elusive later.

VOICE Call sweet, loud *sweet*; two song types: less frequent plaintive, low, sweet *sioo sioo sioo*, more often quick, sharp, ticking accelerating into fast, silvery, vibrant trill, *ti-ti-ti-ti-ti-tik-ik-iki-kirrrrrrrrrrrrrr*.

NESTING Domed grassy nest in dead leaves on ground; 6 or 7 eggs; 1 brood; May–June.

FEEDING Moves through foliage easily, gently, and unobtrusively, picking insects and spiders.

ECSTATIC SONG
The fast, metallic trill seems to take over the whole body of a singing Wood Warbler as it vibrates to its own song.

OCCURRENCE
Local summer visitor breeding in UK, France, and east across Europe except N Scandinavia. In old woods with open space beneath canopy and leaf litter on ground, from April to August. Rare outside breeding areas.

Seen in the UK
J F M **A M J J A S** O N D

SIMILAR SPECIES

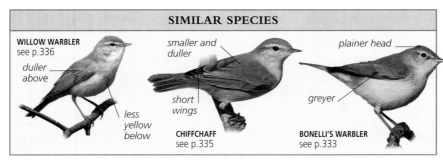

WILLOW WARBLER see p.336 — duller above — less yellow below

smaller and duller — short wings — **CHIFFCHAFF** see p.335

plainer head — greyer — **BONELLI'S WARBLER** see p.333

| Length **13cm (5in)** | Wingspan **19–24cm (7½–9½in)** | Weight **7–12g (¼–7/16oz)** |
| Social **Solitary** | Lifespan **Up to 5 years** | Status **Secure†** |

| Order **Passeriformes** | Family **Sylviidae** | Species *Phylloscopus collybita* |

Chiffchaff 🔊 74

short, round wings

ADULT

IN FLIGHT

short wingtips (longer on Willow Warbler)

tail bobbed down

thin blackish legs (paler on most Willow Warblers)

ADULT

thin pale stripe over eye (longer, sharper, yellower on juvenile)

dark eye-stripe

white crescent under eye

round head

thin bill

pale greenish to yellowish buff underside

olive-tinged green body

ADULT

While the Willow Warbler is more common in many areas, the Chiffchaff is the small greenish warbler that is often the benchmark against which others are judged. It appears commonly in bushy areas by the coast, or close to lakes and reservoirs, during migration, especially quite late in the autumn (when it may sing quite frequently). Sometimes a migrant will appear for a day or so, singing, in a large garden, but in summer it is a bird of taller trees in well-wooded parks or woodland. Distinguishing a Chiffchaff from a Willow Warbler can be a real problem but it is worth persisting and learning their different characters. A frequent downward bob of the tail is a good clue to a Chiffchaff.

VOICE Call almost single syllable, sweet *hweet*; song easy, loud, bright, even-paced repetition of simple notes, *chip-chap-chip-chap-chap-chup-chap-chap-chip*. Slurred *silip* a common late summer call.
NESTING Domed grass nest very low in bush or herbs; 5 or 6 eggs; 1 or 2 broods; April–July.
FEEDING Takes insects and spiders from foliage, slipping easily through without jerky leaps of tits.

ᐱᐱᐱᐱᐱᐱᐱᐱᐱᐱᐱᐱᐱᐱᐱᐱᐱᐱᐱᐱᐱᐱ
FLIGHT: short, low, slow, weak, undulating action.

PERSISTENT SINGER
Early arrivals sing almost constantly before the leaves are on the trees. Chiffchaffs sing again on migration in autumn.

SUBSPECIES

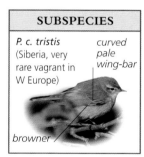

P. c. tristis (Siberia, very rare vagrant in W Europe)

curved pale wing-bar

browner

SIMILAR SPECIES

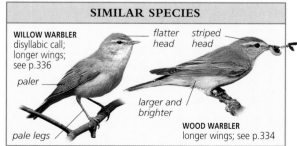

WILLOW WARBLER disyllabic call; longer wings; see p.336

paler

pale legs

flatter head

striped head

larger and brighter

WOOD WARBLER longer wings; see p.334

OCCURRENCE
Breeds in most of Europe except for Iceland; many winter in S Europe, fewer in W Europe. In woods, wooded parks, large gardens, and lower thickets especially on migration (willows near water especially in March).

Seen in the UK
| J | F | M | A | M | J | J | A | S | O | N | D |

| Length **10–11cm (4–4¼in)** | Wingspan **15–21cm (6–8½in)** | Weight **6–9g (7/32–11/32oz)** |
| Social **Solitary** | Lifespan **Up to 5 years** | Status **Secure†** |

| Order **Passeriformes** | Family **Sylviidae** | Species **Phylloscopus trochilus** |

Willow Warbler 🔊 73

plain, round wings

IN FLIGHT

ADULT

strong yellow stripe over eye

brighter, greener back

long wing point

JUVENILE

much yellower underside

narrow pale stripe over eye

flatter head than Chiffchaff's

thin dark eye-stripe

short, thin bill

grey-green to olive-brown upperparts

buff-white to pale yellow underside

longer wingtip than Chiffchaff's

pale yellowish brown legs

ADULT

The *Phylloscopus* warblers are small, slim birds of trees and bushes, able to slip quietly through foliage without the bounce and erratic agility of the small tits or the heavier progress of the larger *Sylvia* warblers. European breeding species are basically pale green and yellowish. The Willow Warbler is generally most common and most widespread, and more conservative in its choice of habitat than most. In spring, it has a fine, simple, and beautifully evocative song. Like other warblers, it is more or less solitary except when feeding young, or when a handful coincidentally feed in the same tree while on migration. Willow Warblers are more grey-brown above and dull white below, less green and yellow, in the far north and northeast of Europe.

VOICE Sweet, simple, double call *hoo-eet*; song lovely, cascading, trilling warble, rising, full notes then falling thinner and fading away with slight flourish.

NESTING Small, domed nest of grass on or near ground in thick cover; 6 or 7 eggs; 1 brood; April–May.

FEEDING Picks insects and spiders from foliage, slipping gently and easily through leaves; catches some flies in air.

FLIGHT: quick, light, bouncy flitting action over short distances.

SIGN OF SPRING
A Willow Warbler's beautiful, fluid cadence is a sure sign of spring: in April, dozens may appear overnight and start singing.

OCCURRENCE
Breeds everywhere north from mid-France and C Europe except for Iceland. Present from April to October; common migrant in S Europe. Prefers light woodland, scrub, and bushes of all kinds, especially birch and willow, but not often gardens.

| Seen in the UK |
| J F M **A M J J A S O** N D |

SIMILAR SPECIES

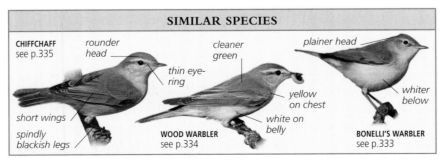

CHIFFCHAFF see p.335

rounder head

thin eye-ring

short wings

spindly blackish legs

cleaner green

yellow on chest

white on belly

WOOD WARBLER see p.334

plainer head

whiter below

BONELLI'S WARBLER see p.333

| Length **11cm (4¼in)** | Wingspan **17–22cm (6½–9in)** | Weight **6–10g (7/32–3/8oz)** |
| Social **Solitary** | Lifespan **Up to 5 years** | Status **Secure** |

| Order **Passeriformes** | Family **Sylviidae** | Species **Sylvia atricapilla** |

Blackcap 🔊 68

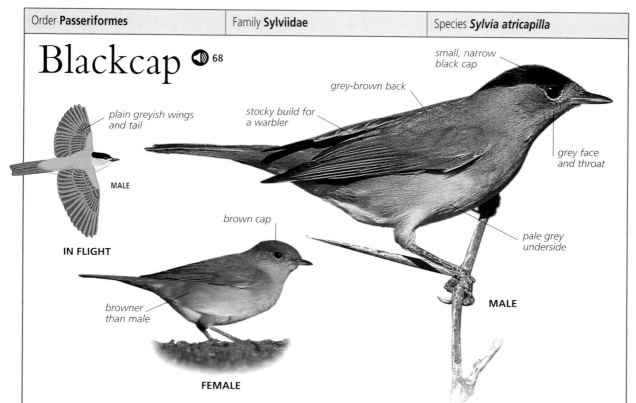

plain greyish wings and tail

MALE

IN FLIGHT

small, narrow black cap

grey-brown back

stocky build for a warbler

grey face and throat

pale grey underside

MALE

brown cap

browner than male

FEMALE

One of the more common Sylvia warblers, mostly found in thick undergrowth or bushy woodland, the Blackcap has a brilliant song and typically hard, unmusical calls. A few stay in northwest Europe for the winter, and many more in south Europe, especially in orchards, vineyards, and olive groves. Blackcaps may visit gardens in autumn to feed on honeysuckle or other berries, and again in winter for food put out on bird-tables. Like other Sylvia warblers, they are not particularly social, but several may feed close together in trees with an abundance of berries.

VOICE Distinct short, hard *tak* call; song brilliant, usually short but sometimes prolonged, fast, varied warbling with bright, clear notes, often accelerating and growing in volume soon after start. *Sylvia* warbler's hard *tak* call contrasts with Willow Warbler's (*Phylloscopus*) softer *hoo-eet* note.

NESTING Small cup of grass and stems in bush; 4 or 5 eggs; 2 broods; April–July.

FEEDING Takes insects from foliage; also feeds on many soft, fleshy berries, especially elder.

FLIGHT: short, quite heavy, flitting, with flurries of quick, flicking wingbeats.

RICH SONG
A male's fast warbling is usually distinct from a Garden Warbler's longer song, but the Blackcap can imitate its close relative.

OCCURRENCE
Breeds in most of Europe except Iceland and N Scandinavia. In summer in N Europe; increasing in UK in winter, more in Spain, Portugal, Italy, and Balkans. In woods, parks, and large bushy gardens, with plenty of thick undergrowth.

Seen in the UK
| J | F | M | A | M | J | J | A | S | O | N | D |

SIMILAR SPECIES

MARSH TIT similar to ♂ ♀; see p.311

larger black cap

black chin

stockier

rounder head

plainer and browner

GARDEN WARBLER similar to ♂ ♀; see p.338

SARDINIAN WARBLER ♂ ♀; see p.344

big black hood of male

long tail

| Length **13cm (5in)** | Wingspan **20–23cm (8–9in)** | Weight **14–20g (½–¹¹⁄₁₆ oz)** |
| Social **Solitary** | Lifespan **Up to 5 years** | Status **Secure** |

Order **Passeriformes**	Family **Sylviidae**	Species **Sylvia borin**

Garden Warbler  67

round head, unlike longer profile of Reed Warbler

short, stubby thick bill

lacks cap, eyestripe, and pale line over eye of Blackcap, Willow Warbler, and Chiffchaff

pale buff-brown upperparts

dull and rather pale

thin pale eye-ring

large dark eye

pale grey patch on sides of neck

pale buff underside

juvenile has sharp pale feather edges

grey legs

IN FLIGHT

ADULT

This small, short-billed, round-faced warbler is obscurely marked but subtly attractive and has a wonderful song. It is generally solitary but twos and threes may gather with other warblers to feed on berries in late summer, putting on fat to fuel the long autumn migration. It appears in gardens and thickets, often near the coast or beside lakes and reservoirs, in autumn, pausing while on migration in areas where it does not nest. The Garden Warbler's movements are a little slower and heavier than a Willow Warbler or a Chiffchaff.

VOICE Call hard, thick, soft *tack*, low *chek-chek, churrr*; like Blackcap, unlike Willow Warbler's; song brilliant outpouring of fast, rather even but varied warbling, very rich, throaty, musical, usually without acceleration and emphasis of Blackcap's.

NESTING Shallow, skimpy cup of grass and moss in bush; 4 or 5 eggs; 1 brood; May–July.

FEEDING Takes insects and spiders from foliage, slipping through with ease; eats many berries and seeds, in autumn, coming to honeysuckle and elder in gardens.

FLIGHT: slightly hesitant; heavy, short flights through trees.

LACK OF PATTERN
The soft grey neck patch shows well here, but there is very little pattern on a Garden Warbler.

OCCURRENCE
Breeds in most of Europe, but absent from Iceland and most of Ireland. Present from April to September in open woodland, tall thickets, shrubs, and trees, and wooded parks, often alongside Blackcaps with little obvious habitat difference.

Seen in the UK
J F **M A M J J A S O** N D

SIMILAR SPECIES

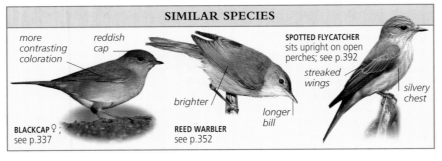

more contrasting coloration

reddish cap

SPOTTED FLYCATCHER sits upright on open perches; see p.392

streaked wings

silvery chest

brighter

longer bill

BLACKCAP ♀; see p.337

REED WARBLER see p.352

Length **14cm (5½in)**	Wingspan **20–24cm (8–9½in)**	Weight **16–23g (⁹⁄₁₆–⁷⁄₈oz)**
Social **Solitary**	Lifespan **Up to 5 years**	Status **Secure**

Barred Warbler

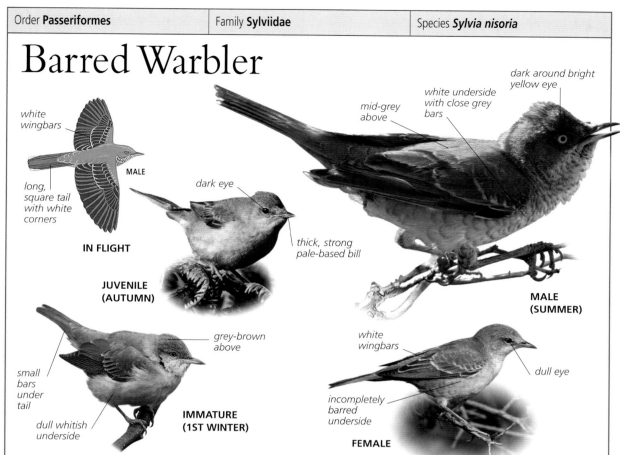

white wingbars

MALE

long, square tail with white corners

IN FLIGHT

dark eye

thick, strong pale-based bill

JUVENILE (AUTUMN)

mid-grey above

white underside with close grey bars

dark around bright yellow eye

MALE (SUMMER)

grey-brown above

small bars under tail

dull whitish underside

IMMATURE (1ST WINTER)

white wingbars

incompletely barred underside

dull eye

FEMALE

One of the larger warblers of Europe, the Barred Warbler is almost Wryneck-like at times, heavily barred beneath and pale-eyed, with a rather severe expression. In autumn, when it is most likely to be seen on migration in northwest Europe, most are pale, almost unbarred juveniles, but they still have a heavy, clumsy, rather aggressive character, and may crash about in low bushes. They are typically skulking birds, not easy to watch, but eventually emerge on the bush tops if conditions are calm and dry.
VOICE Loud, dry, hard rattle, *trrr-r-r-r-rt*; song long, bright, musical warble, like high-pitched Garden Warbler's.
NESTING Substantial nest in thorny bush or scrub; 4 or 5 eggs; 1 brood; May–July.
FEEDING Takes insects and spiders from foliage; tugs at berries.

FLIGHT: low, heavy, long-winged, tail flaunted or spread; high, fluttery song-flight.

AUTUMN MIGRANT
The autumn bird is pale and stocky with dark eyes and pale wingbars.

OCCURRENCE
Breeds in E Europe west to N Italy, in bushy places and woodland clearings; present from April to October. Rare migrant, chiefly in autumn, in NW Europe, on coasts, usually in thickets on dunes or low coastal hills.

Seen in the UK
J F M A M J J **A S O** N D

SIMILAR SPECIES

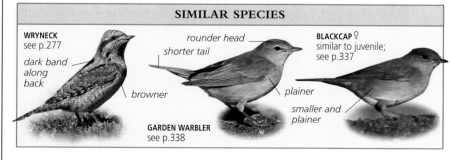

WRYNECK see p.277

dark band along back

browner

rounder head
shorter tail

GARDEN WARBLER see p.338

plainer

BLACKCAP ♀ similar to juvenile; see p.337

smaller and plainer

Length **15–17cm (6–6½in)** | Wingspan **15–20cm (6–8in)** | Weight **12–15g (⁷⁄₁₆–⁹⁄₁₆oz)**
Social **Solitary** | Lifespan **Up to 5 years** | Status **Secure†**

| Order **Passeriformes** | Family **Sylviidae** | Species *Sylvia curruca* |

Lesser Whitethroat

plain brown wings

white sides to dark tail

MALE

IN FLIGHT

grey-brown wings

white chin

FEMALE

broken white eye-ring

pale grey head

JUVENILE

dark legs (pale on Whitethroat)

olive-grey to blue-grey cap

dark smudge through eye

dull grey-brown back

dark patch on cheeks

clear white throat

whitish underside, washed pale pink

MALE

dark grey legs

/\/\/\/\/\/\/\/\/\/\/\/\/\/\

FLIGHT: quick, short, flitting flights with undulating action; fast, whirring wingbeats.

Small, neat, compact, dark-legged, and dark-masked, the Lesser Whitethroat is a secretive warbler of woodland edges and thick, old hedgerows. It is easily located by its song, but often moves to sing again a few metres away before it is seen. In autumn, it can be easy to find on shrubs and trees with berries, sometimes with other warblers but not forming properly coordinated flocks. Young birds at this time are particularly bright and smart.

VOICE Sharp, short, metallic *tak*, very thin *chi*; song in two parts, beginning with low, quiet, muffled warble, becoming short, loud, wooden rattle *chikachikachikachikachikachika*.

NESTING Cup of twigs or grass, lined with hair and roots, in shrub; 4–6 eggs; 1 brood; May–June.

FEEDING Picks insects from foliage; eats many berries in late summer.

WOODEN RATTLE
The male sits upright as he sings, usually moving to a new perch before the next loud rattled phrase.

OCCURRENCE
Summer visitor and breeding bird from April to October in most of Europe west to mid-France and UK; not in Italy, Spain, Portugal, N Scandinavia, and Iceland. In quite tall, dense thickets often at woodland edge, or as part of tall, dense hedges.

Seen in the UK
J F M **A M J J A S O** N D

SIMILAR SPECIES

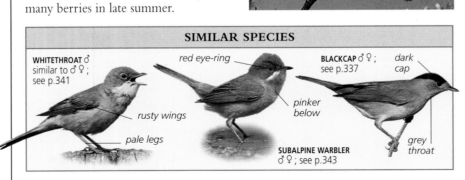

WHITETHROAT ♂ similar to ♂♀; see p.341

rusty wings

pale legs

red eye-ring

pinker below

SUBALPINE WARBLER ♂♀; see p.343

BLACKCAP ♂♀; see p.337

dark cap

grey throat

| Length **13cm (5in)** | Wingspan **17–19cm (6½–7½in)** | Weight **10–16g (3/8–9/16oz)** |
| Social **Solitary** | Lifespan **Up to 5 years** | Status **Secure** |

Order **Passeriformes**	Family **Sylviidae**	Species **Sylvia communis**

Whitethroat

whitish eye-ring

pale bluish grey head

dull brown back

pale underside, washed pink across chest

bright chestnut panel on wings

brown head

clear white throat

MALE

long dark tail, edged white

IN FLIGHT

fidgety tail flicks

pale orange-brown legs (dark on Lesser Whitethroat)

pink-buff below

FEMALE

bright wings with blackish feather centres

JUVENILE (AUTUMN)

MALE

Typically a bird of open spaces with low bushes and scrub, the Whitethroat likes overgrown tracksides, railway embankments, hedgerows and fence-lines with brambles, or thorny thickets around heaths. It sings often, sometimes from a low perch, sometimes from a high wire, but frequently in short, jerky, bouncy song-flights. It is often quite secretive, keeping low down in thick vegetation, but gives itself away by its irritable calls and eventually succumbs to its insatiable curiosity and appears out in the open.

VOICE Harsh, grating *tcharr*, scolding, softer *churr*, sweet, musical *wheet-a-wheet-a-whit;* song often in fluttery song-flight, quick, chattery, rambling warble with dry, scratchy quality.

NESTING Small, neat cup of grass and stems low in thorny shrub; 4 or 5 eggs; 2 broods; April–July.

FEEDING Picks insects from foliage; takes lots of berries and some seeds in late summer and autumn, sometimes visiting gardens.

FLIGHT: low, bouncy or jerky, flitting, undulating, with flicked, untidy tail.

LIVELY SINGER
The male sings with much more vigour than melody, from a bush top or overhead wire.

OCCURRENCE
Breeds almost throughout Europe except for Iceland and much of Scandinavia. In bushy, dry, and heathy places with low, thorny scrub, dense herbs such as nettles, hedges, and thickets; seen from April to October.

Seen in the UK
J F M **A M J J A S O** N D

SIMILAR SPECIES

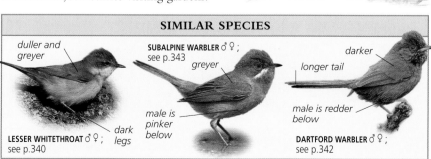

duller and greyer

SUBALPINE WARBLER ♂♀; see p.343

greyer

darker

longer tail

male is pinker below

LESSER WHITETHROAT ♂♀; see p.340

dark legs

male is redder below

DARTFORD WARBLER ♂♀; see p.342

Length **14cm (5½in)**	Wingspan **19–23cm (7½–9in)**	Weight **12–18g (⁷⁄₁₆–⅝oz)**
Social **Solitary**	Lifespan **Up to 5 years**	Status **Secure**

| Order **Passeriformes** | Family **Sylviidae** | Species *Sylvia undata* |

Dartford Warbler

short, spiky, yellow-based bill

red eye and eye-ring

pale spots on dark rust-brown throat

brownish grey back

dark red-brown underside

short, rounded wings

MALE

long, "bouncy" tail

long, slender, dark tail

IN FLIGHT

duller than male (juvenile greyer)

paler underside than male's

FEMALE

MALE

A resident in Europe, the Dartford Warbler is subject to fluctuations in numbers and range according to the severity of winter weather. It prefers warm, flat heaths and slopes with short herbaceous and shrubby growth, including thick heather and clumps of gorse, where it often skulks and is hard to see. It may flick from one bush to another but disappears from sight frustratingly quickly. In warm, still weather, however, it will come to the top and reveal its distinctive colours and shape.
VOICE Very distinctive buzzy call, low *chrrrr* or *djarrr*; song quick, rattling warble with some brighter notes, quite low pitch, little variety, sometimes given in flight.
NESTING Grassy cup lined with finer stems, low down in gorse or heather; 3–5 eggs; 2 or 3 broods; April–July.
FEEDING Finds insects and many spiders in low vegetation.

FLIGHT: quick, undulating with tail flirting; bursts of wingbeats over short distances.

BRIGHT MALE
Only a good view reveals the rich colours of the male.

OCCURRENCE
Breeds very locally in suitable habitat in S Britain, NW, W, and S France, Spain, Portugal, Italy, and on many Mediterranean islands. Found on heaths with heather and gorse and some small bushes, and on warm bushy slopes with few trees but plenty of aromatic and thorny shrubs, all year round.

Seen in the UK
J F M A M J J A S O N D

SIMILAR SPECIES

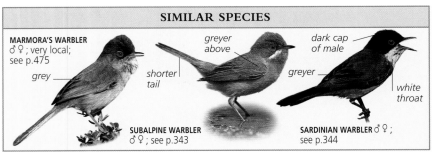

MARMORA'S WARBLER
♂♀; very local; see p.475

grey

shorter tail

greyer above

SUBALPINE WARBLER ♂♀; see p.343

dark cap of male

greyer

white throat

SARDINIAN WARBLER ♂♀; see p.344

| Length **12–13cm (4¾–5in)** | Wingspan **13–18cm (5–7in)** | Weight **9–12g (¹¹⁄₃₂–⁷⁄₁₆oz)** |
| Social **Solitary** | Lifespan **Up to 5 years** | Status **Vulnerable** |

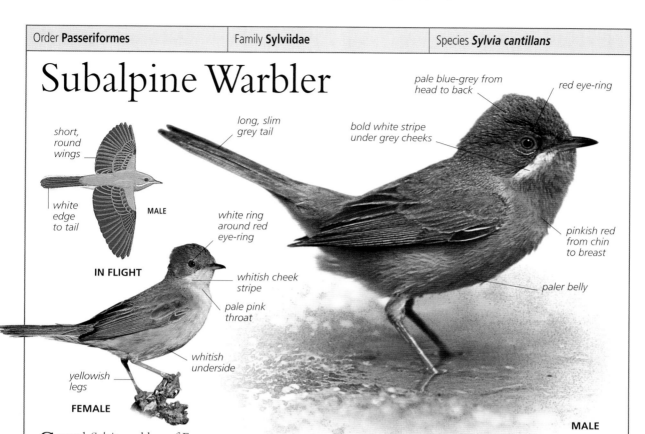

Subalpine Warbler

pale blue-grey from head to back

red eye-ring

bold white stripe under grey cheeks

short, round wings

long, slim grey tail

white edge to tail

MALE

IN FLIGHT

white ring around red eye-ring

whitish cheek stripe

pale pink throat

whitish underside

yellowish legs

FEMALE

pinkish red from chin to breast

paler belly

MALE

Several *Sylvia* warblers of Europe occupy the southern regions, mostly around the Mediterranean; the Subalpine Warbler is typical, inhabiting warm, sun-bathed slopes and fields with rough, tangled hedges and thickets of aromatic shrubs and spiny bushes. It dives out of sight into the spikiest of these and can be frustratingly elusive at times, although it often appears on top and launches into a brief, bouncy song-flight in full view. Females are paler than adult males, which are easy to identify. Females and immatures, which are even paler, are more difficult, especially if they turn up as vagrants farther north in autumn.
VOICE Call sharp, ticking or clicking *tet*, sometimes quickly repeated; song high-pitched, Linnet-like, musical warbling, fast, with quick variation in pitch.
NESTING Small, neat cup nest in low vegetation; 3 or 4 eggs; 2 broods; April–June.
FEEDING Forages in low scrub and herbs, or higher in leafy trees, searching for insects and spiders.

FLIGHT: short flights weak, quick, undulating, with bursts of wingbeats.

WHITE "MOUSTACHE"
Males, and sometimes females, have an obvious white stripe from the bill to the side of the throat.

OCCURRENCE
Breeds on bushy slopes, in low, tangled hedges and thorny thickets, and in open, evergreen oak woods in Spain, Portugal, and Mediterranean Europe, from April to September. Migrants at times appear farther north in low, dense undergrowth near coasts.

Seen in the UK											
J	F	M	A	M	J	J	A	S	O	N	D

SIMILAR SPECIES

darker

more uniform

slimmer tail

DARTFORD WARBLER
♂♀; see p.342

dark head of male

longer tail

SARDINIAN WARBLER ♂♀;
see p.344

greyer

rufous on wings

larger

WHITETHROAT ♂♀;
see p.341

Length **12–13cm (4¾–5in)**	Wingspan **13–18cm (5–7in)**	Weight **9–12g (¹¹/₃₂–⁷/₁₆ oz)**
Social **Solitary**	Lifespan **Up to 5 years**	Status **Secure**

Order **Passeriformes**	Family **Sylviidae**	Species *Sylvia melanocephala*

Sardinian Warbler

IN FLIGHT

short, round wings

long tail

MALE

deep black cap extends onto cheeks

red eye-ring

pale grey back

big white throat

whitish underside

long dark tail with white sides

MALE

orange-red eye-ring

grey head

browner, paler than male

white throat

white edges to tail

FEMALE

One of the Mediterranean warblers, very rare north of its usual range, characteristic of stony places with low, scattered bushes and scrub, the Sardinian Warbler is long-tailed and dark-capped. Typical views may be little more than a glimpse of a small bird with a long tail disappearing into a bush, perhaps not to be seen again. In some places, it frequents taller trees in gardens and orchards. This warbler has a distinctive rapid, rattling call that gives away its presence at frequent intervals, however, and patience is usually rewarded by a better view. It is often seen in pairs or small family groups, but does not form larger flocks.

VOICE Loud, hard, short call and frequent fast, hard rattle, *krr-rr-rr-rr-rr-rr-rr-rr-rr-t*; song fast, unmusical, rattling chatter with calls interspersed.
NESTING Small, neat cup in low bush; 3–5 eggs; 2 broods; April–July.
FEEDING Takes small insects and spiders, mostly low in vegetation or on ground beneath.

∧∧∧∧∧∧∧∧∧∧∧∧∧∧∧∧∧∧∧

FLIGHT: short, bouncy, flitting flights between clumps of cover.

BRIGHT EYE-RING
Even on the browner female Sardinian Warbler, the red eye-ring is a distinct feature.

OCCURRENCE
Resident in most of Spain, Portugal, S France, Mediterranean region, and very rare farther north. In bushy areas, sometimes in open woodland with scrub, more often thickets around buildings, thorny growth over stone walls, and similar areas.

Seen in the UK											
J	F	M	A	M	J	J	A	S	O	N	D

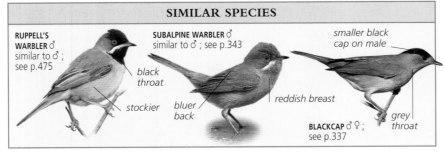

SIMILAR SPECIES

RUPPELL'S WARBLER ♂ similar to ♂; see p.475

black throat

stockier

SUBALPINE WARBLER ♂ similar to ♂; see p.343

bluer back

reddish breast

BLACKCAP ♂ ♀; see p.337

smaller black cap on male

grey throat

Length **13–14cm (5–5½in)**	Wingspan **15–18cm (6–7in)**	Weight **10–14g (³⁄₈–½oz)**
Social **Family groups**	Lifespan **Up to 5 years**	Status **Secure**

Order **Passeriformes**	Family **Sylviidae**	Species **Locustella naevia**

Grasshopper Warbler

blunt wings

rounded tail

IN FLIGHT

finely streaked crown and cheeks

spotted or streaked, pale olive-brown back and rump

scarcely marked, whitish or buff underside

long, broad rounded tail

FLIGHT: low, brief, flitting flights; raises slightly fanned tail as it dives out of sight.

Warblers are split into several different families: this is the most common of the *Locustella* warblers, which are small, streaked, round-tailed, and highly skulking birds, usually hard to see. They have long, trilled, chirping, or rattling songs often likened to the sound of some insects such as crickets; these are usually heard most often at dusk or on warm, still, sultry summer days. These warblers are not to be expected in tall bushes or trees, or openly flitting about in hedgerows.

VOICE Loud call, piercing *psit*; song remarkable, prolonged, unvarying mechanical "reel", fast, hard ticking at close range on one high, sharp note, waxes and wanes as head is turned: *sirrrrrrrrrrrrrrrrrrrrrrrrrrrrrrrrrrrr.*

NESTING Small nest of grass and leaves in dense, low vegetation; 5 or 6 eggs; 2 broods; May–July.

FEEDING Forages in very low, thick vegetation, creeping mouse-like on or near ground, finding mostly small insects and spiders.

HIGH-PITCHED SONG
Many people are unable to hear such high-pitched sounds as the Grasshopper Warbler's song.

OCCURRENCE
Widespread from Ireland, east to Finland and Russia, south to C France, N Spain, and Alps in summer. In marshy areas with grass, low thickets, heathy places, and grassy meadows with thorny bushes where grass grows up through thickets.

Seen in the UK
J F M **A M J J A S O** N D

SIMILAR SPECIES

SEDGE WARBLER see p.350

more obvious stripe over eye

plain back and tail

REED WARBLER see p.352

DUNNOCK see p.390

bigger and stouter

greyer

Length **12.5cm (5in)**	Wingspan **15–19cm (6–7½in)**	Weight **11–15g (³⁄₈–⁹⁄₁₆oz)**
Social **Solitary**	Lifespan **Up to 5 years**	Status **Secure**

| Order **Passeriformes** | Family **Sylviidae** | Species *Locustella luscinioides* |

Savi's Warbler

round wings

rounded tail

IN FLIGHT

long, flat head slopes into long, slender bill

pale throat

brown-buff underside with paler belly

plain brown above

curved outer edge to closed wings

long tail, rounded at tip

long, thick brown undertail coverts

dark under tail

FLIGHT: short, flitting flight between reed clumps. Longer migration flights at night.

An exception to the *Locustella* rule, being plain rather than streaked, Savi's Warbler is more like the plain-backed Reed and Marsh Warblers. It is also, like the Reed Warbler, a bird of dense reedbeds, much less adaptable to various grassy habitats than the more widespread Grasshopper Warbler. It is easily detected by its song (although many may not realize that it is a bird singing at all), especially around dawn and dusk, and patience may reveal its small, dark form, singing from a reed stem. Savi's Warbler is usually rather scarce even in its breeding areas and is rarely seen as a migrant outside its breeding range. It needs to be separated with care from the similarly coloured reed warbler group but the family characteristics, including the round-edged wings, long, thick undertail coverts and rounded tail help in identification, if the bird is seen closely. Savi's Warblers spend the winter in Africa.

VOICE Short, sharp, metallic call; song like Grasshopper Warbler but more slurred, less ticking, faster, lower buzz, *zurrrrrrrrrrrrrrrr*.

NESTING Large, untidy, loose nest of grass in reeds or sedges; 4 eggs; 2 broods; April–June.

FEEDING Forages for insects and spiders in dense vegetation.

OCCURRENCE
Very local; breeding range scattered across Europe from Spain and Portugal to extreme SE England and east into Asia. Appears from April to September in extensive wet reedbeds. Very rare migrant away from this habitat on coast.

| Seen in the UK |
| J F M **A M J J A** S O N D |

SIMILAR SPECIES

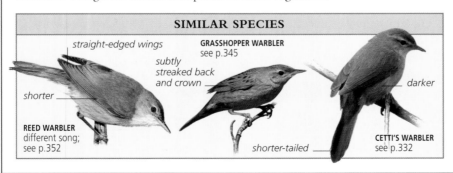

straight-edged wings

GRASSHOPPER WARBLER
see p.345

subtly streaked back and crown

shorter

REED WARBLER
different song;
see p.352

shorter-tailed

darker

CETTI'S WARBLER
see p.332

| Length **14–15cm (5½–6in)** | Wingspan **15–20cm (6–8in)** | Weight **12–15g (7/16–9/16oz)** |
| Social **Solitary** | Lifespan **Up to 5 years** | Status **Secure†** |

Order **Passeriformes**	Family **Sylviidae**	Species *Hippolais icterina*

Icterine Warbler

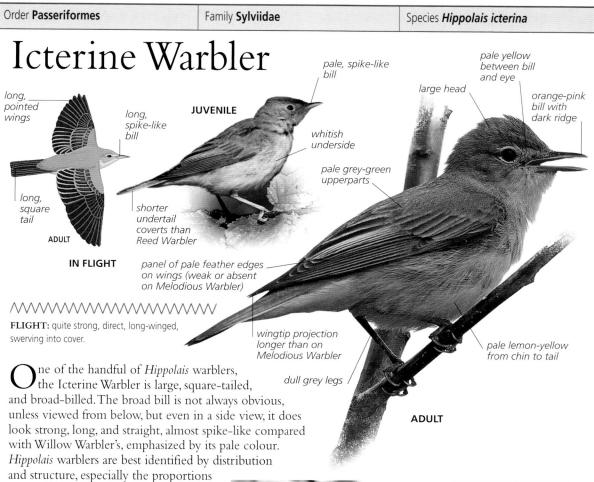

JUVENILE

pale, spike-like bill

long, pointed wings

long, spike-like bill

whitish underside

long, square tail

ADULT

IN FLIGHT

shorter undertail coverts than Reed Warbler

pale grey-green upperparts

pale yellow between bill and eye

large head

orange-pink bill with dark ridge

pale lemon-yellow from chin to tail

panel of pale feather edges on wings (weak or absent on Melodious Warbler)

/\/\/\/\/\/\/\/\/\/\/\/\/\/\/\/\/\/\/\

FLIGHT: quite strong, direct, long-winged, swerving into cover.

wingtip projection longer than on Melodious Warbler

dull grey legs

ADULT

One of the handful of *Hippolais* warblers, the Icterine Warbler is large, square-tailed, and broad-billed. The broad bill is not always obvious, unless viewed from below, but even in a side view, it does look strong, long, and straight, almost spike-like compared with Willow Warbler's, emphasized by its pale colour. *Hippolais* warblers are best identified by distribution and structure, especially the proportions of the various elements of wing, tail, and undertail coverts, which help to separate them from each other and from confusingly similar *Acrocephalus* species.

VOICE Call melodious *ti-ti-looi* or *di-deroi*, hard *tik*; song loud, fast, prolonged, varied warbling with many imitations, short, shrill, nasal notes and *dideroi* call intermixed.

NESTING Deep cup nest suspended from forked branch in tree; 4 or 5 eggs; 1 brood; May–August.

FEEDING Takes insects from foliage; pulls berries from twigs with tug of bill.

GENERIC CHARACTER
Hippolais warblers are heavy, with plain faces and dagger-like bills.

OCCURRENCE
Widespread as summer visitor and breeding bird from E France eastwards and northwards except in N Scandinavia. In open mixed, deciduous, or coniferous woodland between April and September. Spring and especially autumn migrants on NW European coasts including E Great Britain.

Seen in the UK
J F M **A M** J J **A S O** N D

SIMILAR SPECIES

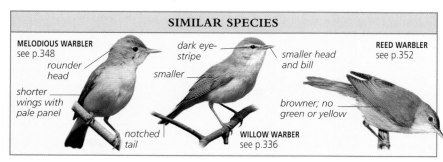

MELODIOUS WARBLER see p.348

rounder head

shorter wings with pale panel

dark eye-stripe

smaller

notched tail

smaller head and bill

WILLOW WARBER see p.336

REED WARBLER see p.352

browner; no green or yellow

Length **13.5cm (5–6in)**	Wingspan **20–24cm (8–9½in)**	Weight **10–14g (³⁄₈–½oz)**
Social **Solitary**	Lifespan **Up to 5 years**	Status **Secure**

Order **Passeriformes**	Family **Sylviidae**	Species *Hippolais polyglotta*

Melodious Warbler

rather long, round wings

ADULT

IN FLIGHT

trace of paler panel on wings (much stronger on Icterine Warbler)

JUVENILE

thick, spiky pale bill

big dark eye in pale face

grey-green upperparts

pale yellow between eye and bill (no dark line)

yellow throat and breast

dull brown legs

ADULT

wingtip projection shorter than on Icterine Warbler

∧∧∧∧∧∧∧∧∧∧∧∧∧∧∧∧∧∧∧∧∧

FLIGHT: rather weak, fluttering flight, usually over short distances.

A replacement of the more easterly Icterine Warbler in southwest Europe, the Melodious Warbler is confusingly similar. Both appear on west European coasts as migrants and require care for reliable separation, especially in autumn when juvenile plumages further confuse the issue. The Melodious and Icterine Warblers are basically green and yellow birds while other *Hippolais* warblers are duller, more pale brown and buff. The Melodious has a plainer wing with a shorter wingtip (primary feather) projection when perched than the Icterine.

VOICE Short clicks and sparrow-like chattering; song fast, rambling, not very accomplished warble, including rattling notes and shrill whistles but generally not much contrast in tone.

NESTING Deep cup in small branches of tree or large bush; 4 or 5 eggs; 1 brood; May–July.

FEEDING Takes insects from foliage; pulls berries from twigs in autumn.

SPRING SONG
The Melodious Warbler's song is a disappointment for a bird with this name: it is a quick, rather uninspired, rambling warble.

OCCURRENCE
Breeds in S and W Europe, France, Spain, Portugal, and Italy. Present from April to October, in light woodland, scrub, hedges, and orchards, with spring (and less often autumn) migrants on W European coasts including S Great Britain.

Seen in the UK
J F M **A M** J J **A S O** N D

SIMILAR SPECIES

ICTERINE WARBLER see p.347

longer wings with more obvious panel

dark eye-stripe

no trace of yellow

shorter bill

WILLOW WARBLER see p.336

GARDEN WARBLER see p.338

Length **12–13cm (4¾–5in)**	Wingspan **18–20cm (7–8in)**	Weight **11–14g (⅜–½oz)**
Social **Solitary**	Lifespan **Up to 5 years**	Status **Secure†**

Moustached Warbler

faintly streaked, blackish cap

black line from bill through eye

bold white wedge-shaped stripe over eye

dark streaked, rust-brown back

short wingtip projection

slight black moustache

white chin and throat

orange-tawny flanks (much paler on Sedge Warbler)

white belly

soft dark streaks on rounded tail

IN FLIGHT

FLIGHT: short, low, flitting flights across reeds.

A streaked *Acrocephalus* warbler, the Moustached Warbler is unusual in that it is a resident in its restricted range in Europe. It is only a very rare vagrant outside its usual range. It is quite distinctive when seen with the Sedge Warbler, its most similar relative, but care is required when identifying potential out-of-range vagrants. Its song is a useful clue in the usual breeding areas. It often tilts over and cocks its tail which the Sedge Warbler does not. Its shorter wingtip is sometimes discernible in a close view, helping to confirm identification.

VOICE Call like Sedge Warbler's but more throaty, *trek* or clicking *trk-tk-tk-tk*; song fast and varied, with frequent Nightingale-like rising whistles.
NESTING Deep grassy nest lined with plant down, in reeds; 5 or 6 eggs; 1 brood; April–June.
FEEDING Eats insects and other small invertebrates, from mud and dense wetland vegetation.

STRIKING HEAD PATTERN
A wedge of white over the eye and a silky white chin are obvious features of the Moustached Warbler in a good view.

SIMILAR SPECIES

paler breast

paler

longer wingtips

SEDGE WARBLER
see p.350

WHINCHAT ♂♀; different habitat and posture; see p.384

different shape

OCCURRENCE
Very local in S Europe, breeding in S and E Spain, S France, Balearics, Italy, and Balkans. In reedbeds and dense waterside sedge or rushes. Resident and very rarely noted outside its breeding range.

Seen in the UK
| J | F | M | A | M | J | J | A | S | O | N | D |

| Length **12–13cm (4¾–5in)** | Wingspan **17–21cm (6½–8½in)** | Weight **10–15g (⅜–⁹⁄₁₆oz)** |
| Social **Solitary** | Lifespan **Up to 5 years** | Status **Secure†** |

| Order **Passeriformes** | Family **Sylviidae** | Species **Acrocephalus schoenobaenus** |

Sedge Warbler 🔊 69

unstreaked, tawny-buff rump

ADULT

IN FLIGHT

blackish and cream streaks on cap

wide silver-white stripe over eye

dark line from bill through eye

ADULT

soft greyish streaks on tawny back

buff across chest (finely streaked on juvenile) and flanks

whitish underside

ADULT

This is one of the *Acrocephalus* warblers, birds of wetlands, especially reedbeds, that fall into two groups, streaked and unstreaked. A small, well-marked, active bird with a loud, fast, varied song, the Sedge Warbler is quite common and widespread but restricted largely to waterside or boggy habitats. It is not exclusively a reedbed warbler, preferring more variety, such as various sedges, nettles, willow, hawthorn scrub, willowherb, and umbellifers intermixed, so sometimes it may be found in hedges beside wet ditches or even drier places with thick, vertical stem growth. It appears as a migrant mostly in similar habitats.

VOICE Call dry, rasping *tchrrrr*, sharper *tek*; song loud, fast, varied, excitable mix of whistles, warbles, clicks, and trills with much mimicry.

NESTING Deep nest of grass mixed with moss, cobwebs, and plant down; 5 or 6 eggs; 1 or 2 broods; April–July.

FEEDING Forages in reeds, sedges, nettles, and bushes, for small insects, spiders, and some seeds.

FLIGHT: short, flitting flights, quite jerky; tail sometimes fanned.

VIGOROUS SONGSTER
A singing bird frequently climbs to the top of a bush or reed stem.

OCCURRENCE
Widespread as breeder except in Iceland. In reeds, from narrow ditches to extensive reedbeds, and associated wetland vegetation such as sedges and reedmace; more rarely in nettles, willowherb, and other rank growth, often with thorn bushes, from April to October.

| Seen in the UK |
| J F M A M J J A S O N D |

SIMILAR SPECIES

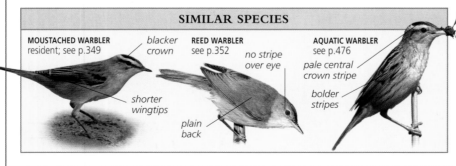

MOUSTACHED WARBLER resident; see p.349

blacker crown

shorter wingtips

REED WARBLER see p.352

no stripe over eye

plain back

AQUATIC WARBLER see p.476

pale central crown stripe

bolder stripes

| Length **13cm (5in)** | Wingspan **17–21cm (6½–8½in)** | Weight **10–13g (³⁄₈–⁷⁄₁₆oz)** |
| Social **Solitary** | Lifespan **Up to 5 years** | Status **Secure†** |

| Order **Passeriformes** | Family **Sylviidae** | Species ***Acrocephalus palustris*** |

Marsh Warbler

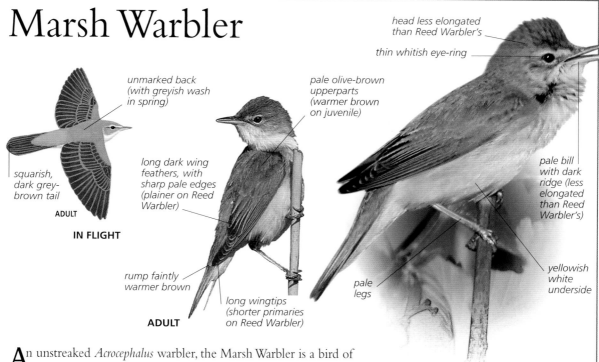

IN FLIGHT

squarish, dark grey-brown tail

ADULT

unmarked back (with greyish wash in spring)

long dark wing feathers, with sharp pale edges (plainer on Reed Warbler)

rump faintly warmer brown

long wingtips (shorter primaries on Reed Warbler)

ADULT

pale olive-brown upperparts (warmer brown on juvenile)

head less elongated than Reed Warbler's

thin whitish eye-ring

pale bill with dark ridge (less elongated than Reed Warbler's)

yellowish white underside

pale legs

ADULT

An unstreaked *Acrocephalus* warbler, the Marsh Warbler is a bird of wet riversides and boggy places with an abundance of rich, thick vegetation; it is not usually a reedbed species. Migrants occur rarely, near the coast, and require patience and close observation for positive identification. Unless the full song is heard, this is a tricky species. Its habitat is always restricted and often rather temporary in nature, so it remains a rare and somewhat erratic breeding bird – one of the last of the summer migrants to arrive in summer.

VOICE Call short, hard *chek* or *chk*; song full of remarkable mimicry (of African as well as European birds), fluent, fast, with twangy, nasal, whistling notes, trills and slower, lower intervals or pauses.

NESTING Quite shallow cup of grass, suspended from tall stems in thick vegetation by "basket handles"; 4 or 5 eggs; 1 brood; June–July.

FEEDING Forages in and under thick plant cover for insects and spiders; also takes some berries.

SUPERB SONGSTER
Marsh Warblers usually sing from bushes, nettles, and other tall, rank vegetation.

ΛΛΛΛΛΛΛΛΛΛΛΛΛΛΛΛΛΛΛΛΛΛΛΛ

FLIGHT: low, short, flitting flights with whirring wingbeats; jerky, bounding action.

OCCURRENCE
Local summer bird, from extreme SE England (where it is rare) across C, SE, and E Europe, and extreme S Scandinavia. Prefers thick wetland vegetation, with or without a mixture of reeds among sedges, willowherb, nettles, and umbellifers. Rare migrant on coasts and islands from May to September.

| Seen in the UK |
| J F M **A M J J A S** O N D |

SIMILAR SPECIES

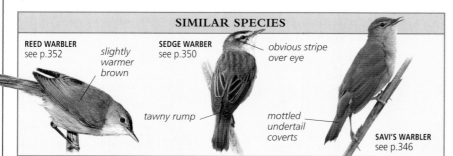

REED WARBLER
see p.352

slightly warmer brown

SEDGE WARBLER
see p.350

obvious stripe over eye

tawny rump

mottled undertail coverts

SAVI'S WARBLER
see p.346

| Length **13–15cm (5–6in)** | Wingspan **18–21cm (7–8½in)** | Weight **11–15g (⅜–⁹⁄₁₆oz)** |
| Social **Solitary** | Lifespan **Up to 5 years** | Status **Secure** |

| Order **Passeriformes** | Family **Sylviidae** | Species ***Acrocephalus scirpaceus*** |

Reed Warbler 🔊 70

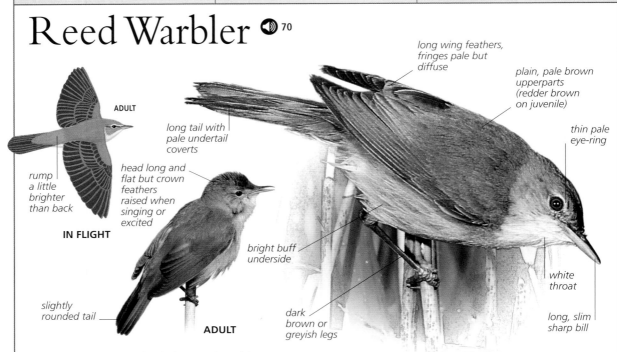

ADULT

rump a little brighter than back

IN FLIGHT

long tail with pale undertail coverts

head long and flat but crown feathers raised when singing or excited

slightly rounded tail

ADULT

bright buff underside

dark brown or greyish legs

long wing feathers, fringes pale but diffuse

plain, pale brown upperparts (redder brown on juvenile)

thin pale eye-ring

white throat

long, slim sharp bill

ADULT

Basically a reedbed bird, the Reed Warbler may sometimes breed away from reeds, in drier spots or in willows growing over shallow water, for example. Its plain colours make it very like some other, rarer, warblers, but unlike the common Sedge Warbler. Its repetitive song is also distinctive as a rule, although subject to some variation. On migration, Reed Warblers may be found in unexpected places, such as thickets and hedgerows, posing identification problems with less closely related species such as Melodious Warblers.
VOICE Call simple, low *churr* or *chk*; song rhythmic, repetitive, low, with occasional high, musical variations, *trrik trrik trrik, chrr chrr chrr chrr, chewe chewe trrrt trrrt trrrt tiri tiri*.
NESTING Deep nest of grass, reedheads, and moss, woven around several upright stems in reedbed; 3–5 eggs; 2 broods; May–July.
FEEDING Forages for insects and spiders on mud and in thick, wet vegetation and foliage of willows; also eats some seeds.

FLIGHT: short, low, jerky flitting flights between reeds or willows, tail low, sometimes spread as it tilts over and dives into cover.

GRASPING REEDS
The Reed Warbler is adept at grasping vertical stems and shuffling through dense reedbeds.

OCCURRENCE
Widespread as breeding bird and summer visitor north to Great Britain and S Scandinavia. In reedbeds, especially extensive, wet ones, but also in reedy ditches and willows beside lakes and rivers. Migrants on coasts between April and October.

| Seen in the UK |
| J F M **A M J J A S O** N D |

SIMILAR SPECIES

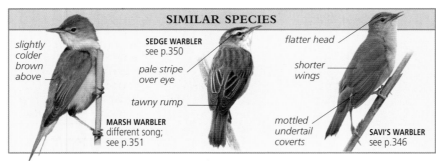

slightly colder brown above

SEDGE WARBLER see p.350

pale stripe over eye

tawny rump

MARSH WARBLER different song; see p.351

flatter head

shorter wings

mottled undertail coverts

SAVI'S WARBLER see p.346

| Length **13–15cm (5–6in)** | Wingspan **18–21cm (7–8½in)** | Weight **11–15g (⅜–⁹⁄₁₆oz)** |
| Social **Solitary** | Lifespan **Up to 5 years** | Status **Secure** |

| Order **Passeriformes** | Family **Sylviidae** | Species ***Acrocephalus arundinaceus*** |

Great Reed Warbler 🔊 71

long, broad tail

bright rump

IN FLIGHT

pale buff stripe from bill to above eye

dark eye-stripe

pale rufous-buff underside

plain warm brown above

big, thick, dark-tipped bill

white throat

long wingtips

broad, slightly rounded, dark brown tail

FLIGHT: low, dashing, thrush-like darts between reed clumps.

Indeed a massive reed warbler, this large, almost thrush-sized warbler typically inhabits reedbeds, but can be found in remarkably small wet spots near rivers or even along ditches and irrigation channels, with just small strips or patches of reed and tall grass. It appears regularly, in very small numbers, north of its usual range, individuals sometimes remaining for a week or two, singing strongly. The song is immediately distinctive: loud and raucous, with a hesitant, frog-like quality.

VOICE Call rolling, harsh *krrrrr* or shorter *tshak*; song remarkably loud, hesitant or full-flowing but repetitive, separate phrases alternately croaking, whistling, warbling, *grik grik grik, jeek jeek chik grrr grrr grrr girik girik girik*.

NESTING Large, deep nest slung in vertical reed stems above water; 3–6 eggs; 1 or 2 broods; May–August.

FEEDING Takes insects and varied invertebrates from foliage and reeds, crashing through stems in search of food.

BOLD SONGSTER
From the top of giant reeds, Giant Reed Warblers create a loud and unmistakable chorus of raucous song.

SIMILAR SPECIES

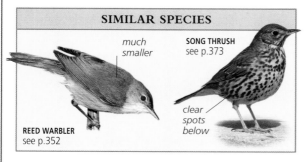

much smaller

REED WARBLER
see p.352

SONG THRUSH
see p.373

clear spots below

OCCURRENCE
Breeds in mainland Europe north to S Scandinavia; local summer visitor. In reedbeds and reedy ditches or strips of reed beside rivers or floods. Present from May to August, when migrants sometimes appear north of usual range.

Seen in the UK
J F M A M J J A S O N D

| Length **16–20cm (6½–8in)** | Wingspan **25–26cm (10in)** | Weight **30–40g (1¹⁄₁₆–1⁷⁄₁₆oz)** |
| Social **Solitary** | Lifespan **Up to 5 years** | Status **Secure†** |

| Order **Passeriformes** | Family **Sylviidae** | Species *Cisticola juncidis* |

Fan-tailed Warbler

dark brown and buff stripes on head

cream and black stripes on back

very short, round wings

small, round tail

IN FLIGHT

short, narrow, often fanned, tail with black and white spots below

unmarked, pale buff underside

thin pink legs

FLIGHT: typically low, fast, whirring, rather weak; song-flight higher, bounding but slow.

Small and insignificant, the Fan-tailed Warbler or Zitting Cisticola is the one European representative of a widespread African and south Asian genus of small, confusingly similar warblers. It is usually revealed by its song, a repetition of a single, sharp, penetrating note given with each bound of a deeply undulating song-flight. Visually, it looks unlike any other European bird despite its basic small, streaky impression, but that itself may make it puzzling if it is silent and skulking in low vegetation. Females may have two or more broods, paired with different males.

VOICE Loud chip call; song usually in deeply undulating song-flight, single short, sharp, penetrating note with each bound, *zeet…zeet…zeet…zeet.*
NESTING Deep, flexible, pear- or flask-shaped nest of grass, feathers, and cobwebs in tall grass; 4–6 eggs; 2 or 3 broods; April–June.
FEEDING Takes insects, spiders, and seeds from rough grass.

DEAD GRASS PATTERN
The pale and dark streaks on the back camouflage this warbler in brown grass and reed stems.

OCCURRENCE
Local breeder around Mediterranean, in Spain, Portugal, and on Atlantic coast of France. In usual range all year but subject to reductions in range in hard winters. In grassy places, marshes, dunes, and sometimes cereal fields with grassy edges.

Seen in the UK
J F M A M J J A S O N D

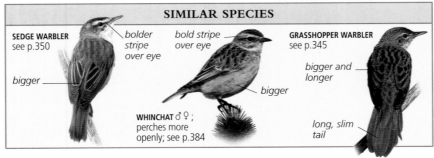

SIMILAR SPECIES

SEDGE WARBLER see p.350

bolder stripe over eye

bigger

bold stripe over eye

bigger

WHINCHAT ♂♀; perches more openly; see p.384

GRASSHOPPER WARBLER see p.345

bigger and longer

long, slim tail

Length **10–11cm (4–4½in)**	Wingspan **12–15cm (4¾–6in)**	Weight **10g (⅜oz)**
Social **Solitary**	Lifespan **Up to 5 years**	Status **Secure†**

| Order **Passeriformes** | Family **Sylviidae** | Species ***Regulus regulus*** |

Goldcrest 🔊 75

neckless shape

blackish wings

ADULT
IN FLIGHT

ADULT

broad white "V"-shaped bar on wings

dull to bright olive-green plumage

black stripe with yellow centre on crown (plain grey-green on juvenile)

whitish around eye

buff to greenish underside

ADULT

Europe's smallest bird, the Goldcrest may sometimes be watched almost at arm's length as it feeds in lower branches of trees. It isn't so much bold or tame as simply oblivious to the presence of people: it generally ignores humans. The Goldcrest's succession of high-pitched, needle-thin but emphatic calls is a feature of many coniferous forests. The song is equally thin but remarkably penetrating, even audible from a fast-passing car. In winter, Goldcrests forage in a variety of places, including hedges, low thickets, and even clumps of gorse or brambles.

VOICE Call high, thin, sibilant but emphasized *see-see-seee*; song high, fast, rhythmic phrase with terminal flourish, complex at close range, at distance *seedli-ee seedli-ee seedli-ee seedli-ee seedli-i-didl-eeoo*.

NESTING Tiny cup of cobwebs, moss, and lichens, slung beneath branch; 7 or 8 eggs; 2 broods; April–July.

FEEDING Picks tiny insects, spiders, and insect eggs from foliage, often hovering briefly.

/\/\/\/\/\/\/\/\/\/\/\/\/\/\/\/\/\/\

FLIGHT: quick, flitting; weak-looking whirr.

PIERCING SONG
The high-pitched song of the Goldcrest penetrates the noise of wind in the trees, and even traffic noise.

OCCURRENCE
Breeds in most of Europe except Iceland, extreme N Scandinavia, and much of Mediterranean Europe. Present all year round in mixed and coniferous woodland, parks, and large wooded gardens; coastal migrants can be in very low scrub.

| Seen in the UK |
| J F M A M J J A S O N D |

SIMILAR SPECIES

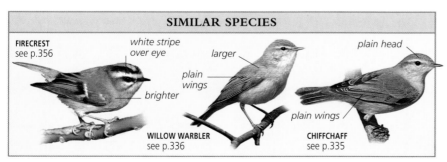

FIRECREST see p.356

white stripe over eye

brighter

larger

plain wings

WILLOW WARBLER see p.336

plain head

plain wings

CHIFFCHAFF see p.335

| Length **8.5–9cm (3¼–3½in)** | Wingspan **13–15.5cm (5–6in)** | Weight **5–7g (³⁄₁₆–¼oz)** |
| Social **Solitary** | Lifespan **2–3 years** | Status **Secure†** |

355

Order **Passeriformes**	Family **Sylviidae**	Species *Regulus ignicapillus*

Firecrest

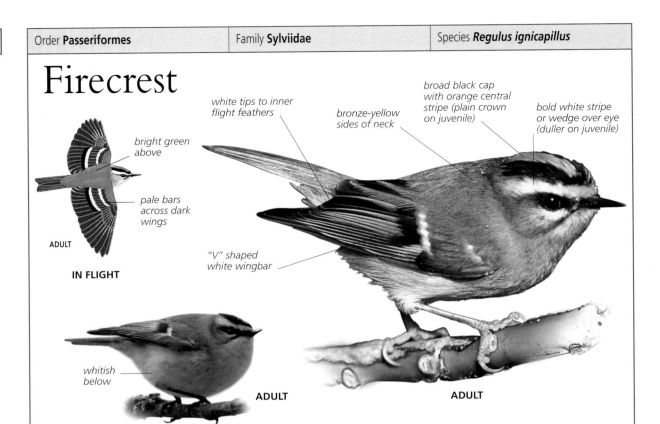

white tips to inner flight feathers

bronze-yellow sides of neck

broad black cap with orange central stripe (plain crown on juvenile)

bold white stripe or wedge over eye (duller on juvenile)

bright green above

pale bars across dark wings

ADULT

IN FLIGHT

"V" shaped white wingbar

whitish below

ADULT

ADULT

Less widespread than the Goldcrest, but in parts of Europe the more likely of the two to be seen, the Firecrest has an obviously close relationship with the slightly duller, plainer species. It has slightly firmer calls and a much less rhythmic, dynamic song, a useful distinction if a bird is seen as a silhouette against the sky at the top of a tall conifer (as so often happens). When it comes lower and allows a close view, it is revealed as one of the brightest of European birds.

VOICE High *zeet*; song sharp, quick, accelerating *zi zi zi zezezeeeee*.

NESTING Moss and lichen cup beneath branch, usually in conifer; 7–11 eggs; 2 broods; April–July.

FEEDING Takes tiny insects and spiders from foliage, slipping through leaves with ease and often hovering briefly.

FLIGHT: short, quick, weak flitting action, usually over very short distances.

SPRUCE NESTER
Firecrests usually nest in conifers; they are typically difficult to see clearly amidst the foliage.

OCCURRENCE
Breeds in extreme S UK, south to Spain and east to Baltic States and Balkans. Found all year in coniferous, broadleaved, and mixed woodland, shrubberies, parks, evergreen scrub, and bushy slopes with many hollies, yews, or similar bushes.

Seen in the UK
J F M A M J J A S O N D

SIMILAR SPECIES

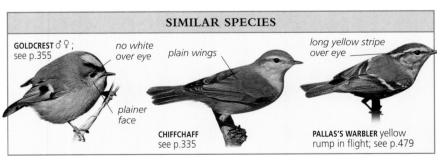

GOLDCREST ♂♀; see p.355

no white over eye

plainer face

plain wings

CHIFFCHAFF see p.335

long yellow stripe over eye

PALLAS'S WARBLER yellow rump in flight; see p.479

Length **9cm (3½ in)**	Wingspan **13–16cm (5–6½ in)**	Weight **5–7g (³⁄₁₆–¼ oz)**
Social **Solitary**	Lifespan **2–3 years**	Status **Secure**

WAXWINGS, WALLCREEPERS, NUTHATCHES, AND TREECREEPERS

THESE ARE ALL BIRDS that find their food while creeping and climbing over hard surfaces: tree bark, walls, or rocks. There are two species pairs and one totally unique species.

WAXWINGS
Upright, crested, short-legged, eye-catching birds, Waxwings are sociable and often very tame. Their numbers in western Europe vary greatly from year to year.

WALLCREEPERS
Nothing else even suggests a Wallcreeper: it is a stunning bird of mountain cliffs and gorges, sometimes coming to quarries, bridges, and large buildings lower down in winter. It is elusive, hard to spot against grey rock, but occasionally gives breathtaking views at close range, which is always a memorable encounter. Wallcreepers creep in a rather crouched stance, bobbing as if mounted on springs, with frequent outward flicks of their wingtips.

NUTHATCHES
Nuthatches are agile, using the strength of their legs and toes to grip, and able to cling underneath branches or overhang, and as easily able to move head-down as right-way up. The Nuthatch is a woodland bird but also feeds on the ground; the Rock Nuthatch is a bird of rocks, walls, and ruins.

TREECREEPERS
Treecreepers literally creep on trees: they can hang beneath a branch, but use the tail as a support, so never descend head-first. Identification is difficult, unless a bird is definitely outside the range of one or other, or it is singing. Treecreeper and Short-toed Treecreeper songs are usually distinctive, but the latter may sometimes sing confusingly like the former. Close examination, preferably of photographs, is necessary to identify a stray outside its normal range. Treecreepers join roving bands of tits in winter, when woods seem empty until, suddenly, trees are "full of birds" for a few minutes, before they move on.

STRONG GRIP
Nuthatches can more or less cling to a branch at any angle, head- down or head-up.

BOMBYCILLIDAE

large crest

WAXWING
p.358

TICHODROMADIDAE

blackish wings with bright red patches

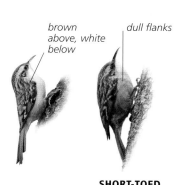

WALLCREEPER
p.359

SITTIDAE

black stripe through eye

NUTHATCH
p.360

CERTHIIDAE

brown above, white below

dull flanks

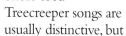

TREECREEPER
p.361

SHORT-TOED TREECREEPER
p.362

| Order **Passeriformes** | Family **Bombycillidae** | Species ***Bombycilla garrulus*** |

Waxwing

ADULT

white bars on black wings

IN FLIGHT

dull black tail with broad yellow tip

pale grey lower back and rump

neat black bib

MALE

blurred bib

yellow tip

thinner yellow stripe

FEMALE

large crest

black line through eye

pale pinkish brown body

waxy red spots on wings

long stripe of yellow or white along closed wingtips

rusty red under tail

MALE

A northern breeding bird, the Waxwing visits western Europe in winter in very variable numbers. The best years follow a summer with good breeding success and high populations, but a poor autumn berry crop will force the Waxwings to move far to the south and west of their usual range in search of food. Although flocks in flight might superficially suggest Starlings, identification is very easy; the birds' tameness helps as they feed in urban areas or gardens.

VOICE Silvery, high, metallic trill on even pitch, *trrreee* or *siirrrrr*.

NESTING Moss-lined nest of twigs in birch or conifer; 4–6 eggs; 1 brood; May–June.

FEEDING Eats insects in summer; in winter some insects, often caught in flight, but mostly large berries such as rowan, hawthorn, and cotoneaster; also eats apples and other fruit.

/\/\/\/\/\/\/\/\/\/\/\/\/\/\

FLIGHT: direct, swooping or swerving, with long, shallow undulations, quick wingbeats; flocks keep formation like waders.

OCCURRENCE
Breeds in conifer forest in extreme NE Europe. In winter, frequent in N Scandinavia, irregular in S Scandinavia and E Europe. Numbers are erratic, sometimes large, in W Europe, big flocks coinciding with high population and lack of food in N Europe.

| Seen in the UK |
| J **F** **M** A M J J A S O **N** **D** |

SIMILAR SPECIES

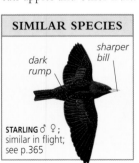

sharper bill

dark rump

STARLING ♂ ♀; similar in flight; see p.365

RESTING FLOCK
Waxwings feed greedily, stripping shrub of berries, and drink a great deal. Between bouts of feeding, flocks rest in undisturbed trees nearby.

| Length **18cm (7in)** | Wingspan **32–35cm (12½–14in)** | Weight **45–70g (1⅝–2½oz)** |
| Social **Flocks** | Lifespan **Up to 5 years** | Status **Secure†** |

Wallcreeper

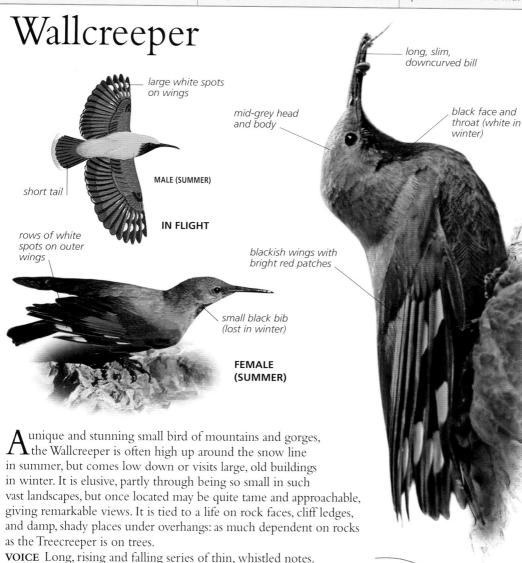

large white spots on wings

MALE (SUMMER)

short tail

IN FLIGHT

rows of white spots on outer wings

long, slim, downcurved bill

mid-grey head and body

black face and throat (white in winter)

blackish wings with bright red patches

small black bib (lost in winter)

FEMALE (SUMMER)

MALE (SUMMER)

A unique and stunning small bird of mountains and gorges, the Wallcreeper is often high up around the snow line in summer, but comes low down or visits large, old buildings in winter. It is elusive, partly through being so small in such vast landscapes, but once located may be quite tame and approachable, giving remarkable views. It is tied to a life on rock faces, cliff ledges, and damp, shady places under overhangs: as much dependent on rocks as the Treecreeper is on trees.

VOICE Long, rising and falling series of thin, whistled notes.
NESTING Untidy nest in hole in cliff or deep in crevice between tumbled rocks; 4 eggs; 1 brood; May–July.
FEEDING Searches rocks, especially wet spots and earthy ledges, and also buildings for insects and spiders, probing with its bill and fluttering constantly.

FLIGHT: hesitant, fluttery, but quite strong, undulating over long distances; remarkable bounding when feeding; moves butterfly-like over short distances.

EXCELLENT CAMOUFLAGE
The red of this juvenile Wallcreeper's wing is not obvious at long range: it often looks dark grey and is easily lost against a rocky background.

OCCURRENCE
Mostly rare and local; breeds in Pyrenees, Alps, and Balkans, typically up close to snow line. Little more widespread in S Europe in winter, coming to lower altitudes. On rocks, cliff faces, in gorges and quarries, but only rarely moving far from breeding range.

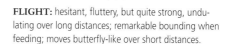

Seen in the UK
| J | F | M | A | M | J | J | A | S | O | N | D |

| Length **15–17cm (6–6½in)** | Wingspan **30–35cm (12–14in)** | Weight **25g (⅞oz)** |
| Social **Family groups** | Lifespan **3–5 years** | Status **Secure†** |

| Order **Passeriformes** | Family **Sittidae** | Species *Sitta europaea* |

Nuthatch 🔊 82

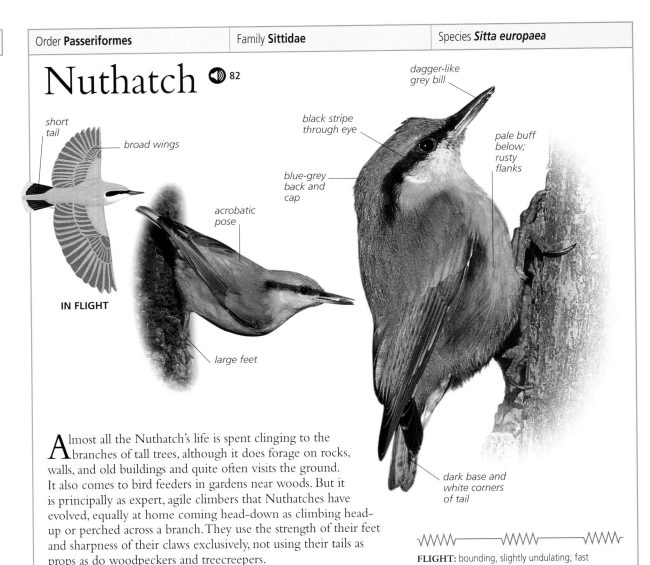

short tail

broad wings

IN FLIGHT

dagger-like grey bill

black stripe through eye

blue-grey back and cap

pale buff below; rusty flanks

acrobatic pose

large feet

dark base and white corners of tail

A lmost all the Nuthatch's life is spent clinging to the branches of tall trees, although it does forage on rocks, walls, and old buildings and quite often visits the ground. It also comes to bird feeders in gardens near woods. But it is principally as expert, agile climbers that Nuthatches have evolved, equally at home coming head-down as climbing head-up or perched across a branch. They use the strength of their feet and sharpness of their claws exclusively, not using their tails as props as do woodpeckers and treecreepers.

VOICE Various loud, full, liquid whistles, *pew pew pew pew*, *chwee chwee* and fast ringing trills, loud *chwit*.

NESTING Uses old woodpecker hole or nest box, lined with bark and leaves, typically plastering mud around entrance; 6–9 eggs; 1 brood; April–July.

FEEDING Eats variety of seeds, berries, nuts, acorns, and beech-mast, often carried to be wedged in bark for easier manipulation; visits peanut baskets.

FLIGHT: bounding, slightly undulating; fast wingbeats.

SIMILAR SPECIES

ROCK NUTHATCH different habitat; see p.481

paler below

plain tail

GROUND FORAGER
Nuthatches frequently drop to the ground to seek fallen nuts and berries, hopping jerkily over the leaf litter.

OCCURRENCE
Breeds in most of Europe, except Iceland, N UK, N Scandinavia, and S Spain. Occupies mixed and deciduous woodland, parkland, and large gardens with big old trees all year round, rarely moving far.

| Seen in the UK |
| J F M A M J J A S O N D |

| Length **12.5cm (5in)** | Wingspan **16–18cm (6½–7in)** | Weight **12–18g (7/16–5/8oz)** |
| Social **Loose flocks** | Lifespan **2–3 years** | Status **Secure** |

Order **Passeriformes**	Family **Certhiidae**	Species **Certhia familiaris**

Treecreeper

whitish wingbar

notched brown tail

IN FLIGHT

fine, curved bill

whitish stripe over eye

silky white underside

large but slender feet

cream and blackish mottling on brown back

rounded black patch on wings; pale bar lacks saw-toothed edge of Short-toed Treecreeper's

pale feather shafts on tail

FLIGHT: weak, low, undulating; typically from high in one tree to base of next.

Even more closely associated with the bark of trees than the Nuthatch, the Treecreeper can cling beneath a branch, hanging freely by its toes, but usually sits upright, propped up on its tail. It sometimes forages on the twigs of smaller bushes, more rarely on stone walls. Typically, Treecreepers work their way upwards on tree trunks and larger branches, often spiralling, before flying down to the next to begin the upward search for food once more. They are often on the edges of mixed tit flocks in autumn and winter.

VOICE Call thin, long, high *seee* and more vibrant *srreee*; song frequent, high, musical, like thin Willow Warbler's in pattern, falling trill with final flourish.

NESTING Untidy nest behind loose bark or ivy stem, sometimes in nest box; 5 or 6 eggs; 1 brood; April–June.

FEEDING Takes insects, spiders, and other tiny items from bark, probing with bill as it shuffles up or around trunks and branches; also forages on walls, and occasionally rocks.

REMARK Subspecies *C. f. familiaris* (Scandinavia) is whiter over eye and very white below.

SIMILAR SPECIES

duller underside

SHORT-TOED TREECREEPER different call and song; see p.363

STRONG GRIP
Treecreepers use their sharp claws and strong toes to grip tightly on rough bark, the tail adding stability and balance.

OCCURRENCE
Breeding bird and year-round resident in Great Britain and Ireland, local in France, N Spain and east across Europe. In mixed, deciduous, or coniferous woods, parks, along tall hedges, and sometimes in well-wooded gardens.

Seen in the UK
J F M A M J J A S O N D

Length **12.5cm (5in)**	Wingspan **18–21cm (7–8½in)**	Weight **8–12g (5/16–7/16oz)**
Social **Mixed flocks**	Lifespan **2–3 years**	Status **Secure**

| Order **Passeriformes** | Family **Certhiidae** | Species *Certhia brachydactyla* |

Short-toed Treecreeper

looks slim and weak in flight

whitish wingbar

IN FLIGHT

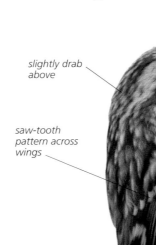

long, slender, slightly downcurved bill

clean white throat

slightly drab above

saw-tooth pattern across wings

plain tail

FLIGHT: quite direct but weak, undulating, with bursts of quick wingbeats.

Few species pairs are as difficult as the two treecreepers: the Short-toed is best told by its calls and song. Even held in the hand, the two can be near impossible to separate on plumage and measurements alone. The Short-toed has more obvious white tips to the wingtip feathers and a slightly different pattern across the closed wing. In general, the Short-toed is a touch duller, a little browner underneath, with a more contrasted white throat; sometimes it may look a little rounder, with its tail angled in more steeply to the bark, but such impressions are of little real value. It is equally dependent on trees, but does clamber on rocks at times.

VOICE Call quite strong, short, clear *tsoit*, sometimes longer *sreet*; song stereotyped, with discrete notes (not flowing like Treecreeper's), *stit-stit-steet, stit-it-steroi-tit*.

NESTING Cup of grass and feathers in crevice, like Treecreeper; 5 or 6 eggs; 1 brood; April–June.

FEEDING Like Treecreeper, creeps about branches and tree trunks, probing and picking (but not chipping away bark) for insects and eggs; sometimes forages on rocks.

SIMILAR SPECIES

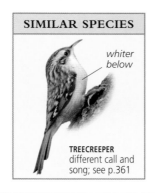

whiter below

TREECREEPER different call and song; see p.361

BARK SPECIALIST
The Short-toed Treecreeper spends its whole life clambering on tree bark searching for food.

OCCURRENCE
Breeds locally in Spain, Italy, France, Low Countries, Italy, and Balkans. Typically in lowland, often deciduous woods, but also in pine forest in hills. Present all year, rare vagrant outside its breeding range.

Seen in the UK
J F M A M J J A S O N D

Length **12.5cm (5in)**	Wingspan **18–21cm (7–8½in)**	Weight **8–12g (5/16–7/16oz)**
Social **Mixed flocks**	Lifespan **2–3 years**	Status **Secure**

WRENS, STARLINGS, AND DIPPERS

THIS GROUP – another convenient grouping rather than a collection of near relatives – includes families that are widespread in the northern hemisphere.

WRENS
Primarily an American family, with just one species in Europe, the wrens are small, brown, barred birds with loud voices and an "irritable" character. The Wren, whose scientific name means "cave dweller", is likely to be found in dark, damp, cobweb-filled places under hedges and around sheds, searching for insects.

STARLINGS
Two species are mostly dark, shiny, quarrelsome birds; the third is paler when juvenile and pink and black when adult. All the starlings are rather squat, sharp-billed, short-tailed birds that walk and run in a quick, shuffling fashion and fly quickly, often in dense flocks. Starlings also gather to roost in woods, reedbeds, and on structures such as piers and bridges, in gigantic flocks, although numbers have recently declined dramatically in many areas.

VIBRANT SONG
A Wren shakes with the effort as it pours out a remarkable song, part of the woodland chorus.

DIPPERS
Superficially wren-like but larger, the Dipper swims, wades, and walks underwater. It is always at the water's edge, even flying along a twisting water course rather than across dry land.

STURNIDAE

glossy black body with green and purple sheen

STARLING
p.365

purple-black body with dull oily sheen

SPOTLESS STARLING
p.366

TROGLODYTIDAE

tiny, rounded, cocked tail

WREN
p.364

CINCLIDAE

bold white chest

DIPPER
p.367

| Order **Passeriformes** | Family **Troglodytidae** | Species ***Troglodytes troglodytes*** |

Wren 🔊 58

dark barring on back and wings

IN FLIGHT

tiny, rounded, cocked tail

long pale stripe over eye

round head, with no neck

pale brown back

fine, long bill

pale buff underside

softly barred flanks

strong pale feet

FLIGHT: low, fast, short, with bursts of quick wingbeats; dives quickly into cover.

One of Europe's smallest birds, the big-voiced Wren uses a remarkable variety of habitats. It is found from sea level to high up in mountain areas, from forest to almost open spaces; subspecies exist in remote island groups. It spends most of the time low down, on or near the ground, often in deep thickets of bramble or bracken or in ornamental shrubberies. Cold winters cause dramatic declines but Wren populations can recover quite quickly.

VOICE Dry, hard calls with irritable, rattling quality, *chit, chiti, tzerrr*; song loud, full-throated, warbling outburst with characteristic low, hard trill and fast, ringing notes.

NESTING Small, loose ball of leaves and grass, tucked in bank, under overhang; 5 or 6 eggs; 2 broods; April–July.

FEEDING Forages in dark, damp places under hedges, around buildings, shrubberies, in ditches, and in patches of dead bracken and similar low, thick cover; finds insects and spiders and feeds on scraps scattered under bushes.

DETERMINED SINGER
With tail raised and bill wide open, a singing Wren puts all its effort into a loud, vibrant song.

OCCURRENCE
Breeds in practically all of Europe except far north; in N and E Europe only in summer. Lives anywhere from open clifftops and heaths to broadleaved and coniferous woodland, parks, gardens, and hedges.

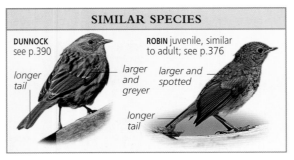

| SIMILAR SPECIES | | SUBSPECIES |

DUNNOCK see p.390

longer tail

ROBIN juvenile, similar to adult; see p.376

larger and greyer

larger and spotted

longer tail

T. t. zetlandicus
(Shetland)

coarsely barred flanks

greyer brown

Seen in the UK
| J | F | M | A | M | J | J | A | S | O | N | D |

| Length **9–10cm (3½–4in)** | Wingspan **13–17cm (5–6½in)** | Weight **8–13g (5/16–7/16oz)** |
| Social **Roosts in flocks** | Lifespan **2–5 years** | Status **Secure** |

Order **Passeriformes**	Family **Sturnidae**	Species ***Sturnus vulgaris***

Starling 🔊 90

ADULT

short, square tail

IN FLIGHT

dull head last to gain adult colours

IMMATURE (MOULTING; AUTUMN)

large, scaly spots around tail

silvery white face with darker mask

body feathers tipped whitish or buff

wing feathers edged bright orange-buff

ADULT (WINTER)

glossy black body with green and purple sheen

blue-based, sharp yellow bill (pink-based on female)

plain brown body

dark bill

JUVENILE

long, strong, red-brown legs

MALE (SPRING)

Generally abundant, Starlings form dense, noisy flocks: no other small bird, apart from shoreline waders, creates such vast flocks that look like clouds of smoke at a distance. However, with widespread declines, flocks have been much reduced. They are found in many habitats, urban, suburban, and rural, many moving west within Europe in winter. In spring, Starlings sing loudly, with characteristic wing-waving actions.

VOICE Loud, slightly grating *cheer*, musical, twangy, whistled *tswee-oo*, variety of clicks, gurgles, squawking notes; song fast mixture of gurgles, rattles, trills, and whistles, some mimicry.

NESTING Loose, bulky nest of grass and stems, lined with roots, moss, wool, and feathers, in tree hole, cavity in building or wall, or large nest box; 4–7 eggs; 1 or 2 broods; April–July.

FEEDING In small to large flocks, finding invertebrates, seeds, and berries on ground; catches flying ants.

FLIGHT: direct, fast, short glides and rapid flicked wingbeats; often in dense flocks, rising and falling.

WINTER FLOCK
Starling flocks in flight are marvellous examples of skilful coordination and close control.

SIMILAR SPECIES

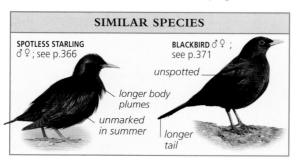

SPOTLESS STARLING
♂♀; see p.366

BLACKBIRD ♂♀; see p.371

unspotted

longer body plumes

unmarked in summer

longer tail

OCCURRENCE
Breeds in all Europe except Spain, Portugal, and S Italy, where it is a winter visitor; only summer in N and E Europe. Breeds in woods, gardens, and towns; in winter, in forest, city centres, and industrial sites, on bridges and piers.

Seen in the UK
J F M A M J J A S O N D

Length **21cm (8½in)**	Wingspan **37–42cm (14½–16½in)**	Weight **75–90g (2⅝–3¼oz)**
Social **Flocks**	Lifespan **Up to 5 years**	Status **Secure**

| Order **Passeriformes** | Family **Sturnidae** | Species *Sturnus unicolor* |

Spotless Starling

ADULT

IN FLIGHT

red-brown tinge on wingtips, seen against light

gradually develops adult colours

JUVENILE (MOULTING)

plain brown back and wings

scattered, small dark spots on underside

JUVENILE

blue-based yellow bill (pink-based on female)

long, loose, pointed plumes around neck and breast

purple-black body with dull oily sheen (duller in winter)

red-pink legs

MALE (SUMMER)

The Iberian equivalent of the Starling, this species has an obvious resemblance and close relationship with the more widespread species: in winter especially, they can be difficult to separate. In summer, however, as groups line the roofs of ancient buildings in Spain, or fly down to feed in the fields, they seem to look and sound a little different, with a subtle character of their own. In winter, Starlings join them across Spain and increasingly they stay to breed south of the Pyrenees, increasing the identification challenge.

VOICE Starling-like squawling and quarrelling notes; song has loud, musical, long-drawn *py-eeeew* and parrot-like sounds.

NESTING Bulky, untidy nest in roof spaces and cavities in walls; 4–7 eggs; 1 brood; April–June.

FEEDING Forages on ground for all kinds of invertebrates and seeds.

FLIGHT: like Starling but tends to look slightly bulkier, broader-winged, slightly slower.

HABITAT
The Spotless Starling is found in similar habitats as the Starling but is more often associated with old buildings and tiled roofs.

SIMILAR SPECIES

STARLING ♂♀ ; larger spots overall in winter; see p.365

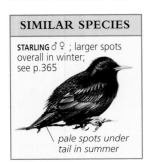

pale spots under tail in summer

SUBTLE SHEEN
Unless seen closely, this bird looks solidly black. It has a more purplish sheen overall than a Starling.

OCCURRENCE
Breeds in Spain, Portugal, and extreme S France, Corsica, Sicily, and Sardinia. All year round in towns and villages, feeding on adjacent farmland; in winter, may mix with common Starlings.

| Seen in the UK |
| J F M A M J J A S O N D |

| Length **21cm (8½in)** | Wingspan **37–42cm (14½–16½in)** | Weight **75–90g (2⅝–3¼oz)** |
| Social **Flocks** | Lifespan **Up to 5 years** | Status **Secure** |

| Order **Passeriformes** | Family **Cinclidae** | Species *Cinclus cinclus* |

Dipper

thickset shape

ADULT

IN FLIGHT

pale feather edges

greyer body

JUVENILE

deep brown head

blackish from back to tail

stout dark bill

bold white chest

chestnut band on belly

thick black legs with large feet

ADULT

Few birds are so strictly confined to one habitat type as the Dipper. In summer, it is essentially a bird of fast-flowing, but often tree-lined rivers in uplands. It remains there if it can in winter, but hard weather may drive it lower down or even to the edges of large lakes and reservoirs or the sea coast. It is quite at home swimming and diving, or simply walking into the water and disappearing underneath as it searches for food. Its springy, bouncing movements and call are also distinctive.

VOICE Sharp, hard, abrupt, and penetrating *dzit* or *djink*; song loud, rich warbling mixed with explosive, grating notes.

NESTING Ball-shaped nest of moss and grass in hole in bank, under overhang or bridge, and behind waterfall; 4–6 eggs; 2 broods; April–July.

FEEDING Unique, walking into water, swimming and diving from surface, or wading into shallows, foraging for caddis-fly and other larvae, small fish, crustaceans, and molluscs.

REMARK Subspecies *C. c. hibernicus* (Scotland and Ireland) has narrow, darker chestnut belly band.

FLIGHT: low, fast, along stream line; fast bursts of wingbeats.

SUBSPECIES

C. c. cinclus (N Europe, N France)

blacker

all-dark belly

CAMOUFLAGE PATTERN
The bright white chest surprisingly serves to render the Dipper less conspicuous in the ripples and reflections of a stony river.

OCCURRENCE
Local in upland areas, absent from Iceland, W France, and NE Europe, but breeding widely elsewhere in suitable habitat: clean, fresh rivers, in moorland areas or tree-lined valleys, or deep in shady gorges. In winter, some move out to larger areas of water, rarely coasts.

Seen in the UK
J F M A M J J A S O N D

| Length **18cm (7in)** | Wingspan **25–30cm (10–12in)** | Weight **55–75g (2–2⅝oz)** |
| Social **Solitary** | Lifespan **Up to 5 years** | Status **Secure†** |

Family **Turdidae**

CHATS AND THRUSHES

Birds in this group are characterized by rather short but strong bills, stout legs, quite large heads, big eyes, and an all-round solid build. Some are common, others very rare; some are resident, others migrate. They occupy a wide range of habitats from gardens to forest, mountain, and moor, and include some of the finest of European bird songsters.

CHATS

Smaller than the thrushes and less stoutly built, the chats are varied: most have different seasonal plumages, with male and female looking different in summer and juveniles looking much like winter adults. The wheatears occupy open places from high, bleak moors to hot Mediterranean heaths. The nightingales live in dense shrubbery in woods or beside heaths. Stonechats are year-round residents on open heath, while Whinchats are summer migrants. The chats demonstrate that there is usually more than one way to exploit a habitat or food supply.

THRUSHES

The spotted thrushes look the same all year and male and female are alike, while the Blackbird and rock thrushes have sexual (and sometimes seasonal) differences in plumage. Superb singers, they make up the bulk of the spring dawn chorus in much of northern Europe. Many are migrants, although in some cases, such as the Blackbird, winter immigrants to western Europe join others of the same species.

INTERMEDIATE
The Nightingale fits neatly between larger thrushes and smaller chats in size and shape.

TURDIDAE

white breast-band

bright orange-yellow bill

black and yellow bill

RING OUZEL
p.370

BLACKBIRD
p.371

FIELDFARE
p.372

"V"-shaped brown-black spots on underside

spots wider, less V-shaped, more evenly spread than on Song Thrush

dull rust-red flanks

SONG THRUSH
p.373

REDWING
p.374

MISTLE THRUSH
p.375

orange-red breast

ROBIN
p.376

soft, pale, buff-brown and greyish olive

THRUSH NIGHTINGALE
p.377

tail often raised

NIGHTINGALE
p.378

blue throat

BLUETHROAT
p.379

black below

BLACK REDSTART
p.380

rich orange-rufous underside

REDSTART
p.381

rich rust-orange underside

ROCK THRUSH
p.382

dark blue body, looks blackish at distance

BLUE ROCK THRUSH
p.383

black streaks on pale straw-brown back

WHINCHAT
p.384

rust-red breast

STONECHAT
p.385

black patch through eye

WHEATEAR
p.386

black face and throat

BLACK-EARED WHEATEAR
p.387

mostly white tail

BLACK WHEATEAR
p.388

| Order **Passeriformes** | Family **Turdidae** | Species ***Turdus torquatus*** |

Ring Ouzel

pale wings

MALE

IN FLIGHT

small head

slender shape

dull, pale breast-band

duller than male

FEMALE

black head

brown-black back

paler wings

white breast-band

long black tail

sooty black underside

MALE

In summer, Ring Ouzels are found in wild, open country with loose rocks, boulders, crags, or drystone walls, more rarely in deeply eroded peat bogs or on steep, bracken-covered slopes. They appear early in spring and occasionally turn up, while migrating, on hills inland or on coasts; in autumn, they are more often found by the sea, particularly on dunes overgrown with berry bushes. They tend to be rather shy and wild, quick to fly off out of sight. They are usually seen with head up, tail cocked, and wings drooped, or a head and bill may just be visible above a skyline rock. Ring Ouzels are declining in areas subject to increased human disturbance on summer weekends.

VOICE Loud, hard, rhythmic *tak-tak-tak*; various chattering and chuckling calls; song loud, wild, simple repetition of short phrases with musical, fluty quality.

NESTING Big, loose cup of grass, twigs, soil, and leaves, in steep bank, rock cavity, or fallen stone wall; 5 or 6 eggs; 2 broods; April–June.

FEEDING Feeds on insects, worms, seeds, and berries; eats berries in bushes on migration.

FLIGHT: fast, direct; often over long distance, recalling Mistle Thrush but lower; frequently flies off over ridge out of sight.

SUBSPECIES

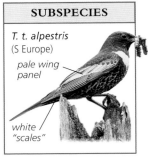

T. t. alpestris (S Europe)

pale wing panel

white "scales"

OCCURRENCE
Breeds locally through Europe, except in Iceland and NE, mostly on high ground, on open moors with rocky places, gullies, exposed tors, and eroded peat bogs. Migrants in early spring and late autumn seen on hills and coasts.

Seen in the UK
J F **M A M J J A S O** N D

SIMILAR SPECIES

BLACKBIRD ♂ similar to ♂ ♀; see p.371

blacker

DIPPER see p.367

white breast

BLUE ROCK THRUSH ♀ similar to ♀ ♂; see p.383

pale below with dark bars

| Length **23–24cm (9–9½in)** | Wingspan **38–42cm (15–16½in)** | Weight **95–130g (3⅜–5oz)** |
| Social **Family groups** | Lifespan **5–10 years** | Status **Secure** |

Blackbird 🔊 66

wingtips subtly paler, especially from below

MALE

IN FLIGHT

brown wings

dull black body

dark bill

IMMATURE MALE (1ST WINTER)

pale streaks on back

gingery brown body

JUVENILE

yellow eye-ring

bright orange-yellow bill

MALE

large, all-black body

dark brown body

dark streaks on throat

raises tail

variably mottled or dark-spotted underside

FEMALE

One of Europe's most familiar birds, the Blackbird is also a common example of a species with very obvious male and female differences. Black adult males are unique while females are always darker than other thrushes, although with a variable tendency to be spotted on the underparts. They range from remote mountain forests to gardens and parks, and are among the most regular garden birds and bird-table visitors in many areas. Blackbirds create most of the spring dawn chorus in suburban and woodland areas.

VOICE Low, soft *chook*, frequent loud, sharp *pink pink pink*; alarm rattle fast, hysterical outburst of sharp rattling notes, high, thin, slightly rough *srreee*; song superb, musical, full-throated, mellow warbling with many variations, phrases often ending in weak, scratchy sounds.

NESTING Grass and mud cup, lined with grass, in shrub, bush, low in tree or hedge; 3–5 eggs; 2–4 broods; March–August.

FEEDING Finds worms, insects, and invertebrates of all kinds on ground, often noisily exploring leaf litter; feeds on fruit and berries in bushes; often visits bird-tables, and eats scattered apples and bread.

FLIGHT: usually quite low, quick, swooping into cover; more undulating over longer range with flurries of wingbeats; raises tail on landing.

OCCURRENCE
Breeds over almost whole of Europe but rare in Iceland. In woods, gardens, parks, and farmland with tall hedges all year, in some areas garden lawn and shrubbery specialist, but essentially bird of woodland with rotting leaf litter on ground.

Seen in the UK
| J | F | M | A | M | J | J | A | S | O | N | D |

SIMILAR SPECIES

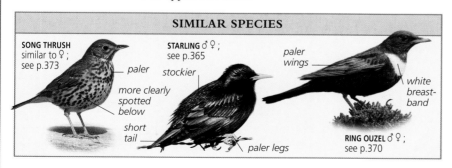

SONG THRUSH similar to ♀; see p.373

paler

more clearly spotted below

short tail

STARLING ♂♀; see p.365

stockier

paler legs

paler wings

white breast-band

RING OUZEL ♂♀; see p.370

| Length **24–25cm (9½–10in)** | Wingspan **34–38cm (13½–15in)** | Weight **80–110g (2⅞–4oz)** |
| Social **Family groups** | Lifespan **Up to 5 years** | Status **Secure** |

| Order **Passeriformes** | Family **Turdidae** | Species ***Turdus pilaris*** |

Fieldfare

blue-grey head with black mask

black and yellow bill

dark brown back (juvenile has brown spots on wing coverts)

mostly white underwings

black tail

ADULT

pale grey rump

IN FLIGHT

black-spotted, pale to deep orange-buff breast

ADULT (SUMMER)

grey rump

whiter flanks

black spots on white flanks

ADULT (WINTER)

A large, striking, and handsome thrush, the Fieldfare has a distinctive call and a unique combination of colours. The white underwing is a useful feature for identification, as is the tendency to move around in flocks and to nest colonially. Flocks in flight keep more or less together but drift along in irregular lines and shapeless packs, less coordinated than, for example, some of the smaller finches.

VOICE Distinctive loud, soft or harder, chuckling *chak-chak-chak* or *tsak-tsak-tsak*, low, nasal Lapwing-like *weeip*; song rather poor, unmusical mixture of squeaks, warbles, and whistles.

NESTING Cup of grass and twigs, in bush or tree, often in loose colonies; 5 or 6 eggs; 1 or 2 broods; May–June.

FEEDING Mostly eats worms and insects on ground; also takes apples, berries, and other fruit from trees, hedges, and bushes.

FLIGHT: quite strong, somewhat undulating, with bursts of wingbeat; irregular glides, quite slow and erratic; often in flocks.

NOMADIC FLOCKS
Fieldfares flock in winter, moving about and feeding together. They often mix with Redwings and interact with thrushes such as Blackbirds and Mistle Thrushes.

OCCURRENCE
Breeds across N and E Europe in wooded regions. Widespread, social winter visitor through all of W and S Europe, in wooded areas, bushy heaths, and farmland (especially old pastures and orchards) with hedges and scattered trees. Visits gardens in severe weather.

Seen in the UK
| J | F | M | A | M | J | J | A | S | O | N | D |

SIMILAR SPECIES

MISTLE THRUSH see p.375

paler back

pale tail

BLACKBIRD ♀; dark underwings in flight; see p.371

more uniform

KESTREL ♂; see p.159

much larger

| Length **25cm (10in)** | Wingspan **39–42cm (15½–16½in)** | Weight **80–130g (2⅞–5oz)** |
| Social **Winter flocks** | Lifespan **5–10 years** | Status **Secure** |

| Order **Passeriformes** | Family **Turdidae** | Species ***Turdus philomelos*** |

Song Thrush 🔊 64

warm orange-buff underwings

plain wings

IN FLIGHT

SINGING

pale eye-ring

plain dark to olive-brown upperparts

pale feather tips on immature (Mistle Thrush has long pale streaks)

streaks under cheeks

"V"-shaped, brown-black spots on underside

yellow-buff underparts, browner on flanks, white on belly

pale pinkish legs

FLIGHT: usually low into nearest cover; higher flight erratic, with swooping glides; bursts of wingbeats.

The classic spotted thrush, the Song Thrush is neatly patterned below, and is rather small, markedly smaller than a Blackbird. It has a marvellously vibrant, varied, full-throated song that is instantly identifiable. A declining bird in many areas, it relies on some woodland or tree cover, or at least big, bushy hedges in farmland. It is equally at home in mixed or deciduous woodland with some clearings and well-wooded gardens or town parks.
VOICE Short, thin, high *stip*; loud alarm rattle; song loud, exuberant, shouted, each separated phrase of 2–4 notes repeated 2–4 times, some musical, some whistled, others harsh or rattled.
NESTING Grassy cup lined with mud and dung, low in bush, hedge, or tree; 3–5 eggs; 2 or 3 broods; March–July.
FEEDING Hops and runs across open ground, stopping to detect and extract earthworms; eats many snails, slugs, other invertebrates, berries, and fruit; shy visitor to bird-tables, but eats scattered scraps and apples.

ELDERBERRY TREAT
Autumn berries provide welcome food for the Song Thrush, in addition to the usual diet of worms and snails.

OCCURRENCE
Breeds in almost all of Europe except Iceland; in summer, only in N and E Europe, resident and winter visitor in S and W. In broadleaved woodland, parkland, farmland with trees and hedges, gardens, parks with lawns, and shrubberies.

Seen in the UK
J F M A M J J A S O N D

SIMILAR SPECIES

MISTLE THRUSH
see p.375

greyer

bigger

pale edges to wing feathers

rounder spots below

REDWING
see p.374

strongly striped head

BLACKBIRD ♀;
see p.371

smaller and darker

bigger

darker

much less sharply spotted

| Length **23cm (9in)** | Wingspan **33–36cm (13–14in)** | Weight **70–90g (2½–3¼oz)** |
| Social **Family groups** | Lifespan **Up to 5 years** | Status **Secure** |

373

| Order **Passeriformes** | Family **Turdidae** | Species *Turdus iliacus* |

Redwing

IN FLIGHT

white spot beside tail

reddish underwing coverts

dark cap

pale stripe over eye

dark cheeks

dark brown back

long, narrow dark spots on breast

silvery white underside

dull rust-red flanks

⋀⋀⋀⋀⋀⋀⋀⋀⋀⋀⋀⋀⋀⋀⋀⋀⋀⋀

FLIGHT: quite quick; in flocks sometimes high, slightly erratic, undulating a little, with in-out flicks of wings; disturbed flocks move to hedge or tree.

Often heard calling during its nocturnal migrations, especially during clear, calm nights in October, the Redwing is a small, social thrush, easily identified by its well-marked head pattern. It moves about in flocks, often mixed with Fieldfares, and, in winter, feeds in loose congregations in fields or in closer groups, feeding on berries in hedgerows. It is not generally a garden bird but will come to larger gardens for food during hard weather, to which it is particularly susceptible. Breeding pairs form small, scattered groups.

VOICE Flight call, often at night, high, thin, simple *seeeeh*, also *chuk, chittuk*; song variable, monotonous repetition of short phrases and chuckling notes with rising or falling pattern.

NESTING Cup of grass and twigs, in low bush or shrub; 4–6 eggs; 2 broods; April–July.

FEEDING Often on ground, in winter in loose flocks advancing across field, finding worms, insects, and seeds; also in hedges, feeding on berries; in hard weather, visits gardens for apples and berries.

BERRY EATERS
Hedgerow berry crops are quickly exhausted by mixed flocks of Redwings and Fieldfares descending on them.

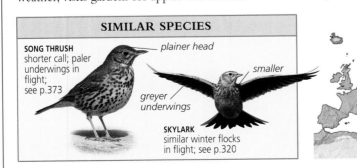

SIMILAR SPECIES

SONG THRUSH
shorter call; paler underwings in flight; see p.373

plainer head

greyer underwings

SKYLARK
similar winter flocks in flight; see p.320

smaller

OCCURRENCE

Breeds in N and E Europe; winters in N and S Europe. Nests in birch woods and conifer forest; in winter, on bushy heaths, in farmland with hedges and old pastures, in larger, undisturbed parks, and in gardens, especially in hard weather.

| Seen in the UK |
| J F M A M J J A S O N D |

| Length **21cm (8½in)** | Wingspan **33–35cm (13–14in)** | Weight **55–75g (2–2⅝oz)** |
| Social **Winter flocks** | Lifespan **Up to 5 years** | Status **Secure** |

Order **Passeriformes**	Family **Turdidae**	Species *Turdus viscivorus*

Mistle Thrush 🔊 65

white underwings

ADULT

pale rump

IN FLIGHT

small, rounded head

slender neck

bold dark eye in plain face

grey-brown back

bold black spots on pale creamy buff underside

ADULT

pale edges to dark wing feathers

very pale head

pale spots on back

whitish side to tail

spots wider, less V-shaped, more evenly spread than on Song Thrush

JUVENILE

ADULT

A large, bold, aggressive thrush, the Mistle Thrush is by far the largest of the "spotted" thrushes and also Europe's biggest true songbird. It is usually found in pairs, but families join up in larger groups during autumn when berries are abundant. In winter, single Mistle Thrushes often defend berry-laden trees against other birds, maintaining a food supply through the colder months. While Song Thrushes often slip away at low level if disturbed, Mistle Thrushes tend to go up to a much greater height and fly off over greater distances.
VOICE Loud, slurred, harsh, rattling chatter *tsairrrk-sairr-sairr-sairrk*; song loud, wild, fluty, not very varied, series of short, repetitive phrases.
NESTING Big, loose cup of roots, leaves, twigs, and grass, often quite exposed high on tree branch; 3–5 eggs; 2 broods; March–June.
FEEDING Bounding hops on ground, searching for worms, seeds, and invertebrates; eats many berries, sometimes coming to larger gardens for fruit.

FLIGHT: strong, direct, sometimes undulating with long swoops between bursts of wingbeats; often high and far-ranging.

PALE THRUSH
Against the dark foliage of a conifer, this bird may look very pale.

OCCURRENCE
Breeds in most of Europe except extreme N; summer visitor in N and E Europe. In parkland, farmland with tall trees, orchards, on edges of moorland near mature forest, woodland clearings, and lower scrub, often feeding on open grassland and large lawns.

Seen in the UK
J F M A M J J A S O N D

SIMILAR SPECIES

SONG THRUSH see p.373
plainer above
smaller
more aligned "V"-spots

FIELDFARE see p.372
grey head
brown back
black tail

BLACKBIRD ♀; see p.371
smaller and darker
no bold spots

Length **27cm (10½in)**	Wingspan **42–48cm (16½–19in)**	Weight **110–140g (4–5oz)**
Social **Winter flocks**	Lifespan **5–10 years**	Status **Secure**

| Order **Passeriformes** | Family **Turdidae** | Species **Erithacus rubecula** |

Robin  60.I, 60.II

big black eye

small, fine
dark bill

bluish grey
sides of neck
and chest

orange-red
breast

soft, warm- to
olive-brown
upperparts

olive-buff
underside

**ADULT
IN FLIGHT**

spindly
brown legs

wingtips
characteristically
drooped

ADULT (SPRING)

mottled
brown
body

red
blotches
appear

JUVENILE

pale orange-red to
washed-out orange
face and breast

ADULT (WINTER)

This is a typical forest or woodland-edge species in most of its range, adapted to follow foraging animals such as wild boar: it picks up food from the earth overturned by the animals. In the UK, it follows gardeners turning the soil, and has become remarkably tame. In most of continental Europe, Robins are much shyer birds. Robins are easily identified (although juveniles have no red at first) and have a distinctive song.

VOICE Sharp, short, abrupt *tik*, series of quick *tik-ik-ik-ik-ik*, high, thin *seeep*; song rich, full, varied warbling in long, musical series, some phrases like Garden Warbler's (see p.338); in autumn/winter more mellow, melancholy.

NESTING Domed nest of leaves and grass in bank, dense bush or hedge, or thick ivy; 4–6 eggs; 2 broods; April–August.

FEEDING Mostly on ground, hopping and flitting in search of spiders, insects, worms, berries, and seeds; comes to feeders and bird-tables for seed mixtures.

FLIGHT: short, flitting darts into cover; longer flights weak, flitting, with bursts of wingbeats.

IN A DIFFERENT LIGHT
In some lights, the white breast spot below the red bib may be very obvious.

OCCURRENCE
Widespread, but absent from Iceland; summer visitor in N and E Europe. In all kinds of forest, especially more open woodland, as well as on bushy heaths, in gardens with hedges and shrubberies, and in town parks.

Seen in the UK
| J | F | M | A | M | J | J | A | S | O | N | D |

SIMILAR SPECIES

DUNNOCK
see p.390

*much greyer
below*

*redder
on tail*

NIGHTINGALE
see p.378

REDSTART ♀
see p.381

*redder
on tail*

| Length **14cm (5½in)** | Wingspan **20–22cm (8–9in)** | Weight **16–22g (9/16– 13/16 oz)** |
| Social **Family groups** | Lifespan **3–5 years** | Status **Secure** |

Order **Passeriformes**	Family **Turdidae**	Species **Luscinia luscinia**

Thrush Nightingale

greyish back

dull tail

IN FLIGHT

pale eye-ring

softly mottled breast

ADULT (FRONT VIEW)

plain head

soft, pale, buff-brown and greyish olive

pale chest, mottled with olive-grey

dull rusty-brown tail

ADULT

FLIGHT: low, short, flitting flights between bushes.

Very like the Nightingale, the Thrush Nightingale replaces the more familiar bird in the north and east of Europe. It is variably mottled beneath and generally duller, without such a rusty-red tail as its southwestern counterpart. Equally difficult to see in dense undergrowth, it has a riveting song, full of loud, fast, rich phrases, which draws attention to it.

VOICE Song rich, throbbing, varied warble of varying pace; lacks the typical crescendo of Nightingale, with more clicking and throaty gurgling notes, less pure, beautiful quality as a whole.

NESTING Nest of grass and stems in dense bush close to ground; 4 or 5 eggs; 1 brood, May–June.

FEEDING Mostly under deep cover on the ground, eating insects, grubs, berries, and a few seeds.

HIDDEN SONGSTER
Males sing from deep within thickets and dense woodland, they bear a strong resemblance to a Nightingale but have a slightly less varied song, with fewer high, pure notes.

OCCURRENCE
Summer visitor to E Europe north to S Scandinavia, Finland, south to Black Sea; in dense, often damp woods and copses, waterside bushes, rarely gardens; vagrant in the UK and W Europe.

Seen in the UK
J F M **A M J J A** S O N D

SIMILAR SPECIES

NIGHTINGALE see p.378
plainer breast
brighter tail

ROBIN juvenile, similar to juvenile; smaller; see p.376
pale spots on back
more mottled below

GARDEN WARBLER greyish neck patch; see p.338
duller tail

Length **15–17cm (6–7in)**	Wingspan **23–26cm (9–10in)**	Weight **18–27g (⅝–1oz)**
Social **Solitary/Family groups**	Lifespan **Up to 5 years**	Status **Secure**

Order **Passeriformes**	Family **Turdidae**	Species *Luscinia megarhynchos*

Nightingale 🔊 61

IN FLIGHT

rufous tail

plain brown wings

ADULT

tail often raised

long undertail coverts give effect of thick rear body

warm brown back

pale ring around large dark eye

strong pinkish legs

ADULT

spotted above

rufous tail

JUVENILE

grey on side of neck

clean grey-buff underside

bright rump

MALE (SINGING)

With one of the finest songs in Europe, the Nightingale is easy to find when singing but otherwise difficult to locate and usually hard to see well. It skulks in thick vegetation, often close to the ground, although with a little patience a clear view can sometimes be obtained. In places, it sings much more openly, but is likely to drop out of sight immediately if approached too closely. Although quite plain in appearance, its identification is usually straightforward.

VOICE Calls include low, mechanical, grating *kerrr*, loud, bright *hweet*; song brilliant but unstructured, very varied, some phrases extremely fast with sudden change from high to low pitch; long, slow, plaintive notes build to sudden throaty trill.

NESTING Cup of grass and leaves in dense bushy cover close to ground; 4 or 5 eggs; 1 brood; May–June.

FEEDING Forages in clear spaces under dense, dark cover, in ditches, and under thickets for worms, larvae, beetles, and berries.

FLIGHT: low, short, flitting, with wings and tail fanned briefly as it dives into cover.

POWERFUL SONG
Males sing more consistently at dawn and dusk, and bursts of song are often intermittent by day.

OCCURRENCE
In Europe from April to August; common in S Europe, scarce in NW as far as England and Germany. Breeds in many kinds of thickets that are dense to ground level, from bushy gullies and overgrown gardens to woodland with bushy (especially coppiced) undergrowth and clumps of dense bushes on heaths.

Seen in the UK
J F M **A M J J A** S O N D

SIMILAR SPECIES

THRUSH NIGHTINGALE see p.377

mottled on chest

duller

ROBIN juvenile, similar to juvenile; see p.376

smaller

GARDEN WARBLER see p.338

duller

smaller

Length **16–17cm (6½in)**	Wingspan **23–26cm (9–10in)**	Weight **18–27g (⅝–¹⁵/₁₆ oz)**
Social **Solitary**	Lifespan **Up to 5 years**	Status **Secure†**

| Order **Passeriformes** | Family **Turdidae** | Species *Luscinia svecica* |

Bluethroat

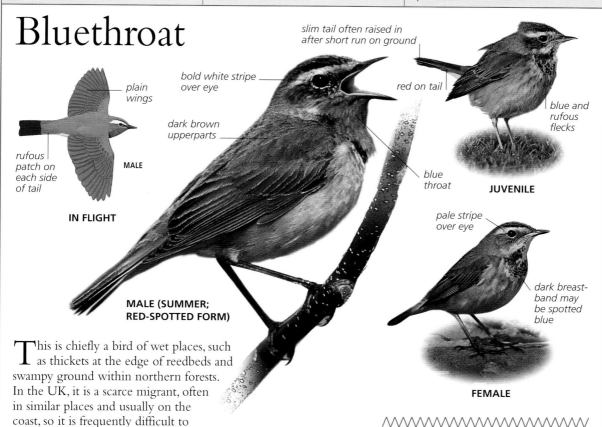

slim tail often raised in after short run on ground

bold white stripe over eye

plain wings

red on tail

dark brown upperparts

blue and rufous flecks

rufous patch on each side of tail

MALE

IN FLIGHT

blue throat

JUVENILE

MALE (SUMMER; RED-SPOTTED FORM)

pale stripe over eye

dark breast-band may be spotted blue

FEMALE

This is chiefly a bird of wet places, such as thickets at the edge of reedbeds and swampy ground within northern forests. In the UK, it is a scarce migrant, often in similar places and usually on the coast, so it is frequently difficult to watch. Not being a regular garden bird it is often overlooked, despite its unusual looks. It also sounds remarkable, its rich song including excellent mimicry. Its general form and actions recall the Robin.

VOICE Sharp, hard *tak*, softer *wheet*, often with hard note as *wheet-turrc*; song powerful, bright, musical, accelerating into melodious outburst with much mimicry.

NESTING Small grassy cup in low bush; 5–7 eggs; 1 brood; May–June.

FEEDING Forages on ground close to cover, picking up seeds, insects, and berries.

REMARK Subspecies *L. s. svecica* (North Europe) has rufous central breast spot; *L. s. magna* (Eastern Europe) has no spot.

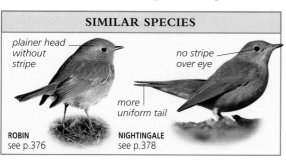

FLIGHT: low, quite quick, flitting, usually short distance into nearby cover.

MUSICAL SONG
Males vigorously sing from concealed perches in dense waterside thickets and low bushes.

OCCURRENCE
Breeds locally in France, Low Countries, Scandinavia, and NE and C Europe. Prefers wet thickets, moist woods, heaths, and bushes on tundra. On migration, a few appear west of usual range, in coastal thickets and reeds during March to October.

| Seen in the UK |
| J F M **A M** J J A **S O** N |

SIMILAR SPECIES

plainer head without stripe

no stripe over eye

more uniform tail

ROBIN
see p.376

NIGHTINGALE
see p.378

SUBSPECIES

L. s. cyanecula (S and C Europe)

white central breast spot

| Length **14cm (5½in)** | Wingspan **20–22cm (8–9in)** | Weight **15–23g (⁹⁄₁₆–¹³⁄₁₆oz)** |
| Social **Solitary** | Lifespan **Up to 5 years** | Status **Secure** |

CHATS AND THRUSHES

| Order **Passeriformes** | Family **Turdidae** | Species *Phoenicurus ochruros* |

Black Redstart 🔊 62

browner head than male

sooty or mousy grey body

dark rust-red tail with darker centre

MALE

IN FLIGHT

pale eye-ring

FEMALE

rust-red above and on sides of tail

pale grey body (Redstart shows more buff)

IMMATURE (WINTER)

grey cap

sooty grey upperparts

white panel on wings

black below

MALE (SPRING)

FLIGHT: quick, agile, darting through small spaces, dashing across roofs, with bursts of flicking wingbeats.

A bird of rocky slopes with scree and crags, or deep gorges, the Black Redstart also readily occupies towns and villages with older buildings that offer holes in which to nest and rough or waste ground where it can feed. It may move into industrial sites and run-down areas of larger cities. In winter, a few use quarries and rocky coves along the coast, but most prefer rough ground, from new building sites to derelict land with brick and concrete rubble.

VOICE Call hard, rattling or creaky notes, short *tsip, tucc-tucc, titititic*; song hesitant warble with strange stone-shaking trills and rattles, carries far in town or on cliff where elusive.

NESTING Grassy nest in hole in building, on ledge, in cavity in cliff or fallen rocks; 4–6 eggs; 2 broods; May–July.

FEEDING Leaps and flies after insects, drops to ground from high perch or boulder to pick up beetles, grubs, worms, some berries, and seeds.

IMMATURE MALE
Males often sing and sometimes breed while still in immature grey plumage.

OCCURRENCE
Breeds in most of Europe except Iceland, N Great Britain (very rare in S), and N Scandinavia, in towns and villages with old buildings, cities and derelict industrial areas, cliffs, gorges, mountain areas, and often in old quarries along coast. In winter, often in quarries or along rocky coast.

Seen in the UK
| J | F | M | A | M | J | J | A | S | O | N | D |

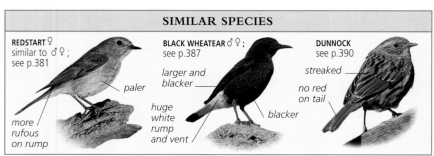

SIMILAR SPECIES

REDSTART ♀ similar to ♂♀; see p.381

paler

more rufous on rump

BLACK WHEATEAR ♂♀; see p.387

larger and blacker

huge white rump and vent

blacker

DUNNOCK see p.390

streaked

no red on tail

| Length **14.5cm (5¾in)** | Wingspan **23–26cm (9–10in)** | Weight **14–20g (½–11⁄16oz)** |
| Social **Family groups** | Lifespan **Up to 5 years** | Status **Secure** |

380

Redstart

dark centre on pale rust-red tail

IN FLIGHT

whitish feather tips obscuring most of dark face

MALE (AUTUMN)

plain head

pale buff underside

rusty tail

FEMALE

bold white forehead

bluish grey from crown to back

black face and throat

slim body

rich orange-rufous underside

slim black legs

rust-red rump

MALE (SPRING)

Robin-like but slim and slender-tailed, the Redstart constantly flickers its tail up and down. Spring males are extremely handsome birds, best located by following up their short, sweet song. A woodland bird, the Redstart prefers to nest in old woods with plenty of space beneath the canopy in which it can feed. As a migrant, it often appears in thickets and bushes, both inland (often in willow thickets beside lakes and reservoirs) and more commonly along the coast.

VOICE Like Chaffinch, clear, rising *wheet* or *hueeee*; *huee-tic*; sharp *tac*; song brief, bright, musical warble, often beginning with several low, rolling notes, finishing in weak trill.

NESTING Grassy nest lined with feathers and hair, in hole or nest box; 5–7 eggs; 1 brood; May–June.

FEEDING In foliage or drops to ground, finding insects, spiders, caterpillars, small worms, and some berries.

FLIGHT: quite quick and agile, flitting from tree to tree; often drops briefly to ground.

AUTUMN PLUMAGE
Autumn females and juveniles have pale wingbars and a bright buffy appearance overall.

OCCURRENCE
Breeds in most of Europe (except Iceland and Ireland), in open woodland or old woods with little undergrowth, some in scattered trees on rocky slopes. Present from April to October, moving to Africa in winter. Migrants are often near the coast or in willow thickets beside reservoirs inland.

Seen in the UK
J F **M A M J J A S O** N D

SIMILAR SPECIES

BLACK REDSTART ♀
similar to ♀; less red on rump; see p.380

darker above

ROBIN
see p.376

plain brown tail

NIGHTINGALE
see p.378

larger

uniform tail

greyer below

Length **14cm (5½ in)**	Wingspan **20–24cm (8–9½ in)**	Weight **12–20g (7/16–11/16 oz)**
Social **Solitary**	Lifespan **Up to 5 years**	Status **Vulnerable**

| Order **Passeriformes** | Family **Turdidae** | Species *Monticola saxatilis* |

Rock Thrush

IN FLIGHT

- white on back (dark on female)
- dark brown wings
- MALE
- rust-orange tail with thin dark centre

JUVENILE

- pale bars above
- dark spots on rufous breast

MALE (SPRING)

- stout, spiky bill
- pale spots wear off in summer
- powder-blue head and neck
- rich rust-orange underside (orange buff on female with narrow dark bars)
- white bars wear off in summer
- strong dark legs

∧∧∧∧∧∧∧∧∧∧∧∧∧∧∧∧∧∧∧∧

FLIGHT: strong, direct, quick, with bursts of wingbeats; fluttery song-flight.

A small thrush with a short tail and stocky body, the Rock Thrush is characteristic of high alpine pastures, rocky slopes, and small upland fields with stone walls. It perches on boulders, poles, overhead wires, and other prominent places and so may be relatively easy to see. Its song-flight also catches the eye, although finding a small bird in a habitat that is generally so open and expansive can be difficult. Identifying it is usually simple enough: males especially are quite striking.

VOICE Squeaky *whit* and hard *chak*; song fluty, soft, Blackbird-like rich, musical warble, descending, often in song-flight.

NESTING Grassy cup in hole in wall or cavity among rocks or scree; 4 or 5 eggs; 1 brood; May–June.

FEEDING Looks for food from high perch, dropping down onto insects, small reptiles, and worms; also eats berries and seeds.

SUMMER COLOURS
Fresh feathers have whitish tips, but by mid-summer, these wear off to create a more uniform appearance.

SIMILAR SPECIES

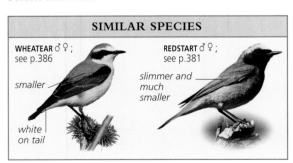

WHEATEAR ♂♀; see p.386
- smaller
- white on tail

REDSTART ♂♀; see p.381
- slimmer and much smaller

OCCURRENCE
Found from March to September in S Europe north to Alps and Pyrenees, very rare outside this range. Breeds in high alpine meadows and on grassy slopes with boulders and crags; also on cliffs and in deep gorges.

| Seen in the UK |
| J F M A M J J A S O N D |

| Length **17–20cm (6½–8in)** | Wingspan **30–35cm (12–14in)** | Weight **50–70g (1¾–2½oz)** |
| Social **Family groups** | Lifespan **Up to 5 years** | Status **Declining†** |

Order **Passeriformes**	Family **Turdidae**	Species **Monticola solitarius**

Blue Rock Thrush

brownish wings

MALE

IN FLIGHT

dark brown tail with no trace of rufous

close, fine brown bars on pale face

plain brown back

close dark bars on pale underside

FEMALE

rich, bright blue head

long, thick, spike-like bill

dark blue body, looks blackish at distance

quite short, stout legs

JUVENILE

MALE

Dark, but intensely blue in good light and at close range, the Blue Rock Thrush is mostly found around cliffs and deep mountain gorges; it also occurs around coastal villages and developments, perching freely on roofs and wires. In more remote places, it is often best located by its loud song. Females are more problematical than males but equally distinctive if seen well. The thrush-like silhouette, with a particularly long, thick, spike-like bill, is a useful feature.

VOICE Deep, thrush-like *chook*, higher squeaky notes; song rich, musical warbling, melancholy, Blackbird-like, carrying far across gorges and cliffs.
NESTING Grassy cup in hole in wall or cliff or under rocks; 4 or 5 eggs; 1 or 2 broods; May–July.
FEEDING Picks insects, spiders, worms, lizards, berries, and seeds from ground.

FLIGHT: swift, direct; flappy over long distance, recalling Blackbird.

ELUSIVE ON CLIFFS
The blue male is handsome when seen well but is surprisingly difficult to spot against rocks.

OCCURRENCE
Breeds in Spain, Portugal, and Mediterranean area, where mostly resident. In gorges, rocky areas with crags and boulders in mountains, and also around buildings and quarries, often near coasts.

Seen in the UK
J F M A M J J A S O N D

SIMILAR SPECIES

BLACKBIRD ♂ ♀; ♀ more uniform; see p.371

blacker

ROCK THRUSH juvenile, similar to juvenile, ♀; see p.382

smaller and brighter

shorter rufous tail

SPOTLESS STARLING ♂ ♀; see p.366

stockier

dark and glossy

shorter tail

Length **21–23cm (8½–9in)**	Wingspan **35–40cm (14–16in)**	Weight **60–80g (2⅛–2⅞oz)**
Social **Family groups**	Lifespan **5–10 years**	Status **Vulnerable†**

Order **Passeriformes**	Family **Turdidae**	Species **Saxicola rubetra**

Whinchat

FEMALE (SPRING)

IN FLIGHT

white triangle on each side of tail

almost black cheeks, edged white

deep apricot breast

pale yellowish underside

MALE (SPRING)

bold pale stripe over eye

dark cap

black streaks on pale straw-brown back

pale throat

MALE (AUTUMN)

buff stripe over eye (stronger than any Stonechat's)

yellow-buff underside

IMMATURE (1ST WINTER)

buff stripe over eye

streaked cap and cheeks

FEMALE (SPRING)

Declining in much of its range as its preferred habitats are altered, the Whinchat is a bird of grassland with tall, woody stems or small bushes which give it an open perch quite close to the ground. Young conifer plantations serve it well for a few years but "rough", uncultivated ground is often not tolerated in agricultural or suburban areas and so the Whinchat is squeezed out. It is a summer visitor to Europe, unlike the similar but resident Stonechat.

VOICE Loud, short, *tictic* or *tuc-tuc-tuc*, *wheet* or *wheet-tuk*; song varied, at best Robin-like but with clicks, rattles, and grating notes mixed in.

NESTING Grassy nest low in tussock, bush, or ground; 5 or 6 eggs; 1 or 2 broods; May–July.

FEEDING Watches from perch and drops to ground to pick up insects and worms; also feeds on some seeds and berries.

FLIGHT: short, low, flitting, quite quick, usually on to isolated, slender upright stem or fence.

SPRING PERFECTION
A male in spring is a superbly patterned, crisply coloured bird.

OCCURRENCE
Present from April to September in open places with heather, grass, scattered taller stems, or young trees. Widespread but increasingly scarce and localized; absent from Iceland. Migrants typically rest near coasts in open grassy or marshy places.

Seen in the UK
J F M **A M J J A S O** N D

SIMILAR SPECIES

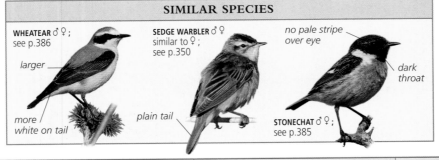

WHEATEAR ♂♀; see p.386

larger

more white on tail

SEDGE WARBLER ♂♀ similar to ♀; see p.350

plain tail

no pale stripe over eye

dark throat

STONECHAT ♂♀; see p.385

Length **12.5cm (5in)**	Wingspan **21–24cm (8½–9½in)**	Weight **16–24g (⁹⁄₁₆–⁷⁄₈oz)**
Social **Family groups**	Lifespan **Up to 5 years**	Status **Secure**

Stonechat 🔊 63

large white wing patch

pale rump

IN FLIGHT

MALE (SUMMER)

pale over eye and paler throat may recall Whinchat

dark-edged, less well-marked lower throat than Whinchat

black streaks on brown back (back duller in winter)

stocky shape

FEMALE

blackish head and throat

white patch on sides of neck

rust-red breast

paler belly

short blackish tail (Whinchat's has white shades)

slim black legs

MALE (SUMMER)

pale throat

mottled chest

JUVENILE

Small, chunky, and upright, the Stonechat often perches on the tops of bushes or on overhead wires in otherwise rather open places. Heaths, upland moors, and stretches of grassland with gorse thickets above coastal cliffs are perfect for it; in winter, it may move to the coast to escape hard weather, and inland breeding populations may temporarily disappear after bad winters. Migrant Siberian Stonechats are more like pale Whinchats in some respects but resident Stonechats are usually easily identified.

VOICE Hard, scolding *tsak* or *tsak-tsak*, sharp *wheet*, often *wheet-tak-tak*; song sometimes in flight, rapid, chattery warble with some chattering and grating notes, less musical than Whinchat's.

NESTING Grassy cup, lined with hair and feathers, often in dense grass with entrance tunnel; 5 or 6 eggs; 2 broods; May–July.

FEEDING Drops to ground to pick up insects, spiders, worms, and some seeds; catches some insects in flight, returning to eat them on a perch.

FLIGHT: low, quick, direct, whirring, onto prominent perch.

SUBSPECIES

S. t. maura (Siberia); clear rufous rump

pale line over eye

white feather fringes

SIMILAR SPECIES

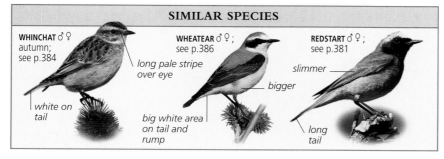

WHINCHAT ♂♀ autumn; see p.384

long pale stripe over eye

white on tail

WHEATEAR ♂♀; see p.386

bigger

big white area on tail and rump

REDSTART ♂♀; see p.381

slimmer

long tail

OCCURRENCE

Breeds in most of Europe except Iceland, Scandinavia, and NE Europe; rather scarce inland in NW Europe. Likes open places with gorse, heather, and bushes, on heaths or above coastal cliffs, and dunes.

Seen in the UK											
J	F	M	A	M	J	J	A	S	O	N	D

Length **12.5cm (5in)**	Wingspan **18–21cm (7–8½in)**	Weight **14–17g (½–⅝oz)**
Social **Family groups**	Lifespan **Up to 5 years**	Status **Declining†**

Order **Passeriformes**	Family **Turdidae**	Species **Oenanthe oenanthe**

Wheatear

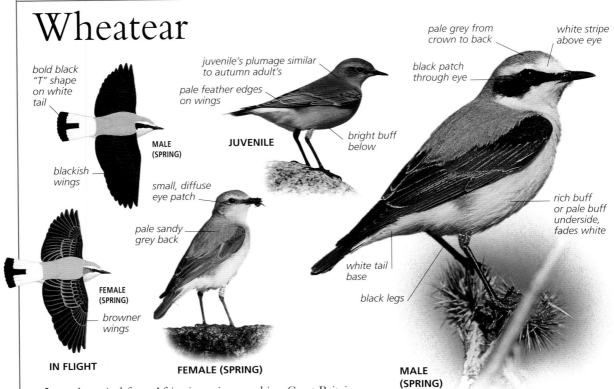

bold black "T" shape on white tail

blackish wings

juvenile's plumage similar to autumn adult's

pale feather edges on wings

JUVENILE

bright buff below

pale grey from crown to back

white stripe above eye

black patch through eye

rich buff or pale buff underside, fades white

small, diffuse eye patch

pale sandy grey back

white tail base

black legs

MALE (SPRING)

FEMALE (SPRING)

browner wings

IN FLIGHT

FEMALE (SPRING)

MALE (SPRING)

An early arrival from Africa in spring, reaching Great Britain in early March, the Wheatear breeds in open areas with grassy places on which it feeds adjacent to scree, stone walls, crags, or, more rarely, holes in sandy ground in which to nest. It is a frequent migrant outside its breeding areas, turning up along coasts, on farmland, and on grass beside reservoirs. Very much a terrestrial bird, it usually avoids trees and bushes. It frequently flies ahead of people, not going far, and revealing its distinctive white rump each time it moves.

VOICE Hard *chak-chak*, bright *wheet-chak-chak*; song long, rambling, quick warble with rolling, scratchy notes, often in song-flight.

NESTING Grassy cup, in hole in ground, rabbit burrow, or under fallen rocks, in stone wall; 5 or 6 eggs; 1 or 2 broods; April–July.

FEEDING Bouncy hops, short runs, on open ground, picking up insects and spiders; sometimes catches flies in flight or with sudden short, fluttery leap.

FLIGHT: low, flitting, quite strong, undulating; sometimes swoops up to perch.

ROCK HOPPER
The Wheatear has strong legs and feet, ideal for the rocky habitat in which it is often found.

SUBSPECIES

O. o. leucorhoa
(Greenland; W Europe in spring)

richer colours

larger

SIMILAR SPECIES

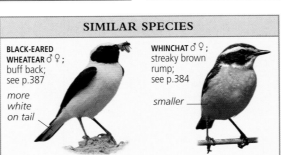

BLACK-EARED WHEATEAR ♂♀; buff back; see p.387

more white on tail

WHINCHAT ♂♀; streaky brown rump; see p.384

smaller

OCCURRENCE
Breeds almost throughout Europe but very local, where open grassy or heathy ground is mixed with scree, boulders, and cliffs, typically at rather high altitude or in hilly areas. Migrants encountered on fields, grassy areas, dunes, and golf courses, from March to October.

Seen in the UK
J F **M A M J J A S O** N D

Length **14.5–15.5cm (5¾–6in)**	Wingspan **26–32cm (10–12½in)**	Weight **17–30g (⅝–1¹⁄₁₆oz)**
Social **Small flocks**	Lifespan **Up to 5 years**	Status **Secure**

Order **Passeriformes**	Family **Turdidae**	Species ***Oenanthe hispanica***

Black-eared Wheatear

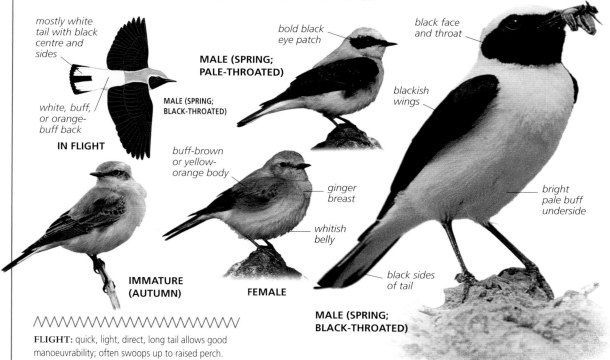

mostly white tail with black centre and sides

MALE (SPRING; PALE-THROATED)

bold black eye patch

black face and throat

white, buff, or orange-buff back

MALE (SPRING; BLACK-THROATED)

IN FLIGHT

blackish wings

buff-brown or yellow-orange body

ginger breast

bright pale buff underside

whitish belly

IMMATURE (AUTUMN)

FEMALE

black sides of tail

MALE (SPRING; BLACK-THROATED)

FLIGHT: quick, light, direct, long tail allows good manoeuvrability; often swoops up to raised perch.

Slimmer, lighter, and less solid than a Wheatear, the Black-eared Wheatear occurs in two forms – black-throated and pale-throated – and requires careful identification in plumages other than the spring male's, which is bold and striking. It combines some of the actions of the smaller chats with the typical behaviour of ground-feeding wheatears. This southern European wheatear is common on warm, stony Mediterranean slopes, readily perching on low bushes and tall stems.

VOICE Wheezy *tssch*, hard *tack*; song fast, rattling warble, quite bright and explosive, sometimes with mimicry.

NESTING Grassy cup in hole, under boulders or stones, or at base of bush; 4 or 5 eggs; 1 or 2 broods; April–June.

FEEDING Watches from bush top or stone, drops to ground, or chases after insects; eats some small seeds.

REMARK Subspecies *O. h. hispanica* (SW Europe) has less black on face and throat and yellower back; *O. h. melanoleuca* (S Italy, Balkans) has more black on face, whiter back, and longer wingtips. Both have black- and pale-throated forms.

SLENDER FORM
This is a slim, elegant wheatear, often perching on bushes, using its tail to maintain its balance.

OCCURRENCE
Very locally in Spain, Portugal, and Mediterranean countries, in variety of open, often barren places with scattered bushes, rocks, and high stony pastures, from March to October. Only rare vagrant farther north in spring or autumn.

Seen in the UK
J F M A M J J A S O N D

SIMILAR SPECIES

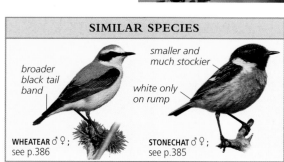

broader black tail band

smaller and much stockier

white only on rump

WHEATEAR ♂♀; see p.386

STONECHAT ♂♀; see p.385

Length **13.5–15cm (5¼–6in)**	Wingspan **25–30cm (10–12in)**	Weight **15–25g (⁹⁄₁₆–⅞ oz)**
Social **Family groups**	Lifespan **Up to 5 years**	Status **Vulnerable**

Order **Passeriformes**	Family **Turdidae**	Species *Oenanthe leucura*

Black Wheatear

wings look paler in flight

ADULT

black "T"-shaped mark on white tail

IN FLIGHT

black body

stout black bill

body duller and browner than male's

FEMALE

white vent

strong blackish legs

mostly white tail

FLIGHT: quite strong, often low; fast up and down slopes or across cliff faces.

MALE (SPRING)

One of the larger wheatears of the region, the Black Wheatear is also more of a resident than the others. It is declining in some northern parts of its range. It prefers rocky or stony ground, and is often on or around the base of sheer cliffs, being surprisingly inconspicuous in the strong light and shade of scree slopes or boulders. If it flies, however, its very large and striking white rump and tail become immediately obvious.

VOICE Bright, whistled *pewp*, hard *tet-tet*; song low, rich or lighter, harsh twittering, sometimes in song-flight.

NESTING Grassy cup in hole in ground, rabbit burrow, under fallen rocks, or in stone wall; 5 or 6 eggs; 1 or 2 broods; April–July.

FEEDING Forages on ground, moving up and down slopes in short flights; swoops from perches on insects and spiders.

UPRIGHT STANCE
Like all wheatears and chats, the Black Wheatear has a bold, upright posture on strong legs, and moves with quick, leaping hops.

OCCURRENCE
Breeds on slopes with rocks and scree from Pyrenees southwards through Spain and Portugal. Resident there but does not move outside this restricted range. Often seen perched on cliffs, crags, and boulders or feeding on patches of short grass.

Seen in the UK
J F M A M J J A S O N D

SIMILAR SPECIES

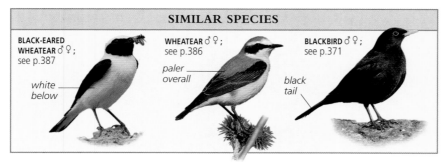

BLACK-EARED WHEATEAR ♂♀; see p.387

white below

WHEATEAR ♂♀; see p.386

paler overall

BLACKBIRD ♂♀; see p.371

black tail

Length **16–18cm (6½–7in)**	Wingspan **30–35cm (12–14in)**	Weight **25–35g (⅞–1¼oz)**
Social **Family groups**	Lifespan **Up to 5 years**	Status **Endangered**

Family **Prunellidae**

ACCENTORS

SMALL, SLIM–BILLED, shuffling ground birds, accentors are often overlooked. The Dunnock, however, is common in many habitats and a frequent garden songster that deserves to be a more popular favourite.

ALPINE ACCENTOR
Accentors make up a very small family of sharp-billed birds, found only in Europe and Asia. The Alpine Accentor is a high mountain bird.

PRUNELLIDAE

black streaks on back

broad red-brown streaks on flanks

DUNNOCK
p.390

ALPINE ACCENTOR
p.391

Family **Muscicapidae**

FLYCATCHERS

THERE ARE TWO GROUPS OF flycatchers: one includes basically brown ones (Spotted, Red-breasted), the other, birds in which the summer male is black and white (the "pied" group). They are short-billed, upright, long-winged, short-tailed birds. Male and female Spotted Flycatchers are alike, but in the others summer males are very different from female and winter plumages. All are migrants, the Spotted Flycatcher being one of the last to arrive in spring. They mainly catch flies on the wing, although the pied group also drop to the ground; Spotted Flycatchers characteristically fly out and return to the same perch.

MUSCICAPIDAE

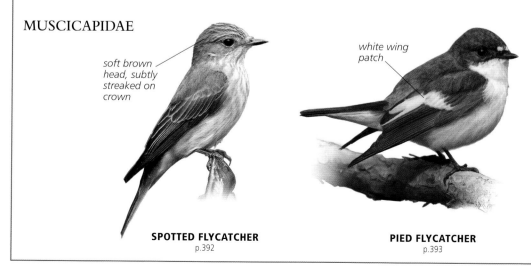

soft brown head, subtly streaked on crown

white wing patch

SPOTTED FLYCATCHER
p.392

PIED FLYCATCHER
p.393

389

Order **Passeriformes** | Family **Prunellidae** | Species *Prunella modularis*

Dunnock  59

black streaks on
rich brown back
and wings

brown cheeks

streaked
brown cap

brown eye

fine dark
bill

black and brown
streaks on back

mid-grey
breast

grey
throat

soft, warm
brown streaks
on flanks

IN FLIGHT

line of pale
spots across
wings

ADULT

uneven, grey
and brown
streaking below

JUVENILE

orange-
brown
legs

ADULT

The generally quite dull and unobtrusive looks of the Dunnock bely its unusual mating habits: it forms one-male-two-female or even one-female-two-male trios when breeding. Easily overlooked, it is widely distributed in a great variety of places, like the Wren. Its sharp calls and fast, high-pitched but slightly "flat" song call attention to it. If disturbed, it generally flies at near ground-level into the nearest thick bush, and is sometimes taken for some rare, vagrant warbler.

VOICE Loud, high, penetrating *pseeep*, thin, vibrant *teeee*; song quick, slightly flat, high-pitched, fast warble with little contrast or variation in pitch.

NESTING Small grassy cup, lined with hair and moss, in bush or hedge; 4 or 5 eggs; 2 or 3 broods; April–July.

FEEDING Forages on ground, shuffling, crouched, often in or around bushes, close to cover; picks up small insects and seeds; feeds on scraps beneath bird-tables and grated cheese under shrubberies.

FLIGHT: short, flitting, whirring; wings round, tail quite long.

SHUFFLING FEEDER
Dunnocks creep forward, flicking their tails, picking food from the ground, without the bouncy hop of House Sparrows.

OCCURRENCE
Breeds throughout Europe except in Iceland. Present only in summer in N and E Europe; in some parts of S Europe in winter. Widespread, on heaths and moors with low, dense scrub and exposed coastal areas as well as in higher forest, bushy gardens, ornamental flowerbeds, and parks.

Seen in the UK
J F M A M J J A S O N D

SIMILAR SPECIES

ROBIN juvenile;
see p.376

no black
streaks

more buff
below

WREN
see p.364

short
tail

barred

MEADOW PIPIT
see p.406

paler

smaller

streaked
buff
underside

Length **14cm (5½in)** | Wingspan **19–21cm (7½–8½in)** | Weight **19–24g (¹¹⁄₁₆–⁷⁄₈oz)**
Social **Family groups** | Lifespan **Up to 5 years** | Status **Secure**

Alpine Accentor

dark midwing panel, edged with white spots

grey head

yellow-based dark bill

dark streaks on pale grey-brown back

dark band on closed wings

short, rounded wings

IN FLIGHT

dark tail

streaked, dull buffish or grey underside

broad red-brown streaks on flanks

strong colours obvious only in close views

Where it is moderately common, the Alpine Accentor can be found by searching high mountain slopes with mixed pastures and rock, or in almost entirely rocky places at high altitude. In areas where it is more thinly spread, locating it can be quite difficult. In winter, accentors move to lower levels and turn up quite regularly at traditional sites outside the breeding range, usually hilltops, but also centred on old buildings such as castle complexes on rocky outcrops. They can be quite tame in winter, but are often shy and elusive in summer in the mountains. In shape and behaviour, they resemble large Dunnocks.

VOICE Short, trilling or rolling calls, *trru, tschirr, drrp*; song erratic, uneven series of trills and squeaky notes, sometimes in flight.

NESTING Grassy nest in rock crevice or under rocks; 3–5 eggs; 2 broods; May–August.

FEEDING Forages on ground, finding insects, spiders, and seeds.

FLIGHT: quite strong, lark- or thrush-like with flurries of wingbeats, quick swoops.

OCCURRENCE
Breeds at high altitude mostly in Pyrenees and Alps, and locally in Italy and Balkans, usually on wide open slopes with short grass and plenty of boulders, or almost wholly rocky places. Sparse at lower altitude in winter, a few moving outside breeding range to Mediterranean islands.

Seen in the UK
J F M A M J J A S O N D

SIMILAR SPECIES

DUNNOCK greyer on breast; see p.390

plainer wings

less reddish flanks

CHARACTERISTIC MARKING
This Alpine Accentor on rocky ground reveals the obvious dark wing panel that is usually the most distinctive feature at a distance.

Length **15–17cm (6–6½in)**	Wingspan **22cm (9in)**	Weight **25g (⅞oz)**
Social **Family groups**	Lifespan **Up to 5 years**	Status **Secure**

Order **Passeriformes**	Family **Muscicapidae**	Species **Muscicapa striata**

Spotted Flycatcher

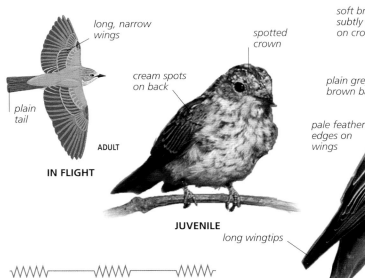

IN FLIGHT

long, narrow wings

plain tail

ADULT

cream spots on back

spotted crown

JUVENILE

long wingtips

bold dark eye

soft brown head, subtly streaked on crown

quite thick bill

plain grey-brown back

pale feather edges on wings

soft, pale grey-brown streaks on breast

silvery white underside

short black legs

ADULT

long, plain brown tail held downwards

FLIGHT: strong, quick, agile; swooping over long distances with bursts of wingbeats; catches flies with rapid twists, returning quickly to perch.

Many birds take the occasional fly in mid-air but flycatchers specialize in it: not in continuous flight, like swallows, but flying out from a perch and back again. This gives the Spotted Flycatcher a sharp-eyed, constantly alert appearance that is very appealing even if its plumage lacks strong colours or pattern. This slim, upright, short-legged bird is a late spring arrival from Africa, spreading out into places with "edge" habitats, such as woodland clearings, allotments, parks, churchyards, and tennis courts, when open space meets cover for nesting and somewhere to perch.

VOICE Short, slightly metallic or scratchy, unmusical *tzic* or *tzee*, *tzee-tsuk tsuk*; song of similar quality, short, scratchy, weak warble.

NESTING Cup of grass, leaves, moss, and feathers in creeper, old nest, cavity in wall, or open-fronted nest box; 3–5 eggs; 1 or 2 broods; June–August.

FEEDING Mostly catches insects in air, after flight from perch (from near ground level to treetop height); usually returns to same perch.

OPEN PERCH
A perch with a view of insects that catch the light as they cross sunlit clearings lets this flycatcher find a good meal.

SIMILAR SPECIES

GARDEN WARBLER
see p.338

shorter tail

PIED FLYCATCHER ♀;
see p.393

plainer wing

less upright

white stripe on wings

OCCURRENCE
Breeds almost throughout Europe except in Iceland, present from May to September and often a late arrival in spring. In open woodland, parkland, gardens with bushes and trees, and similar places.

Seen in the UK
J F M A **M J J A S O** N D

Length **14cm (5½ in)**	Wingspan **23–25cm (9–10in)**	Weight **14–19g (½–11⁄16 oz)**
Social **Solitary**	Lifespan **3–5 years**	Status **Declining**

Pied Flycatcher

MALE (SUMMER)

blackish wings with bold white patch

white sides to tail

IN FLIGHT

bold dark eye

white wing patch

dull brown and buffish white body

FEMALE

one or two spots on forehead

black and white plumage (in autumn, like female but retains forehead spots)

black tail with white sides

white underparts

dull brown back

white wing patch

pale throat; often with darker edge

blacker wings

short darker legs

JUVENILE

MALE (SUMMER)

Neither a garden bird nor as universally distributed as the Spotted Flycatcher, the Pied Flycatcher is a forest species, preferring space under the canopy in which it can feed, flying out for flies or dropping to the ground. Such places often have Redstarts (see p.382), Wood Warblers (see p.334), and Pied Flycatchers as a characteristic trio of small birds in summer. All become elusive after nesting, Pied Flycatchers almost "disappearing" for a time, although they are frequent migrants in coastal areas later in autumn.

VOICE Sharp *whit* or *whit-tic, wheet*; song brief, simple, musical phrase with notes clearly separated, slightly hesitant, ending with trill, *see, see, see sit, see-sit sitip-seweee.*

NESTING Cup of leaves and moss, in tree hole, old woodpecker hole or, by preference, nest box; 5–9 eggs; 1 brood; April–May.

FEEDING Catches flies in air and picks insects from foliage and from ground; also eats seeds and berries.

REMARK Subspecies *F. h. iberiae* (Spain) has pale rump, bigger forehead spot, bigger wingbar and primary patch.

FLIGHT: strong, quite bounding; often drops to ground, catches flies in air but usually goes on to different perch.

NEST BOX
A male feeds his chicks in a box specially provided: numbers are increased by nest box schemes in suitable woods.

OCCURRENCE
Breeds in UK and across most of mainland Europe. Seen from April to August in broadleaved woodland with clear space beneath canopy, often where there are Wood Warblers. Migrants often on coasts in autumn.

Seen in the UK
J F M **A M J J A S O** N D

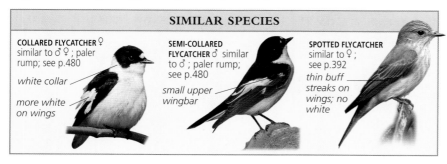

SIMILAR SPECIES

COLLARED FLYCATCHER ♀
similar to ♂♀; paler rump; see p.480

white collar

more white on wings

SEMI-COLLARED FLYCATCHER ♂ similar to ♂; paler rump; see p.480

small upper wingbar

SPOTTED FLYCATCHER
similar to ♀; see p.392

thin buff streaks on wings; no white

Length **13cm (5in)**	Wingspan **21–24cm (8½–9½in)**	Weight **12–15g (7/16–9/16oz)**
Social **Solitary**	Lifespan **3–5 years**	Status **Secure**

Family **Passeridae**

SPARROWS

Related more closely to the African weavers than to the similar finches, sparrows are adaptable and widespread birds, the House Sparrow being closely associated with people wherever they are.

Male and female plumages are markedly different in House and Spanish Sparrows, but identical in the Tree Sparrow (both looking more like a House Sparrow male). Rock Sparrows, in another genus altogether, are quite different, but have no obvious variations in their plumage.

All sparrows are social. Spanish Sparrows are especially gregarious where they are common.

House Sparrows used to gather in hundreds to feed on grain and weed seeds in winter stubbles, but have declined in most places in recent years. Breeding birds may be in loose colonies, using a variety of sites including the base of White Stork nests, thickets, and dense creepers on walls, but House Sparrows mostly occupy holes or cavities of some sort in roofs and under eaves. Renovation of wooden eaves with plastic cladding has often

FEEDING FLOCK
Sparrows burst from a stubble field with a loud roar of wings. They feed in tight-packed flocks, often mixed with Greenfinches and Linnets, which also pack closely together in many situations, while other finches and buntings form loose groups.

been blamed for local declines. Even inside a small cavity, the nest is a mass of grass and other stems, feathers, and scraps, with an overall rounded effect that recalls the weavers, but lacking the neatness and precision of construction that characterizes their nests. Both House and Tree Sparrows are suffering a widespread decline in western Europe.

PASSERIDAE

black bib (bigger in spring)

HOUSE SPARROW
p.395

rich brown cap

TREE SPARROW
p.397

large white cheek patch

brown tail tipped with whitish spots

SPANISH SPARROW
p.396

ROCK SPARROW
p.398

Order **Passeriformes**	Family **Passeridae**	Species ***Passer domesticus***

House Sparrow  91

IN FLIGHT

whitish wingbar

MALE

pale greyish rump

grey cap with red-brown sides

red-brown back

thick black bill

black bib (bigger in spring)

unmarked grey underside (grey-buff on female)

MALE (SUMMER)

plain plumage

IMMATURE (AUTUMN)

yellow-brown cap

pale stripe behind eye

plain brown tail

FEMALE

Recent declines have seen numbers in gardens and town parks reduce and farmland flocks disappear but House Sparrows remain almost as widespread as ever. They are familiar in towns and gardens and even in farmland concentrate around buildings. Males are easily recognized, although they look quite like Tree and Spanish Sparrows. Females lack the bold patterns and can be taken for some finches, but in reality are equally distinct. House Sparrows are typically social and noisy birds.

VOICE Basic lively, simple *chirrup*, *chilp*; loud chattering chorus from flocks; series of chirps forms simple song.

NESTING Untidy nest of grass and feathers, in roof space, cavity in wall, in House Martin nest (see p.326), tree cavity, or more openly in creepers and around wires; 3–7 eggs; 1–4 broods; April–August.

FEEDING Usually on ground, taking seeds, buds, roots, berries, and many insects for young; catches some insects clumsily in flight; visits bird-tables for nuts, seed mixtures, and kitchen scraps.

FLIGHT: fast, whirring, with bursts of rapid wingbeats; often in noisy flocks.

WINTER MALE
In winter, the red-brown band behind the eye and dark bib are partly obscured by pale feather tips.

OCCURRENCE
Resident almost throughout Europe, but only very local in Iceland. Thrives in villages, around farms, and in all kinds of habitats close to human habitation, but has declined recently on farmland, in towns, and in suburban gardens.

Seen in the UK
J F M A M J J A S O N D

SIMILAR SPECIES

CHAFFINCH ♀
similar to ♀ ;
see p.412

duller

white wingbars

SPANISH SPARROW ♂
similar to ♂ ; see p.396

big white cheeks

streaked below

SUBSPECIES

P. d. italiae
(Italy, Crete)

brown crown

white cheeks

Length **14cm (5½in)**	Wingspan **20–22cm (8–9in)**	Weight **19–25g (¹¹⁄₁₆–⁷⁄₈oz)**
Social **Flocks**	Lifespan **2–5 years**	Status **Secure**

Order **Passeriformes**	Family **Passeridae**	Species *Passer hispaniolensis*

Spanish Sparrow

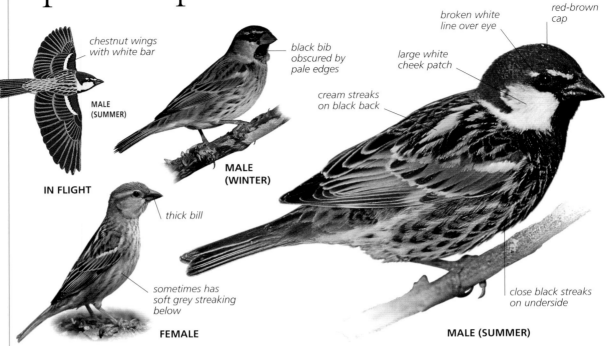

chestnut wings with white bar

MALE (SUMMER)

IN FLIGHT

black bib obscured by pale edges

MALE (WINTER)

cream streaks on black back

broken white line over eye

red-brown cap

large white cheek patch

thick bill

sometimes has soft grey streaking below

FEMALE

close black streaks on underside

MALE (SUMMER)

Much like the House Sparrow but, in summer, more brightly and strikingly patterned, the Spanish Sparrow is essentially a southeast European bird, and not actually common in Spain. Where House Sparrows are sparse it may take over that species' role in towns, but in many areas, it is a bird of farmland and damp places with willow thickets. It is highly social, sometimes found in large flocks and breeding in colonies. In Italy, the sparrows seem to be a constant form of hybrid between Spanish and House Sparrows.

VOICE Slightly higher and more metallic than House Sparrow's; loud chirruping, fast chorus from flock or colony.

NESTING Bulky nest of grass, in thicket, or in base of stick nest of stork or heron, often in tall willows or other wetland thicket, in colony; 3–7 eggs; 1 or 2 broods; April–July

FEEDING Eats seeds and berries, mostly from ground; feeds insects to chicks.

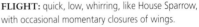

FLIGHT: quick, low, whirring, like House Sparrow, with occasional momentary closures of wings.

DIFFICULT TEST
This female is clearly streaked, but most female Spanish Sparrows are very difficult to distinguish from female House Sparrows.

OCCURRENCE
Local breeding bird in Spain, Sardinia, and Sicily, more frequent (but in summer only) in Balkans. In farmland, villages, and wet places with willow thickets and tall trees.

Seen in the UK
J F M A M J J A S O N D

SIMILAR SPECIES

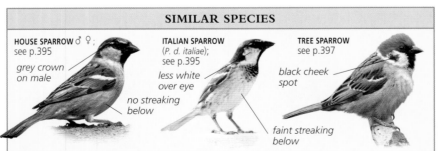

HOUSE SPARROW ♂ ♀; see p.395

grey crown on male

no streaking below

ITALIAN SPARROW (*P. d. italiae*); see p.395

less white over eye

TREE SPARROW see p.397

black cheek spot

faint streaking below

Length **14–16cm (5½–6½in)**	Wingspan **20–22cm (8–9in)**	Weight **20–25g (11/16–7/8oz)**
Social **Flocks**	Lifespan **2–5 years**	Status **Secure†**

Order **Passeriformes**	Family **Passeridae**	Species *Passer montanus*

Tree Sparrow

rich brown cap (duller on juvenile)

square black patch on white cheeks (less distinct on juvenile)

white collar

black mask and bib

two pale wingbars (buff on juvenile)

black and brown streaks on back

buffish rump

unmarked grey-buff underside

plain brown tail, often cocked

IN FLIGHT

ADULT

FLIGHT: quick, direct, undulating, with occasional momentary closures of wings.

The history of the Tree Sparrow has seen widespread declines and increases through Europe. By the end of the 20th century it was, in many areas, in the depths of a severe decline, and is now absent from large regions where it was quite recently common. Unlike House and Spanish Sparrows, male and female Tree Sparrows look alike. It is sometimes a suburban bird but is most suited to woodland with scattered clearings and farmland with mature trees.

VOICE Loud chirruping and cheeping calls like House Sparrow; a disyllabic *tsu-wit*; hard, short *tek tek* in flight.

NESTING Rounded or domed nest of straw and grass, in hole in tree or building, or in nest box; 4–6 eggs; 2 or 3 broods; April–July

FEEDING Mostly picks seeds from ground; also eats some insects, buds, and scraps around farms; visits bird-tables and feeders for seeds and nuts.

FLUFFIER IN WINTER
The Tree Sparrow keeps warm by fluffing its flank feathers over its wings; details such as its wingbars are then obscured.

SIMILAR SPECIES

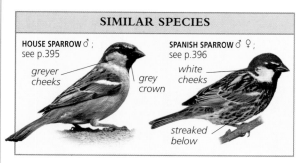

HOUSE SPARROW ♂; see p.395
greyer cheeks
grey crown

SPANISH SPARROW ♂ ♀; see p.396
white cheeks
streaked below

OCCURRENCE
Breeds in most of Europe but very local in UK, and absent from Iceland and N Scandinavia. Bird of farmland with scattered trees, parks, woodland, and woodland edge, but also town bird in much of S and E Europe.

Seen in the UK
J F M A M J J A S O N D

Length **14cm (5½in)**	Wingspan **20–22cm (8–9in)**	Weight **19–25g (11/16–7/8oz)**
Social **Flocks**	Lifespan **2–5 years**	Status **Secure**

Order **Passeriformes**	Family **Passeridae**	Species **Petronia petronia**

Rock Sparrow

IN FLIGHT

thin white wingbars

pale spots on tail

pale crown

sandy buff streaks on back

pale wingbar

bold cream and blackish stripes on head

big pale bill

dull brown body with dark streaks

brown tail, tipped with whitish spots

whitish underside with long, even, grey stripes

Rock Sparrows need broken ground with cavities in which to nest. This requirement may be met by cliffs and gorges in mountains, lower slopes with scattered rocks, farmed land with earth cliffs, road cuttings, old buildings, or hollow poles which provide nesting places. The birds are best located by following up their distinctive nasal calls but can be elusive as they remain perched on rocky ledges, inconspicuous with their dull, pale colours and lack of any strong pattern.

VOICE Distinctive twangy, nasal, oftrepeated note, *tyeoo*, *tee-vit*, or *peoo*.
NESTING Domed nest of grass and feathers in cavity in old building, hollow post, and earth bank, or in crag, often loosely colonial; 5 or 6 eggs; 2 or 3 broods; May–July.
FEEDING Finds seeds and invertebrates on ground, in grass, or among stones and boulders on open ground.

FLIGHT: low, quick, whirring, undulating over a distance; tail often fanned.

DRINKING POOL
A small pool in a dry region is usually a good place to sit in wait for Rock Sparrows and other birds that come to drink.

OCCURRENCE
Breeds in Spain, Portugal, S France, S Italy, and Balkans; typically in dry, stony, or sandy areas with cliffs or road cuttings; in rocky places, gorges, and mountainous regions, present all year.

Seen in the UK
J F M A M J J A S O N D

SIMILAR SPECIES

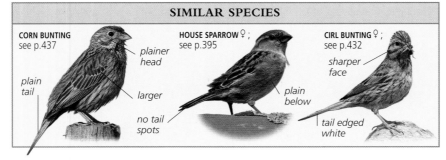

CORN BUNTING see p.437

plain tail

plainer head

larger

no tail spots

HOUSE SPARROW ♀; see p.395

plain below

CIRL BUNTING ♀; see p.432

sharper face

tail edged white

Length **15–17cm (6–6½in)**	Wingspan **21–23cm (8½–9in)**	Weight **20–28g (¹¹⁄₁₆–1oz)**
Social **Small flocks**	Lifespan **2–5 years**	Status **Secure**

WAGTAILS and PIPITS

W̲HILE SIMILAR TO LARKS, these are smaller, slimmer birds with longer tails and a more steeply undulating flight. They lack the larks' prolonged song-flights but the pipits have more ritualized song-flight patterns and less varied songs.

WAGTAILS

More boldly patterned or more colourful than pipits, the wagtails are often associated with water or wet meadowland. Pied and White Wagtails, however, are more likely than almost any other bird to be seen on tarmac or concrete in urban areas and even the Grey Wagtail, which breeds beside fast-flowing streams, is a regular bird on urban rooftops in winter.

Male and female plumages are often different and winter plumages are duller than summer ones; juveniles are also recognizably different. Some species are resident in Europe, others migrate to Africa for the winter.

PIPITS

Streaky brown is the typical description of a pipit: species can be hard to tell apart. Calls help, as does the time of year, habitat, and location.

GREY OR YELLOW?
Although called the Grey Wagtail, this bird confuses many people with its yellow coloration.

Similar species pairs may have different lifestyles, such as Meadow Pipits (moorland in summer, lowlands in winter) and Tree Pipits (woodland edge in summer, Africa in winter). There is little plumage variation between sexes and seasons.

MOTACILLIDAE

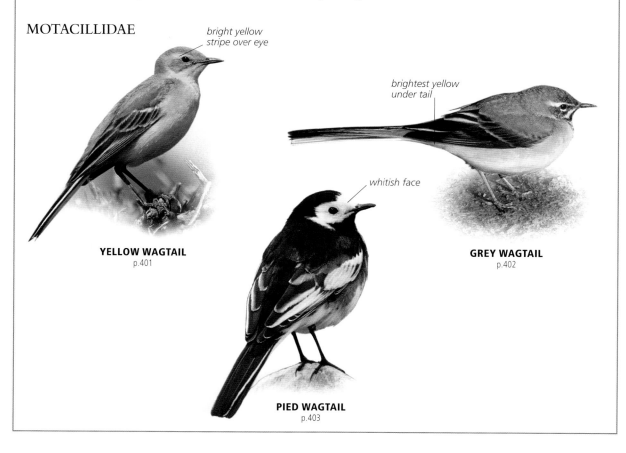

bright yellow stripe over eye

brightest yellow under tail

whitish face

YELLOW WAGTAIL
p.401

GREY WAGTAIL
p.402

PIED WAGTAIL
p.403

MOTACILLIDAE *continued*

dark spots on wing covers

TAWNY PIPIT
p.404

buff-yellow underside

TREE PIPIT
p.405

slim, weak, all-dark bill

MEADOW PIPIT
p.406

pinkish to brick-red face, throat, and upper breast

RED-THROATED PIPIT
p.407

dark legs

ROCK PIPIT
p.408

long white stripe over eye

WATER PIPIT
p.409

Yellow Wagtail 🔊 56

two white bars on blackish wings

MALE (SPRING)

IN FLIGHT

green crown

bright yellow stripe over eye

green back

MALE (SPRING)

white sides to black tail

long, spindly black legs

bright yellow underparts

pale stripe over eye

white lines on wings

tail shorter than Pied Wagtail's

buff underside

JUVENILE (AUTUMN)

pale line over eye

grey-green back

FEMALE (SPRING)

E legant and colourful, the Yellow Wagtail, particularly the summer male, is a highly distinctive bird. However, autumn birds, especially juveniles, cause confusion with rarer species and also juvenile Pied Wagtails, which can appear strongly yellowish. The call always helps to identify it. In summer, it lives around pools and reservoirs and damp, grassy fields where horses and cattle disturb the insects it eats. In winter, it is found near big mammal herds on African plains.
VOICE Call distinctive, loud, full, flat, or rising *tsli*, or *tsweep* or *tswi-eep*; song repetition of brief, chirping phrases.
NESTING Grassy cup in vegetation on ground; 5 or 6 eggs; 2 broods; May–July.
FEEDING Forages on ground, skipping and leaping after flies in short flycatching sallies; eats insects and other invertebrates.

FLIGHT: strong but undulating, with long, sweeping bounds; flurries of quick wingbeats.

FEEDING
The Yellow Wagtail is usually found around livestock in damp fields and pastures: it eats insects dislodged from the grass by the grazing cattle and horses.

SUBSPECIES

M. f. flava (C Europe)

long white stripe over eye

blue-grey crown and cheeks

glossy black crown and cheeks

M. f. feldegg (SE Europe)

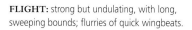

OCCURRENCE
Widespread in summer, breeding throughout Europe except for Ireland and Iceland. Often near water, in wet fields and pastures with livestock. Migrant flocks often on muddy reservoir edges or adjacent grass.

Seen in the UK
J F M **A M J J A S O** N D

SIMILAR SPECIES

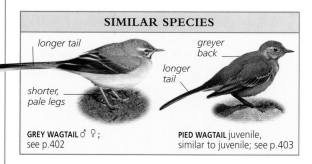

longer tail

shorter, pale legs

greyer back

longer tail

GREY WAGTAIL ♂ ♀; see p.402

PIED WAGTAIL juvenile, similar to juvenile; see p.403

Length **17cm (6½in)**	Wingspan **23–27cm (9–10½in)**	Weight **16–22g (⁹/₁₆–¹³/₁₆oz)**
Social **Small flocks**	Lifespan **Up to 5 years**	Status **Secure**

| Order **Passeriformes** | Family **Motacillidae** | Species **Motacilla cinerea** |

Grey Wagtail 🔊 57

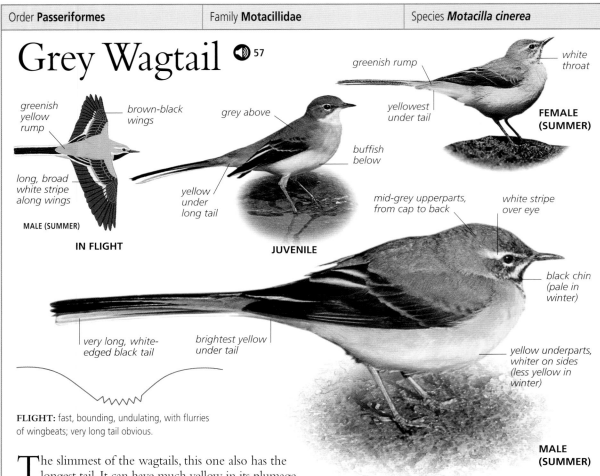

greenish yellow rump

brown-black wings

grey above

greenish rump

yellowest under tail

white throat

FEMALE (SUMMER)

long, broad white stripe along wings

MALE (SUMMER)

IN FLIGHT

yellow under long tail

buffish below

JUVENILE

mid-grey upperparts, from cap to back

white stripe over eye

black chin (pale in winter)

very long, white-edged black tail

brightest yellow under tail

yellow underparts, whiter on sides (less yellow in winter)

MALE (SUMMER)

FLIGHT: fast, bounding, undulating, with flurries of wingbeats; very long tail obvious.

The slimmest of the wagtails, this one also has the longest tail. It can have much yellow in its plumage and can be confused with the Yellow Wagtail; it may also be mistaken for the Pied Wagtail which often inhabits the same areas of clear, fast-flowing, rocky streams in summer and shallow pools in winter, and has a similar call. In winter, the Grey Wagtail may feed on almost any puddle, even on a flat rooftop in a city centre or at a garden pond. Visits to such places by this shy bird are invariably short but revealed by its calls.
VOICE Sharp, explosive, metallic *tchik* or *zi*, or *zi-zi*; song penetrating, metallic, sharp trills and warbles.
NESTING Grassy cup in hole in bank, wall, tree roots, or under bridge; 4–6 eggs; 2 broods; April–August.
FEEDING Catches flies and small invertebrates on ground or in air; quite active, bouncy feeder.

SUBTLE BEAUTY
Even a female or young Grey Wagtail has a lovely combination of smoky grey, buff, yellow, white, and black.

SIMILAR SPECIES

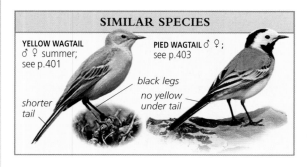

YELLOW WAGTAIL
♂ ♀ summer;
see p.401

shorter tail

PIED WAGTAIL ♂ ♀;
see p.403

black legs

no yellow under tail

OCCURRENCE

Widespread breeder north to Great Britain, Ireland, and S Scandinavia, along clean, often tree-lined rivers or more open upland streams. Very widespread near water in winter, briefly at puddles in towns and cities.

| Seen in the UK |
| J F M A M J J A S O N D |

| Length **18–19cm (7–7½in)** | Wingspan **25–27cm (10–10½in)** | Weight **15–23g (⁹⁄₁₆–¹³⁄₁₆oz)** |
| Social **Family groups** | Lifespan **Up to 5 years** | Status **Secure†** |

Pied Wagtail 🔊 55

blackish rump

white streaks on wings

MALE (SUMMER)

IN FLIGHT

greyer back than male's

greyer head and upperparts than adult male's

JUVENILE

buffish below

FEMALE (PIED)

black cap, chin, and throat (white chin and throat outside breeding season)

whitish face

black back

black breast

sooty flanks

white belly

MALE (SUMMER; PIED)

long, white-edged black tail

FLIGHT: quick, direct, with long undulating bounds and bursts of wingbeats.

Widespread and familiar, the Pied (or White) Wagtail is frequently seen in and around towns, often feeding on areas of tarmac, concrete, or stone slabs. It is also frequently seen on roofs, from which it typically calls before moving off: its call is a useful indicator of its presence. In summer, it can be found anywhere from builder's yards and woodsheds to remote quarries and natural cliffs and along stony river or lake sides. Although creating few identification problems, its non-breeding plumages are quite complex.

VOICE Calls loud, musical *chrip, chuwee, chrruwee,* and variants, merging into harder, unmusical *tissik* or *chiswit*; song mixes similar calls and trills.

NESTING Grassy cup in cavity in bank, cliff, or woodpile, in outbuilding or under bridge; 5 or 6 eggs; 2 or 3 broods; April–August.

FEEDING Feeds very actively on ground, roofs, or waterside mud or rocks, walking, running, leaping up or sideways, or flying in pursuit of flies; takes insects, molluscs, and some seeds.

SUBSPECIES *M. a. alba*

White Wagtail
The form found on mainland Europe; it is greyer and whiter than the Pied (*M. a. yarrellii*).

greyer cap

pale grey back

JUVENILE

browner wings with white bars

white flanks

MALE

SIMILAR SPECIES

GREY WAGTAIL ♂ ♀; yellow rump; see p.402

see p.402

yellow under tail

YELLOW WAGTAIL juvenile, similar to juvenile; different call; see p.401

see p.401

browner

OCCURRENCE

Breeds throughout Europe; found only in summer in N and E Europe but widespread in winter. Very varied habitat, often near water and in built-up areas, feeding on car parks, pavements, and roof-tops, but not usually in gardens.

Seen in the UK
| J | F | M | A | M | J | J | A | S | O | N | D |

| Length **18cm (7in)** | Wingspan **25–30cm (10–12in)** | Weight **19–27g (¹¹⁄₁₆– ⁷⁄₈ oz)** |
| Social **Winter flocks** | Lifespan **Up to 5 years** | Status **Secure** |

Order **Passeriformes**	Family **Motacillidae**	Species *Anthus campestris*

Tawny Pipit

IN FLIGHT

sparsely streaked pale back

ADULT

dark tail with white sides

long pale line over eye

pale sandy- or grey-brown back

dark spots on wing coverts

quite long, spike-like, pale-based bill

faint markings on breast

pale cream-buff underparts

JUVENILE

streaked back

dark stripe between eye and bill

plain sandy back

slender, pinkish or yellowish brown legs

WORN ADULT

ADULT

A large, stout-bodied, long-tailed, and rather wagtail-like pipit, the Tawny Pipit is widespread in mainland Europe and a scarce but annual visitor to the UK. It can easily be located by its spring song, although it is often difficult to see as it sings high in a clear sky. It prefers dry, stony, or sandy areas, such as warm, rocky Mediterranean slopes with scattered bushes and aromatic shrubs, or sand dunes by the sea. Pipits are often difficult to identify, especially out of their usual range, but a summer Tawny in a typical situation is usually quite easy to pick out with confidence.

VOICE Sparrow-like *schilp*, more grating, emphatic *tsee-i*, short *chup*; song in high undulating flight loud repetition of ringing, low-high double note *tchu-veee tchu-veee*.

NESTING Grass-lined cup in short vegetation on ground; 4 or 5 eggs; 1 or 2 broods; April–June.

FEEDING Catches and eats mostly insects on ground.

DARK SPOTS AND STREAKS
A row of dark spots and feather centres relieve an otherwise nearly uniform pale buff appearance.

FLIGHT: fast, direct; undulating with bursts of quick wingbeats; flies off long distance, going quite high.

OCCURRENCE
Breeds on bushy, stony slopes, in dry cultivated land with much stony soil, in grassland, and in dunes. Widespread in Europe north to Baltic but most typical of warmer areas in S Europe. Rare migrants farther north likely near coast.

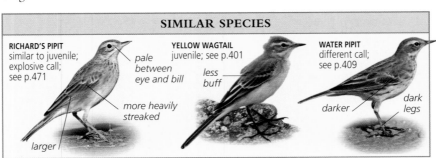

SIMILAR SPECIES

RICHARD'S PIPIT
similar to juvenile; explosive call; see p.471

pale between eye and bill

more heavily streaked

larger

YELLOW WAGTAIL
juvenile; see p.401

less buff

WATER PIPIT
different call; see p.409

darker

dark legs

Seen in the UK
J F M A M J J A S O N D

Length **15–18cm (6–7in)**	Wingspan **28–30cm (11–12in)**	Weight **35g (1¼oz)**
Social **Solitary**	Lifespan **Up to 5 years**	Status **Vulnerable**

Tree Pipit  54

blackish tail with white sides

neat black stripes on pale back (plumage similar to juvenile)

strong pale stripe over eye

upperparts browner in summer

pinkish-based bill

dark spots across wings

thin blackish streaks on chest

plain, pale yellowish flanks (streaked on Meadow Pipit)

buff-yellow underside

thin, pale pink legs with short claws

ADULT (SPRING)

IN FLIGHT

ADULT (AUTUMN)

FLIGHT: quite strong, direct, slightly undulating with bursts of quick wingbeats; often flies up into tree; less hesitant than Meadow Pipit.

One of the small, streaky pipits, the Tree Pipit is very like a Meadow Pipit but is more confident-looking and sleeker, although thickset: small points of character rather than plumage features separate these little brown birds. It has a superb song, rich and musical, in a distinctive song-flight, so summer males are not so difficult to identify. For autumn migrants, a call-note may be necessary for positive identification. While they do often occur within sight of each other, Tree and Meadow Pipits have different habitats, Tree Pipits occurring mostly on woodland edges and Meadow Pipits on heaths or moorlands. Tree Pipits do not form flocks.

VOICE Distinctive calls, including low, hissy buzz, *teeess* or *teaze*, thin, sharp *tzit*; loud, sweet song, with long series of notes and fast trills ending in loud, slow *sweee-sweee-sweee*, from perch or in flight ending on tree or bush.

NESTING Grassy cup on ground in thick grass; 4–6 eggs; 1 or 2 broods; April–July.

FEEDING Takes small insects from ground.

STRIKING SONG
Whether from a tree or in flight, the Tree Pipit's song is rich and musical with Canary-like trills.

OCCURRENCE
Breeds in most of Europe except in Ireland and Iceland. Occurs from spring to autumn only, usually in open woodland, woodland glades, or around edges of plantations, on bushy heaths and moors with scattered trees; migrants in more open areas at coasts.

Seen in the UK
J F M **A M J J A S O** N D

SIMILAR SPECIES

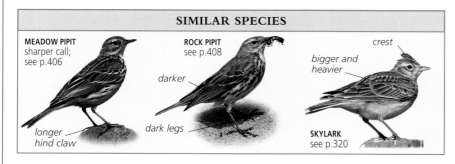

MEADOW PIPIT
sharper call;
see p.406

longer hind claw

ROCK PIPIT
see p.408

darker

dark legs

crest

bigger and heavier

SKYLARK
see p.320

Length **15cm (6in)**	Wingspan **25–27cm (10–10½in)**	Weight **20–25g (¹¹⁄₁₆ –⁷⁄₈ oz)**
Social **Solitary**	Lifespan **Up to 5 years**	Status **Secure**

| Order **Passeriformes** | Family **Motacillidae** | Species *Anthus pratensis* |

Meadow Pipit 🔊 53

pale stripe over eye

dark streak or patch on sides of neck

slim, weak, all-dark bill

soft blackish streaks on greyish, olive-, or yellowish brown back (darker on juvenile)

evenly streaked chest and flanks

olive-buff or creamy underside (yellower on juvenile)

pale orange-brown legs

very long hind claw

ADULT

dark tail with broad white sides

ADULT

IN FLIGHT

A small, streaky brown bird, the Meadow Pipit is worth looking at closely for the subtleties and beautiful patterns of its plumage. It gives the impression of constant nervous energy and worry; its calls may have a slightly hysterical quality. The Meadow Pipit often forms flocks, and winter flocks make shorter calls, more like the "pip-it" of their name. In summer, Meadow Pipits prefer heaths and wide open moors, often quite boggy places up on the hills. Their tinkling songs are characteristic of summer days in the open hills. In winter, many move to open farmland. These common, widespread birds often fall prey to Merlins and are parasitized by Cuckoos.

VOICE Sharp, weak, squeaked *pseeep* or *tsee*, frenetic repetition in alarm; winter flocks have short, quiet *pip*, *pi-pip* calls, short *tit*; song long series of simple repeated notes and trills, in parachuting song-flight starting and ending on ground.

NESTING Nest lined with fine stems in grass on ground; 4 or 5 eggs; 2 broods; May–July.

FEEDING Shuffles about on ground, picking up insects and other tiny invertebrates; eats some seeds.

FLIGHT: quite slow, erratic bounds and slight undulations with bursts of wingbeats; flies up weakly, jerkily.

DELICATE PATTERN
A close view of this streaky brown bird reveals a beautiful, intricate pattern.

OCCURRENCE
Breeds widely in NW, N, and E Europe; in winter, in W, SW, and S Europe. Nests on heaths, moorland, coasts, dunes, and bogs from sea level to high hills; in winter, mostly on lowland farmland and marshy places near coasts.

Seen in the UK
J F M A M J J A S O N D

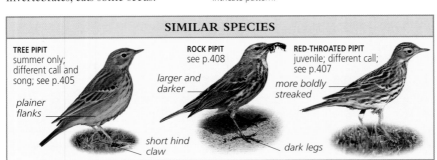

SIMILAR SPECIES

TREE PIPIT summer only; different call and song; see p.405
plainer flanks
short hind claw

ROCK PIPIT see p.408
larger and darker

RED-THROATED PIPIT juvenile; different call; see p.407
more boldly streaked
dark legs

| Length **14.5cm (5¾in)** | Wingspan **22–25cm (9–10in)** | Weight **16–25g (9/16–7/8oz)** |
| Social **Flocks** | Lifespan **Up to 5 years** | Status **Secure** |

Order **Passeriformes**	Family **Motacillidae**	Species *Anthus cervinus*

Red-throated Pipit

streaked dark cap

strong pale stripe over eye

boldly streaked, dull whitish underside

IMMATURE (1ST WINTER)

streaked rump

ADULT (SUMMER)

short, white-edged tail

IN FLIGHT

finely streaked crown

pinkish to brick-red face, throat, and upper breast

blackish brown and cream stripes on back

yellow base to bill

short-tailed, stocky shape

even, broad black stripes on white underparts

ADULT (SUMMER)

duller face than in summer

ADULT (WINTER)

In summer, this is a pipit of the far north; in spring it may be found in damp, grassy places and around pools and salt pans in southeast Europe. It is a regular but rare autumn migrant in west Europe. Although it is just another "streaky pipit", it does have a call note that, once heard, is remarkably distinctive, instantly revealing the presence of a calling bird flying over or flushed from the grass.

VOICE Call distinctive: high, slightly explosive, penetrating, fading out *psseeeeee*, also harder *chup*; song rhythmic repetition of sharp, fine, ringing notes and buzzy trills.

NESTING Grassy cup on ground in vegetation; 4 or 5 eggs; 1 brood; May–June.

FEEDING Takes insects and other invertebrates from ground; eats some small seeds.

FLIGHT: stronger than Meadow Pipit, less hesitant, more direct like Tree Pipit.

VARIABLE RED
This adult Red-throated Pipit is at the reddest-faced, least-streaked extreme of summer.

SIMILAR SPECIES

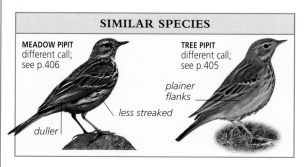

MEADOW PIPIT
different call;
see p.406

duller

TREE PIPIT
different call;
see p.405

plainer flanks

less streaked

OCCURRENCE
Breeds in extreme N Scandinavia on tundra, mountains, and in willow swamp. Widespread migrant in damp places in C and E Europe, but rare in W; likely on open ground, swampy areas, coastal dunes, and islands.

Seen in the UK
J F M A M J J A S O N D

Length **14–15cm (5½–6in)**	Wingspan **22–25cm (9–10in)**	Weight **16–25g (9/16–7/8 oz)**
Social **Small flocks**	Lifespan **Up to 5 years**	Status **Secure†**

| Order **Passeriformes** | Family **Motacillidae** | Species **Anthus petrosus** |

Rock Pipit

SUMMER

grey-edged, dark tail

IN FLIGHT

dark back

dull underside

WINTER

diffusely streaked, dusky olive back

dark legs

weak pale stripe over eye

pale eye-ring

long, strong, all-black bill

grey-brown streaks on yellowish to dull white underside

SUMMER

dark brown to blackish legs (pale pink on Meadow Pipit)

FLIGHT: fairly strong; bursts of wingbeats between glides.

One of the smaller pipits, this is a stocky, relatively heavily built, and quite dark bird, with distinctive dark legs. It is a coastal rather than a cliff bird, breeding in rocky places by the sea and feeding along shorelines of all kinds. In summer, it is mostly seen around cliffs and rocky islands, but in winter it moves out to open sand and shingle beaches and even the deeply incised muddy creeks of big salt marshes. Its song-flight and song are very similar to those of the Meadow Pipit.

VOICE Call rather full, more slurred than Meadow, *feest* or *pseeep*, usually singly; song richer, stronger trill in similar song-flight.

NESTING Hair-lined nest on ground, in cavity in rocks; 4 or 5 eggs; 1 or 2 broods; April–July.

FEEDING Forages on grass and rocks above cliffs in summer, more often on weedy and stony beaches in winter; picks up insects, sandhoppers, small periwinkles, and similar creatures.

TRICKY IN WINTER
The Scandinavian subspecies looks very like the Rock Pipit in winter but turns up inland more often.

OCCURRENCE
Breeds on all rocky coasts of Scandinavia, Shetlands, N and W Great Britain, Ireland, and NW France. Winters widely on softer coasts and is common around salt-marsh creeks and muddy inlets, in S to W Spain, with Scandinavian birds moving south.

| Seen in the UK |
| J F M A M J J A S O N D |

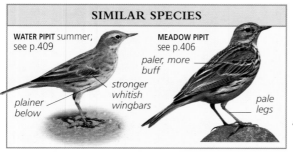

| SIMILAR SPECIES | SUBSPECIES |

WATER PIPIT summer; see p.409

plainer below

stronger whitish wingbars

MEADOW PIPIT see p.406

paler, more buff

pale legs

L. a. littoralis (Scandinavia) spring

greyer back

less streaked

| Length **16.5cm (6½in)** | Wingspan **23–28cm (9–11in)** | Weight **20–30g (¹¹⁄₁₆–1¹⁄₁₆oz)** |
| Social **Small flocks** | Lifespan **Up to 5 years** | Status **Secure** |

Order **Passeriformes**	Family **Motacillidae**	Species ***Anthus spinoletta***

Water Pipit

IN FLIGHT

two strong whitish bars on dark wings

SUMMER

dark tail with white edges

greyish head

long white stripe over eye

weakly streaked, warm brown back

little marking on chin and throat

pink-flushed whitish underparts

dark to reddish brown legs (pale pink on Meadow Pipit)

SUMMER

white stripes on brown head

dark brown back

streaked flanks

white bib

two white wingbars

white underparts

dark brown to blackish legs

WINTER

Unusual in Europe, Water Pipits breed in high mountain areas and move down in winter, which takes many of them north-wards rather than south. These winter birds visit muddy edges of reservoirs, muddy places around reedbeds, and salt-marsh pools, very unlike their summer territories – alpine pastures and boulder-strewn slopes around the snow line. Migrants are generally shy and not easy to watch closely; care must be taken to separate them from migrant Rock Pipits of Scandinavian origin.

VOICE Call between squeaky Meadow Pipit and fuller Rock Pipit, quite strong, thin *fist*; song strong series of trills in high song-flight like Rock Pipit's.

NESTING Grass-lined cup on ground among grass; 4 or 5 eggs; 2 broods; May–July.

FEEDING Takes small insects and other invertebrates from ground.

FLIGHT: strong, with bursts of wingbeats; often flies off to considerable height and distance; drops to ground in long, fast dive.

SHY BIRD
Water Pipits are large, wary birds, and are not very easy to spot and identify.

OCCURRENCE
Breeds locally at high altitude in Pyrenees, Alps, Italy, and Balkans, most often on high pastures with scattered boulders. In winter, spreads widely across W and S Europe, in marshy areas, coastal marshes, and lagoons with muddy edges.

Seen in the UK
J F M A M J J A S O N D

SIMILAR SPECIES

ROCK PIPIT
see p.408

duller

less white below

more buff

SKYLARK
see p.320

streaked breast

WHEATEAR ♀;
white rump;
see p.386

short tail

Length **17cm (6½ in)**	Wingspan **23–28cm (9–11in)**	Weight **20–36g (¹¹⁄₁₆–1⁵⁄₁₆ oz)**
Social **Small flocks**	Lifespan **Up to 5 years**	Status **Vulnerable**

FINCHES

THERE ARE TWO MAIN groups of finches, the *Fringilla* species (Chaffinch and Brambling) and the Cardueline finches. The Chaffinch and Brambling are clearly closely related, their different colours arranged in very similar patterns; they have the same basic shape and behaviour, and frequently mingle in winter.

The others, however, are a very diverse group. Their shapes and behaviour are strongly determined by their food. The crossbills have remarkable beaks with crossed, hooked tips, specially adapted to extracting seeds from conifer cones. The Hawfinch has a massive bill for cracking open tough seeds and stones, such as olive and cherry, while the Bullfinch has a softer, rounded bill for manipulating buds and soft fruits. The Greenfinch has a big bill, used for dealing with strong seeds and for tearing open tough fruits such as hips, while the Goldfinch and Siskin have delicate, pointed bills for extracting seeds from fruiting plants such as thistles and teazels and from cones of trees such as larch and alder.

SOCIAL FEEDERS
Goldfinches feed on seeds that tend to be abundant in small areas, so can afford to gather in large flocks.

Some species, such as the redpolls, are acrobatic and light enough to feed while perching on plants, while others are much less agile and feed while standing on the ground by pulling seedheads down with their bills.

Most finches are gregarious but flock behaviour varies: Chaffinches form loose aggregations, while Twites, Redpolls, and Siskins dash about in flight in tight, coordinated parties.

FRINGILLIDAE

blue-grey head and bill

CHAFFINCH
p.412

bright yellow-orange breast and shoulder

BRAMBLING
p.413

yellow stripe

GREENFINCH
p.414

bright yellow forehead

SERIN
p.415

pale grey sides of neck and chest

CITRIL FINCH
p.416

bold black, white, and red patterning on head

GOLDFINCH
p.417

black cap

SISKIN
p.418

pink-red chest

LINNET
p.419

unmarked tawny-buff face and throat

TWITE
p.420

pale to deep pink on breast

LESSER REDPOLL
p.421

conspicuously hooked bill

CROSSBILL
p.422

brighter red rump

PARROT CROSSBILL
p.423

cherry-red head and breast (loses red in winter)

SCARLET ROSEFINCH
p.424

strawberry red and pale grey

PINE GROSBEAK
p.425

pale grey back

BULLFINCH
p.426

pale tawny brown cap

HAWFINCH
p.427

| Order **Passeriformes** | Family **Fringillidae** | Species *Fringilla coelebs* |

Chaffinch 🔊 92

greenish rump

two white wingbars

dark wings with yellowish feather edges

MALE (SUMMER)

IN FLIGHT

blue-grey head and bill

brown back

dark patch in front of eye

head with ochre-brown smudges

MALE (WINTER)

brownish pink cheeks and throat

white patch

pink underside, whiter on belly and under tail

dark tail with broad white sides

MALE (SUMMER)

greyish olive from head to back

duller plumage than male's but similar white bars

FEMALE

pale greyish underside (Brambling has white belly)

One of Europe's most abundant birds, the Chaffinch forms an obvious species pair with the Brambling. The two often feed together outside the breeding season; their general shape, pattern, and behaviour are very similar. Chaffinches breed in separate territories, proclaimed by males singing loudly from prominent perches, but they are social birds at other times. They are often very tame, coming for food in car parks and picnic sites and frequently visiting gardens.

VOICE Flight call short, single, soft *chup*, frequent loud *pink!* in spring, loud *hweet, jilip*; song bright, cheery, rattled phrase with a flourish, *chi-chip-chip, chirichirchiri cheeip-tcheweeoo.*

NESTING Neat, perfectly camouflaged cup of grass, leaves, moss, cobweb, and lichen, against trunk of tree or bush; 4 or 5 eggs; 1 brood; April–May.

FEEDING Eats insects in summer, mostly caterpillars from foliage; otherwise takes seeds, shoots, and berries; visits bird-tables for seed mixtures, especially sunflower seeds.

FLIGHT: direct, quite quick, undulating; bursts of wingbeats between glides with wings closed.

SINGING MALE
The cheerful, rattling, and far-carrying song of the Chaffinch is an early sign of spring.

OCCURRENCE
Summer visitor in N and E Europe; all year in W and S (absent from Iceland); breeds in woods (extensive pine forest or conifer plantations to deciduous stands), parks, and gardens. Under beeches, in fields, and gardens in winter.

| Seen in the UK |
| J F M A M J J A S O N D |

SIMILAR SPECIES

BRAMBLING ♀ winter, similar to ♀; white rump; see p.413

orange upper wingbar

BULLFINCH ♂ similar to ♂ summer; white rump; see p.426

HOUSE SPARROW ♀ similar to ♀; see p.395

thick black bill and black cap

no white on tail

| Length **14.5cm (5¾in)** | Wingspan **25–28cm (10–11in)** | Weight **19–23g (¹¹⁄₁₆–¹³⁄₁₆oz)** |
| Social **Flocks** | Lifespan **2–5 years** | Status **Secure** |

Order **Passeriformes**	Family **Fringillidae**	Species *Fringilla montifringilla*

Brambling

oval white centre on rump

whitish or buff lower wingbar

broad orange-buff upper wingbar

MALE (WINTER)

IN FLIGHT

dark back

dark spots on flanks

"scaly" head

yellow bill

pale chin and throat

bright yellow-orange breast and shoulder

clean white belly

MALE (WINTER)

dark sides to pale panel on back of head

orange-buff shoulder

pale orange breast (pale feather edges wear off in summer)

whiter belly

FEMALE (WINTER)

black head

black back

MALE (SUMMER)

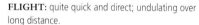

FLIGHT: quite quick and direct; undulating over long distance.

Generally less common and not nearly so ubiquitous as the Chaffinch, the Brambling can nevertheless gather in gigantic flocks in winter, especially in central Europe. In the west, their numbers fluctuate from year to year with the food supply, especially tree seeds such as beech-mast. Spring males can often be seen in fine summer plumage in their winter quarters before they migrate.

VOICE Flight call slightly harder than Chaffinch, single *tchek*, distinctive nasal *tsweek*; song includes deep, nasal, buzzing *dzeeee* note like Greenfinch.

NESTING Cup of lichen, bark, roots, and stems, lined with hair and feathers; 5–7 eggs; 1 brood; May–June.

FEEDING Eats insects in summer, seeds at other times; often on ground feeding on beech-mast.

VERY LARGE FLOCKS
Bramblings may form huge flocks in winter. Millions have been noted in central Europe but scores or hundreds are more usual in most areas.

SIMILAR SPECIES

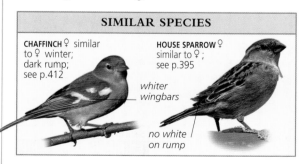

CHAFFINCH ♀ similar to ♀ winter; dark rump; see p.412

HOUSE SPARROW ♀ similar to ♀; see p.395

whiter wingbars

no white on rump

OCCURRENCE
Breeds in Scandinavia and NE Europe, in northern forest. In winter, throughout Europe, in farmland, parks, especially areas with beech, birch, and spruce, at times in gigantic concentrations where tree seeds are abundant.

Seen in the UK
| J | F | M | A | M | J | J | A | S | O | N | D |

Length **14.5cm (5¾in)**	Wingspan **25–28cm (10–11in)**	Weight **19–23g (¹¹⁄₁₆–¹³⁄₁₆oz)**
Social **Flocks**	Lifespan **2–5 years**	Status **Secure**

| Order **Passeriformes** | Family **Fringillidae** | Species *Carduelis chloris* |

Greenfinch 🔊 95

grey wings, yellow along edges of flight feathers

MALE

IN FLIGHT

plainer head than House Sparrow

thicker bill than Siskin

browner than adult

subtle streaks on back and along flanks

JUVENILE

duller than male

FEMALE

dark patch between bill and eye

apple-green plumage

stout pale bill

MALE (SUMMER)

yellow stripe

yellow sides to tail

MALE (WINTER)

This large, stocky, thick-billed finch breeds in loose groups in tall trees, old hedges, overgrown gardens, and orchards, and is a frequent garden visitor in winter, either to feeders or to shrubs with berries. Where common, Greenfinches gather into large feeding flocks, flying up together in a rush, more like Linnets and sparrows than Chaffinches. Adults are easily identified, but duller juveniles can be more troublesome at times.

VOICE Flight call fast, light, tinny chatter, *tit-it-it-it-it*, loud, nasal *tzoo-eee*, hard *jup-jup-jup*; fine song series of staccato trills of varying pace and quality, some metallic and thin, others full, musical, with droning, buzzy *dzweee* intermixed; often in flight *chup-chup-chup, chip-ipipip chr'r'r'r'r'r, tit-it-it-it-it chup-up*.

NESTING Bulky nest of grass and twigs, lined with finer stems, hair, and feathers, in thick bushes or trees; 4–6 eggs; 1 or 2 broods; April–July.

FEEDING Eats seeds, from trees to short plants, many taken from ground; also feeds on berries and nuts; visits bird-tables and feeders.

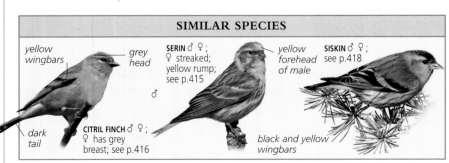

FLIGHT: fast, bounding, undulating, with bursts of quick wingbeats between closed-winged swoops; display flight slower, with stiff, wavy wingbeats.

GARDEN VISITOR
The Greenfinch often visits bird-tables and feeders for sunflower seeds, seed mixes, and peanuts.

OCCURRENCE
In all of Europe except Iceland, but in summer only in extreme north. Does best in open deciduous woods, parks, big gardens of country houses, bushy areas, orchards, or around farmsteads with tall, dense hedges.

| Seen in the UK |
| J F M A M J J A S O N D |

SIMILAR SPECIES

yellow wingbars

grey head

dark tail

CITRIL FINCH ♂ ♀; ♀ has grey breast; see p.416

SERIN ♂ ♀; ♀ streaked; yellow rump; see p.415

♂

yellow forehead of male

SISKIN ♂ ♀; see p.418

black and yellow wingbars

| Length **15cm (6in)** | Wingspan **25–27cm (10–10½in)** | Weight **25–32g (⅞–1⅛oz)** |
| Social **Flocks** | Lifespan **2–3 years** | Status **Secure** |

Order **Passeriformes**	Family **Fringillidae**	Species **Serinus serinus**

Serin

bright yellow forehead

dark crescent around cheeks

yellow rump

two short, narrow pale wingbars

streaked green back

stubby bill

MALE

pale wingbars

IN FLIGHT

forked tail

black-streaked flanks

MALE

paler yellow on face than male

less yellow below than male

FEMALE

MALE

A tiny, bouncy, colourful finch with sharp, spluttering calls, the Serin is characteristic of many Mediterranean areas. Males sing from the tops of spindly conifers, or in a fast, fluttery song-flight. Although superficially like other green and yellow finches, the Serin is generally easily identified in its usual range. However, various possible escaped cage birds have to be ruled out when identifying a potential out-of-range vagrant, including dull, streaky young Canaries.

VOICE Silvery, rapid trill, *zirr-r-r-r-r-r*; rising *tuweee*; song very quick, sharp, jingling or breaking glass quality, trills and twitters, often in stiff-winged song-flight.

NESTING Tiny, hair-lined cup of grass and moss in tree or bush; 4 eggs; 2 or 3 broods; May–July.

FEEDING Eats tiny seeds, mostly from ground or on low-growing plants.

FLIGHT: light, buoyant, deeply undulating; song-flight slower, on stiff, outstretched wings.

SIZZLING SONG
Males drop their wings to show off their yellow rumps as they sing their fast, high-pitched song.

OCCURRENCE
Resident in Spain, Portugal, S and W France, and Mediterranean area; summer visitor north to Baltic; only rare vagrant outside this range in W Europe. In villages, orchards, vineyards, olive groves, town parks, gardens, wooded areas, and along roadsides.

Seen in the UK
J F M A M J J A S O N D

SIMILAR SPECIES

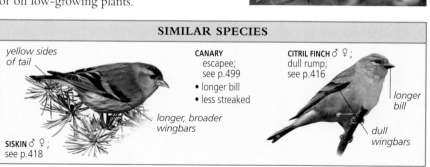

yellow sides of tail

CANARY escapee; see p.499
• longer bill
• less streaked

CITRIL FINCH ♂ ♀; dull rump; see p.416

longer bill

longer, broader wingbars

SISKIN ♂ ♀; see p.418

dull wingbars

Length **11–12cm (4¼–4¾ in)**	Wingspan **18–20cm (7–8in)**	Weight **12–15g (7/16–9/16 oz)**
Social **Flocks**	Lifespan **2–3 years**	Status **Secure**

| Order **Passeriformes** | Family **Fringillidae** | Species **Serinus citrinella** |

Citril Finch

IN FLIGHT

pale yellow-green rump

MALE

yellow wingbars

unstreaked flanks

duller plumage than male's

FEMALE

black and yellow crossbars on wings

MALE (WINTER)

dark tail

grey cap and hindneck

pale, greenish yellow face

pale grey sides of neck and chest

dull green back

thin bill

yellow-green breast

MALE (SUMMER)

A small, neat finch, with a combination of soft grey, pale lemon-yellow, and apple-green on its body and boldly barred wings, the Citril Finch is a bird of high altitude forest-edge habitats. It feeds on the ground or in trees in clearings or around grassy Alpine meadows within easy reach of spruce trees. It is usually found in small groups or family parties, looking puzzlingly like subtly marked Siskins or small, dull Greenfinches at first.

VOICE Various quick flight calls, short *tek* or *te-te-te*; song quick, varied, rambling warble with wheezy notes and buzzy trills.

NESTING Nest of grass and lichens, lined with plant down, high in tree; 4 or 5 eggs; 1 or 2 broods; May–July.

FEEDING Feeds on seeds, both from trees and on ground beneath.

FLIGHT: light, fast flight with bouncy undulations.

MOUNTAIN SPECIALIST
Citril Finches can be spotted high up, close to the tree line around rocky Alpine pastures.

OCCURRENCE
Bird of mountain forests and adjoining high level pastures, and spruce woods at tree line, in N Spain, S France, Alps, Corsica, and Sardinia, rarely moving far from breeding areas.

| Seen in the UK |
| J F M A M J J A S O N D |

SIMILAR SPECIES

SERIN ♂ ♀; yellower rump; see p.415

smaller bill

yellow tail sides

black cap on male

yellow streak on wing edges

bigger

SISKIN ♂ ♀; yellower rump; see p.418

GREENFINCH ♂ ♀; see p.414

| Length **11–12cm (4¼–4¾in)** | Wingspan **18–20cm (7–8in)** | Weight **12–15g (⁷⁄₁₆–⁹⁄₁₆oz)** |
| Social **Flocks** | Lifespan **2–3 years** | Status **Secure** |

Goldfinch  94

broad yellow panel on black wings

tawny back

bold black, white, and red patterning on head

sharp pale bill

black tail with white spots at tip

ADULT

yellow on closed wings

squarish tawny-chestnut patch on each side of breast

IN FLIGHT

grey head

duller wings than adult's

pale underside

ADULT

broad buff tip to tail

JUVENILE

Although it is widespread in Europe, even in cool, damp climates, the Goldfinch seems most at home in the hot, sunny summer of the Mediterranean. Its bouncy, lively actions and flashing colours go well with the bright, dry conditions and surroundings of brightly flowering plants on the seeds of which it feeds. It is, however, also found farther north in farmland with scattered woods and plenty of rough, open ground. Such places tend to be labelled "waste" and are all too often tidied up and stripped of the seed-bearing herbs and shrubs on which so many finches depend.

VOICE Calls are highly distinctive variations on usual finch theme: chattering, skipping flight call, *skip-i-lip* or *tililip* with liquid, lilting quality, rough *tschair*; song musical and varied, mixture of call notes and liquid trills.

NESTING Neat nest of roots, grass, and cobwebs, in tree or shrub; 5 or 6 eggs; 2 broods; May–July.

FEEDING Feeds on soft, half-ripe seeds on low-growing to medium-height plants, less often on ground; also eats tree seeds from alder and birch.

FLIGHT: particularly light and skipping, dancing, bouncy action with flurried beats of long wings.

FLASHING WINGS
A flock of Goldfinches in flight becomes a blur of yellow stripes; these, plus the bouncy, undulating action, make identification simple.

SIMILAR SPECIES

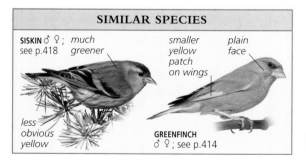

SISKIN ♂ ♀; see p.418 | **much greener** | **smaller yellow patch on wings** | **plain face**

less obvious yellow

GREENFINCH ♂ ♀; see p.414

OCCURRENCE
Breeds in most of Europe except Iceland and N Scandinavia; only in summer in NE Europe, resident elsewhere, common in S. Likes weedy places with tall, seed-bearing flowers such as thistles, teasels; also alders and larch.

Seen in the UK
| J | F | M | A | M | J | J | A | S | O | N | D |

Length **12.5–13cm (5in)**	Wingspan **21–25cm (8½–10in)**	Weight **14–17g (½–⅝oz)**
Social **Flocks**	Lifespan **2–3 years**	Status **Secure†**

Order **Passeriformes**	Family **Fringillidae**	Species **Carduelis spinus**

Siskin

yellow rump

broad yellow bar on black wings

yellow patch on each side of black tail

whitish belly

MALE

IN FLIGHT

dark streaks on green back

black cap

black chin

lime-green to yellowish breast

MALE

looks like greyer, washed-out female

paler and greyer head than male's

black streaks on white underside

JUVENILE

FEMALE

A tree-seed feeder, the Siskin is particularly associated with conifers, but also feeds in birch and alder trees in winter. It visits gardens to eat peanuts and sunflower seeds, but is not usually a ground-feeder. In winter, it associates in flocks, which share a bounding, tight-packed sociability with the Lesser Redpoll. Males sometimes separate out from the flocks in spring to sing from treetops. When feeding, these tiny finches are acrobatic, almost tit-like in their actions.

VOICE Flight calls loud, whistled, clear, with slightly squeaky or metallic quality, *tsy-zee* or *tsu-ee*; feeding birds give low, hoarse buzz or purr; song mixes calls and fast trills with hard twittering notes, from tree or in flight.

NESTING Tiny nest of twigs and stems, lined with plant down and hair, high in tree; 4 or 5 eggs; 1 or 2 broods; May–July.

FEEDING Eats seeds of pine, larch, and various other trees.

FLIGHT: dashing, darting, undulating; often in tight-packed, coordinated flocks.

NUT BASKET FEEDER
Siskins come to gardens in spring, when natural supplies of seed are low; they like the reddish peanuts especially.

OCCURRENCE
Breeds locally in N and E Europe, Alps, Pyrenees, UK, and Ireland, in forest of spruce and pine. In winter, more widespread and found especially in stands of larch, spruce, and alders along rivers; visits gardens for peanuts.

Seen in the UK
J F M A M J J A S O N D

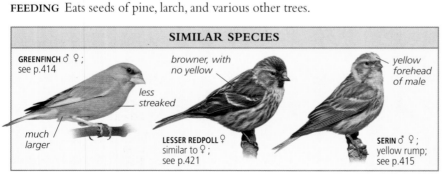

SIMILAR SPECIES

GREENFINCH ♂ ♀; see p.414

less streaked

much larger

browner, with no yellow

LESSER REDPOLL ♀ similar to ♀; see p.421

yellow forehead of male

SERIN ♂ ♀; yellow rump; see p.415

Length **12cm (4¾in)**	Wingspan **20–23cm (8–9in)**	Weight **12–18g (⁷/₁₆–⁵/₈oz)**
Social **Flocks**	Lifespan **2–3 years**	Status **Secure**

| Order **Passeriformes** | Family **Fringillidae** | Species ***Carduelis cannabina*** |

Linnet  93

whitish streaks on dark wings

MALE (SUMMER)

dark tail with white side streaks

IN FLIGHT

browner head

less pink-red on chest

MALE (WINTER)

short black legs

pale red forehead, deeper in spring

pale grey head

plain pale orange-brown back

pale cheek spot

pink-red chest

MALE (SUMMER)

pale cheek spot

triangular grey bill

streaked brown body

tawny-buff chest

white belly

FEMALE

A small, lively, sociable finch, the Linnet breeds in small colonies and feeds in flocks all year round. Flocks move together, tightly coordinated, unlike the looser aggregations formed by Chaffinches. They are ground feeders, while Redpolls and Siskins are mostly tree feeders and Goldfinches feed on tall herbs; at times most finches can be found together in mixed flocks. Linnets prefer waste ground with plentiful seeding plants and bushes, or hedgerows, in which to nest.
VOICE Light, twittering, chattering flight call *tidit tidititit*, nasal *tseeoo*; song musical, quite rich, varied warbling with chattering intermixed, often chorus from flocks.
NESTING Neat little nest of stems and roots, lined with hair; 4–6 eggs; 2 or 3 broods; April–July.
FEEDING Often feeds in groups all year, on seeds, taken from ground; young fed on insects; rarely comes to gardens.

FLIGHT: light, dancing, jerky undulations; flurries of wingbeats; sudden drop to ground to feed.

SWIRLING FLOCK
Linnets fly in tight, lively, bouncy flocks, with well-coordinated movements.

SIMILAR SPECIES

streaked tawny body

buff throat

TWITE ♂ ♀;
see p.420

streaked tawny-brown body

pale wingbar

LESSER REDPOLL ♂ ♀;
see p.421

OCCURRENCE
Found locally on heaths, rough grassland, commons, farmland, and upland meadows, in most of Europe except N Scandinavia and Iceland. Present only in summer in N and E Europe, but resident elsewhere.

| Seen in the UK |
| J F M A M J J A S O N D |

| Length **12.5–14cm (5–5½in)** | Wingspan **21–25cm (8½–10in)** | Weight **15–20g (9/16–11/16 oz)** |
| Social **Flocks** | Lifespan **2–3 years** | Status **Secure** |

| Order **Passeriformes** | Family **Fringillidae** | Species *Carduelis flavirostris* |

Twite

IN FLIGHT

narrow pale bar and long whitish streaks on wings

dark pink rump

MALE (SUMMER)

yellow bill

MALE (WINTER)

blackish streaks on tan-brown back

grey bill

white streaks on wing like Linnet, but pale buff bar like Redpoll

unmarked tawny-buff face and throat

buff underside with black streaks

lacks pink on rump

white side streak on tail often striking

FEMALE

MALE (SUMMER)

Unusual in that it feeds its young on seeds, the Twite needs plentiful flowers and herbs going to seed all year round. The loss of many flowery meadows has caused widespread declines and contractions in its range. Twites resemble Linnets but share characteristics with the Redpoll. They are, however, ground feeders, not treetop feeders like Redpolls. Like other smaller finches, they move in tight, coordinated flocks, rising from the ground, and circling and dropping again as one.

VOICE Flight call little harder than Linnet's, main call nasal, twanging, rising *twa-eeet*; song quick with buzzing notes and trills intermixed.

NESTING Deep cup of twigs, grass, and moss, lined with hair, in bush or bank; 4–6 eggs; 1 or 2 broods; May–June.

FEEDING Eats seeds; unlike most finches, also feeds young on seeds.

FLIGHT: bounding, fast, energetic with deep undulations; dives rapidly into cover.

ATTRACTED TO WATER
Twites use shallow pools for drinking and bathing, and are often easier to see then than when they are feeding in tall weeds.

SIMILAR SPECIES

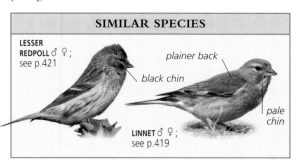

LESSER REDPOLL ♂ ♀; see p.421

plainer back

black chin

LINNET ♂ ♀; see p.419

pale chin

OCCURRENCE
Breeds in N Great Britain and Scandinavia, in weedy fields, at edges of moorland, and around upland farms. Winters around North Sea and Baltic coasts, mostly on salt marshes along coasts but declining. Scarce inland.

| Seen in the UK |
| J F M A M J J A S O N D |

| Length **12.5–14cm (5–5½in)** | Wingspan **21–25cm (8½–10in)** | Weight **15–20g (9/16–11/16oz)** |
| Social **Flocks** | Lifespan **2–3 years** | Status **Secure** |

Lesser Redpoll

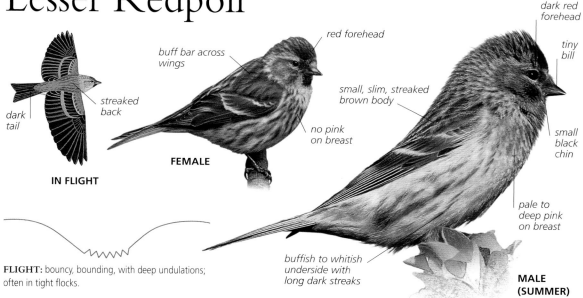

red forehead

buff bar across wings

dark tail

streaked back

IN FLIGHT

no pink on breast

FEMALE

dark red forehead

tiny bill

small, slim, streaked brown body

small black chin

pale to deep pink on breast

FLIGHT: bouncy, bounding, with deep undulations; often in tight flocks.

buffish to whitish underside with long dark streaks

MALE (SUMMER)

Typically a treetop bird, the Redpoll can also be found with Linnets in weedy fields, or feeding on the ground under birches where vast amounts of seeds have fallen. Most often, however, Redpoll groups feed, frequently with Siskins, in trees and move between treetops in noisy, well-coordinated flocks, circling together and often returning to the same tree after being disturbed. They may effectively "disappear" instantly on settling, becoming quiet and unobtrusive as they feed.

VOICE Flight call particularly hard, staccato chattering, metallic *chuchuchuchuchuchuch*, loud twangy *tsooeee*; song in flight combines chatter with fast, thin, reeling trill, *trreeeeee*.

NESTING Cup of twigs and grasses, lined with hair or wool, in bush or tree; 4–6 eggs; 1 or 2 broods; May–July.

FEEDING Mostly feeds in trees, on seeds, such as birch, alder, and larch, but also on or near ground in weedy fields and under birch trees.

C. flammea

Common Redpoll
A N European redpoll, scarce visitor to the UK and SW Europe.

whiter band over eye

paler

FEEDING FLOCK
The outer twigs of seeding birch trees can be decorated with Redpolls, which can hang upside down and perch at all angles in their efforts to reach the seeds.

SIMILAR SPECIES

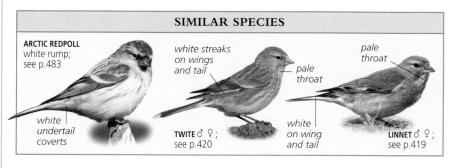

ARCTIC REDPOLL
white rump;
see p.483

white undertail coverts

white streaks on wings and tail

pale throat

TWITE ♂ ♀;
see p.420

white on wing and tail

pale throat

LINNET ♂ ♀;
see p.419

OCCURRENCE
Breeds with fluctuating numbers and range in Iceland, Ireland, Alps, Great Britain, Low Countries, and NE through Scandinavia. In winter to S France and Italy. In birch woods, larch, and bushy heaths; on ground under birch in spring.

Seen in the UK
J F M A M J J A S O N D

Length **11–14.5cm (4¼–5¾in)**	Wingspan **20–25cm (8–10in)**	Weight **10–14g (⅜–½oz)**
Social **Flocks**	Lifespan **2–3 years**	Status **Secure†**

Order **Passeriformes**	Family **Fringillidae**	Species *Loxia curvirostra*

Crossbill 🔊97

dark wings, rarely with narrow pale bars

MALE

bright pinkish rump

IN FLIGHT

brown wings

greenish body

FEMALE

small dark eye

conspicuously hooked bill

MALE

red underside

streaked below

rich red rump

dark tail

JUVENILE

Several species of crossbills occur in Europe, with the three plain-winged ones – the common, Parrot, and Scottish Crossbills (see p.484) – being the most difficult to separate. The common Crossbill feeds on spruce seeds but also survives quite well in areas where larch or pine predominate (trees favoured by the smaller Two-barred and larger Scottish or Parrot Crossbills). It is subject to periodic irruptions when large numbers travel far and wide in search of food: almost any clump of pines may then host Crossbills for a time. They feed quietly but may burst out of a treetop with loud flight calls.

VOICE Loud, abrupt calls, similar to young Greenfinch but louder, more staccato, *jup-jup-jup* or *chip-chip-chip*; quiet conversational notes while feeding; song mixes buzzy notes, calls, and bright warbles and trills.

NESTING Small nest of twigs, moss, and bark, lined with hair or wool; 3 or 4 eggs; 1 brood; January–March.

FEEDING Eats seeds of spruce, larch, pine, and other conifers, using crossed bill to prise them from cones on twigs; also eats some berries, buds, and insects.

FLIGHT: strong, direct, bursting from treetops in sudden noisy flurry; fast wingbeats between glides with closed wings.

THIRSTY FINCH
Crossbills eat dry seeds and need easy access to pools for frequent bouts of drinking.

OCCURRENCE
Scattered over most of Europe except in Iceland, but erratic, not present in most years in many areas. Core areas in extensive woods of spruce, larch, and pine, with variety of more or less distinct local populations.

Seen in the UK
J F M A M J J A S O N D

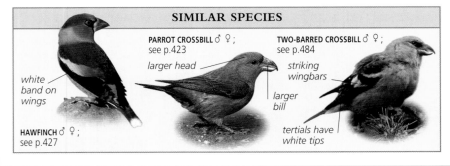

SIMILAR SPECIES

white band on wings

HAWFINCH ♂ ♀; see p.427

PARROT CROSSBILL ♂ ♀; see p.423

larger head

TWO-BARRED CROSSBILL ♂ ♀; see p.484

striking wingbars

larger bill

tertials have white tips

Length **16 cm (6½ in)**	Wingspan **27–30cm (10½–12in)**	Weight **34–38g (1³⁄₁₆–1⁶⁄₁₆ oz)**
Social **Small flocks**	Lifespan **2–5 years**	Status **Secure**

Parrot Crossbill

bright rump

dark wings

bull neck

forked "fish-tail"

IN FLIGHT

dark tail

brighter red rump

often greyer around bulky neck

heavy, bulging, hooked bill

dull red with browner wings

small eye in large head

ADULT MALE

FLIGHT: fast, bounding, with long undulations, usually low over tree tops.

O ne of three exceedingly similar crossbill species in northern Europe, the Parrot Crossbill is the largest and biggest-billed. It is adapted to pine forests with tough pine cones, but others also feed in pines (especially the Scottish Crossbill) and identification is seldom easy. Like the Common Crossbill, it may 'erupt' from breeding areas in years when food is scarce and numbers are high, and moves south and west. Small groups are then sometimes briefly seen in unexpected places, including small clumps of pines in coastal areas.

VOICE Like Common Crossbill but a tendency to be deeper, harder, less clicking: *toop-toop-toop*; song includes call notes, warbles, trills, and buzzing sounds.

NESTING Untidy nest of thin twigs and moss, high in a pine tree; 3 or 4 eggs; 1 brood; March or April.

FEEDING Eats pine seeds, prised from cones with the crossed bill; often drinks at nearby pools on ground; also a few berries, insects, and other seeds.

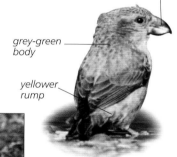

bulging bill

grey-green body

yellower rump

FEMALE

THIRSTY WORK
Like other crossbills, the Parrot Crossbill drinks frequently between bouts of feeding on dry pine seeds, repeatedly coming to favourite pools or rainwater puddles in the forest.

SIMILAR SPECIES

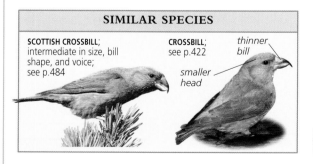

SCOTTISH CROSSBILL; intermediate in size, bill shape, and voice; see p.484

CROSSBILL; see p.422

thinner bill

smaller head

OCCURRENCE
Mostly resident in old coniferous forest in Scandinavia; rare visitor to the UK and probably breeds regularly in tiny numbers in Scotland. True status hard to define due to identification problems.

Seen in the UK
| J | F | M | A | M | J | J | A | S | O | N | D |

| Length **16–17cm (6½in)** | Wingspan **30cm (12in)** | Weight **35–40g (1¼–1⁷⁄₁₆oz)** |
| Social **Small flocks** | Lifespan **2–5 years** | Status **Secure** |

| Order **Passeriformes** | Family **Fringillidae** | Species ***Carpodacus erythrinus*** |

Scarlet Rosefinch

JUVENILE
brown wings

MALE (SUMMER)

dull brown back

red rump

IN FLIGHT

two pale buff wingbars

soft streaks below

JUVENILE

cherry-red head and breast (loses red in winter)

bold black eye

brown back

short, rounded bill

fine streaks on crown and cheeks

mid-brown back

hint of narrow wingbars

pale underside with fine dark streaks

FEMALE

MALE (SUMMER)

Several species of Rosefinches are widespread across Asia but only this one breeds in Europe. It is a bright, sturdy finch with a thick, bulky bill and small dark eyes in a plain face, giving a distinctive expression in all plumages. It has shown a tendency to spread westwards in recent years, with sporadic breeding even in Great Britain; singing males may turn up in early summer in unexpected places. This may or may not lead to long-term colonization; other species, such as the Serin, have shown a similar pattern without properly establishing themselves.

VOICE Short, ascending whistle, *vuee* or *tsoee*; song soft, rhythmic, whistling sequence.
NESTING Small neat grass nest low in bush; 4 or 5 eggs; 1 or 2 broods; May–July.
FEEDING Feeds on seeds, buds, shoots, and some insects, found in bushes or on ground.

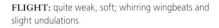

FLIGHT: quite weak, soft; whirring wingbeats and slight undulations.

JUVENILE
The pale wingbars and dark, round eyes are conspicuous on this juvenile.

OCCURRENCE
Present from May to August from C Europe eastwards, north to Scandinavia, breeding in deciduous woodland and bushy places, often in wetland areas near lakes or rivers. In autumn, rare migrant on W European coasts.

Seen in the UK
J F M A **M** J J **A S O** N D

SIMILAR SPECIES

BULLFINCH ♂ ♀; see p.426

dark cap

grey back

grey cap

CHAFFINCH ♂ ♀; see p.412

white wingbars

CROSSBILL ♂ ♀; different habitat; see p.422

thicker bill

larger

Length **15cm (6in)**	Wingspan **22–26cm (9–10in)**	Weight **21–27g (¾–¹⁵/₁₆oz)**
Social **Small flocks**	Lifespan **2–3 years**	Lifespan **2–3 years**

Order **Passeriformes**	Family **Fringillidae**	Species ***Pinicola enucleator***

Pine Grosbeak

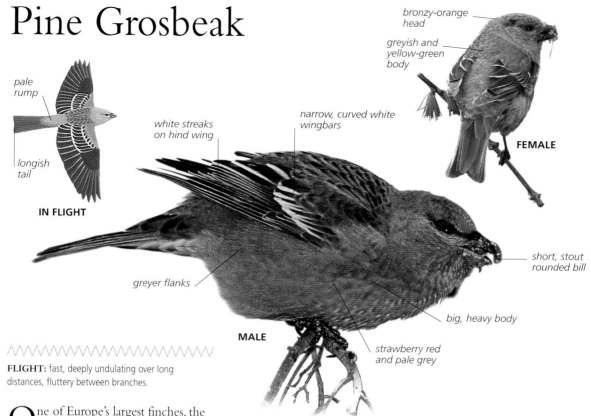

bronzy-orange head

greyish and yellow-green body

FEMALE

pale rump

longish tail

IN FLIGHT

white streaks on hind wing

narrow, curved white wingbars

short, stout rounded bill

greyer flanks

big, heavy body

MALE

strawberry red and pale grey

FLIGHT: fast, deeply undulating over long distances, fluttery between branches.

One of Europe's largest finches, the Pine Grosbeak gives the impression of being a gentle, but fearless bird, often almost oblivious of human presence. It is hard to find in summer when it breeds in northern forests, but may be seen feeding close to houses, even in ornamental trees in town centres, at other times, when it sometimes allows itself almost to be touched. It is restricted to forests and trees with berries and soft seeds all year round, so is unlikely to be seen outside its usual range.
VOICE Clear, single or multi-syllabic ringing notes with a loud, fluty quality; short, clear song.
NESTING Close to the trunk of a conifer, typically a spruce; 3–4 eggs; 1 brood; May–July.
FEEDING Eats seeds, buds, and shoots; some insects in summer, from trees, bushes, and shorter herbaceous plants, often in shady forest undergrowth.

BERRY SPECIALIST
Soft berries and buds are essential to the winter survival of this large, confiding finch; it often finds a good supply in towns and suburbs.

SIMILAR SPECIES

TWO-BARRED CROSSBILL; see p.484
smaller
bolder white wingbars

SCARLET ROSEFINCH; usually less red; see p.424
plainer wings
much smaller

OCCURRENCE
Breeds in old, extensive mixed or coniferous forest in extreme NE Europe and C Scandinavian mountain forests; in winter a little more widespread, rare in southern Scandinavia.

Seen in the UK
J F M A M J J A S O N D

Length **19–22cm (8–9in)**	Wingspan **30–35cm (12–15in)**	Weight **50–65g (1¾–2⅜oz)**
Social **Small flocks**	Lifespan **Up to 5 years**	Status **Secure**

Bullfinch 🔊 96

bright white rump

grey-white band on dark wings

MALE

IN FLIGHT

no dark cap

same plumage as female's

JUVENILE

dull brownish back

beige-grey underside

FEMALE

distinct hood formed by black cap, bill, and chin

pale grey back

thick, stubby bill

vivid red-pink underside

short dark legs

black tail

white under tail

MALE

FLIGHT: quite slow, but direct, slightly undulating over a distance.

A pest in some areas, but seriously declining in many, the Bullfinch is a quiet, inconspicuous bird despite its bold plumage. It uses its round bill to feed on soft buds, flowers, and shoots rather than hard seeds, usually feeding in pairs or family groups. If disturbed, it moves out of sight through a thicket or hedge. Its whistled calls are then highly distinctive. It does not visit bird-tables or feeders, although it may come to gardens in spring to raid flowering fruit trees.
VOICE Call low, soft, clear whistles, slightly descending, *peuuw, deu,* or *phiu*; song infrequent, creaky pea-whistle quality, with calls intermixed.
NESTING Cup of twigs, lined with moss and grass, in bush or tree; 4 or 5 eggs; 2 broods; April–June.
FEEDING Eats soft buds, seeds, berries, shoots, and some invertebrates, from low bushes and shrubs, occasionally on ground.

MALE CALLING
Usually inconspicuous, despite its bright colours, the Bullfinch's piping, whistled call is the best clue to its presence in a shrubbery or hedgerow.

SIMILAR SPECIES

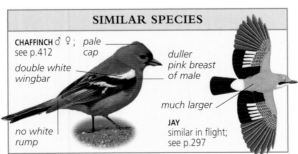

CHAFFINCH ♂ ♀; see p.412

pale cap

double white wingbar

no white rump

duller pink breast of male

much larger

JAY similar in flight; see p.297

OCCURRENCE
Breeds in most of Europe except Iceland, most of Spain, Portugal, and S Balkans; visits S Spain and Greece in winter. In woodland, farmland with hedges, thickets, parks, gardens with thick shrubberies and similar dense, low cover.

Seen in the UK											
J	F	M	A	M	J	J	A	S	O	N	D

Length **15cm (6in)**	Wingspan **22–26cm (9–10in)**	Weight **21–27g (¾–15/16 oz)**
Social **Small flocks**	Lifespan **2–3 years**	Status **Secure**

| Order **Passeriformes** | Family **Fringillidae** | Species ***Coccothraustes coccothraustes*** |

Hawfinch

broad buff-white wingbar

MALE (SUMMER)

IN FLIGHT

wing greyer than male

FEMALE

large yellow-brown bill (blue and black in summer)

small black bib

pale tawny brown cap

large head

grey nape

MALE (WINTER)

scaly back

JUVENILE

bars beneath

bright to dull tawny brown body

broad, clear white tip to short tail

In much of Europe, this is one of the most elusive finches, but in southern Europe it can be remarkably approachable, even in suburban trees and orchards or clumps of cherries or olives. It is not, even then, very obvious, but can be located by its quiet, clicking calls. Where the Hawfinch is more typically shy, it may provide little more than a glimpse as it flies up through trees and away over the canopy. It may sometimes be seen perched high on treetops, its size and stocky build then unlike other finches except for the Crossbill.

VOICE Call Robin-like, short, sharp, metallic *tik* or *tzik*, thin *tzree, tikitik*; weak, unmusical, scratchy song.

NESTING Nest of twigs, roots, and moss, lined with rootlets, in old tree; 4 or 5 eggs; 1 brood; April–May.

FEEDING Mostly takes large tree seeds, berries, cherries, and other fruit stones from trees; also picks hornbeam, sycamore, beech, and other seeds from ground in late winter.

FLIGHT: strong, fast, direct, swooping with undulations and bursts of powerful, whirring wingbeats.

SEED CRUSHER
The big bill of the Hawfinch deals easily and neatly with small seeds and can cope with tough stones of cherries and olives.

OCCURRENCE
Widespread but very localized, breeding in deciduous woodland, parks, large gardens, orchards, and olive groves. In winter, in similar places with plentiful tree seeds. Absent from Iceland, Ireland, and N Scandinavia.

Seen in the UK
| J | F | M | A | M | J | J | A | S | O | N | D |

SIMILAR SPECIES

CHAFFINCH ♂ ♀; see p.412

smaller bill

double wingbar

slim tail

CROSSBILL similar shape on high perch; see p.422

different colour

BULLFINCH ♂ ♀; large white rump; see p.426

small bill

| Length **18cm (7in)** | Wingspan **29–33cm (11½–13in)** | Weight **48–62g (1¹¹⁄₁₆–2¼oz)** |
| Social **Small flocks** | Lifespan **2–5 years** | Status **Secure** |

BUNTINGS

A BUNTING LOOKS MUCH like a finch: in general, buntings are a little slimmer and longer-tailed, and the structure of the bill is more constant, with a small upper mandible fitting neatly into a deeper, broader lower one that has a curiously curved cutting edge.

Most buntings have dark tails with white sides, but some, such as the Corn Bunting, have plainer tails. They show a variety of head patterns. Males are much like females in winter, with these patterns obscured by dull feather edges, but the dull colours crumble away in spring to reveal striking breeding plumage colours.

Females and juveniles, lacking these patterns, are more difficult to identify and some require care. Habitat, location, and time of year may be useful. Calls also help: several much rarer species visit western Europe in the autumn and look rather like Reed Buntings, but a hard, sharp "tik" call concentrates attention, as the Reed Bunting does not have any corresponding call note. Songs are mostly brief, not especially musical, and repetitive, although some, such as the Yellowhammer's

PRE-ROOST GATHERING
Corn Buntings get together before flying to a roost in a thicket or reedbed. Their flight calls often give them away.

all-summer-long song phrase, have a particularly pleasing and evocative character.

Most buntings are seed-eaters outside the breeding season and have suffered declines in areas where intensive modern farming has reduced the opportunities for birds to find weed seeds in winter. The Cirl Bunting has also declined with a lack of grasshoppers, which it feeds to its young in summer.

EMBERIZIDAE

red-brown breast sides

SNOW BUNTING
p.429

black cap, face, and breast

LAPLAND BUNTING
p.430

yellow head with dusky stripes

YELLOWHAMMER
p.431

streaked, dark greenish cap

CIRL BUNTING
p.432

grey head and chest

ROCK BUNTING
p.433

pale eye-ring

ORTOLAN BUNTING
p.434

rust-red crown and face

LITTLE BUNTING
p.435

white stripe under cheeks

REED BUNTING
p.436

streaked, pale brown back

CORN BUNTING
p.437

| Order **Passeriformes** | Family **Emberizidae** | Species **Plectrophenax nivalis** |

Snow Bunting

IMMATURE

small white wing patch

black tail with white sides

white wings

IN FLIGHT

black wingtips

MALE (WINTER)

dark grey head and back

JUVENILE

stocky body with short legs

red- or orange-brown cap and cheeks

white head

black back and wingtips

white below

MALE (SUMMER)

dark-tipped yellow bill

red-brown breast sides

black and brown streaks on back (plumage greyer in summer)

brown cap

MALE (WINTER)

sandy brown back

short black legs

white underside

FEMALE (WINTER)

In summer, Snow Buntings are in the far north or on the highest peaks, usually where snow is still present. In winter, they roam widely over high ground, from ski resorts to barren, exposed mountainsides, but are more easily seen where they winter on the coast. Flocks prefer shingle banks and sheltered, muddy or gravelly marshes just inland of the beach, sometimes mixed with other buntings, finches, and larks. Their complex face and chest patterns may be confusing but the extensive white areas seen when they fly are good clues.

VOICE Loud call deep, clear *pyew* or *tsioo*, frequent lighter, trilling, rippling *tiri-lil-il-il-il-ip*; song short, clear, ringing phrase.

NESTING Nest of moss, lichen, and grass stems in cavity among rocks; 4–6 eggs; 1 or 2 broods; May–July.

FEEDING Takes insects in summer, mainly seeds and strandline invertebrates on beaches in winter.

FLIGHT: bouncy, erratic, as if swept by wind, with flurries of wingbeats and deep undulations; long wings.

ATTRACTED BY SEEDS

Snow Buntings can be attracted to patches of seeds scattered on the ground at the edge of a shingle beach in winter.

SIMILAR SPECIES

REED BUNTING ♀ similar to ♂ ♀ winter; see p.436

LAPLAND BUNTING juvenile; see p.430

streaked below

OCCURRENCE

Breeds very locally in N Scotland, Iceland, and N Scandinavia, on tundra or similar mountain top habitat. In winter, on coasts in S to N France and inland E Europe, at fringes of breeding range.

Seen in the UK
J F M A M J J A S O N D

| Length **16–17cm (6½in)** | Wingspan **32–38cm (12½–15in)** | Weight **30–40g (1¹⁄₁₆–1⁷⁄₁₆oz)** |
| Social **Flocks** | Lifespan **2–3 years** | Status **Secure†** |

| Order **Passeriformes** | Family **Emberizidae** | Species **Calcarius lapponicus** |

Lapland Bunting 🔊 95

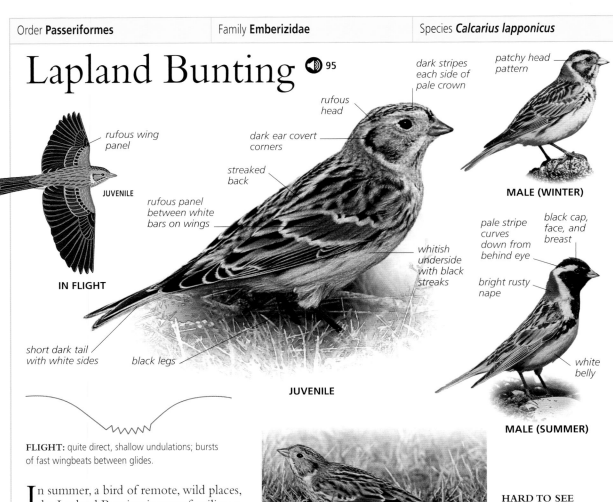

rufous wing panel

JUVENILE

rufous panel between white bars on wings

IN FLIGHT

short dark tail with white sides

black legs

dark stripes each side of pale crown

rufous head

dark ear covert corners

streaked back

whitish underside with black streaks

JUVENILE

patchy head pattern

MALE (WINTER)

pale stripe curves down from behind eye

black cap, face, and breast

bright rusty nape

white belly

MALE (SUMMER)

FLIGHT: quite direct, shallow undulations; bursts of fast wingbeats between glides.

In summer, a bird of remote, wild places, the Lapland Bunting is more familiar as a winter bird or autumn migrant. It appears near the coast, on grassy places such as golf courses, in dunes, and around grassier parts of salt marshes. It tends to creep inconspicuously until flushed from almost underfoot, or is noticed flying overhead with its distinctive calls. Its plumage patterns are superficially like those of a Reed Bunting, although more complex and richly marked, but its shape and actions recall a Snow Bunting.

VOICE Calls typically hard, quick, staccato rattle and clear whistle, *t-r-r-r-r-ik teu* or *tikikikiktik teu*; song in flight like short bursts of Skylark's song.

NESTING Nest of moss, lichens, and grass on ground, in hollow in tussock, or among rocks; 5 or 6 eggs; 1 brood; May–June.

FEEDING Shuffles on ground, finding seeds; feeds on insects in summer.

HARD TO SEE
This female Lapland Bunting, feeding among long grassy vegetation at the edge of coastal salt marshes, is hard to spot.

OCCURRENCE
Breeds in N Scandinavia, in tundra and high plateaux. In winter, mostly on salt marsh and short, wet grassland close to coasts, around North Sea and Baltic. Quite rare inland.

| Seen in the UK |
| J F **M** A M J J A **S O N D** |

SIMILAR SPECIES

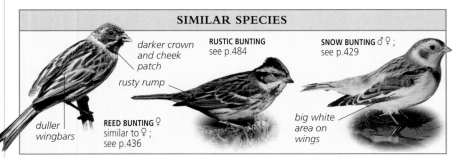

darker crown and cheek patch

RUSTIC BUNTING
see p.484

SNOW BUNTING ♂♀;
see p.429

rusty rump

duller wingbars

REED BUNTING ♀
similar to ♀;
see p.436

big white area on wings

| Length **14–15cm (5½–6in)** | Wingspan **25–28cm (10–11in)** | Weight **20–30g (11/16–1 1/16 oz)** |
| Social **Small flocks** | Lifespan **2–3 years** | Status **Secure†** |

| Order **Passeriformes** | Family **Emberizidae** | Species *Emberiza citrinella* |

Yellowhammer 🔊 98

yellow head with dusky stripes

black and rufous streaks on back

MALE (SUMMER)

rufous rump

IN FLIGHT

black tail with white sides

MALE (SPRING)

yellow underside with fine dark streaks

mixed rufous, buff, and black back

MALE (SUMMER)

less yellow on head

pale cheek spot

darker back

rusty flanks

more streaked below

FEMALE

The common bunting of farmland and bushy heaths, the Yellowhammer is typical of warm, sunny days when the males sing non-stop. In winter, they gather in small groups, or mix with other buntings and finches, roaming weedy fields or ploughed land, searching for seeds. Small parties of Yellowhammers draw attention to themselves by their sharp calls. In flight, they show the typical long, white-edged black tail of buntings.

VOICE Call sharp, quick, metallic, spluttering *tsik*, *tzit*, or *twitik*; song sharp, thin, metallic trill with one or two longer, higher or lower notes at end, *ti-ti-ti-ti-ti-ti-ti-ti-teee-tyew*, or simpler quick trill.

NESTING Hair-lined nest of grass and straw on ground in base of bush or below bank; 3–5 eggs; 2 or 3 broods; April–July.

FEEDING Eats some insects in summer, otherwise mostly takes seeds from ground.

FLIGHT: undulating, quite fast; steep rise from ground when disturbed; bursts of wingbeats.

SINGING MALE
The Yellowhammer's sharp song is characteristic of warm summer days on gorsy heaths.

WINTER FLOCK
Seeds attract Yellowhammers in winter, and they feed in dense groups where they are still common.

OCCURRENCE
Breeds in most of Europe except for Iceland, N Scandinavia, S Spain, and S Portugal. Widespread in S and W Europe in winter. On upland pastures, heaths, farmland with hedges, and coastal grassland.

| Seen in the UK |
| J F M A M J J A S O N D |

SIMILAR SPECIES

REED BUNTING ♀ similar to ♀; see p.436

ORTOLAN BUNTING ♂ ♀; dull rump; see p.434

white eye-ring

pale pink bill

yellow body

CIRL BUNTING ♂ ♀; dull olive rump; different call; see p.432

browner

darker, less yellow

| Length **16cm (6½in)** | Wingspan **23–29cm (9–11½in)** | Weight **24–30g (⅞–1¹⁄₁₆oz)** |
| Social **Flocks** | Lifespan **2–3 years** | Status **Secure†** |

| Order **Passeriformes** | Family **Emberizidae** | Species *Emberiza cirlus* |

Cirl Bunting

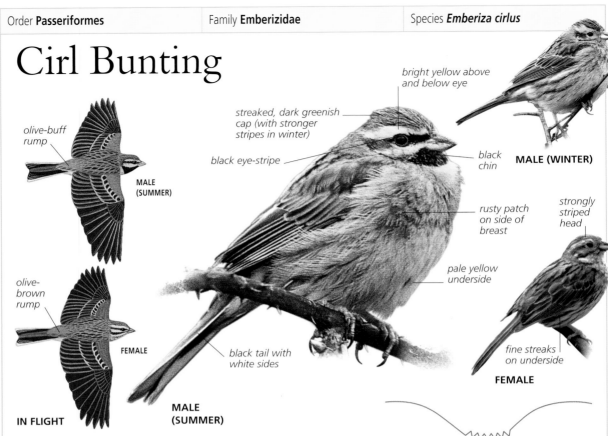

olive-buff rump

MALE (SUMMER)

olive-brown rump

FEMALE

IN FLIGHT

streaked, dark greenish cap (with stronger stripes in winter)

black eye-stripe

bright yellow above and below eye

black chin

MALE (WINTER)

rusty patch on side of breast

strongly striped head

pale yellow underside

black tail with white sides

MALE (SUMMER)

fine streaks on underside

FEMALE

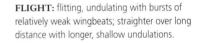

FLIGHT: flitting, undulating with bursts of relatively weak wingbeats; straighter over long distance with longer, shallow undulations.

With a song that recalls both Yellowhammer and Bonelli's Warbler (see p.333), the Cirl Bunting is a common bird of open, bushy slopes and well-treed farmland with hedges and thickets. It needs old, unimproved grassland, especially with a great many grasshoppers, so is suffering in the face of agricultural intensification. Males sing from bush tops but also from inconspicuous perches part way up trees, quite difficult to spot and likely to sit quite still for minutes on end.

VOICE Call very simple, short, high, thin *sip*; song fast, rattling trill on one note, or slower, lighter, more bubbling variant, *t-r-r-r-r-r-r-r-r-r* or *ti-ti-ti-ti-ti-ti-ti-ti-ti-ti-ti-ti*.

NESTING Rough nest of grass and stalks, low in shrub or hedge; 3 or 4 eggs; 2 broods; April–July.

FEEDING Needs to eat grasshoppers and similar insects in summer; otherwise takes seeds from ground.

FEMALE BIRD
The female Cirl Bunting has a strongly striped pale yellow and blackish head. The wings have rusty patches and the breast is pale yellowish.

SIMILAR SPECIES

YELLOWHAMMER ♂♀; rusty rump; see p.431

pale chin

REED BUNTING ♀ similar to ♂♀; see p.436

bolder head pattern

more rufous

OCCURRENCE
Breeds in extreme SW England, France, Spain, Portugal, and east to Balkans. Found all year on warm, bushy, often stony slopes, around tall, leafy orchard edges, and in olive groves. In winter, in weedy or grassy fields and gardens.

Seen in the UK
| J | F | M | A | M | J | J | A | S | O | N | D |

| Length **15–16cm (6–6½in)** | Wingspan **22–26cm (9–10in)** | Weight **21–27g (¾–¹⁵/₁₆oz)** |
| Social **Small flocks** | Lifespan **2–3 years** | Status **Secure** |

| Order **Passeriformes** | Family **Emberizidae** | Species ***Emberiza cia*** |

Rock Bunting

rusty rump
broad wings
MALE
IN FLIGHT

duller head stripes

buffish grey chest

rusty brown streaks on back

MALE (WINTER)

black stripes on crown, through eye, and below cheek
slim body
grey head and chest
orange-brown underside

less grey on breast than male

FEMALE

white-edged, black tail

MALE (SUMMER)

This is a small, slim, colourful bunting but can be frustratingly difficult to see. It tends to sit somewhere on a stony slope, often among thick bushes, calling frequently with a short, very thin, high note that is hard to pinpoint. In winter, it gathers in small flocks, often in grassy or weedy fields. It may be seen foraging beside roads in stony cuttings, or around archaeological sites with plenty of fallen stones and rough grass.

VOICE Call very thin, short, high *sip*, monotonous and elusive; song high, clear, erratic warbling phrase.

NESTING Nest of grass, roots, and bark on ground in thick cover; 4–6 eggs; 2 broods; April–June.

FEEDING Feeds on insects in summer; takes seeds from or near ground at other times.

FLIGHT: quite slow, low, erratic, with bursts of wingbeats; also flitting between bushes.

GROUND FEEDER
Rock Buntings feed on grassy clearings, among low rocks and shrubs, around tumbled boulders, and often along roadside cuttings.

OCCURRENCE
Breeds throughout Spain and Portugal, in Mediterranean region, and locally in Alps and C Europe. In rocky areas with dry, bushy slopes and crags and boulders, in alpine meadows and grassy places alongside road cuttings. Seen all year in majority of range.

| Seen in the UK |
| J F M A M J J A S O N D |

SIMILAR SPECIES

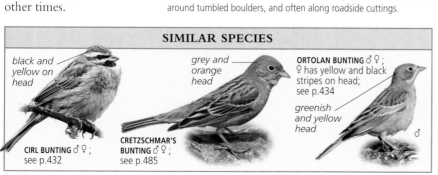

black and yellow on head

CIRL BUNTING ♂♀; see p.432

grey and orange head

CRETZSCHMAR'S BUNTING ♂♀; see p.485

ORTOLAN BUNTING ♂♀; ♀ has yellow and black stripes on head; see p.434

greenish and yellow head ♂

| Length **15cm (6in)** | Wingspan **22–26cm (9–10in)** | Weight **21–27g (^{3}/4–^{15}/16oz)** |
| Social **Small flocks** | Lifespan **2–3 years** | Status **Vulnerable** |

Ortolan Bunting

pale eye-ring

pink bill

brighter buff than female

JUVENILE

yellow eye-ring

green head

yellow moustache

pale green chest

sharp, triangular bill

pale eye-ring

greyer than juvenile

orange underside

FEMALE

olive-buff rump

MALE (SUMMER)

IN FLIGHT

black tail with white sides

MALE (SUMMER)

M ale Ortolans sing from bushes or trees on warm open slopes or in areas of upland pasture with hedges, walls, and copses. Their persistence makes up for a lack of real musical quality. Ortolan Buntings are also scarce but regular migrants in many coastal areas. They are usually quite shy and quick to fly off but tend to feed in open, grassy places where they can be watched from a distance. They are slim, pale buntings with sharp pink bills and obvious pale eye-rings.
VOICE Call thick, metallic, *dl-ip* and *chu*; song fluty, simple, ringing phrase, often repeated several times then changed to higher pitch, *sia sia sia si sia sru sru sru sru*.
NESTING Simple, hair-lined nest of grass and straw, on or near ground; 4–6 eggs; 2 or 3 broods; April–July.
FEEDING Eats insects in summer, and seeds from ground at other times, often from short grass clearings in dunes or fields.

FLIGHT: flitting, undulating, with bursts of relatively weak wingbeats; straighter over long distance with longer, shallow undulations.

STREAKY FEMALE
The general colour and pattern of the male are evident but subdued and faintly streaked on the female.

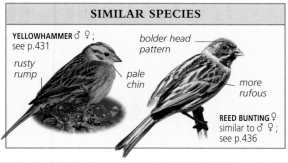

SIMILAR SPECIES

YELLOWHAMMER ♂ ♀; see p.431

rusty rump

pale chin

bolder head pattern

more rufous

REED BUNTING ♀ similar to ♂ ♀; see p.436

OCCURRENCE
Breeds across most of Europe except for UK, Iceland, and N Scandinavia, in variety of places from warm, bushy, stony slopes to semi-alpine pasture. Migrates to Africa in winter; rare on NW European coasts in spring and autumn.

Seen in the UK											
J	F	M	A	M	J	J	A	S	O	N	D

Length **15–16cm (6–6½in)**	Wingspan **22–26cm (9–10in)**	Weight **21–27g (¾ –$^{15}/_{16}$oz)**
Social **Small flocks**	Lifespan **2–3 years**	Status **Vulnerable†**

| Order **Passeriformes** | Family **Emberizidae** | Species **Emberiza pusilla** |

Little Bunting

IN FLIGHT

greyish forewings (rufous on Reed Bunting)

MALE (SUMMER)

pale crown stripe

greyish shoulders

MALE (WINTER)

white edge of tail

buff, brown, and black streaks on back

blackish corners of cheeks

pale eye-ring; weak or absent on Reed Bunting

blackish sides of crown

rust-red crown and face

sharp, straight-edged bill; bulbous on Reed Bunting

white underside with dark streaks

rufous face

pale legs

FEMALE

MALE (SUMMER)

Rather like a small Reed Bunting (and requiring great care when identifying migrants in autumn), the Little Bunting is a bird of the far north. It breeds in the vast taiga zone with mixed coniferous and birch forest. Rarely, one or two may winter in western Europe. It is, like many buntings, very much a ground bird most of the time, scuttling about on or very close to ground level even when disturbed. Generally rather quiet and unobtrusive, it is easy to overlook.

VOICE Call short, sharp, ticking *zik*; song short, high, mixed warble with clicking, rasping, and whistled notes.

NESTING Nest of grass and moss, in hollow on ground under bush; 4 or 5 eggs; 1 brood; May–June.

FEEDING Eats insects in summer; picks seeds from ground in autumn.

FLIGHT: quick, weak, light flitting action with flicked tail and short bursts of wingbeats.

SPRING MALE
The rufous colouring over the entire crown, face, and cheeks makes a summer adult eye-catching.

OCCURRENCE
Breeds in extreme NE Europe in open spaces in conifer forest. Autumn migrants rare on NW European coasts and islands, even rarer inland, often in damp places with weedy growth.

| Seen in the UK |
| J F M A M J J A S O N D |

SIMILAR SPECIES

REED BUNTING ♀ similar to ♂ ♀; see p.436

weak eye-ring

rounded bill

rusty rump and flank streaks

RUSTIC BUNTING ♂ ♀; see p.484

LAPLAND BUNTING juvenile; see p.430

bigger

short dark legs

| Length **12–13cm (4¾–5in)** | Wingspan **18–20cm (7–8in)** | Weight **15–18g (⁹⁄₁₆–⅝oz)** |
| Social **Small flocks** | Lifespan **2–3 years** | Status **Secure†** |

| Order **Passeriformes** | Family **Emberizidae** | Species ***Emberiza schoeniclus*** |

Reed Bunting

IN FLIGHT

rufous forewings

wide white tail sides

MALE (SUMMER)

MALE (SUMMER)

black and white neck and head

buff-brown back with black streaks

white stripe under cheeks

streaked whitish underside

pale red-brown legs

bold black tail with wide white sides

obscured head pattern

MALE (WINTER)

cream stripe over eye

hint of pale collar

cream and black streaks on back

FEMALE

long, notched tail

One of the more common buntings, especially in any damp or wet landscape, the Reed Bunting is easy to find and identify in summer. Males sing monotonously from low perches in the wetland vegetation. In winter, when males are far less striking, Reed Buntings are not so easily identified and also spread widely over all kinds of open ground and in thickets of willow, young conifers, and farmland hedgerows. They visit gardens at times.

VOICE Typical call quite full, loud, high *tseeu* or *psiu*, high, thin, pure *sweee*; *zi zi*; song short, stereotyped, simple, jangly phrase, two or three groups of notes clearly separated, *srip srip srip sea-sea-sea stitip-itip-itipip*.

NESTING Bulky nest of grass, sedge, and other stems, lined with roots and hair, on or close to ground in thick cover; 4 or 5 eggs; 2 broods; April–June.

FEEDING Mostly feeds on insects in summer, seeds at other times, taken low in bushes or on ground, often on open grass near water.

FLIGHT: slightly erratic, bounding, with flicking tail; dives into cover with flourish.

WINTER DRABNESS
Brown feather tips in winter obscure the male's head pattern. However, in spring, they wear away to reveal the full colours.

OCCURRENCE
Breeds in C and N Europe except for Iceland; seen in winter only in S Europe. Inhabits wet places with reeds, sedge, rushes, willow thickets, and fringes of lakes and rivers; also drier heathy slopes and heathland bogs. Sometimes visits gardens in winter, especially in hard weather.

Seen in the UK
J F M A M J J A S O N D

SIMILAR SPECIES

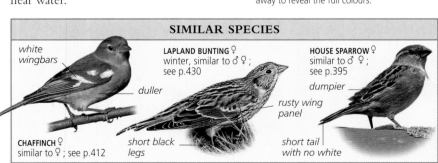

white wingbars

CHAFFINCH ♀
similar to ♀; see p.412

duller

LAPLAND BUNTING ♀
winter, similar to ♂ ♀;
see p.430

short black legs

rusty wing panel

HOUSE SPARROW ♀
similar to ♂ ♀;
see p.395

dumpier

short tail with no white

Length **15cm (6in)**	Wingspan **21–26cm (8½–10in)**	Weight **15–22g (⁹⁄₁₆ –¹³⁄₁₆ oz)**
Social **Small flocks**	Lifespan **2–3 years**	Status **Secure**

Order **Passeriformes**	Family **Emberiza**	Species **Emberiza calandra**

Corn Bunting  99

weak head
pattern

dark stripes
on crown

dark eye with
thin pale ring

streaked, pale
brown back

large, triangular pale
yellowish bill; thicker
than Skylark's

dark lower
edge to
cheeks

row of dark
spots on
wing coverts

dark streaks below
often merge into
central spot

SINGING

pale breast

streaks extend
well down body
unlike Skylark

stocky
shape

plain
tail

plain
wings

IN FLIGHT

plain brown
tail

A large bunting, the Corn Bunting is superficially like a Skylark, being a similarly pale, streaky brown. It is, however, plain on both wings and tail, and it perches on wires, fence posts, clumps of earth, or bushes, singing a short, simple phrase repeated with little variation. It feeds on the ground like other buntings, hopping and creeping rather than walking like a lark. It can often be seen flying over in small groups, calling distinctively, towards dusk, heading for communal roosts which may be scores or even hundreds strong.

VOICE Call short, abrupt, clicking *plip* or *quit*; song jangling, dry, fast rattled phrase like rattled keys or broken glass, *ti-ti-ti-tchee-iriririrr*.

NESTING Nest of grass and roots, lined with finer material, on ground; 3–5 eggs; 1 or 2 broods; April–June.

FEEDING Picks insects and seeds in summer, seeds in winter, from ground.

FLIGHT: long undulations; powerful bouts of wingbeats between looping glides with closed wings; in display, sometimes flies off with legs lowered.

WINTER FLOCKS
Where they remain common, Corn Buntings feed in small groups or even larger flocks in winter, resorting to hedges when disturbed.

OCCURRENCE
Breeds locally in UK, across Europe north to Baltic, most commonly in S Europe. Around meadows, cereal prairies, and farmland with hedges and scattered trees. Present all year except in E Europe, where summer visitor only but declining in many areas.

Seen in the UK
J F M A M J J A S O N D

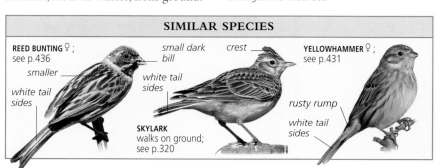

SIMILAR SPECIES

REED BUNTING ♀;
see p.436

smaller

white tail
sides

small dark
bill

white tail
sides

SKYLARK
walks on ground;
see p.320

crest

YELLOWHAMMER ♀;
see p.431

rusty rump

white tail
sides

Length **18cm (7in)**	Wingspan **26–32cm (10–12½ in)**	Weight **38–55g (1⅜–2oz)**
Social **Small flocks**	Lifespan **2–3 years**	Status **Secure†**

RARE SPECIES

Europe has a remarkably wide variety of habitats and extends over a huge geographical spread from the Arctic to the Mediterranean, from the Atlantic to the Black Sea. There are regular European species that breed only in very small areas of this range (for example, the Pied Wheatear along the Black Sea shores), or appear only as migrants in small parts of the continent (for instance, the Great Shearwater that regularly sweeps past the southwest of Ireland on its oceanic migrations). These are, nevertheless, seen every year in the right areas. Some, for example, Mediterranean Shearwaters that are numerous in the Mediterranean in summer, but rare elsewhere, are relatively numerous. Others, with abbreviated treatment here, such as the Pied-billed Grebe from North America, are really rarities, properly belonging to the avifauna of other continents. A few individuals stray far from their breeding range to turn up in Europe, some

species every year, others not so regularly. They are always, however, recorded in very small numbers.

The birds on the following pages include some that are rare everywhere in Europe and always unpredictable, most of which one cannot really plan to see, and others that are rare or restricted in range, but easily seen if one visits the right place at the right time of year.

SUMMER SPECIAL
Sooty Shearwaters from the southern hemisphere appear off northwest European coasts during their "winter" travels, in our summer and autumn.

Family **Gaviidae**	Species **Gavia adamsii**

White-billed Diver

If anything, this massive diver is even bigger than a Great Northern Diver (see p.105), similarly chequered in summer but with an uptilted, yellowish white bill. The bill lacks a complete dark ridge and tip in winter, when the sides of the face are also paler than a Great Northern Diver's. In flight, its heavy head and longer feet are sometimes noticeable.

OCCURRENCE Rare in summer in Arctic Europe; in winter, very few south into North Sea.

VOICE Silent in winter; loud wailing and laughing notes in summer.

uptilted, yellowish white bill

WINTER

pale cheeks

Length **80–90cm (32–35in)**	Wingspan **1.35–1.5m (4½–5ft)**

Family **Podicipedidae**	Species **Podilymbus podiceps**

Pied-billed Grebe

This stocky, big-headed grebe is like a large Little Grebe (see p.106) with a much stouter bill, which is plain yellowish in winter, but uniquely white with a black band in summer. In summer, it also has a black throat. Juveniles have dark head stripes. Rare visitors from North America may remain for some weeks on a suitable lake or reservoir. They tend to keep quite close to well-vegetated shores.

OCCURRENCE Rare in W Europe in autumn/winter, from Americas.

VOICE Silent when not breeding.

stout white bill with black band

ADULT (SUMMER)

black throat

Length **31–38cm (12–15in)**	Wingspan **50cm (20in)**

Family **Procellariidae**	Species ***Puffinus gravis***

Great Shearwater

Breeding in the southern oceans and migrating north in the northern summer and autumn, the Great Shearwater is a master of its challenging marine environment. It banks steeply and often to a great height, gliding at speed with few flaps. Its dark brown cap looks black at a distance, and the brown back, narrow white collar, white over the tail, and dark patches under the wing all aid identification.

OCCURRENCE Rare to moderately common well out at sea off W Europe, from August to October.

VOICE Silent.

brown back

dark brown cap

narrow white collar

Length **43–51cm (17–20in)**	Wingspan **1.05–1.22m (3½–4ft)**

Family **Procellariidae**	Species ***Puffinus baroli***

North Atlantic Little Shearwater

The Little (or Macaronesian) Shearwater is like a small, slightly dumpy Manx Shearwater (see p.114), often with a paler grey inner wing/black outer wing contrast and with more extensive white on its face quite easy to see at moderate range. It must nevertheless be watched with care in order to prove its identity, especially outside its normal range. It flies with rather fast wingbeats and few, short glides.

OCCURRENCE Breeds in Azores, Madeira, and Canaries; rare off NW Europe, in summer and autumn.

VOICE Rhythmic laughing notes at colony at night.

dark upperparts

extensive white on face

Length **25–30cm (10–12in)**	Wingspan **58–67cm (23–26in)**

Family **Procellariidae**	Species ***Puffinus yelkouan***

Mediterranean Shearwaters

Mediterranean "Manx" shearwaters are now separated as a full species or even split into two species. East Mediterranean birds (Yelkouan) are like small, more flappy Manx Shearwaters with feet projecting slightly beyond the tail, while the western ones (Balearic) are browner both above and below, but paler and smaller than Sooty Shearwaters. These birds may be seen on or low over the sea off Mediterranean shores in summer.

OCCURRENCE Breeds in coast and islands in Mediterranean; a few north to North Sea.

VOICE Strangled, yodelling notes over colonies at night.

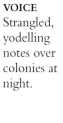

brownish above

Length **34–39cm (13½–15½in)**	Wingspan **78–90cm (31–35in)**

Family **Procellariidae**	Species ***Puffinus griseus***

Sooty Shearwater

One of the southern ocean seabirds that migrates north in the European summer, the Sooty Shearwater is regularly seen off some headlands and ferry routes in West European seas. It is slightly pot-bellied, with long, narrow, angular wings, and appears all dark except for a variably pale underwing panel that typically looks like a soft, silvery white central patch. The Sooty Shearwater is quite noticeably larger than a Manx Shearwater (see p.114) when they are seen together, and can resemble a dark skua at times.

OCCURRENCE Biscay, Irish and British coasts, from August to October.

VOICE Silent.

all-dark appearance

Length **40–50cm (16–20in)**	Wingspan **0.95–1.1m (3–3½ft)**

| Family **Procellariidae** | Species *Oceanites oceanicus* |

Wilson's Storm-petrel

Abundant in Antarctic seas, Wilson's Storm-petrels rarely stray north of the equator. They remain well out at sea, sometimes with Storm Petrels (see p.115), feeding on floating offal and sometimes approaching fishing vessels or following ships. The white rump is very broad, the upperwing has a pale band but the underwing is all-dark. The long wings and legs give a particularly buoyant action.

OCCURRENCE Very rare off NW Europe in late summer.

VOICE Silent.

long legs

| Length **16–18cm (6½–7in)** | Wingspan **38–42cm (15–16½in)** |

| Family **Procellariidae** | Species *Oceanodroma castro* |

Madeiran Storm-petrel

Very like Leach's Petrel (see p.116), the Madeiran Storm-petrel is distinguished with difficulty by its broader white rump, extending well around the sides, and less forked tail. It is an entirely marine bird except when visiting nesting colonies at night; it is rather solitary at sea and does not follow ships. It breeds in burrows and crevices on rocky islands.

OCCURRENCE Breeds off Portugal and in Madeira; rare at sea north of this range.

VOICE Cooing purrs and squeaky notes from burrow at night.

pale band on inner wings

| Length **19–21cm (7½–8½in)** | Wingspan **43–46cm (17–18in)** |

| Family **Pelecanidae** | Species *Pelecanus onocrotalus* |

White Pelican

Huge and contrasted black and white, the White Pelican is rose-pink in summer (the juvenile is duller). It has an orange-yellow bill pouch and a dark eye in a patch of pink. Overhead it shows black trailing edges and tips to the wings, like a White Stork (see p.134), but it lacks the stork's long legs and slender neck. Flocks circle and soar in a more coordinated fashion than storks.

OCCURRENCE Breeds in Balkans and E Europe, on large lakes and marshes.

VOICE Various grunts at nest.

dark eye in pink area
orange-yellow bill pouch
black under-wings
ADULT

| Length **1.4–1.75m (4½–5¾ft)** | Wingspan **2.45–2.95m (8–9¾ft)** |

| Family **Pelecanidae** | Species *Pelecanus crispus* |

Dalmatian Pelican

Globally rare and endangered, the Dalmatian Pelican is one of the world's largest birds. Its obvious pelican form, greyish head and body, reddish bill pouch in summer, and dull wings (with no sharp black and white contrast) identify it. Close views reveal a pale eye in a whitish area (dark on pink on the White Pelican). In flight, it is a magnificent sight, soaring effortlessly in warm air.

OCCURRENCE Breeds in Greece and Danube Delta, on large reedy lakes and swamps.

VOICE Silent.

ADULT
reddish bill pouch
pale eye in whitish area
greyish body (juvenile duller)

| Length **1.6–1.8m (5¼–6ft)** | Wingspan **2.7–3.2m (8¾–10ft)** |

Family **Phalacrocoracidae**	Species *Phalacrocorax pygmeus*

Pygmy Cormorant

A typical cormorant, the Pygmy Cormorant is nevertheless stocky, round-headed, short-billed, and thick-necked. Close views reveal a brown head and neck; in winter the throat is pale and the head less brown. The juvenile is paler below. When perched, or in flight, a rather long and rounded tail is evident. Groups often swim amongst vegetation or perch in overhanging trees or reeds, sometimes visiting coasts in winter.
OCCURRENCE Balkans and Black Sea coasts, on rivers and deltas.
VOICE Croaks and grunts at colonies.

long, rounded tail

ADULT (SUMMER)

Length **45–55cm (18–22in)**	Wingspan **75–90cm (30–35in)**

Family **Ardeidae**	Species *Egretta gularis*

Western Reef Egret

Like a thick-billed Little Egret (see p.127), the Western Reef Egret is typically dark grey with a white chin in West Africa, but white with pale grey or dark irregular marks in the Red Sea. Its legs are dull and brownish, the bill brown or blackish, with a yellower base. It is generally seen on the coast (but Little Egrets also frequent rocky shores), and often on piers and quays and associated structures.
OCCURRENCE Sinai and Red Sea; very rare in Morocco.
VOICE Silent.

white chin

dark grey body

Length **55–68cm (22–27in)**	Wingspan **88–112cm (35–44in)**

Family **Threskiornithidae**	Species *Plegadis falcinellus*

Glossy Ibis

Extremely slender but round-winged in flight, and elegant, round-bodied, but long-necked on the ground, the Glossy Ibis looks almost black unless seen closely in good light. Then it shows reflections of bronze and deep coppery red. Its slim, downcurved bill is distinctive as it wades and probes for food at the water's edge. Flocks tend to fly in wavy lines.
OCCURRENCE Rare in S Europe; more in Balkans and Middle East.
VOICE Mostly silent.

coppery red and bronze body

slim, curved bill

Length **55–65cm (22–26in)**	Wingspan **88–105cm (35–41in)**

Family **Anatidae**	Species *Anser caerulescens*

Snow Goose

Appearing especially brilliant white, even in the company of swans, the Snow Goose has a grey patch adjacent to bold black wingtips, a thick reddish bill, and deep pink legs. Some are grey-brown, bluer on the wings, with a white head, sometimes called "blue geese". Hybrid Canada x Greylag Geese are much bigger but may have a similar pattern, and "farmyard" white geese occasionally fly free.
OCCURRENCE Rare visitor to NW Europe from N America, or escapee.
VOICE Soft, rising, cackling notes.

ADULT

reddish bill

brilliant white body (juvenile duller and greyer)

black wingtips

Length **65–75cm (26–30in)**	Wingspan **1.33–1.56m (4¼–5ft)**

| Family **Anatidae** | Species *Anser erythropus* |

Lesser White-fronted Goose

Now extremely rare, numbers of this goose are being boosted by birds "fostered" under other geese, blurring the true wild pattern. In a winter goose flock, its faster action, long wingtips, neat round head, and very short, shocking-pink bill help to identify it; other useful features are the bold white blaze over the crown and bright yellow eye-ring. Juveniles lack the white blaze.

OCCURRENCE Very rare breeder in N Scandinavia; rare in W Europe in winter.

VOICE High, quick, yelping notes.

large white forehead blaze

ADULT

long wingtips

| Length **56–66cm (22–26in)** | Wingspan **1.15–1.35m (3¾–4½ft)** |

| Family **Anatidae** | Species *Branta ruficollis* |

Red-breasted Goose

Uniquely patterned black, white, and deep red, the Red-breasted Goose is an easy bird to identify but is sometimes surprisingly difficult to spot in a dense flock of slightly larger Brent or Barnacle Geese (see pp.68, 67). Strong sunlight makes even White-fronted Geese (see p.64) very contrasty, and the Red-breasted can be hard to find among them.

OCCURRENCE Large flocks in winter around Black Sea; very rare in W Europe.

VOICE Loud, sharp, double *pik-wik*.

white spot on face

ADULT

striking black, white, and red plumage

| Length **54–60cm (21½–23½in)** | Wingspan **1.1–1.25m (3½–4ft)** |

| Family **Anatidae** | Species *Tadorna ferruginea* |

Ruddy Shelduck

Clearly a shelduck in shape and actions, the Ruddy Shelduck is instantly identifiable by its mostly rich rust-orange plumage. The males have a pale head and narrow black collar; females have whiter faces. In flight, the forewing is strikingly white. Other shelduck species that may escape from collections look similar but differ in head and neck details. Periodic appearances of Ruddy Shelducks in the UK in late summer may involve truly wild birds.

OCCURRENCE Rare in E Greece and Turkey; vagrant in W Europe.

VOICE Nasal honking calls.

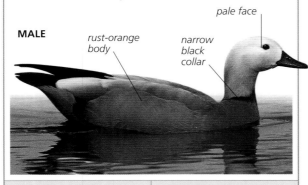

pale face

MALE

rust-orange body

narrow black collar

| Length **58–70cm (23–28in)** | Wingspan **1.1–1.35m (3½–4½ft)** |

| Family **Anatidae** | Species *Alopochen aegyptiaca* |

Egyptian Goose

Introduced as an ornamental bird to England, the Egyptian Goose has established itself in the wild but not spread far beyond eastern England. It looks a little like a pale brown Shelduck (see p.69), with a short pink bill, a brown eye patch, and big white wing patches. Some look quite rufous, others greyer.

OCCURRENCE E England; occasional wanderers elsewhere.

VOICE Loud, raucous, staccato cackling.

brown eye patch

ADULT

large white wing patch

long pink legs

| Length **63–73cm (25–29in)** | Wingspan **1.1–1.3m (3½–4¼ft)** |

Family **Anatidae**	Species *Aix galericulata*

Mandarin

Suitably exotic-looking for a duck of Southeast Asian origin, the Mandarin has been introduced into south Great Britain. Drakes have bushy orange "whiskers" and triangular orange "sails" on the back, and a black chest; females are dark grey-brown, mottled paler on the sides, with fine white "spectacles". They often perch in trees near freshwater lakes and rivers.
OCCURRENCE Locally in Great Britain, around wood-fringed lakes and slow rivers.
VOICE Short, rising whistling note.

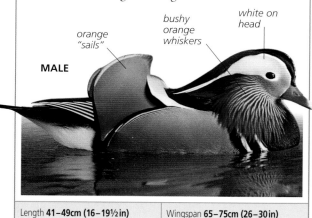

white on head

bushy orange whiskers

orange "sails"

MALE

Length **41–49cm (16–19½in)**	Wingspan **65–75cm (26–30in)**

Family **Anatidae**	Species *Aix sponsa*

Wood Duck

Rather like the Mandarin, the Wood Duck has escaped into the wild but is far less well-established. Males have a long, dark, drooped crest, bold white face marks, and a white band between the dark chest and orange flanks; females look like female Mandarins but with a dark-tipped (not pale-tipped) bill and shorter and broader "spectacles".
OCCURRENCE Very rare in Iceland, from North America; scattered escapees in UK.
VOICE Mostly silent.

dark, drooping crest

MALE pale orange flanks

Length **43–51cm (17–20in)**	Wingspan **68–78cm (27–31in)**

Family **Anatidae**	Species *Anas rubripes*

Black Duck

Clearly related to the Mallard (see p.73), the Black Duck is like a plainer, darker female Mallard with a contrasted pale head, blue hindwing patches lacking the white edges shown by a Mallard, and a bold white underwing that is viewed when flying or flapping its wings. The bill is greenish yellow and the legs rich orange. Various farmyard Mallard derivatives may look superficially similar.
OCCURRENCE Rare vagrant from North America in NW Europe.
VOICE Mallard-like quacks.

pale bill

ADULT

dark overall

plain dark wings

Length **53–61cm (21–24in)**	Wingspan **80–90cm (32–35in)**

Family **Anatidae**	Species *Anas americana*

American Wigeon

Bearing an obvious resemblance to the European Wigeon (see p.70), the male American Wigeon is moderately easy to identify by the white forehead and broad dark green band across a pale, speckled face and a dusky pinkish body. Females, however, are very difficult, but a contrasted greyer head and dark eye patch sometimes help in identification; white "wingpits" are diagnostic if seen clearly.
OCCURRENCE Regular but rare visitor from North America to W Europe, in autumn/winter.
VOICE Drake has Wigeon-like whistle.

dark eye patch

MALE

dusky pinkish body

Length **48–56cm (19–22in)**	Wingspan **75–85cm (30–34in)**

Family **Anatidae**	Species *Anas discors*

Blue-winged Teal

This small, long-billed surface-feeding duck is usually seen in immature plumage which is dark, mottled, and Teal-like (see p.72), with a broken pale line over the eye and whitish spot near the bill, pale blue forewings, and yellowish legs. Males have a bold, vertical white blaze on the face and bright blue on the wings; females are duller. The face pattern is echoed by summer male Shovelers (see p.76), a species with which Blue-winged Teals at times associate. Occasionally one may remain for some time on a suitable lake in Europe.

OCCURRENCE Rare autumn/winter vagrant from North America.

VOICE Mostly silent.

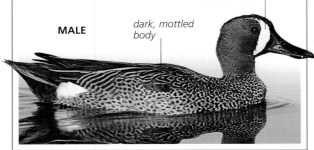

vertical white blaze on face

MALE

dark, mottled body

Length **37–41cm (14½–16in)**	Wingspan **55–65cm (22–26in)**

Family **Anatidae**	Species *Marmaronetta angustirostris*

Marbled Duck

Rare and local, the Marbled Duck is a pale, mottled grey-brown bird with a distinctive dark mask running into a slight tuft on the nape. The bill looks dark, the tail and rear end of the bird pale. In flight, the wings show little pattern except for darker tips and an almost-white trailing edge. Pale spots on the flanks are distinctive if seen at close range. Most Marbled Ducks found in NW Europe are suspected to be escaped birds from collections.

OCCURRENCE Very rare, in S Spain, Morocco, and Turkey.

VOICE Silent.

slight tuft at nape

dark mask

ADULT

pale spots on flanks

Length **39–42cm (15½–16½in)**	Wingspan **63–70cm (25–28in)**

Family **Anatidae**	Species *Aythya nyroca*

Ferruginous Duck

A fast-declining bird, the Ferruginous Duck is a sleek, rich mahogany-red diving duck, with broad, dazzling white wing stripes. Drakes have white eyes and dark grey bills fading to whitish before a black tip. Females are duller and brown-eyed. All have a pure white patch under the tail and peaked heads. Hybrid diving ducks may look very similar and require close attention to features such as eye and bill colours.

OCCURRENCE Declining breeder in C and E Europe; rare visitor in NW Europe.

VOICE Quiet; occasional purring growls.

white eye

dark grey bill with black tip

white patch under tail

MALE

Length **38–42cm (15–16½in)**	Wingspan **60–67cm (23–26½in)**

Family **Anatidae**	Species *Aythya collaris*

Ring-necked Duck

This is a close relative of the Tufted Duck (see p.78), identified by a more pointed head shape with no tuft, and grey wingbars. Drakes have grey flanks with a white "peak" at the front, brown females a Pochard-like (see p.77) pale face and white "spectacle". Hybrid ducks with similar patterns cause identification problems. A whitish ring behind the black bill tip is the sign of a true Ring-necked Duck.

OCCURRENCE Very rare but regular vagrant from North America to W Europe.

VOICE Low growling notes.

whitish ring behind black bill tip

pointed head

white "peak" on grey flanks

MALE

Length **37–46cm (14½–18in)**	Wingspan **65–75cm (26–30in)**

Family **Anatidae**	Species *Aythya affinis*

Lesser Scaup

A black-fronted, pale-bodied diving duck (resembling Tufted Duck and Scaup, see pp.78, 79), this rare bird has a rounded head with a very slight bump on the nape but no tuft. Its bill is pale blue-grey with a tiny black tip. The back is quite coarsely marked with wavy grey bands (greyer and more thickly marked than Scaup). The white flanks are sullied with pale grey, and faintly barred, unlike the pure white of an adult Scaup. Females are like female Scaup with a peaked nape.

OCCURRENCE Very rare vagrant to W Europe in autumn/winter, from N America.
VOICE Mostly silent.

blue-grey bill with small black tip

small bump on nape

wavy grey bands on back

MALE

Length **38–45cm (15–18in)**	Wingspan **70cm (28in)**

Family **Anatidae**	Species *Somateria fischeri*

Spectacled Eider

A rare and elusive northern species, this is a large duck but smaller than the Eider (see p.80); the drake is similarly white above and black below, and has a wedge-shaped face with a pale green head marked by a large whitish disc around the eye. The brown female has a pale buff version of this pattern. Unlike the King Eider, this species has not been found accompanying Eider flocks in NW Europe and remains a very difficult bird to see.

OCCURRENCE Breeds in Siberia and Alaska; very rare in Norway.
VOICE Silent in winter.

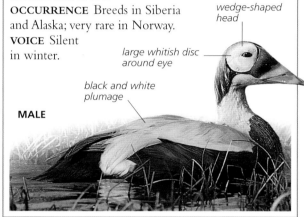

wedge-shaped head

large whitish disc around eye

black and white plumage

MALE

Length **50–58cm (20–23in)**	Wingspan **80–95cm (32–37in)**

Family **Anatidae**	Species *Polysticta stelleri*

Steller's Eider

This is a small and unusual eider, with a "normal" head and bill shape. The drake is largely pale, with a black stern and collar and bold black eye-spot. Females and immatures are dark, with two narrow white bars on the hindwing and white under the wing; the thick bill is grey, the head rather square with a slight bump on the nape.

OCCURRENCE Arctic breeder, regular in N Norway, rare in Baltic in winter.
VOICE Mostly silent.

slight bump on nape

thick grey bill

black and white wing feathers

MALE

Length **42–48cm (16½–19in)**	Wingspan **68–77cm (27–30in)**

Family **Anatidae**	Species *Melanitta perspicillata*

Surf Scoter

Big, bulky, and almost Eider-like (see p.80) in its deep bill and wedge-shaped head, the Surf Scoter is worth looking for in large scoter flocks offshore. Very similar to Coots (see p.168) when asleep, drakes show a bold white nape patch and multi-coloured bill, but females are difficult to identify, looking like Velvet Scoters (see p.84) with all-dark wings. The deep, dark bill is distinctive only at close range.

OCCURRENCE Rare but regular vagrant from North America; exceedingly rare inland.
VOICE Silent.

multi-coloured bill

white nape patch

MALE

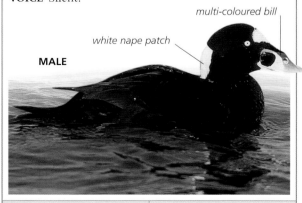

Length **45–56cm (18–22in)**	Wingspan **85–95cm (34–37in)**

Family **Anatidae**	Species *Netta rufina*

Red-crested Pochard

This large, bulky duck behaves more like a surface-feeder than a diving duck. Drakes have obvious "fuzzy" ginger heads, red bills, and black chests; females are plain brown with a dull whitish lower face. Both have very broad white wingbars. They tend to turn up amongst flocks of Tufted Ducks (see p.78) and Pochards (see p.77) on fresh water.

OCCURRENCE Breeds locally in S and E Europe; elsewhere occasional (usually escapees).
VOICE Various quiet barking notes.

MALE (SUMMER)
ginger head
red bill
white flanks

Length **53–57cm (21–22½in)**	Wingspan **85–90cm (34–35in)**

Family **Anatidae**	Species *Histrionicus histrionicus*

Harlequin Duck

Harlequin Ducks prefer rushing rivers, moving to coasts and lakes in winter, but rarely travelling far. Drakes are boldly patterned but look essentially dark, with strange white stripes and spots on the head, neck, and chest. Females are dark brown, dumpy diving ducks, with a diffuse white face patch and bright white ear-spot.

OCCURRENCE Iceland; exceedingly rare vagrant in NW Europe.
VOICE Mostly silent.

MALE
blue-grey body with white stripes
brown flanks
white face patch
pointed tail

Length **38–45cm (15–18in)**	Wingspan **63–70cm (25–28in)**

Family **Anatidae**	Species *Bucephala islandica*

Barrow's Goldeneye

Like a large Goldeneye (see p.85), this Icelandic species has a squarer, bigger head, marked on the drake by a long, kidney-shaped white patch where the Goldeneye has a rounder spot. The back is more solidly black. Females are much harder to tell, with a rounder crown, bulkier nape, and more yellow on the bill in summer.

OCCURRENCE Breeds and winters in Iceland; rare vagrant elsewhere.
VOICE Deep growling notes from female.

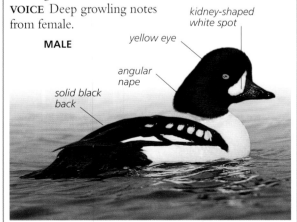

MALE
kidney-shaped white spot
yellow eye
angular nape
solid black back

Length **42–53cm (16½–21in)**	Wingspan **67–82cm (26–32in)**

Family **Anatidae**	Species *Oxyura leucocephala*

White-headed Duck

Stiff-tailed ducks include the introduced Ruddy Duck (see p.89) and native White-headed Duck in Europe. The latter is bulkier, paler, and less rich red in colour: drakes have more white on the head and swollen, pale sky-blue bills, while females have grey bills with a swollen base and black and whitish bands across the cheek. Immatures have black heads.

OCCURRENCE Rare in S Spain and Turkey, on large freshwater lakes.
VOICE Mostly silent.

MALE (SUMMER)
stiff tail
white head
sky-blue bill

Length **43–48cm (17–19in)**	Wingspan **60–70cm (23½–28in)**

Family **Accipitridae**	Species *Gypaetus barbatus*

Lammergeier

One of Europe's most spectacular birds, the Lammergeier is a massive, long-tailed vulture, flying with occasional deep, slow wingbeats but mostly with long, flat-winged, magnificent glides. Adults have white heads and rusty underparts and look shiny charcoal-grey above. Immatures are more uniformly grey or dark-hooded and paler-bellied. The long, wedge-shaped tail is most obvious on males.
OCCURRENCE Rare in Pyrenees, Crete, and Balkans; reintroduced in Alps.
VOICE Silent.

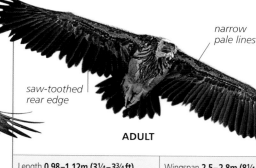

flat wings

white head

ADULT

diamond-shaped tail

Length **1.05–1.25m (3½–4ft)**	Wingspan **2.35–2.75m (7¾–9ft)**

Family **Accipitridae**	Species *Torgos tracheliotus*

Lappet-faced Vulture

A massive vulture, the Lappet-faced Vulture resembles the Black Vulture (see p.143) but has a paler body and shows narrow pale lines across the underwing. Close views reveal a bluish white head and very deep pale bill. The wings are broad, deeply fingered, and bulging on the rear edge; the tail is extremely short. The wings are held flat or slightly arched in flight, which consists mostly of a series of long glides and high, circling soaring.
OCCURRENCE Very rare, in Middle East.
VOICE Silent.

deeply fingered wings

narrow pale lines

saw-toothed rear edge

ADULT

Length **0.98–1.12m (3¼–3¾ft)**	Wingspan **2.5–2.8m (8¼–9¼ft)**

Family **Accipitridae**	Species *Aquila heliaca*

Imperial Eagle

This large, dark eagle has a pale grey base to the tail and white marks on the shoulders. Immatures are paler, with a bold pale rump, white upperwing bands, and contrasted underwings with a pale patch behind the angle. They have buffish bodies, with heavy, dark streaks. The wings are held flat or drooped in a glide, unlike the Golden Eagle (see p.154).
OCCURRENCE Rare in upland forests in Balkans.
VOICE Loud, barking notes.

dark brown streaks on breast

dark hindwings

pale window on inner primaries

JUVENILE

Length **70–80cm (28–32in)**	Wingspan **1.75–2.05m (5¾–6¾ft)**

Family **Accipitridae**	Species *Aquila adalberti*

Spanish Imperial Eagle

One of Europe's big eagles, the Spanish Imperial Eagle is a bird of lowlands and forested areas. It flies on rather flat wings, unlike the Golden Eagle (see p.154), adults showing a bold white front edge, a pale head, and a two-tone, dark-tipped tail. Juveniles are ginger-brown with dark wingtips, hindwings, and tail, a pale rump, and a whitish band along the upperwing.
OCCURRENCE Rare resident in C and S Spain.
VOICE Deep, barking notes.

dark and pale bands on wings

JUVENILE

Length **72–85cm (28–34in)**	Wingspan **1.8–2.1m (6–7ft)**

Family **Accipitridae**	Species **Aquila pomarina**

Lesser Spotted Eagle

This eagle migrates in large flocks into Africa for the winter. It is plain brown when adult except for a pale mark at the base of the primary feathers; the forewing is paler than the hindwing (often uniform, or reversed, on Spotted Eagle). Juveniles have a single line of white across the upperwings, a white band above the tail, and pale patches on the outer wings.

OCCURRENCE Breeds in SE Europe north to E Baltic; migrates to Africa through Middle East.

VOICE High-pitched yapping barks.

pale spots on wings

plain brown body

JUVENILE

Length **55–65cm (22–26in)**	Wingspan **1.43–1.68m (4¾–5½ft)**

Family **Accipitridae**	Species **Aquila clanga**

Spotted Eagle

Of the big brown eagles, this is the stockiest and broadest-winged in silhouette. Adults are very dark except for a small pale patch at the base of the primaries and a paler patch above the tail; immatures are marked by rows of pale feather tips across the wings and a white crescent over the tail. The strong legs are heavily feathered.

OCCURRENCE Rare in summer in E Europe.

VOICE Occasional low barking notes.

strong, yellow-based bill

pale feather tips across wings

IMMATURE

Length **59–69cm (23–27in)**	Wingspan **1.53–1.77m (5–5¾ft)**

Family **Accipitridae**	Species **Aquila nipalensis**

Steppe Eagle

One of the massive, heavy eagles of Asia, migrating into Africa for the winter, the Steppe Eagle is closely related to the Tawny Eagle. It flies on flat or drooped wings, and immatures have a broad white band along the middle of the underwings, gradually lost over several years until the all-dark adult plumage is attained. It has a particularly long, protruding head compared with the chunkier spotted eagles.

OCCURRENCE Migrant in Middle East.

VOICE Silent on migration.

broad white band on underwings

long, protruding head

IMMATURE

Length **62–74cm (24–29in)**	Wingspan **1.65–1.9m (5½–6¼ft)**

Family **Accipitridae**	Species **Elanus cearuleus**

Black-shouldered Kite

A medium-sized, blunt-headed, broad-winged bird of prey with a short, narrow tail, the Black-shouldered Kite is often seen hovering towards dusk, like a big, clumsy Kestrel (see p.159). It is uniquely grey with black shoulder patches and a white underside, the wingtips grey above but black beneath. Juveniles are duller with pale scaly feather edges on a darker grey back.

OCCURRENCE Rare in S and W Spain and Portugal, common in Egypt, near marshland.

VOICE Sharp, high *kree-ak* sound.

broad head

black shoulder patch

ADULT

Length **31–36cm (12–14in)**	Wingspan **71–85cm (28–34in)**

Family **Accipitridae**	Species *Circus macrourus*

Pallid Harrier

Of the harriers, the male Pallid Harrier is the palest and most ghost-like, identified by its white breast and narrow dark wedge-shaped wingtip patch. Females are like Montagu's Harriers (see p.149) with darker hindwings and a pale collar, while juveniles have a bolder whitish collar beneath dark cheeks and a dark band around the hind neck. All have white rumps.
OCCURRENCE Rare migrant in SE Europe.
VOICE High, whinnying, trilling chatter.

very pale grey body

dark, wedge-shaped wingtip patch

MALE

Length **40–50cm (16–20in)**	Wingspan **0.97–1.18m (3¼–3¾ft)**

Family **Accipitridae**	Species *Buteo rufinus*

Long-legged Buzzard

A big, bright buzzard with a pale cinnamon or rusty tail and whitish flight feathers with narrow black tips, the Long-legged Buzzard often hovers over open ground. It has a long-winged, eagle-like appearance. The belly or at least flank patches are dark and the upperwing has a dark wrist patch against a paler outer mark. The tail is not black-tipped.
OCCURRENCE Breeds in Greece, Turkey, and N Africa; moves south for winter.
VOICE Mostly silent.

pale head

dark flanks

Length **50–60cm (20–23½in)**	Wingspan **1.3–1.5m (4¼–5ft)**

Family **Accipitridae**	Species *Accipiter brevipes*

Levant Sparrowhawk

Clearly a long-tailed, broad-winged, bird-eating hawk, Levant Sparrowhawk is more sociable than the Sparrowhawk (see p.151) and migrates in flocks. Males have black-tipped, rather pointed wings, which are mostly white beneath; females also show dark wingtips. Both have dark eyes (Sparrowhawk's are yellow) and a black chin stripe, while males have grey cheeks (rusty on Sparrowhawk).
OCCURRENCE Breeds in Balkans and E Europe; migrates to Africa in autumn.
VOICE Shrill, repeated *ke-wik*.

grey cheeks

MALE

black-tipped, pointed wings

orange underside (female white with dark bars)

Length **30–37cm (12–14½in)**	Wingspan **63–76cm (25–30in)**

Family **Falconidae**	Species *Falco vespertinus*

Red-footed Falcon

Small, delicate, slightly rounded in its contours, the Red-footed Falcon is between a Hobby and Kestrel (see pp.161, 159) in shape and behaviour. It hovers, but also swoops gracefully in pursuit of insects. Old males are smoky grey with paler wingtips, young ones grey with dark wingtips and reddish belly patches. Females are barred grey and brown, and pale buff on the crown and underside.
OCCURRENCE Breeds in E Europe, regular but rare in spring/summer in W Europe.
VOICE High, quick, staccato chatter.

smoky grey body

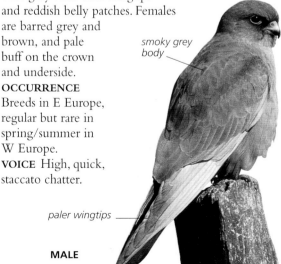

paler wingtips

MALE

Length **28–34cm (11–13½in)**	Wingspan **65–76cm (26–30in)**

Family **Falconidae**	Species *Falco eleonorae*

Eleonora's Falcon

This is a large, rakish, long-tailed, sharp-winged falcon of Mediterranean regions. One form is all-dark and blackish, and another has a white collar, dark moustache, and rufous underside. The underwing is two-toned, dark in front. Juveniles are plainer with narrow bars and pale cheeks. Eleonora's Falcons catch small migrant birds over the sea or big insects over lakes and marshes.

OCCURRENCE Scarce on Mediterranean islands and coasts.

VOICE Sharp, nasal, grating chatter.

white neck

ADULT (PALE FORM)

Length **37–42cm (14½–16½in)**	Wingspan **87–104cm (34–41in)**

Family **Falconidae**	Species *Falco rusticolus*

Gyr Falcon

The biggest and most heavily built falcon, the Gyr Falcon may be dark brownish (juveniles), slaty grey, or almost pure white according to age and location: the grey birds breed in N Europe, while the white ones visit mostly in late winter or spring from Greenland. The outer wing has paler areas underneath and the forewing is rather darker than the trailing edge.

OCCURRENCE Rare vagrant in W Europe, and rare breeder in Iceland and N Norway.

VOICE Deep, hoarse, rattling chatter.

white body and wings with black spots

ADULT (PALE FORM)

Length **53–63cm (21–25in)**	Wingspan **1.09–1.34m (3½–4½ft)**

Family **Falconidae**	Species *Falco cherrug*

Saker

A massive, powerful, broad-winged falcon, equal to a male Gyr Falcon in size, and paler and browner than a Peregrine (see p.162), the Saker has a pale buff head with only a thin dark moustache, and usually dark thigh feathers. The underwing is strongly two-toned. Juveniles are darker, blackish on the flanks, and dark under the tail unlike a young Lanner.

OCCURRENCE Rare in SE Europe, in hills, forests, and semi-arid grassland.

VOICE Loud, harsh, ringing chattering calls.

pale buff head

ADULT dark thigh feathers

brownish upperparts

Length **47–55cm (18½–22in)**	Wingspan **1.05–1.29m (3½–4¼ft)**

Family **Falconidae**	Species *Falco biarmicus*

Lanner

This is one of the big falcons, longer-tailed and slimmer-winged than a Peregrine (see p.162), and darker and greyer than a Saker. It has a paler breast than a Peregrine and its head is marked with buff or rufous. The juvenile is browner, heavily striped below, but paler under the tail. All have a dark forewing band underneath the wing, most obvious on young birds.

OCCURRENCE Rare in S Italy and Balkans, in semi-arid areas and mountains.

VOICE Harsh, loud, rasping chatter.

buff or rufous head

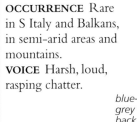

blue-grey back

ADULT

Length **43–50cm (17–20in)**	Wingspan **95–105cm (37½–41¼in)**

Family **Phasianidae**	Species *Alectoris graeca*

Rock Partridge

A rare partridge of mountains and rocky slopes, best identified by distribution and habitat, the Rock Partridge has a pure white throat, black curving down beside the bill, and very little white behind the eye. It is a plain-backed, barred-flanked bird, like a Chukar or Red-legged Partridge (see below, p.98) in general appearance, with a short red bill and red legs and a rufous tail showing in flight.

OCCURRENCE Scarce on alpine slopes, in Alps, Italy, and Balkans.

VOICE Short, hard, choking notes in long series, more varied than Chukar's.

black stripe through eye

barred flanks (less regular on juvenile)

ADULT

Length **33–36cm (13–14in)**	Wingspan **46–53cm (18–21in)**

Family **Phasianidae**	Species *Alectoris barbara*

Barbary Partridge

A rare bird very restricted in range in Europe, the Barbary Partridge has striped flanks, a mostly whitish grey face without a dark eye-stripe, and a spotted, reddish brown collar. Its breast is grey, the belly pale orange, and legs pale reddish. A dark central stripe shows on the crown. It is very like a Red-legged Partridge (see p.98) when seen flying off and its general behaviour is similar.

OCCURRENCE Gibraltar, Sardinia, Atlantic islands, and N Africa.

VOICE Series of quick, rhythmic, hoarse notes.

whitish grey face

reddish brown collar

striped flanks (less neatly patterned on juvenile)

ADULT

Length **32–34cm (12½–13½in)**	Wingspan **46–49cm (18–19½in)**

Family **Phasianidae**	Species *Alectoris chukar*

Chukar

A large, pale, rather plain partridge with bold flank stripes, the Chukar is characterized by black on the forehead but not beside the bill, a creamy throat sometimes spotted at the base, and a broad pale line behind the eye. Only when introduced birds are encountered do these subtle points matter too much as distribution is usually sufficient to identify it.

OCCURRENCE Common in Middle East; rare in NE Greece.

VOICE Loud, rhythmic, hollow *cha-cha-cha-chaker chaker chaker.*

broad white line behind eye

bold stripes on flanks

ADULT

Length **32–35cm (12½–14in)**	Wingspan **47–52cm (18½–20½in)**

Family **Phasianidae**	Species *Chrysolophus amherstiae*

Lady Amherst's Pheasant

This striking pheasant is very difficult to see in dense undergrowth beneath dark conifer forest. Males are uniquely patterned black and white with yellow on the rump; they have long red feathers beside the extremely long tail. Females are dark rufous, closely barred black, with a paler, unbarred belly unlike the Golden Pheasant; at 60–80cm (23½–32in) long, they are much smaller than the males.

OCCURRENCE Introduced but rare resident in C England.

VOICE Loud, strident *aaahk-aik-aik* at dusk.

striking black and white plumage

very long tail

long red feathers

MALE

Length **1.05–1.2m (3½–4ft)**	Wingspan **70–85cm (28–34in)**

451

Family **Phasianidae**	Species *Chrysolophus pictus*

Golden Pheasant

Introduced but not spreading far from old release sites, the Golden Pheasant is difficult to see, despite its bright colours. Males are strikingly red and yellow, with long, marbled, golden-brown tails. Females, which are much smaller at 60–80cm (23½–32in) long, are pale brown with black barring all over, and much less spotted than a Pheasant (see p.100).
OCCURRENCE Rare; very local in S Scotland and S and E England.
VOICE Loud, shrieking *eh-aik*.

MALE
red and yellow plumage

long golden-brown tail

Length **90–105cm (35–41in)**	Wingspan **65–75cm (26–30in)**

Family **Turnicidae**	Species *Turnix sylvatica*

Small Button-quail

This tiny bird is an enigma and may even be extinct in Europe, but is common in Africa. It may survive in dry, heathy places with palmetto scrub. If flushed, it reveals a tiny, Quail-like form (see p.97) with obvious pale upperwing patches. On the ground, its pale greyish face with no dark stripes would be distinctive. The Small Button-quail calls at dusk and dawn.
OCCURRENCE Very rare in S Spain; scarce in Morocco.
VOICE Deep, booming *hoooo hoooo hoooo* notes.

spots on orange breast

ADULT

Length **15–17cm (6–6½in)**	Wingspan **25–30cm (10–12in)**

Family **Rallidae**	Species *Porzana parva*

Little Crake

A tiny, elusive bird of dense waterside vegetation and ditches, sometimes emerging onto open mud or weed, the Little Crake looks like a tiny Water Rail (see p.164) with a short bill. Males are blue-grey and unmarked below, and brown with blackish streaks and a few long buff lines above. Females are pale brown, buff beneath, with a few blackish streaks on top.
OCCURRENCE Sporadic across C and E Europe; vagrant in W.
VOICE Nasal, yapping notes accelerate into fast trill.

long wings and tail

blackish streaks on pale brown back

red base to bill

FEMALE

Length **17–19cm (6½–7½in)**	Wingspan **34–39cm (13½–15½in)**

Family **Rallidae**	Species *Porzana pusilla*

Baillon's Crake

Compared with the Little Crake, this is a rounder, dumpier, short-winged, and short-tailed bird. Sexes are alike: brown above with black-edged white flecks, grey below with white bars on the flanks; the bill is green (red at base on Little and Spotted Crakes, see p.165) and the legs greenish. Juveniles are greyer and more barred than paler young Little Crakes.
OCCURRENCE Rare and very local in W Europe; vagrant in NW.
VOICE Quiet, short, soft, dry rattle.

short wings

green bill with no red

ADULT

Length **16–18cm (6½–7in)**	Wingspan **33–37cm (13–14½in)**

Family **Rallidae**	Species *Fulica cristata*

Crested Coot

Rare in Europe, this bird is very like a Coot (see p.168). It is best distinguished by a duller bill against the white facial shield, a rather more rounded shape to the black face against the bill base, often a "bump" near the tail when swimming, and a plainer upperwing with no pale trailing edge. The small red knobs on the forehead which give it its name are usually hard to see but can be quite obvious in spring.

OCCURRENCE Very rare in SW Spain, Morocco.

VOICE Shrill double note, unlike Coot, and hollow, *nasal ka-hah*.

ADULT (SPRING)

red knobs on forehead

bulging rump

greyish body

Length **39–44cm (15½–17½in)**	Wingspan **75–85cm (30–34in)**

Family **Rallidae**	Species *Porphyrio porphyrio*

Purple Gallinule

This huge Moorhen-like (see p.167) bird is often elusive in thick, reedy swamps, but comes into the open at times. It is immediately obvious: large and purplish blue (in Egypt, with a green back and turquoise face). The massive red bill and shield and long pink-red legs are easy to see, as is the bold white patch under the short tail.

OCCURRENCE SW Spain, Sardinia, Egypt.

VOICE Loud, abrupt, bleating and hooting calls, hardly bird-like in tone.

red shield

purplish blue body

large red bill

bold white patch under tail

long pink-red legs

ADULT

Length **45–50cm (18–20in)**	Wingspan **90–100cm (35–39in)**

Family **Gruidae**	Species *Anthropoides virgo*

Demoiselle Crane

Obviously a crane, this large grey bird is not always easy to tell from the common Crane (see p.170), especially in flight. Good views reveal a white head plume, long black breast feathers, and neat, narrow feathers (not bushy) cloaking the tail; in flight, the upperwing is less contrasted than on the Common Crane but wear increases the contrast as a pale grey "bloom" fades, and the smaller size is not obvious. Demoiselles are mainly Middle Eastern birds, regular in Cyprus, but very rare farther west.

OCCURRENCE Rare in extreme E Europe.

VOICE High, sharp trumpeting notes.

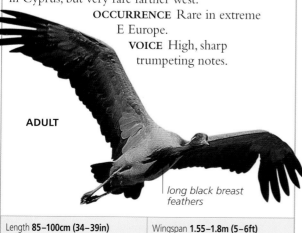

ADULT

long black breast feathers

Length **85–100cm (34–39in)**	Wingspan **1.55–1.8m (5–6ft)**

Family **Burhinidae**	Species *Burhinus senegalensis*

Senegal Thick-knee

This is a close relative of the Stone-curlew (see p.177) and can be distinguished only with care in a close view by a broad greyish band across the closed wing (narrow black and white bands on Stone-curlew). In flight, the white wingtip spots are slightly larger. Senegal Thick-knees are often seen on buildings or in groups on muddy riverbanks, unlike Stone-curlews.

OCCURRENCE In delta, along Nile, and Cairo, Egypt.

VOICE Loud, ringing whistles varying in pitch and volume.

ADULT

broad greyish band

Length **38–45cm (15–18in)**	Wingspan **76–88cm (30–35in)**

Family **Glareolidae**	Species *Cursorius cursor*

Cream-coloured Courser

This is a desert bird, only rarely straying beyond this harsh environment. It is hard to spot on the ground, on which it moves in quick, jerky runs, head high, but striking in flight with black wingtips above and solidly black underwings. Pale grey-buff or pinkish buff, except for a grey nape and black and white stripes behind the eye onto the nape, it has long, pale whitish-grey legs. Juveniles have soft dark mottles above and the head stripes are duller, the nape pale grey-brown.
OCCURRENCE Breeds in Middle East, North Africa; rare vagrant farther north.
VOICE Short, high, sharp flight calls.

long pale legs

ADULT

Length **24–27cm (9½–10½in)**	Wingspan **70cm (28in)**

Family **Glareolidae**	Species *Glareola nordmanni*

Black-winged Pratincole

Pratincoles are beautiful, elegant, aerial birds, although they spend much time hunched on the ground. The Black-winged Pratincole is rather dark, with little red on the bill, quite extensive black on the face, and a tail shorter than the wingtips (unlike Collared Pratincole, see p.213). It is easier to identify in flight, but lighting effects demand care: the underwings are solidly blackish, and the upperwing dark with no pale trailing edge (Collared shows a white line).
OCCURRENCE Breeds around Black Sea; rare migrant/vagrant in W Europe mostly in summer.
VOICE Hard, grating flight call, *kettek* or *kit-i-kit*.

ADULT (SUMMER)

Length **24–28cm (9½–11in)**	Wingspan **60–70cm (23½–28in)**

Family **Glareolidae**	Species *Glareola maldivarium*

Oriental Pratincole

While obviously a pratincole, this is not an easy species to identify, sharing characteristics with both Collared Pratincole (see p.213) and Black-winged Pratincole. It looks short-tailed (with only a shallow fork and no streamers), and combines the dark upperwing and lack of a white trailing edge of Black-winged with the chestnut-red underwing coverts of Collared. A worn or moulting Collared is the likeliest source of confusion. It is similarly elegant in the air, and a proficient hunter of flying insects.
OCCURRENCE Very rare vagrant to W Europe, from Asia, mostly in late summer.
VOICE Strident and tern-like.

ADULT

short tail

Length **23–27cm (9–10½in)**	Wingspan **50–60cm (20–23½in)**

Family **Charadriidae**	Species *Charadrius semipalmatus*

Semipalmated Plover

Very much like the Ringed Plover (see p.185), it is unlikely that a vagrant Semipalmated Plover will usually be noticed. In summer, it has less white behind the eye and a thinner black breast-band than a Ringed. In winter or immature plumages, the shorter bill, narrow breast-band, and fractionally smaller size are useful; suspicions need to be confirmed by the call.
OCCURRENCE Very rare vagrant to W Europe, from North America.
VOICE Bright, rising, double whistle, *chi-weee*, more clearly articulated than a Ringed Plover's.

dark bill

ADULT (WINTER)

narrow black breast-band

Length **16–17cm (6½in)**	Wingspan **33–38cm (13–15in)**

| Family **Charadriidae** | Species *Charadrius vociferus* |

Killdeer

The Killdeer is larger than a Ringed Plover (see p.185) and has a long tail, a tall but horizontal stance, and an obvious double black breast-band. Its legs are dull, the rather long bill black. In flight, it is striking because of its long tail and contrastingly rusty rump, unlike any other ringed plover type, and bold white stripe on almost black wings. It may turn up almost anywhere on open, flat, sandy or wet or derelict ground.
OCCURRENCE Rare vagrant to W Europe in autumn or winter, from North America.
VOICE Loud, fluty whistle, *klu-eee* or *kil-deeee*.

double black breast-band

long black bill

long tail

ADULT

| Length **23–26cm (9–10in)** | Wingspan **45–50cm (18–20in)** |

| Family **Charadriidae** | Species *Charadrius mongolus* |

Lesser Sand Plover

In summer, this species has a more solidly dark reddish breast-band than the Greater Sand Plover. It is also stockier, with a broad, round head that is not so disproportionately large, and a slightly shorter, more tapered bill. It has blackish or dark grey-green legs. In winter and immature plumages, it shows a neat dark patch on each side of the breast and a rather narrow pale stripe over the eye.
OCCURRENCE Rare vagrant in Europe, from Asia, nowhere regular.
VOICE Hard, short, trilled or repeated *trrrk* or *tirrik*.

ADULT (WINTER)

narrow pale stripe over eye

blackish or dark grey-green legs

thick black bill

| Length **17–19cm (6½–7½in)** | Wingspan **45–58cm (18–23in)** |

| Family **Charadriidae** | Species *Charadrius leschenaultii* |

Greater Sand Plover

Much bigger than a Ringed Plover (see p.185), the Greater Sand Plover is also longer-legged, more upright, larger-billed, and has a big, broad, bulbous head. In summer, it is rusty red on the head and chest; in winter, it is plain with dusky chest sides on the white underside. Young birds have pale scaly feather edges on the back. Care is needed to separate this from the Lesser Sand Plover.
OCCURRENCE Rare vagrant in Europe; regular in Israel and Egypt.
VOICE Trilling *trr-rr*, often repeated quickly.

JUVENILE

large black bill

pale, scaly feather edges

dusky chest sides

long, dull green legs

| Length **19–22cm (7½–9in)** | Wingspan **57–64cm (22½–25in)** |

| Family **Charadriidae** | Species *Charadrius asiaticus* |

Caspian Plover

This small- to medium-sized plover is long-legged, small-billed, and elegant. Pale brown above and white below, it has a broad chestnut breast-band in summer; winter birds and immatures have pale earthy buff across the chest, more extensively dark than on a sand plover. The legs are greenish, the wings show a white stripe, and the rump is all-dark in flight.
OCCURRENCE Rare vagrant in Europe from Asia; scarce migrant in Middle East.
VOICE Short *chup*.

ADULT (WINTER)

long legs

| Length **19–21cm (7½–8½in)** | Wingspan **57–64cm (22½–25in)** |

Family **Charadriidae**	Species *Pluvialis fulva*

Pacific Golden Plover

More similar to the Golden Plover (see p.181) than the American Golden Plover, this slightly smaller, longer-legged plover is hard to detect. In summer, it has bolder black chequering above and more black below than the Golden Plover. In winter, it is similar to the American species but less grey, longer-billed, and longer-legged. The dusky underwing is visible in flight. It has longer tertials than the American bird.

OCCURRENCE Very rare vagrant from Siberia to W Europe, mostly late summer.
VOICE Sharp, whistled *chu-wit* like Spotted Redshank (see p.191).

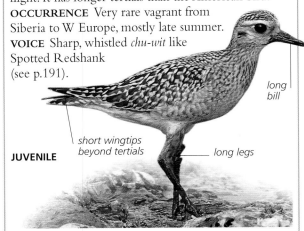

long bill

JUVENILE

short wingtips beyond tertials

long legs

Length **21–25cm (8½ –10in)**	Wingspan **45–50cm (18–20in)**

Family **Charadriidae**	Species *Pluvialis dominica*

American Golden Plover

Difficult to find in Golden Plover flocks (see p.181), a winter American Golden Plover tends to look greyer, with a bolder head pattern, slightly longer legs, and longer wingtips. In summer, it is less yellow above, more extensively black beneath, with bold white chest sides. In flight, the dusky grey underwing is a crucial clue. Grey Plovers (see p.182) are larger and much bigger-billed; Pacific Goldens are more difficult to separate.

OCCURRENCE Regular but very rare vagrant in W Europe, from North America, in autumn-winter.
VOICE *Klu-i*, stressed on first syllable.

white over eye

black ear-spot

long wingtips beyond tertials

white underparts

JUVENILE

Length **24–27cm (9½–10½in)**	Wingspan **50–55cm (20–22in)**

Family **Charadriidae**	Species *Vanellus spinosus*

Spur-winged Lapwing

A big and boldly marked plover, scarcely reaching Europe, this is a common species along riversides and on sandy places in the Middle East. It is easily distinguished by its black cap and breast, bold white neck, and grey-brown back. It often stands in pairs or forms noisy groups. In flight, the wings show black tips and a broad white diagonal band on top.
OCCURRENCE Rare in Greece; common in Israel, Egypt, especially along Nile.
VOICE Loud, metallic, repeated, high *titi-tirik* and similar notes.

black cap

white cheeks

black breast

grey-brown back

ADULT

Length **25–28cm (10–11in)**	Wingspan **60–65cm (23½–26in)**

Family **Charadriidae**	Species *Vanellus gregarius*

Sociable Lapwing

A rather large, bulky plover, typically associating with Lapwings, the Sociable Lapwing looks grey with a dark belly and a black and white striped face in summer. In winter, the body is more uniform sandy grey and the head less boldly marked, but still showing a dark cap and pale stripes over the eye. In flight, the wings reveal a broad white triangular patch and black tips; the tail has a black band.
OCCURRENCE Rare vagrant to W Europe from Asia, sometimes in winter; rare migrant in SE Europe.
VOICE Harsh, chattering notes in flight, but usually silent.

dark cap

white over eye

JUVENILE

Length **27–30cm (10½–12in)**	Wingspan **60–65cm (23½–26in)**

Family **Charadriidae**	Species ***Vanellus leucurus***

White-tailed Lapwing

More upright, long-legged, and elegant than other lapwings, this species is identified by its white tail with no black band, and long yellow legs that extend well behind the tail in flight. It may show a dark grey breast-band against a whiter belly. The wings have a neat black-edged white band and large black tips. Young birds are spotted above.
OCCURRENCE Very rare in Romania; very rare vagrant in W Europe in late summer.
VOICE Usually silent.

pale grey-brown head

dark grey breast-band

long yellow legs

ADULT

Length **26–29cm (10–11½in)**	Wingspan **60cm (23½in)**

Family **Scolopacidae**	Species ***Limicola falcinellus***

Broad-billed Sandpiper

A small, short-legged wader, this bird has a contrasting dark back and white belly. In spring, it has a "frosty" look, fading to darker brown with coppery edges in late summer, with long whitish stripes on the back. In winter, it is much paler and plainer grey. The best feature is then the two white lines over each eye that become bolder in summer. The bill is rather heavy, faintly kinked down, and thinner at the tip.
OCCURRENCE Breeds in Scandinavia, migrates through E Europe; rare in west, mostly in late spring.
VOICE A high, buzzing trill, *bree-eeet.*

double white line over eye

dark back

ADULT (SUMMER)

Length **15–18cm (6–7in)**	Wingspan **30–34cm (12–13½in)**

Family **Scolopacidae**	Species ***Xenus cinereus***

Terek Sandpiper

Disproportionately large-billed, this rare, short-legged sandpiper has a rather low, forward-leaning carriage that is exaggerated as it lurches and runs over muddy shores. It is plain dull greyish brown with a whiter underside, marked by a white trailing edge to the wing (but the rump is grey). In summer, it has blackish bands along the back. The legs are pale to rich orange-yellow.
OCCURRENCE Rare migrant in E Europe, very rare vagrant in W Europe, from Asia.
VOICE Soft notes in short, rapid series in flight.

large, upturned bill

ADULT (SUMMER) *yellow legs*

Length **22–25cm (9–10in)**	Wingspan **38–40cm (15–16in)**

Family **Scolopacidae**	Species ***Calidris subminuta***

Long-toed Stint

A small, long-legged, long-toed, slightly curve-billed stint, the Long-toed Stint tends to creep about like a tiny crake, or stretch upwards and look very upright. It is like a bright Least Sandpiper or a tiny Wood Sandpiper (see p.208), with yellowish legs, a dark cap reaching the bill, dark cheeks, a pale bill base, and bright upperpart "V"s on autumn juveniles.
OCCURRENCE Very rare vagrant from E Siberia.
VOICE Short trill, *chrrip.*

ADULT *cream stripes on back*

short bill

pale legs

Length **14–15cm (5½–6in)**	Wingspan **25–30cm (10–12in)**

Family **Scolopacidae**	Species *Calidris minutilla*

Least Sandpiper

Rather like a tiny Pectoral Sandpiper (see p.460), the Least Sandpiper is distinguished from the Little Stint (see p.200) by its pale legs and from Temminck's (see p.196) by its streakier back, darker breast, and fine pale "V" on the upperparts on juveniles. Pale legs instantly indicate something quite rare; the tiny (barely sparrow) size and angular shape are also distinctive. The Long-toed Stint is very similar but even rarer.

OCCURRENCE Rare vagrant in W Europe, from North America.

VOICE Sharp, abrupt *keek, ki-keek* or *tree-eep*.

streaked back

ADULT

pale legs

tiny bill

Length **13–14cm (5–5½in)**	Wingspan **25–30cm (10–12in)**

Family **Scolopacidae**	Species *Calidris tenuirostris*

Great Knot

With some resemblance to the Knot (see p.193), especially in winter, the Great Knot is nevertheless larger, with a small head, longer bill, slightly longer legs, and a longer, more tapered rear end. Juveniles are browner than young Knots, with dark scaly upperparts; they have dark breasts with rows of neat dark spots. Winter adults are grey but have darker breast spots, unlike a Knot. The bill is thick-based, tapered, and slightly downcurved; the rump is grey, like a Knot's.

OCCURRENCE Very rare vagrant in NW Europe and Middle East, from Siberia.

VOICE Vagrants mostly silent.

tapered rear end

short legs

spotted underside

ADULT (SUMMER)

Length **24–27cm (9½–10½in)**	Wingspan **40cm (16in)**

Family **Scolopacidae**	Species *Calidris bairdii*

Baird's Sandpiper

Of the vagrant small North American sandpipers, Baird's Sandpiper is the buffiest and longest-tailed, looking low-slung and tapered, with a short black bill and short blackish legs. Immatures have particularly well-marked pearly white, scaly feather edges on the upperparts and a buff breast above very white underparts. In flight, the rump is mostly dark and the wings have a thin white stripe.

OCCURRENCE Rare vagrant from North America, mostly in autumn.

VOICE Short, purring trill, *trreeet*.

short black bill

scaly back

JUVENILE

very long wings

short legs

Length **14–17cm (5½–6½in)**	Wingspan **30–33cm (12–13in)**

Family **Scolopacidae**	Species *Calidris fuscicollis*

White-rumped Sandpiper

Almost as slim and tapered as Baird's Sandpiper, the juvenile White-rumped Sandpiper in autumn is told by its less scaly upperside, with more rufous and black, a tiny pale bill base, a whiter stripe over the eye, and white "V" lines on the back. Adults are plain and grey in winter, also with long, tapered wingtips. In flight, the white patch above the tail is an obvious feature.

OCCURRENCE Very rare in W Europe, in autumn, from North America.

VOICE Thin, sharp, squeaky *tzeet*.

ADULT (WINTER)

short bill

long wings

black legs

Length **14–17cm (5½–6½in)**	Wingspan **30–33cm (12–13in)**

Family **Scolopacidae**	Species *Calidris pusilla*

Semipalmated Sandpiper

This is like a dull Little Stint (see p.200) with less clear pale "V"s above, a thicker, blunter bill, and with tiny webs between the toes. Bright juveniles have some gingery rufous on the head and back, others are greyer with a little buff around the neck and chest sides. Western Sandpipers are very similar, with slightly longer bills and more distinct streaks on the sides of the breast.

OCCURRENCE Rare vagrant from North America.
VOICE Short, muffled *tchrp*, stint-like *tip*.

JUVENILE

greyish or gingery rufous back

thick, blunt bill

partly webbed toes

Length **13–15cm (5–6in)**	Wingspan **25–30cm (10–12in)**

Family **Scolopacidae**	Species *Calidris mauri*

Western Sandpiper

Very rare in Europe, and very difficult to tell from a Semipalmated Sandpiper or Little Stint (see p.200), the Western Sandpiper is a tiny, stint-like wader with long legs and a slim, quite long, slightly curved bill. It has a slight pale "V" on its back in autumn, with a band of rufous feathers on each side, a broad pale band over the eye, and tiny webs between its toes.

OCCURRENCE Very rare vagrant in W Europe, in autumn, from North America.
VOICE High, thin, rough *jeet*.

ADULT

long, slightly curved bill

tiny webs between toes

Length **14–17cm (5½–6½in)**	Wingspan **28–31cm (11–12in)**

Family **Scolopacidae**	Species *Calidris ruficollis*

Red-necked Stint

In summer, the largely rust-red neck and upper chest of the Red-necked Stint are distinctive, although larger Sanderlings (see p.197) look similar in late summer. However, autumn vagrants are confusingly like Semipalmated Sandpipers and Little Stints (see p.200), requiring close study for identification. The very short bill, unwebbed feet, short legs, and deep body may help; grey wings against the brighter back, and a lack of white "V"s above are also useful.

OCCURRENCE Very rare in Europe, from E Siberia.
VOICE High, hoarse *chrit*.

ADULT (SUMMER)

rust-red neck

short bill

Length **13–16cm (5–6½in)**	Wingspan **25–30cm (10–12in)**

Family **Scolopacidae**	Species *Tryngites subruficollis*

Buff-breasted Sandpiper

Although often near water on migration, this sandpiper is as likely to be seen on open, dry ground where it runs in short, quick bursts. It is small, rounded, with long yellow legs and a short black bill, rather like a tiny Ruff (see p.194). The upperparts are spangled dark with scaly pale fringes. The whole neck and breast area is a warm, rich buff, slightly spotted at the sides. In flight, the rump looks dark, the wings have just a diffuse paler central band.

OCCURRENCE Regular but rare vagrant to NW Europe, from North America.
VOICE Mostly rather silent.

JUVENILE

scaly pattern on back

buff breast

long yellow legs

Length **18–20cm (7–8in)**	Wingspan **35–37cm (14–14½in)**

Family **Scolopacidae**	Species *Calidris melanotos*

Pectoral Sandpiper

One of the commoner North American birds in Europe (but still rare), this Ruff-like (see p.194) wader is quite small, with bright buff feather edges above and a white "V" on the back, a dark cap, and, most usefully for identification, a closely streaked breast sharply defined against the white belly. In flight, the rump shows oval white sides. The legs are yellow.

OCCURRENCE Rare but regular in W Europe in autumn, from North America.

VOICE Short, quite deep, throaty trill, *trrr't*.

bright buff feather edges on upperparts

closely streaked breast

dark cap

JUVENILE

defined breast-band

Length **19–23cm (7½–9in)**	Wingspan **38–44cm (15–17½in)**

Family **Scolopacidae**	Species *Calidris acuminata*

Sharp-tailed Sandpiper

A smallish wader, the Sharp-tailed Sandpiper is like the rather less rare Pectoral Sandpiper without the closely streaked breast-band. Adults have a cap that is streaked dark and pale rufous, and a streaked breast petering out in a scattering of loose spots; immatures have a plain peachy breast, finely marked at the sides. All have greenish yellow legs and a yellowish base to the short, slightly curved bill.

OCCURRENCE Very rare vagrant in NW Europe, from NE Asia (E Siberia), early autumn.

VOICE Soft, quiet *wheep*.

streaked rufous cap

short bill

JUVENILE

diffuse breast-band

Length **17–21cm (6½–8½in)**	Wingspan **40cm (16in)**

Family **Scolopacidae**	Species *Tringa flavipes*

Lesser Yellowlegs

This looks like a small, delicate, greyer, slender Redshank or small Greenshank, and is also similar to the smaller, rounder Wood Sandpiper (see pp.209, 206, 208). It has long, bright, pale yellow to orange-yellow legs and, in flight, reveals plain wings and a square white rump. The bill is thin and straight (Greater Yellowlegs has a thicker bill, faintly upturned). It often wades quite deeply and so may be confused with Wilson's Phalarope (see p.462).

OCCURRENCE Regular but rare vagrant from North America to W Europe.

VOICE High, clear *tew*, sometimes repeated at same pitch.

thin, straight bill

yellow legs

JUVENILE

Length **23–25cm (9–10in)**	Wingspan **45–50cm (18–20in)**

Family **Scolopacidae**	Species *Tringa melanoleuca*

Greater Yellowlegs

More Greenshank-like (see p.206) than the Lesser Yellowlegs, the Greater Yellowlegs is nevertheless not always easy to distinguish from the Lesser. The bill is thicker, slightly paler-based, and faintly upturned, and, in most birds, there are more obvious white spots on the upperparts (which are browner than a summer Greenshank's or Lesser Yellowlegs'). The square white rump differs from the "V"-shaped wedge on a Greenshank.

OCCURRENCE Very rare vagrant to W Europe, from North America.

VOICE Loud, often with three notes, with third note at lower pitch, *tew-tew-tew*, but sometimes very like Lesser.

square white rump

JUVENILE

Length **30–35cm (12–14in)**	Wingspan **53–60cm (21–23½in)**

Family **Scolopacidae**	Species *Tringa solitaria*

Solitary Sandpiper

A small *Tringa* sandpiper, much like the Green Sandpiper (see p.204), this species is distinguished by its dark rump. It has a less obvious white face stripe but a bolder white eye-ring than the Green Sandpiper, and is darker, duller, and shorter-legged than a Wood Sandpiper (see p.208). Its long, tapered rear end is often bobbed up and down, as with the Green Sandpiper and the smaller, browner Common Sandpiper (see p.203).

OCCURRENCE Rare vagrant to NW Europe, from North America.

VOICE Rich *tewit-weet*.

dark rump

dark, dull plumage

bold white eye-ring

ADULT

Length **18–21cm (7–8½in)**	Wingspan **35–39cm (14–15½in)**

Family **Scolopacidae**	Species *Actitis macularia*

Spotted Sandpiper

Very like a Common Sandpiper (see p.203), the shorter-tailed Spotted Sandpiper is distinguished in summer by a scattering of small or large black spots underneath. In autumn or winter, adults are plainer, greyer, with brighter yellow legs and a sharper call. Juveniles are plainer, less spotted on the edges of the longer wing feathers, but more contrastingly barred on the wing coverts.

OCCURRENCE Very rare visitor from North America, sometimes winters.

VOICE Sharp, thin *peet* or *peet-weet*.

short tail

plain greyish upperparts

ADULT (WINTER)

Length **18–20cm (7–8in)**	Wingspan **32–35cm (12½–14in)**

Family **Scolopacidae**	Species *Bartramia longicauda*

Upland Sandpiper

An unusual, slim-necked, long-tailed wader of dry ground, the Upland Sandpiper looks rather like a slim, young Ruff (see p.194) with shorter legs, a thin bill, and a dark-capped crown. The dark eyes stand out well on its pale face. The Upland Sandpiper is all-dark on wings and tail, and the underwing is dark and closely barred.

OCCURRENCE Very rare in autumn, from North America.

VOICE Whistling, bubbling note, *quip-ip-ip-ip* in flight.

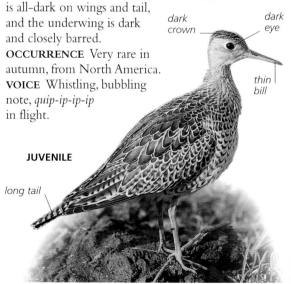

dark crown

dark eye

thin bill

JUVENILE

long tail

Length **28–32cm (11–12½in)**	Wingspan **50–55cm (20–22in)**

Family **Scolopacidae**	Species *Micropalama himantopus*

Stilt Sandpiper

In water, when its long green legs are hidden, this species is often mistaken for a grey Ruff, Redshank (see pp.194, 209), or some other medium/small wader in winter plumage. Its long, thick, slightly downcurved bill is a useful clue. In summer, the barred underparts and a rusty cheek patch are obvious. Juveniles are marked with rufous above, streaked on the flanks, and show a dark cap and pale stripe over the eye (dowitcher -like). A square white rump shows in flight.

OCCURRENCE Rare vagrant to NW Europe from North America.

VOICE Soft, chirrupy *trrr-p*.

pale stripe over eye

long, thick bill

ADULT (WINTER)

Length **18–23cm (7–9in)**	Wingspan **37–42cm (14½–16½in)**

Family **Scolopacidae**	Species *Limnodromus scolopaceus*

Long-billed Dowitcher

This wader looks something like a cross between a Redshank and a Snipe (see pp.209, 212): it is typically quite pale and rather plain in autumn or winter, with a dark cap edged by a bold white line over each eye (an obvious "V" from the front). The bill is long, thick, and snipe-like, the legs short and greenish. In flight, it reveals a white wedge or long oval on the back and a whitish trailing edge to the wing. Good views reveal broad black and narrow white bars on the tail.

OCCURRENCE Regular but still rare vagrant to W Europe from North America.

VOICE Short, sharp *kik* or *keek*, often repeated in series such as *kip-ip-ip-ip*.

close dark bars on tail

relatively short, green legs

long, thick bill

JUVENILE

Length **27–30cm (10½–12in)**	Wingspan **42–49cm (16½–19½in)**

Family **Scolopacidae**	Species *Gallinago media*

Great Snipe

Difficult to identify when out of its usual range, the Great Snipe is a big, dark, heavy snipe with a rather thick bill and heavy, dark underpart barring. The closed wing shows lines of white feather tips. In flight, the adult shows big white tail sides, reduced on a juvenile. The upperwing is a better clue, with a central dark band, edged with white, right across to the outer edge. Its low, heavy, quiet flight is also a fair clue.

OCCURRENCE Breeds in Scandinavia, Baltic area, and eastwards; rare migrant in E Europe, vagrant in W.

VOICE Occasional deep croaks.

ADULT

thick bill

dark barring on underparts

white tips to coverts

Length **26–30cm (10–12in)**	Wingspan **43–50cm (17–20in)**

Family **Scolopacidae**	Species *Phalaropus tricolor*

Wilson's Phalarope

The largest phalarope, and most prone to feed on mud, Wilson's is slender but short-legged, less elegant on land than when swimming. In summer, a striking dark band curves back from the eye and down the neck. In winter, the back is pale grey, and there is only a hint of the neck stripe. Immatures have dark feathers with buff edges on the upperparts, and yellowish legs. All have a long, fine, straight bill and a white rump in flight. Yellowlegs and Wood Sandpipers (see p.208) are very similar.

OCCURRENCE Vagrant to W Europe from North America, in late spring and autumn.

VOICE Short, nasal *vit* in flight.

long, fine bill

ADULT (WINTER)

pale legs

Length **22–24cm (9–9½in)**	Wingspan **38–44cm (15–17½ in)**

Family **Laridae**	Species *Chroicocephalus genei*

Slender-billed Gull

This looks very like the Black-headed Gull (see p.237), except for the fact that it has a white head: there is no trace of a hood. It has a rather thick, long, orange-red to blackish bill and long, orange-red legs. Juveniles have weak brown markings on the wings and a narrow tail band. A close view reveals a pale eye (unlike any similar gulls), but this is hard to see at any distance.

OCCURRENCE Rare breeder, locally along Mediterranean coastal area; vagrant elsewhere.

VOICE Low, strained version of Black-headed Gull calls.

white head

ADULT (SUMMER)

long, orange-red legs

long red to black bill

Length **37–42cm (14½–16½in)**	Wingspan **90–102cm (35–40in)**

| Family **Laridae** | Species *Chroicocephalus philadelphia* |

Bonaparte's Gull

Resembling a small Black-headed Gull (see p.237), this gull has a slim, black bill like a Little Gull's (see p.238) and a light, airy flight like a tern's. Its upperwing is like a Black-headed's (a white outer triangle and black trailing edge), but the underwing is pure white except for a sharp, thin, black edge towards the tip. Adults in summer have slaty-black hoods; in winter and on immatures, the head is white with a dark ear-spot. Juveniles have a darker diagonal band across the wing coverts and darker trailing edge than a Black-headed Gull.

OCCURRENCE Rare vagrant to W Europe from North America.
VOICE High, sharp, tern-like notes and squealing calls.

black hood
slim black bill
ADULT (SUMMER)

| Length **31–34cm (12–13½in)** | Wingspan **79–84cm (31–33in)** |

| Family **Laridae** | Species *Larus delawarensis* |

Ring-billed Gull

Like a stocky Common Gull (see p.240), but paler above, the Ring-billed Gull has less white between the grey of the back and black wingtip, a pale eye (looking dark at a distance), and a thicker bill, with a black band near the tip. The legs are greenish to yellow. One-year olds are more spotted on the neck and flanks, the black tail band less clear-cut.

OCCURRENCE Rare vagrant in NW Europe, regular in SW Britain, from North America.
VOICE Raucous, squealing calls.

black band on thick bill
pale grey back
pale iris
yellowish legs
ADULT (SUMMER)

| Length **41–49cm (16–19½in)** | Wingspan **1.12–1.24m (3¾–4ft)** |

| Family **Laridae** | Species *Ichthyaetus audouinii* |

Audouin's Gull

Once very rare, now increasing, Audouin's Gull is slimmer-winged, narrower-tailed, and stubbier-billed than the Herring Gull (see p.242). Adults are very pale grey, their wingtips extensively black with tiny white spots on the feather tips. They have grey or greenish legs and dark red bills, banded black and tipped yellow. The dark eye and long white face shape are distinctive. Young birds are dark, with mostly black tails, very long wings, and blackish legs.

OCCURRENCE Breeds in E Spain, Balearics, Morocco, and Mediterranean islands, rare in Atlantic.
VOICE Low, nasal calls.

dark eye
ADULT
grey legs

| Length **44–52cm (17½–20½in)** | Wingspan **1.17–1.28m (3¾–4¼ft)** |

| Family **Laridae** | Species *Ichthyaetus ichthyaetus* |

Great Black-headed Gull

In summer, this huge gull has a black hood, pale grey back, white outer wings, and small black wingtips, set off by yellow legs and a long yellow bill banded black and red. In winter, the hood is lost and immature birds have a grey smudge through the eye. Often confused with Herring Gulls (see p.242), the flat forehead and long bill profile are useful for identification.

OCCURRENCE Very rare in Europe, regular in Middle East outside breeding season.
VOICE Deep, nasal, throttled call in flight but mostly silent.

jet-black hood
pale grey back
yellow legs
ADULT (SUMMER)

| Length **58–67cm (23–26in)** | Wingspan **1.46–1.62m (4¾–5¼ft)** |

| Family **Laridae** | Species *Xema sabini* |

Sabine's Gull

This rare autumn migrant is brought close inshore in northwest Europe by Atlantic gales. It resembles a juvenile Kittiwake (see p.236) but the wing pattern is composed of three sharp triangles, dark (grey on adults, grey-brown on juveniles) in front, black at the tip, and pure white at the back, with no diagonal black band. Unlike immature Kittiwakes, which lose most of their black and look dull and scruffy, Sabine's Gull looks very neat.

OCCURRENCE Regular in autumn off W Europe, rare in North Sea.

VOICE Tern-like calls unlikely to be heard from migrants.

short dark bill with yellow tip

dark head (pale in autumn, winter)

ADULT (SUMMER)

black legs

| Length **30–36cm (12–14in)** | Wingspan **80–87cm (32–34in)** |

| Family **Laridae** | Species *Leucophaeus pipixcan* |

Franklin's Gull

A small, dark, short-legged gull, Franklin's Gull looks like a Laughing Gull but has a white band crossing the wing near the black and white tip. The bill is typically short, less tapered, and less drooping in appearance than a Laughing Gull's, but some are difficult to separate on the ground. Young birds have white on the breast and flanks where Laughing Gulls are dark, and dark outer primaries unlike the much paler Black-headed Gulls (see p.237). Common Gulls are paler and much larger (see p.240).

OCCURRENCE Very rare in NW Europe, from North America.

VOICE Soft nasal calls, but usually rather silent.

dark hood (jet-black in summer)

pale eye-ring

ADULT (WINTER)

| Length **32–36cm (12½–14in)** | Wingspan **80–87cm (32–34in)** |

| Family **Laridae** | Species *Leucophaeus atricilla* |

Laughing Gull

A long-winged, sharp-featured gull with a long black bill and black legs, the Laughing Gull is usually easy to identify. Summer adults have jet-black hoods with thin white eyelids. In winter, the head has only dusky smudges. The back is a deep mid-grey. Young birds are browner on the wings with black along the hind edge, have black tail bands, and are smoky grey across the breast and along the flanks, looking very contrasty with their white rump and underside.

OCCURRENCE Rare vagrant in W Europe, from North America.

VOICE Loud, squealing notes.

black bill

IMMATURE (1ST WINTER)

long wings

black legs

| Length **36–41cm (14–16in)** | Wingspan **0.98–1.1m (3¼–3½ft)** |

| Family **Laridae** | Species *Pagophila eburnea* |

Ivory Gull

Almost pigeon-like with its dumpy form and short legs, the Ivory Gull is longer and more tapered at the back and longer-winged in flight. It has black legs, dark eyes, and a grey bill with a yellow tip. Juveniles are lightly spotted with black and smudged dark on the face. Albino Kittiwakes and Common Gulls (see pp.236, 240) may cause identification problems: the bill colours are then important.

OCCURRENCE Rare vagrant in W Europe; breeds in Svalbard and high Arctic islands.

VOICE Loud, shrill tern-like calls but mostly silent in winter.

white body with dark spots

IMMATURE (1ST WINTER)

| Length **41–47cm (16–18½in)** | Wingspan **1–1.13m (3¼–3¾ft)** |

Family **Laridae**	Species *Rhodostethia rosea*

Ross's Gull

A rare Arctic species, Ross's Gull is pigeon-like in form, with rather broad-based but long wings and a wedge-shaped tail. Its very short bill is black, the short legs red or pinkish. Summer adults are flushed bright pink and have a thin black collar; winter birds are duller and the black is reduced or replaced by smudges of grey. Young birds have a Little Gull-like (see p.238) dark zigzag pattern but the hindwing is all white; they show a dark ear-spot and a smoky grey hindneck.

OCCURRENCE Rare vagrant, mostly in winter, sometimes spring, from Arctic.

VOICE Mostly silent.

thin black collar

ADULT (SUMMER)

no black on wings

Length **29–32cm (11½–12½in)**	Wingspan **73–80cm (29–32in)**

Family **Sternidae**	Species *Thalasseus maximus*

Royal Tern

This is a large, magnificent tern, almost the size of a Caspian Tern (see p.229) but more elegant. A very pale bird, it has white underwings marked only by narrow dark feather tips (Caspian Tern has a big black patch) and a white rump (the smaller Lesser Crested Tern is greyer). It is white-headed with a black nape, but has a black cap in summer; the bill is dagger-like, and rich orange. Immatures have dark primaries and hindwing bars, like a young Common Gull (see p.240).

OCCURRENCE Very rare vagrant in NW Europe, from North America and/or Africa.

VOICE Rather weak, strident, scratchy calls.

long orange bill

ADULT (WINTER)

black legs

Length **42–49cm (16½–19½in)**	Wingspan **86–92cm (34–36in)**

Family **Sternidae**	Species *Sterna bengalensis*

Lesser Crested Tern

Large and elegant, like a slightly darker-backed Sandwich Tern (see p.232), the Lesser Crested has a grey rump (hard to see), black legs, and a long, slim, dagger-like, bright orange bill. In summer, it has a ragged black crest, in winter a white forehead and crown. It is difficult to separate in isolation from the Royal Tern which is bigger, thicker-billed, and whiter-rumped. Immatures have dark wing markings like a young Common Gull (see p.240), but less bold than on a Royal Tern.

OCCURRENCE Occasional birds in Sandwich Tern colonies in summer, but a rare vagrant.

VOICE Loud, grating, Sandwich Tern-like *kirrik*.

orange bill

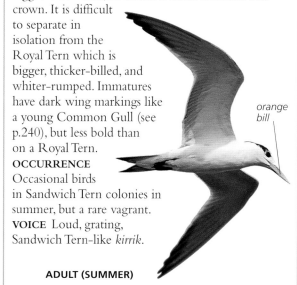

ADULT (SUMMER)

Length **33–40cm (13–16in)**	Wingspan **76–82cm (30–32in)**

Family **Sternidae**	Species *Onychoprion fuscatus*

Sooty Tern

A large black and white tern, the Sooty Tern is a tropical bird, spending most of its time well out over the sea. It looks boldly pied, with a long and deeply forked tail, the streamers tipped with a blob of white; feather textures differ, causing slight variations in some lights. The forehead has a deep, wide white patch that reaches to just above the eye. Young birds are dark above with pale bars, and mostly very dark brown below.

OCCURRENCE Rare vagrant in summer and autumn; breeds in Red Sea, Caribbean.

VOICE Mostly silent away from breeding areas.

white forehead

ADULT

deeply forked tail

Length **42–45cm (16½–18in)**	Wingspan **72–80cm (28–32in)**

Family **Sternidae**	Species ***Onychoprion anaethetus***

Bridled Tern

Like the Sooty Tern, the Bridled Tern is a tropical seabird that looks very dark above and white below, with a long, forked, white-edged, dark tail. Good views reveal a contrast between the brownish grey back and black cap and wingtips; the white forehead tapers back into a point behind the eye. Care is required to rule out the larger, blacker Sooty Tern, which is also a rare vagrant in Europe. Juveniles are dark and barred above, paler below.

OCCURRENCE Rare vagrant mostly in late summer; breeds in Red Sea, W Africa.

VOICE Mostly silent.

white over eye

brown back

forked, white-edged, dark tail

ADULT

Length **37–42cm (14½–16½in)**	Wingspan **65–72cm (26–28in)**

Family **Sternidae**	Species ***Sterna forsteri***

Forster's Tern

Very like a Common Tern (see p.233), Forster's has paler wingtips (all frosty-white or pearly grey in winter) and white underparts in summer. In winter, it shows a bold blackish mask and a black bill (like the Gull-billed Tern, see p.228); the bill is shorter than on a Sandwich Tern (see p.232) and slimmer than a Gull-billed's. Juveniles have a black mask and dusky grey centres to the tertials. The shape, size, and behaviour are much the same as for the Common Tern.

OCCURRENCE Rare vagrant to W Europe, mostly late autumn or winter, from North America.

VOICE Mostly silent in winter.

black ear patch

ADULT (WINTER)

Length **33–36cm (13–14in)**	Wingspan **64–70cm (25–28in)**

Family **Sternidae**	Species ***Chlidonias leucopterus***

White-winged Black Tern

The three *Chlidonias* terns, Black, Whiskered (see pp.231, 230), and White-winged Black, are the "marsh" terns; this is the smallest and dumpiest, with a slightly quicker flight action than the other two. In summer, it is boldly black with white wings and tail; autumn juveniles have a dark back, pale wings, whitish rump, and clean white breast sides (no dark spot as on Black); winter adults are much paler, with whitish rumps.

OCCURRENCE Breeds in E Europe, rare migrant elsewhere, mostly in autumn.

VOICE Short, simple, squeaky *kek* notes in flight.

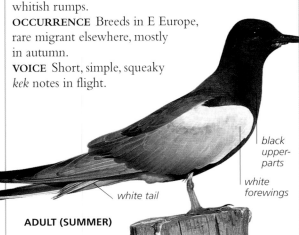

black upper-parts

white forewings

white tail

ADULT (SUMMER)

Length **20–24cm (8–9½in)**	Wingspan **50–56cm (20–22in)**

Family **Alcidae**	Species ***Uria lomvia***

Brünnich's Guillemot

Extremely like a Guillemot (see p.222), Brünnich's is more thickset and thicker-billed, with a white streak along the gape and a pointed white breast against the dark foreneck (rounded on Guillemot). Its flanks are pure white, not streaked. In winter, the head is dark to below the eye, without the dark eye-stripe of a Guillemot or the white patch behind the eye of a Razorbill (which it otherwise resembles, see p.220). It looks the most solid, short-necked, and deep-bellied of the auks in flight.

OCCURRENCE Breeds in Iceland, far N Scandinavia; vagrant in NW Europe.

VOICE Crow-like growling notes.

ADULT (SUMMER)

white stripe on black bill

Length **40–44cm (16–17½in)**	Wingspan **64–75cm (25–30in)**

Family **Pteroclididae**	Species *Pterocles orientalis*

Black-bellied Sandgrouse

Sandgrouse are pigeon-like in form, partridge-like in plumage, with tiny bills, short legs, long, tapered wings, and long tails. The Black-bellied has a short tail, and is chunky, with an obvious bold black underside and white underwings with black tips. On the ground, the male's plain face and breast (one narrow bar at the lower edge) are distinctive; the back is spotted on the male and closely barred in a complex pattern on the spotted female. Black-bellied Sandgrouse are shy; they fly long distances each morning to drink.

OCCURRENCE Breeds in C and S Spain, Turkey, in wide open, dry plains.

VOICE Rolling, bubbling flight call, fast trill slowing to stutter.

FEMALE

black underside

Length **30–35cm (12–14in)**	Wingspan **60–65cm (23½–26in)**

Family **Pteroclididae**	Species *Pterocles alchata*

Pin-tailed Sandgrouse

Slimmer than the Black-bellied Sandgrouse, the Pin-tailed has a fine tail spike, a bright white belly, and white underwings with bold black tips (like Black-bellied). On the ground, the face looks rufous with a black eye-stripe, the breast narrowly banded with black, the back beautifully spotted (male) or barred (female) with buff in a delicate and complex pattern. Large flocks visit pools to drink each morning.

OCCURRENCE Rare breeder in C and S Spain, very rare in S France, in arid, stony areas.

VOICE In flight, rhythmic, grating, descending *cata-cata* or *rrria-rrria*.

spotted back

MALE

Length **28–32cm (11–12½in)**	Wingspan **55–63cm (22–25in)**

Family **Cuculidae**	Species *Clamator glandarius*

Great Spotted Cuckoo

This unusual-looking bird is quite unlike the Cuckoo (see p.253). In flight, it is long and slim, with broad-based, tapered, slightly rounded wings and a long, narrow tail, its head small and held up, the chest rather deep. Adults are grey above, with white spots, whitish below, and grey on the crown. Juveniles have a black cap, darker back with pale spots, and a rusty outer wing. The yellowish chest and white belly are striking. It perches with wings and tail drooped.

OCCURRENCE Breeds in Spain, Portugal, and S France; rare in Italy; vagrant farther north.

VOICE Loud, rattling or cackling trill often given from cover.

white below

ADULT

white-spotted, grey upperparts

Length **35–39cm (14–15½in)**	Wingspan **55–65cm (22–26in)**

Family **Cuculidae**	Species *Coccyzus americanus*

Yellow-billed Cuckoo

A small, neat, white-breasted cuckoo, the Yellow-billed Cuckoo is pale brown above with rusty wingtips and black tail sides with big white spots. The white tail feather tips may be obvious when perched. The small eyes are dark, and the bill short and slightly curved. In flight the long wings and tail and the slender, raised head give a typically cuckoo appearance: it usually swoops upwards to a perch. Yellow-billed Cuckoos rarely survive more than a day or so when they reach Europe.

OCCURRENCE Very rare vagrant to NW Europe in late autumn, from North America.

VOICE Vagrants are silent.

black-tipped yellow bill

pale brown above

white spots on tail when spread

Length **29–32cm (11½–12½in)**	Wingspan **48–52cm (19–20½in)**

Family **Psittacidae**	Species *Psittacula krameri*

Ring-necked Parakeet

This introduced bird to Europe and the Middle East thrives in very localized areas. The typical long-tailed parakeet form, with a short bill and bright green plumage, as well as the loud calls, are distinctive (but other species escape at times and could be confused). Males have a black chin and dark neck-ring (red on the nape); females have a plain green head.

OCCURRENCE Various populations from SE England to Istanbul and Cairo, mostly in towns and suburbs.

VOICE Loud, squealing, squawking calls.

red nape

black chin of male

MALE — *long tail*

Length **39–43cm (15½–17in)**	Wingspan **42–48cm (16½–19in)**

Family **Strigidae**	Species *Strix nebulosa*

Great Grey Owl

Sitting upright in a tree this owl looks enormous; in flight it is very impressive. It is very large-headed with a grey facial disc and white marks between the eyes. The outer wings have buff patches and dark bars. Young birds look dark grey; watching them may attract dangerous attacks from angry parents.

OCCURRENCE Breeds in boggy northern forests, in NE Sweden and Finland; rare vagrant elsewhere.

VOICE Deep, growling notes; song is series of slow, booming hoots.

white crescents between small yellow eyes

broad grey facial disc

ADULT *massive body*

Length **59–68cm (23–27in)**	Wingspan **1.28–1.48m (4¼–4¾ft)**

Family **Strigidae**	Species *Strix uralensis*

Ural Owl

This big, grey-brown owl has a longer tail than a Tawny (see p.261) or Great Grey Owl, a plainer, pale grey-buff face, and small dark eyes (with no white crescents). The back has two lines of pale spots; the underside is white with dark streaks. The wings are more uniform than a Great Grey's, although the tips are closely barred. Young birds are paler-faced and browner than young Great Greys, distinguished by their eye colour.

OCCURRENCE Rare breeder in mountain forest in E Scandinavia, NE Europe, and E Europe.

VOICE Deep, cooing hoots, two, then two, then three with pauses.

dark streaks on pale underside

ADULT — *long tail*

Length **50–59cm (20–23in)**	Wingspan **1.03–1.24m (3½–4ft)**

Family **Strigidae**	Species *Nyctea scandiaca*

Snowy Owl

A huge owl, big-headed and broad-winged, with a tiny hint of ear tufts, the Snowy Owl is nearly all-white (male) or white with narrow dark bars (female). Young birds are mostly grey-brown. Barn Owls (see p.256) in car headlights can look big and white but Snowy Owls are giants, with a distinctive fast-up, slow-down wingbeat and a liking for remote landscapes.

OCCURRENCE Rare breeder in Iceland and N Scandinavia; vagrant elsewhere.

VOICE Grating, chattering, and barking notes.

yellow eye

all-white body

MALE

Length **53–65cm (21–26in)**	Wingspan **1.25–1.5m (4–5ft)**

Family **Strigidae**	Species **Surnia ulula**

Hawk Owl

As it perches on a treetop or, more especially, flies across a forest clearing, this owl gives a hawk-like effect. It has a much bigger head, with a bold black "frame" to the white face and fierce yellow eyes; the back is dark, broadly smudged and spotted with white. Young Hawk Owls are darker and more uniformly grey. No other owl has the same big-headed, long-tailed, upright shape of the Hawk Owl.
OCCURRENCE Breeds in N Scandinavia in boggy areas or clearings in forest.
VOICE Long, rapid, bubbling song and fast, chattering calls.

black "frame" to white face

white underside with grey bars

long tail

ADULT

Length **35–43cm (14–17in)**	Wingspan **69–82cm (27–32in)**

Family **Caprimulgidae**	Species **Caprimulgus ruficollis**

Red-necked Nightjar

This nightjar is restricted in range and habitat and best separated from the common Nightjar (see p.265) by its call. Its rusty collar is distinctive in good light; both sexes have white wing and tail spots and usually a large white throat patch. The shape and actions are the same as a Nightjar's, very short- and broad-headed, long-tailed, and long-winged; the tail is often fanned and broad, almost like a third wing.
OCCURRENCE Breeds almost throughout Spain and Portugal, rare in S France; rare vagrant elsewhere.
VOICE Repetitive, wooden, hollow, slightly squeaky double tap, *ko-tok ko-tok ko-tok ko-tok*.

rusty collar

Length **30–34cm (12–13½in)**	Wingspan **60–65cm (23½–26in)**

Family **Apodidae**	Species **Apus caffer**

White-rumped Swift

Overhead the White-rumped Swift is a fast-moving, narrow-winged, shapely swift, very dark except for paler marks on the hindwing and sometimes a paler midwing band. It has narrow, swept-back, scythe-shaped, stiff wings and a narrow, forked tail, often held closed in a single spike. The white rump is a narrow, curved band barely visible from below (unlike the broad, easily seen rump of the Little Swift).
OCCURRENCE Breeds very locally in SW Spain; winters in Africa.
VOICE Short, hard notes running together as fast trill.

stiff dark wings

crescentic white rump

pale underwing

narrow, forked tail

Length **14–15cm (5½–6in)**	Wingspan **33–37cm (13–14½in)**

Family **Apodidae**	Species **Apus affinis**

Little Swift

With straight, blade-like wings and a short, square tail, the Little Swift looks paler-winged and darker-bodied overhead, with a dull white throat, but in any other view the square white rump is very obvious. In comparison, House Martins (see p.326) are less stiff- and narrow-winged, and white beneath as well as on the rump; White-rumped Swifts have scythe-like wings, a narrow white rump, and a deeply forked tail. Flocks of Little Swifts are often seen over towns and villages in areas where this species is common.
OCCURRENCE Very rare vagrant from N Africa and Middle East.
VOICE Fast, twittering sounds.

dull white throat

broad white rump

square tail

Length **12–13cm (4¾–5in)**	Wingspan **32–34cm (12½–13½in)**

Family **Picidae**	Species *Dendrocopos syriacus*

Syrian Woodpecker

This is the most similar woodpecker to the common Great Spotted Woodpecker (see p.280). Male (red nape), female (black nape), and juvenile (red crown) variations are like Great Spotted but the red under the tail is replaced by pale pink-red on Syrian. The black face-stripe does not join the black hind-neck (hard to tell when the bird is hunched), and the bill is rather long.

OCCURRENCE Common in SE Europe, scarce but spreading northwest into E Europe.

VOICE Soft *kik;* drumming slightly longer and less abrupt than Great Spotted's.

pale pink-red under tail

FEMALE

Length **23–25cm (9–10in)**	Wingspan **38–44cm (15–17½in)**

Family **Picidae**	Species *Dendrocopos leucotos*

White-backed Woodpecker

A rare, large woodpecker of undisturbed forest, this obvious pied type has barred wings (no white shoulder patch) and a white rump. Close views are difficult, but show a red crown on males and buffish underparts deepening to pinkish red under the tail. A white band between the dark back and cheek patches is distinctive. Old woodland is important for this declining species.

OCCURRENCE Very rare in Pyrenees; scattered resident in E Europe and S Scandinavia.

VOICE Dull, throaty *kik,* sometimes repeated; drumming accelerates.

red crown

pinkish red under tail

MALE

Length **25–28cm (10–11in)**	Wingspan **40–45cm (16–18in)**

Family **Alaudidae**	Species *Calandrella rufescens*

Lesser Short-toed Lark

This small, pale, streaky lark is much like the Short-toed Lark (see p.316), but distinguished by a gorget of fine, dark streaks; it also has more uniform wings (less contrasted rows of dark feather centres). In some ways it looks more like an undersized Skylark (see p.320), but lacks the pale hind edge to the wings. Close views should show a longer wingtip point. The calls are important in helping to make identification certain.

OCCURRENCE Rare in S and E Spain and Turkey, Middle East, and N Africa.

VOICE Dry, buzzing trill, *drrrrt;* song rich, varied, quick, with buzzy calls intermixed.

streaked breast

long wing point

Length **13–14cm (5–5½in)**	Wingspan **24–32cm (9½–12½in)**

Family **Alaudidae**	Species *Chersophilus duponti*

Dupont's Lark

In its hot, sandy, open, often saline habitat, Dupont's Lark is usually difficult to see: it prefers to run rather than fly when disturbed. This streaky lark stands upright, showing a closely streaked breast and white underside. In flight, the wings look plain above, pale below (unlike Skylark or Calandra Lark, see p.320, 315), and the tail a little longer than on Crested Lark (see p.317).

OCCURRENCE Rare and local in E Spain and North Africa, in short, sparse, vegetation.

VOICE Call thin *choo-chee;* song short, whistling, fluty notes.

flattish crown

curved bill

long, narrow neck

Length **17–18cm (6½–7in)**	Wingspan **30cm (12in)**

Family **Motacillidae**	Species *Anthus richardi*

Richard's Pipit

This is a large pipit, Skylark-like (see p.320) in its size, bulk, and general plumage. It often stands upright, breast pushed out, on long, thick legs, its bold, strong bill quite distinct. There is no trace of a crest. The face is marked by a bold whitish area around the eye with a dark mark beneath; there is a thick black line on each side of the throat (less marked on a Tawny Pipit, see p.404). The long tail is often bobbed.

OCCURRENCE Regular, but rare, late autumn migrant in NW Europe, from Asia.

VOICE Loud, rasping *schreep* and quieter variations.

long blackish tail with white sides

very long hind claws

Length **17–20cm (6½–8in)**	Wingspan **29–33cm (11½–13in)**

Family **Motacillidae**	Species *Anthus godlewski*

Blyth's Pipit

Only recently observed with any regularity in Europe, Blyth's Pipit is difficult to identify, resembling Richard's and juvenile Tawny Pipits (see p.404). It is fractionally smaller than Richard's, with a shorter tail, a slightly richer underside colour, a slightly shorter, pointed bill, and shorter hind claws. It may look more like a small pipit species and more wagtail-like than Richard's, but only close observation and several clearly heard calls can separate them for certain.

OCCURRENCE Very rare vagrant in NW Europe, from Asia.

VOICE Slightly higher than Richard's, less explosive, with fading, breathy quality, *psh-eee*.

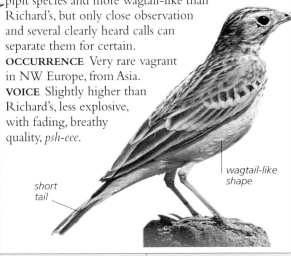

wagtail-like shape

short tail

Length **15–17cm (6–6½in)**	Wingspan **28–30cm (11–12in)**

Family **Motacillidae**	Species *Anthus hodgsoni*

Olive-backed Pipit

Looking rather dark and uniform above or bright and streaked in front in a brief view, this pipit reveals a subtle pattern on closer examination. It is greenish, with very soft streaking above, and has a dark cap, a broad, bright cream stripe above the eye, a dark stripe through the eye, and a cream spot on the ear coverts. The underside is bright yellow-buff to buff with bold black streaks. It frequently walks in longish vegetation, bobbing its tail, but flies into trees if disturbed.

OCCURRENCE Rare vagrant in NW Europe from Asia, mostly in late autumn.

VOICE Tree Pipit-like hoarse or buzzing *spees* or *tees*.

bold dark streaks on pale underside

Length **14–15cm (5½–6in)**	Wingspan **24–27cm (9½–10½in)**

Family **Motacillidae**	Species *Anthus gustavi*

Pechora Pipit

Slim and streaky like most pipits, the Pechora Pipit has bolder pale stripes on the back (edged black), striking white wingbars, a buff breast, a white belly streaked with black, and a pinkish-based bill. It is difficult to see well and crouches when disturbed. The breast/belly contrast, wingbars, and longer wingtips help separate it from a juvenile Red-throated Pipit (see p.407).

OCCURRENCE Rare vagrant in autumn in NW Europe, from Asia.

VOICE Short, slightly buzzed, clicking *dzep*, not often heard.

bold streaks

two wingbars

Length **14–15cm (5½–6in)**	Wingspan **23–25cm (9–10in)**

| Family **Motacillidae** | Species ***Motacilla citreola*** |

Citrine Wagtail

In spring, male Citrine Wagtails look like Yellow Wagtails (see p.401), but with a greyer back and a narrow black band between the back and yellow head. Females lack the black, have less yellow, but show yellow around dark cheeks (which have a paler centre). Autumn juveniles are like young Pied Wagtails (see p.403), with a plainer chest; they have pale-centred cheeks, and are pale buffish between the eye and bill.

OCCURRENCE Regular in Middle East, rare in SE Europe, vagrant (mostly autumn) in NW Europe.

VOICE Like Yellow, but harsh, distinctly buzzed *tzsip*.

black collar

yellow underside

MALE (SPRING)

| Length **16–17cm (6½in)** | Wingspan **24–27cm (9½–10½in)** |

| Family **Turdidae** | Species ***Cercotrichas galactotes*** |

Rufous Bush Robin

Bush robins are small, long-tailed chats, often raising and fanning their tails and bounding along in leaping hops on the ground. This species is bright buff, with a ginger-orange tail tipped with narrow black and big white spots. A bold white line over the eye and black eye-stripe are distinct. In SE Europe and the Middle East, birds have dull sandy brown on the head and back, with a contrasted rufous rump. The tail spots are especially obvious in flight.

OCCURRENCE Scarce breeder in S Spain, Balkans, Middle East, and North Africa.

VOICE Clicking and buzzing calls; high, clear, thrush-like song.

bold white line

rufous tail with white spots

| Length **15–17cm (6–6½in)** | Wingspan **22–27cm (9–10½in)** |

| Family **Turdidae** | Species ***Tarsiger cyanurus*** |

Red-flanked Bluetail

This rare bird inhabits dense northern forests. Males are slaty blue above, brighter on the crown, with a white chin and pale underside edged orange. Females and immatures are browner, with dusky chests and orange flanks; the tail is dull steel-blue. The white throat stands out as a well-defined wedge. In general, the form and behaviour resemble the Robin (see p.376), but it is a shy bird. It frequently flicks its wings and tail.

OCCURRENCE Breeds in extreme NE Europe; very rare vagrant in NW Europe in autumn.

VOICE Short whistle and hard, short *tak*; song short, bright, melancholy phrases.

steel-blue tail

orange flanks

FEMALE

| Length **13–14cm (5–5½in)** | Wingspan **21–24cm (8½–9½in)** |

| Family **Turdidae** | Species ***Oenanthe isabellina*** |

Isabelline Wheatear

A pale, female or young Wheatear (see p.386), especially of the large Greenland race, can look very much like this rarer species. Greyish brown above, with slightly darker and weakly marked wings, it has a tapering whitish stripe over the eye (less broad and flared than on Wheatear). The tail has a broad black band with a very short central "T" stem, leaving a squarish, slightly creamy white rump.

OCCURRENCE Breeds in NE Greece, Turkey, and Middle East; rare vagrant in NW Europe.

VOICE Short, hard call; song includes short, repeated whistles.

thick, long bill

dark spot on wings

creamy buff underside

black tail

| Length **15–16cm (6–6½in)** | Wingspan **27–31cm (10½–12in)** |

Family **Turdidae**	Species *Oenanthe cypriaca*

Cyprus Wheatear

Earlier regarded as a race of the Pied Wheatear, this bird is now treated as a species. It likes rocky areas with short, often sparse, vegetation. Males have dusky grey caps outlined with white, against a largely dark body, except for buff and white underparts and a vivid white rump and tail base; females are browner, rich buff beneath with a dark throat. The tail has a broad black band at the tip.

OCCURRENCE Except as a migrant through the Middle East to Africa, confined to Cyprus, March to October.

VOICE A sharp *tsak* and upswept *tsuee*, often in combination; buzzy, repetitive song.

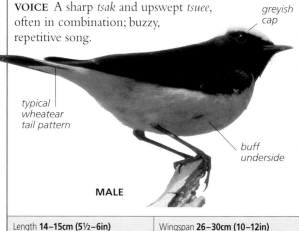

greyish cap

typical wheatear tail pattern

buff underside

MALE

Length **14–15cm (5½–6in)**	Wingspan **26–30cm (10–12in)**

Family **Turdidae**	Species *Catharus minimus*

Grey-cheeked Thrush

Several North American thrushes are small, like softly marked miniature Song Thrushes (see p.374). The Grey-cheeked Thrush is dull olive-brown, with a greyer face marked by a thin, weak eye-ring and cold greyish chest marked by rounded, blurry, dark spots. The underwing (usually hard to see) has several dark and whitish bands. This is a shy bird, usually remaining on or close to the ground in thick bushes.

OCCURRENCE Very rare late autumn vagrant in NW Europe, from North America.

VOICE Shrill *tsee*.

blurry dark spots

olive-brown body

AUTUMN

Length **15–17cm (6–6½in)**	Wingspan **28–32cm (11–12½in)**

Family **Turdidae**	Species *Zoothera dauma*

White's Thrush

Only a juvenile Mistle Thrush (with bright, pale speckling above and pale wingbars; see p.376) looks much like a White's Thrush. The latter is a difficult bird to observe, being wary and flighty and keeping to low cover where it creeps and shuffles secretively. It is a pale, sandy buff bird with black crescents below; the back has black-edged buff spots. In flight, the tail looks dark with pale sides and the underwing flashes striking black and white stripes (Mistle Thrush has plain white underwings).

OCCURRENCE Rare vagrant, typically in late autumn, in NW Europe, from Asia.

VOICE Silent bird on migration.

black-edged, buff spots on back

black crescents below

Length **27–31cm (10½–12in)**	Wingspan **40–45cm (16–18in)**

Family **Turdidae**	Species *Zoothera sibirica*

Siberian Thrush

Like the other *Zoothera* species, White's Thrush, this bird has bold black and white bands under the wings, sometimes visible as it flies off. Males are slate-grey, blacker on the face, with a white stripe over each eye, a white belly, and white tips to the outer tail feathers; immature males are duller. Females are brown, barred with black crescent-shaped marks below, with a buff line over the eye and a broader buff band under the cheeks. The white tail corners and banded underwings are distinctive.

OCCURRENCE Very rare vagrant in NW Europe in autumn/winter, from Asia.

VOICE Thin, simple call, *tsee*.

buff band under cheeks

dark crescent-shaped marks below

FEMALE

Length **20–21cm (8–8½in)**	Wingspan **34–36cm (13½–14in)**

| Family **Turdidae** | Species *Turdus ruficollis* |

Dark-throated Thrush

The two forms, the Black-throated and the Red-throated, are both Blackbird-like (see p.372) with pale grey-brown backs and dull white underparts with dusky streaking. The face and chest are black on male Black-throateds and dusky rust-red on Red-throateds. Immatures are streaked on the face and breast, with a hint of the black or red colour developing in males. The rump is a paler grey than the tail.

OCCURRENCE Both very rare in W Europe, in autumn/winter, from Asia.

VOICE Fieldfare-like (see p.373) chacking calls.

grey back

dark chest

JUVENILE MALE (BLACK-THROATED)

| Length **23–26cm (9–10in)** | Wingspan **37–40cm (14½–16in)** |

| Family **Turdidae** | Species *Turdus obscurus* |

Eyebrowed Thrush

A neat, smallish thrush, the Eyebrowed Thrush shares the pale stripe over the eye with a Redwing (see p.375). It has a white spot under the eye, and orange flanks and breast-band. The bill is yellowish at the base and the legs look dull orange-yellow. The underside is more uniformly orange than any Redwing's and entirely lacks streaks or spots. Adults have a grey hood, more contrasted than the autumn immatures.

OCCURRENCE Rare autumn vagrant in NW Europe, from Siberia.

VOICE Thin, Redwing-like *tseeee* note.

pale stripe over eye

white spot under eye

AUTUMN

| Length **21–23cm (8½–9in)** | Wingspan **36–38cm (14–15in)** |

| Family **Turdidae** | Species *Turdus naumanni* |

Dusky Thrush

The two forms of this species are very different: the "Dusky Thrush" has rich, dark upperparts, bold black and white face stripes, and white underparts spotted black (concentrated in a breast-band and flank spots), and "Naumann's Thrush" has orange-buff on the face, orange spots on the underside, and a rusty rump and tail. Both are obvious medium-large thrushes, bold and upright like a Song Thrush (see p.374), but shy and likely to fly off low and fast if approached.

OCCURRENCE Very rare vagrant in W Europe, from Siberia, mostly in autumn/winter.

VOICE Rather hard, sharp alarm note.

white stripes on face

DUSKY FORM

| Length **20–23cm (8–9in)** | Wingspan **36–39cm (14–15½in)** |

| Family **Sylviidae** | Species *Sylvia hortensis* |

Orphean Warbler

A big warbler of olive groves and bushy slopes, the Orphean Warbler is rather like a massive Lesser Whitethroat (see p.340). Males have dark grey heads, somewhat blacker on the cheeks, with big white throats and whitish eyes. Females and immatures are slightly browner above and slightly buffer on the chest (not so white beneath or pure grey above as the male), and have dark eyes. The long dark tail has white sides, unlike a Blackcap's (see p.337).

OCCURRENCE Breeds in Mediterranean region and N to C France; very rare outside this range.

VOICE Hard *tak*; simple/repetitive in SW, more prolonged in SE Europe.

pale eye

buff chest

long tail

MALE

| Length **15–16cm (6–6½in)** | Wingspan **20–25cm (8–10in)** |

Family **Sylviidae**	Species *Sylvia melanothorax*

Cyprus Warbler

A typical *Sylvia* warbler in its spiky-billed, capped appearance with an inquisitive expression, bulky body, and slim, sometimes raised tail, this is a Sardinian Warbler-like (see p. 344) bird with dusky markings beneath. Males show a white streak between the black head and dark-spotted throat; the underparts are spotted blackish. Females are greyer, the chest more subtly mottled. Both have a red orbital ring and a thin white outer eye-ring. All have dark spots under the white-edged tail.

OCCURRENCE Breeds in Cyprus; rare migrant through Middle East.

VOICE Dry, ticking notes and rattling alarm call.

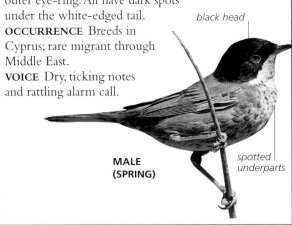

black head

spotted underparts

MALE (SPRING)

Length **13cm (5in)**	Wingspan **15–18cm (6–7in)**

Family **Sylviidae**	Species *Sylvia rueppelli*

Rüppell's Warbler

Bigger than a Sardinian Warbler (see p.344), Rüppell's Warbler is even more striking, with its black face, grey nape, white moustachial stripe, and big black throat. A red eye-ring is also obvious. Females have a grey face and a pale throat with darker spots. Juveniles are paler still, grey-faced, with a faint reddish eye-ring; they share the adults' dark-centred, pale-edged wing feathers and at least a hint of a white moustache. All have pale reddish brown legs.

OCCURRENCE Breeds in SE Europe; very rare vagrant elsewhere; winters in Africa.

VOICE Sharp, hard *zak* and rattle.

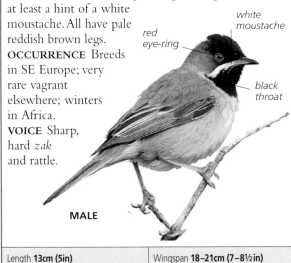

white moustache

red eye-ring

black throat

MALE

Length **13cm (5in)**	Wingspan **18–21cm (7–8½in)**

Family **Sylviidae**	Species *Sylvia conspicillata*

Spectacled Warbler

This warbler looks very like a small Whitethroat (see p.341). The male has a grey head with a white throat (greyer in the centre) and a black patch between the eye and the bill; the eye is encircled by white. The back is grey-brown, and the wings have a rusty patch; the breast is darker pinkish than a Whitethroat's. The female is browner-headed, like the female Whitethroat, but the wing is more uniform rust-brown, the wingtip projection shorter. Juveniles in autumn are similar.

OCCURRENCE Breeds in Spain east to Italy and in the Canaries; very rare vagrant farther north.

VOICE Distinctive buzzing *dzz* or *d-rrr*, sometimes longer, dry, hesitant rattle.

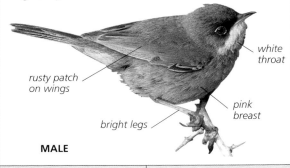

white throat

rusty patch on wings

pink breast

bright legs

MALE

Length **12–13cm (4¾–5in)**	Wingspan **14–17cm (5½–6½in)**

Family **Sylviidae**	Species *Sylvia sarda*

Marmora's Warbler

Split into two species, Marmora's and Balearic (*S. balearica*) Warblers, both similar to a Dartford Warbler (see p.342) in size, shape, behaviour, and general character, but much more restricted in range. It is a greyer bird, the male rather plain smoky grey except for a paler throat, red bill base, red eye, and red legs. Females are paler, duller but equally grey, with no trace of brown; immatures are just slightly tinged with olive-brown above, a little less so on the wings than a juvenile Dartford Warbler, and a little paler, purer grey, on the throat than the commoner bird.

OCCURRENCE Scarce residents in Corsica and Sardinia; and in the Balearics; very rare vagrants elsewhere.

VOICE Dull, buzzy *tshek* or *tsak* notes; song quite soft, fast rattling warble with brighter trill at end.

MALE (SUMMER)

Length **13–16cm (5–6½in)**	Wingspan **13–17.5cm (5–7in)**

Family **Sylviidae**	Species *Acrocephalus paludicola*

Aquatic Warbler

Rare and declining, the Aquatic Warbler is much like a bright, strongly patterned Sedge Warbler (see p.350), especially the immature Sedge that tends to have a pale crown stripe and some faint breast streaks resembling the marks on Aquatic. A peach-buff bird with streaks of black and cream, it has a striped head with a sharply defined pale crown stripe. Fine streaks on the chest and flanks are distinctive in summer. A cream "V" on the back is characteristic.

OCCURRENCE Rare breeder south of Baltic; regular but very rare migrant in reedbeds in W Europe.

VOICE Short, metallic *tak*; song varied, less energetic than that of Sedge Warbler.

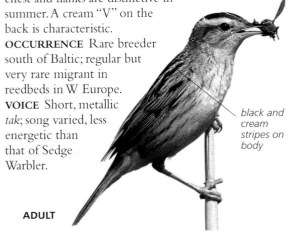

black and cream stripes on body

ADULT

Length **12–13cm (4¾–5in)**	Wingspan **17–19cm (6½–7½in)**

Family **Sylviidae**	Species *Locustella fluviatilis*

River Warbler

Small warblers with rounded tails, very long undertail coverts, and rounded outer edges to the closed wings, the *Locustella* species have a thick, tapered rear end and a sharp bill. Their songs are insect-like, prolonged trillings: that of the River Warbler is the best way to locate it. This is a dark brown bird, with whitish tail feather tips and a gorget of soft, brown streaks (unlike Reed, Savi's, or Cetti's Warblers; see pp.352, 346, 332) but a plain back (unlike Grasshopper or Sedge Warblers; see pp.345, 350).

OCCURRENCE Breeds in NE Europe from Baltic south to Black Sea; rare in summer in W Europe.

VOICE Song has rhythmic, fast, hissing, mechanical quality, *tsi-tsi-tsi-tsi-tsi-tsi-tsi-tsi*.

plain back

ADULT

rounded edge of wings

Length **15–16cm (6–6½in)**	Wingspan **19–22cm (7½–9in)**

Family **Sylviidae**	Species *Locustella lanceolata*

Lanceolated Warbler

The most-streaked *Locustella* species, this is a particularly secretive bird, creeping and hiding in the sparsest cover and very difficult to watch. It looks like a small, dark Grasshopper Warbler (see p.345) with striped underparts (or in autumn at least a gorget of diffuse streaks) and quite distinct, narrow dark spots under the tail (softer, longer marks on Grasshopper). Narrower, more defined pale edges to the tertials may aid identification but, while typical ones are distinctive, there is a problem with overlapping features with these two species.

OCCURRENCE Very rare but regular vagrant in NW Europe, from Siberia, in autumn.

VOICE Short clicking call note, rarely heard.

pale edges to tertials

ADULT

Length **12cm (4¾in)**	Wingspan **15–16cm (6–6½in)**

Family **Sylviidae**	Species *Acrocephalus dumetorum*

Blyth's Reed Warbler

This species is dull and plain in appearance, with markedly uniform wings. It has a long bill and a pale stripe from the bill to the eye (less clear behind the eye). Its wingtips are short and its legs dark (Marsh Warbler's are long and pale, respectively; see p.351). The wingtips are plain dark (Marsh has sharp pale feather edges) and the bill has a pale base, the lower mandible tipped darker (completely pale on Marsh).

OCCURRENCE Breeds in NE Europe, rare vagrant in W Europe in summer and autumn.

VOICE Short, hard, clicking call; song rich, varied, each phrase slowly repeated several times.

pale stripe from bill to eye

short wingtips

Length **13–14cm (5–5½in)**	Wingspan **17–19cm (6½–7½in)**

| Family **Sylviidae** | Species *Acrocephalus agricola* |

Paddyfield Warbler

This small, pale, reed-type warbler is marked by a strong pale stripe over the eye, edged darker above and below. Its bill is quite short and pale, with a dark tip. A rufous rump may be obvious (less so on autumn juveniles). The short wings, with quite strongly patterned tertials (dark centres and pale edges), help to separate it from Blyth's Reed Warbler. The unrelated Booted Warbler can also look very similar.

OCCURRENCE Breeds around Black Sea; very rare vagrant in W Europe.

VOICE Short, hard *tack* and *chek* notes; fast, varied song with much mimicry.

short wingtips

strong pale stripe over eye

| Length **12–13cm (4¾–5in)** | Wingspan **15–17cm (6–6½in)** |

| Family **Sylviidae** | Species *Hippolais olivetorum* |

Olive-tree Warbler

The largest of the *Hippolais* group, the Olive-tree Warbler is one of Europe's largest warblers, typically seen moving heavily through olive, almond, or holm oak foliage. It has a pale, strong, dagger-like bill, thick dark legs, and a long, square, white-edged dark tail. Its wingtip projection is particularly long. It appears very grey (less so on more olive-hued juveniles) with a marked pale wing panel. There is a short white line from the bill to just above (not behind) the eye.

OCCURRENCE Scarce breeder in Balkans and Middle East; winters in Africa.

VOICE Short, hard *tack*; harsh, grating, repetitive song.

long wingtips

dagger-like pale bill

thick dark legs

| Length **16–18cm (6½–7in)** | Wingspan **24–26cm (9½–10in)** |

| Family **Sylviidae** | Species *Hippolais languida* |

Upcher's Warbler

A rather large *Hippolais* warbler, with a pale, broad-based, dagger-like bill, Upcher's Warbler has short undertail coverts and long, square tail with whitish sides. The wings have a paler central panel when closed, formed by pale feather edges. Strong dark legs and a habit of swaying its tail from side to side are helpful identifying features. It is slightly larger and rounder-headed than the very similar Olivaceous Warbler.

OCCURRENCE Rare summer visitor in Middle East; winters in Africa.

VOICE Hard, metallic *tack* note; energetic warbling song.

pale sandy grey above

paler below

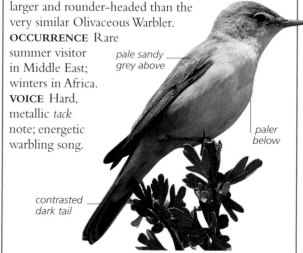

contrasted dark tail

| Length **14–15cm (5½–6in)** | Wingspan **20–23cm (8–9in)** |

| Family **Sylviidae** | Species *Hippolais pallida* |

Olivaceous Warblers

Split into Eastern and Western Olivaceous Warblers, these neat, long, flat-headed warblers are distinguished by their lack of features as much as anything else. They have an all-pale lower mandible and greyish legs. The wings are plain although the feather tips are paler. The wingtip point is short (longer on Icterine, p.347 shorter on Booted). They frequently dip their tail, like a Chiffchaff (see p.335).

OCCURRENCE Western scarce in Spain, Eastern more common in SE Europe. Rare vagrants elsewhere.

VOICE Sparrow-like twitter and hard, dry *tack* calls; song unmusical, fast warble with recurring pattern.

narrow whitish sides of tail

| Length **12–14cm (4¾–5½in)** | Wingspan **18–21cm (7–8½in)** |

Family **Sylviidae**	Species *Iduna caligata*

Booted Warbler

This *Hippolais* warbler is like a Willow Warbler (see p.336) in general form, and also recalls the Paddyfield Warbler in pattern. Pale sandy grey or warmer brown, it has a spiky, dark-tipped pale bill and a slight dark stripe through the eye and a thin pale line above it, sometimes quite marked. The broad-based bill, short undertail coverts, and long, slim, square-tipped tail (bobbed upwards but not flicked down) help to identify it.

OCCURRENCE Breeds in extreme NE Europe; rare vagrant in W Europe in autumn.

VOICE Dry, hard, tapping *tak* or *tek*.

dark sides to crown

long, slim tail

pale sandy grey body

Length **11–12cm (4¼–4¾in)**	Wingspan **18–20cm (7–8in)**

Family **Sylviidae**	Species *Phylloscopus fuscatus*

Dusky Warbler

Stockier than a Chiffchaff (see p.335), the Dusky Warbler is browner, tinged grey or olive, with a broad, long pale stripe over the eye (whitest in front, buff behind), pale orange-brown legs, and pale underparts with slightly brighter, buffer flanks. The dark eye-stripe often sharpens the effect of the line over the eye.

OCCURRENCE Rare but annual vagrant in NW Europe, from Asia, in late autumn.

VOICE Hard *tchuk* or *tak* notes, or repeated *tek tek*.

dark eye-stripe

broad pale stripe over eye

fine, pointed bill

pale underparts

Length **11–12cm (4¼–4¾in)**	Wingspan **14–20cm (5½–8in)**

Family **Sylviidae**	Species *Phylloscopus swarzi*

Radde's Warbler

A rare and highly prized warbler, skulking and elusive but calling often, Radde's Warbler looks dark, bulky, and strong-billed. Its legs are thick and bright pinkish. A long, slightly kinked pale stripe runs over each eye, edged darker above and below. The back is olive green, less brown than the Dusky Warbler's, and the underside more brightly flushed with orange-buff on the flanks and rusty-buff on the undertail coverts.

OCCURRENCE Very rare but regular late autumn vagrant to NW Europe, from Asia.

VOICE Soft, repeated, slightly chuckling *chup* or *chep*.

pale stripe over eye

thick pale legs

Length **12cm (4¾in)**	Wingspan **15–20cm (6–8in)**

Family **Sylviidae**	Species *Phylloscopus borealis*

Arctic Warbler

The *Phylloscopus* warblers, including the Willow Warbler (see p.336), are greenish, delicate, and quick-moving birds. The Arctic Warbler is rather thickset and solid with a stout bill (pale lower mandible with a dark tip), a thick, dark stripe through the eye and a long cream line above it, and a thin cream wingbar (sometimes two). Larger than the Greenish Warbler, it has longer wingtips, and pinker legs in autumn.

OCCURRENCE Breeds in extreme N Scandinavia; very rare migrant/vagrant in autumn in NW Europe.

VOICE Call hard, sharp *dzit*; song low, fast trill.

thin wingbar

AUTUMN

long wingtips

Length **12–13cm (4¾–5in)**	Wingspan **16–22cm (6½–9in)**

Family **Sylviidae**	Species *Phylloscopus trochiloides*

Greenish Warbler

This delicate, fast-moving warbler is grey-green above and silvery white below. The yellowish stripe over the eye typically reaches the top of the bill (unlike Arctic Warbler). A single narrow cream wingbar is usual, a second short one occasional. Whiter flanks help to give it a lighter look than the Arctic Warbler; but it may be confused with some Chiffchaffs with a pale wingbar.

OCCURRENCE Breeds in NE Europe; rare migrant in late summer; occasional in late spring in W Europe.

VOICE Loud, sweet, disyllabic *schu-weet* or *tshi-li*; quick, trilling song.

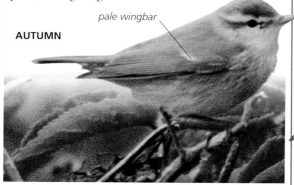

pale wingbar

AUTUMN

Length **10cm (4in)**	Wingspan **15–21cm (6–8½in)**

Family **Sylviidae**	Species *Phylloscopus inornatus*

Yellow-browed Warbler

A very small, strongly patterned, beautiful warbler, the Yellow-browed Warbler is clear grey-green or olive-green above, whiter beneath, with black-centred, white-tipped tertials and two yellowish cream wingbars; the upper one is thin and short, the lower longer and broad, edged dark green and black, catching the eye in the briefest view. It is an elusive, active little bird, often hard to see well against the sky in leafy trees.

OCCURRENCE Rare but regular migrant in NW Europe, in late autumn, from Asia.

VOICE Penetrating, sharp, rising *sweeet* or *chi-weet*.

long cream stripe over eye

white tips

two pale wingbars

AUTUMN

Length **9–10cm (3½–4in)**	Wingspan **14–20cm (5½–8in)**

Family **Sylviidae**	Species *Phylloscopus humei*

Hume's Leaf Warbler

A very close and very similar relative of the Yellow-browed Warbler, Hume's is another tiny, but slightly duller, warbler. It is dusky grey-green, whiter below, with a long, cream stripe over the eye, an all-dark bill (no pale base), and two pale wingbars, one broad and obvious, the upper one short, weak, and often inconspicuous. Its tertials are dull with pale tips, less blackish than on the Yellow-browed.

OCCURRENCE Very rare vagrant in NW Europe, from Asia.

VOICE Loud, cheeping *tsee-oo*, falling at end, or flatter *tsweeet*.

dark bill

AUTUMN

Length **9–10cm (3½–4in)**	Wingspan **14–20cm (5½–8in)**

Family **Sylviidae**	Species *Phylloscopus proregulus*

Pallas's Warbler

The tiniest and most boldly patterned leaf warbler, Pallas's Warbler is eye-catching but not easy to separate from the Yellow-browed unless the crown and rump are visible. It is brighter green and neckless, its head boldly striped yellow and dark green-black, with a long central crown stripe. Its rump is pale, lemon-yellow or cream, best seen as it hovers briefly while feeding.

OCCURRENCE Very rare but regular late autumn vagrant in NW Europe, from Asia.

VOICE Rising, soft *chuee* call.

boldly striped head

broad lower wingbar, edged darker

AUTUMN

Length **9cm (3½in)**	Wingspan **12–16cm (4¾–6½in)**

| Family **Muscicapidae** | Species *Ficedula parva* |

Red-breasted Flycatcher

This delightful, tiny flycatcher is best identified by its black tail with a long rectangle of white on each side at the base. Males have a grey hood and a small orange-red throat patch. Females and juveniles have plainer heads, with marked pale eye-rings, and smudgy marks beside the throat. The legs are short and black, the tail often cocked upwards. Migrants can be very tame.

OCCURRENCE Breeds in E and NE Europe; rare but regular autumn migrant in NW Europe.

VOICE Short, dry, Wren-like *trr-r-rt* and *tut*; song high, sharp, rhythmic, falling away in purer cadence.

pale eye-ring

long wings

JUVENILE

white on tail

| Length **11–12cm (4¼–4¾in)** | Wingspan **18–21cm (7–8½in)** |

| Family **Muscicapidae** | Species *Ficedula albicollis* |

Collared Flycatcher

This is very similar to the Pied Flycatcher (see p.393), plumages other than the adult male's being difficult. Females are greyer than Pied, with a greyish rump; the wing patch is thinner, but there is a larger white patch on the primaries. Juveniles may have a short upper wingbar. Confusions arise with hybrids, as well as Semi-collared Flycatchers.

OCCURRENCE Breeds in E Europe, north to Baltic islands; rare vagrant in W Europe in spring.

VOICE Thin *tseeeep* and short *tek*; song slow, harsh whistles of varying pitch.

bold white patch near bill

white collar

MALE large white wing patch

| Length **12–13cm (4¾–5in)** | Wingspan **22.5–24.5cm (8¾–9¾in)** |

| Family **Muscicapidae** | Species *Ficedula semitorquata* |

Semi-collared Flycatcher

A difficult bird of restricted range, the Semi-collared Flycatcher is best identified when breeding, by the adult males: the white throat hooks around under the ear coverts in a half collar, the wings have a lot of white with a very big primary patch, and the rump is pale grey; Iberian Pied Flycatchers (see p.393), however, look very similar. Females have very thin white wing marks, but a strong upper wingbar.

OCCURRENCE Breeds in Balkans and Turkey; migrates through Middle East in spring and autumn.

VOICE Call low, piping whistle; song slow, like Collared Flycather's, more rhythmic like Pied's.

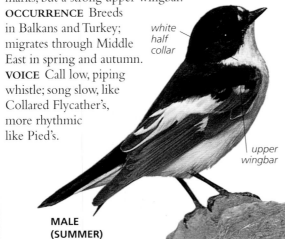

white half collar

upper wingbar

MALE (SUMMER)

| Length **12–13cm (4¾–5in)** | Wingspan **23–24cm (9–9½in)** |

| Family **Paridae** | Species *Parus cyanus* |

Azure Tit

A rare bird of remote places, this is a little-known species in Europe. It has a basic Blue Tit-like look (see p.306), but lacks yellow and green entirely, and has a longer tail with broad white sides. The crown is all white (no blue cap), the back pale grey, and wings blue with broad white bands. The underside is all white. Hybrids of Azure and Blue Tits have a pale blue cap and bluer tails with less white at the corners.

OCCURRENCE Breeds in extreme NE Europe in damp woods and willow stands.

VOICE Most calls very like Blue Tit's; song more stuttering.

all-white crown

long tail

| Length **12–13cm (4¾–5in)** | Wingspan **19–21cm (7½–8½in)** |

Family **Paridae**	Species ***Parus cinctus***

Siberian Tit

With a pattern recalling a Willow Tit (see p.310), this bigger bird also has a richer, brighter brown back and rusty flanks, contrasting with grey wings, white cheeks, and a dark brownish cap and bib. The cap is often dull, washed grey or grey-brown. From the front, the bib makes a broad wedge under the large, almost bulbous cheeks. In very cold conditions in northern conifer forest, it is often fluffed out, looking unexpectedly large.
OCCURRENCE Breeds in N Scandinavia in old, remote forest and birch wood.
VOICE Basic *zi-zi-tah tah tah* much like Willow but less emphatic; song more nasal and buzzing.

grey wings

brown flanks

Length **13–14cm (5–5½in)**	Wingspan **20–21cm (8–8½in)**

Family **Paridae**	Species ***Parus lugubris***

Sombre Tit

The same size as a Great Tit (see p.307), the Sombre Tit's basic pattern is more like a Willow Tit's (see p.310). It has a deep, grey-black cap, a very large, wide black bib, and a restricted wedge of white between these across the cheeks. It is a rather heavy little bird, with a thick bill. Relatively quiet, it does not draw attention to itself unless calling.
OCCURRENCE Breeds in woods or on bushy slopes, in Balkans, Turkey, and Middle East.
VOICE Like Blue Tit (see p.306), plus Great Tit-like grating churrs.

white wedge on cheeks

grey back

pale streak on wings

pale, dull grey below

Length **13–14cm (5–5½in)**	Wingspan **21–23cm (8½–9in)**

Family **Sittidae**	Species ***Sitta whiteheadi***

Corsican Nuthatch

A tiny nuthatch of pine forest in Corsica, this unmistakable bird has a typical stout, tapered nuthatch form, with a spike-like bill, short legs but strong feet, and short, square tail held well clear of its perch. Males have a blackish cap and eye-stripe separated by a long white line. Females have the black replaced by grey, but the white line over the eye is equally distinctive.
OCCURRENCE Only found in Corsica, in mature pine forest in high mountain valleys.
VOICE Throaty, harsh call, repeated at intervals; song an even, fast, high trill.

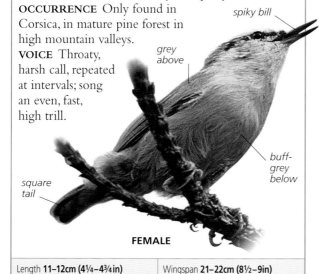

spiky bill

grey above

square tail

buff-grey below

FEMALE

Length **11–12cm (4¼–4¾in)**	Wingspan **21–22cm (8½–9in)**

Family **Sittidae**	Species ***Sitta neumayer***

Rock Nuthatch

The best places to see this bird are archaeological sites in Greece and Turkey, where it often flits about the ruins, drawing attention to itself by its bold behaviour and voice. It looks like a large, washed-out Nuthatch (see p.360), but has a plain grey tail (no dark and white marks) and whitish flanks (no rust-orange). It often perches very upright, breast pushed out, bobbing, almost like a slim Dipper (see p.367).
OCCURRENCE Breeds in Balkans and Turkey, in rocky areas.
VOICE Rising and falling, far-carrying, strident whistling notes.

long black eye-stripe

long bill

plain tail

Length **14–15cm (5½–6in)**	Wingspan **23–25cm (9–10in)**

Family **Laniidae**	Species *Lanius isabellinus*

Isabelline Shrike

A close relative of the Red-backed Shrike (see p.287), the Isabelline Shrike is similar in basic appearance but paler, more sandy-brown, with a pale, rusty tail. Males have a black mask and dark wings; females are plainer. Immatures look like rusty-tailed, sandy-backed young Red-backed Shrikes, with paler, plainer upperparts; the tail may be entirely pale rufous or darker with rufous edges and a rufous rump.
OCCURRENCE Very rare vagrant in W Europe, from Asia, mostly in late autumn.
VOICE Migrants are generally silent.

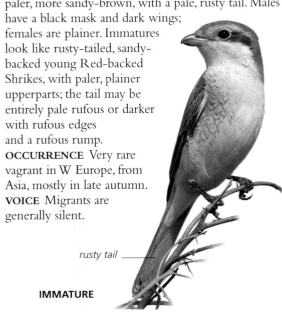

rusty tail

IMMATURE

Length **16–18cm (6½–7in)**	Wingspan **26–28cm (10–11in)**

Family **Laniidae**	Species *Lanius nubicus*

Masked Shrike

The small, slender, and well-marked Masked Shrike is mostly black and white with peachy orange flanks and is easily identifiable. Males are brighter than females. Juveniles look like young Woodchat Shrikes (see p.290), but have slim, blacker tails, greyer upperparts with at least some scaly whitish shoulder marks making more of a patch than on Woodchat, and a bigger white patch on the primaries.
OCCURRENCE Breeds in Balkans and Turkey; winter migrant in Africa.
VOICE Scratchy, hoarse note and rough rattling call.

white forehead

black eye-stripe

large white shoulder patch

orange flanks

MALE

Length **17–18cm (6½–7in)**	Wingspan **24–26cm (9½–10in)**

Family **Corvidae**	Species *Cyanopica cyanus*

Azure-winged Magpie

This striking bird is a magpie in form (see p.296), with an upright, short body, short wings, long tail, and stout bill and legs, but quite different from the familiar pied bird in plumage. It has a deep black cap and a white throat; the back is pale fawn, the underside a paler shade, while the wings and tail are pale, dusty blue. Small flocks roam through pine woods, calling frequently.
OCCURRENCE Breeds in S Spain and Portugal; resident.
VOICE High, nasal, calls with bright, whining quality, and deeper, grating notes.

pale blue wings

fawn body

long, pale blue tail

Length **31–35cm (12–14in)**	Wingspan **38–40cm (15–16in)**

Family **Sturnidae**	Species *Sturnus roseus*

Rose-coloured Starling

This exotic-looking starling is usually seen in western Europe in immature plumages, but adults occur in southeast Europe at times. Adults are pale pink (dusky in winter) with a black hood, wings, and tail; whitish feather tips obscure some of the black in winter. Immatures are like sandy grey young Starlings (see p.365) except for a shorter yellowish bill, more contrast between pale body and dark wings, and a pale rump.
OCCURRENCE Rare migrant in SE Europe, vagrant in W Europe, in summer and autumn.
VOICE Short, harsh calls and varied, unmusical, rattling song.

dusky pink body

ADULT (WINTER)

Length **19–22cm (7½–9in)**	Wingspan **37–40cm (14½–16in)**

Family **Vireonidae**	Species *Vireo olivaceus*

Red-eyed Vireo

Vireos are like rather large, stocky warblers with quite thick bills. The Red-eyed Vireo has a strong head pattern (grey cap edged black, broad white stripe over the eye, and black eye-stripe), a dark red eye, and a basic green above, whitish below pattern. The greenish tail is short and square, helping to give it a slightly different look from European warblers with green upperparts. It tends to be quite slow and heavy in its movements through foliage.

OCCURRENCE Very rare but annual vagrant in NW Europe, in autumn, from North America.
VOICE Short, nasal *chway*.

white over red eye

Length **14cm (5½in)**	Wingspan **23–25cm (9–10in)**

Family **Parulidae**	Species *Dendroica striata*

Blackpoll Warbler

In summer, this is a black and white bird, but in autumn immatures are dull-looking: greenish with soft dark streaks on the back and pale greyish yellow underparts with soft dusky streaks. They have a pale stripe above and pale marks below the eye, white under the tail, and dark wings with two long, narrow, diagonal wingbars. The white under the tail and streaked chest are important points to eliminate even rarer species.

OCCURRENCE Very rare vagrant in NW Europe in late autumn, from North America.
VOICE Liquid, sharp, short *chip*.

two curved white wingbars

IMMATURE (AUTUMN)

Length **12–13cm (4¾–5in)**	Wingspan **15cm (6in)**

Family **Parulidae**	Species *Dendroica coronata*

Yellow-rumped Warbler

While some autumn plumages are very difficult to identify, most North American wood warblers are striking birds, at least in summer. Yellow-rumped Warblers are small, neat, active birds, distinguished by a yellow patch on each side of the chest, and a bright yellow rump. They are streaky brown above, and whitish with blacker streaks below in autumn. Two long, diagonal wingbars are clearly visible. The head is not strongly patterned but has white crescents above and below the eye.

OCCURRENCE Very rare vagrant in NW Europe in late autumn, from North America.
VOICE Frequent, hard, single notes, *chik* or *twip*.

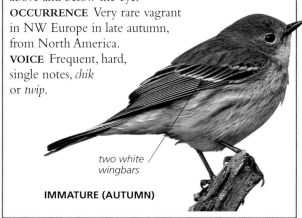

two white wingbars

IMMATURE (AUTUMN)

Length **12–13cm (4¾–5in)**	Wingspan **15cm (6in)**

Family **Fringillidae**	Species *Carduelis hornemanni*

Arctic Redpoll

Redpolls are split into several races or species; the Arctic Redpoll is usually recognized as a species but can be hard to distinguish. Males have a large white rump, white underparts, and unmarked undertail coverts; a broad whitish wingbar, small red forehead patch, and a short yellow bill aid identification. Females and immatures are more streaked, but the rump is unstreaked white, the wingbar white, and the undertail area unmarked white.

OCCURRENCE Breeds in extreme N Scandinavia; winters in Scandinavia and vagrant in W Europe.
VOICE Calls like Redpoll's, not safely distinguishable.

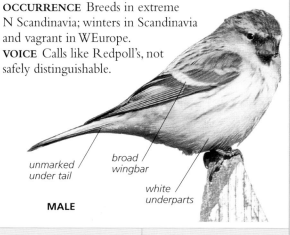

unmarked under tail

broad wingbar

white underparts

MALE

Length **12–14cm (4¾–5½in)**	Wingspan **21–27cm (8½–10½in)**

| Family **Fringillidae** | Species *Loxia leucoptera* |

Two-barred Crossbill

A smallish crossbill, the Two-barred Crossbill is typically marked by two broad white wingbars (unlike the Crossbill, see p.422, which may only occasionally have wingbars) and white spots on the tertial tips. Males are rather deep cherry-red with blacker wings (Crossbills tend to be more orange-red with browner wings); females are green and brown, with yellow-green rumps. Dark back spots are distinctive. Juveniles are duller, browner, with much thinner wingbars, and are difficult to identify with absolute certainty.

OCCURRENCE Rare breeder in extreme NE Europe; vagrant elsewhere.

VOICE Weaker than Crossbill's.

MALE *two broad wingbars*

| Length **16cm (6½in)** | Wingspan **26–29cm (10–11½in)** |

| Family **Fringillidae** | Species *Loxia scotica* |

Scottish Crossbill

If Parrot and common Crossbills (see p.422) are hard to distinguish, the Scottish Crossbill, being intermediate between the two, can seem impossible to identify. It is restricted to pine forest in Scotland, but the other two can be found there, too. It has bulging cheeks, a thick neck, and a deep, hefty bill; its plumages are like those of the other species and its calls are confusingly difficult unless recorded and analysed technically. Research continues into the relationship of all these forms.

OCCURRENCE Resident in N Scotland, presumably does not move elsewhere.

VOICE Much as Crossbill's, sometimes deeper, thicker notes.

bulging cheeks

becomes increasingly red with age

IMMATURE MALE

| Length **16–17cm (6½in)** | Wingspan **27–37cm (10½–14½in)** |

| Family **Passeridae** | Species *Montifringilla nivalis* |

Snowfinch

The Snowfinch is large, long-winged, and short-legged with black-tipped white wings recalling a Snow Bunting (see p.429). The head is grey (black face and bib on summer male), the back dull grey-brown, and the tail white with a narrow black centre. In summer, the male has a black bill; otherwise the bill is yellow.

OCCURRENCE Sparse breeder, at high altitude, in Pyrenees, Alps, Italy, and Balkans.

VOICE Wide variety of sharp, hoarse, or mewing calls; trilling, sparrow-like song.

black bill

dull grey-brown back

MALE (SUMMER)

| Length **17–19cm (6½–7½in)** | Wingspan **34–38cm (13½–15in)** |

| Family **Emberizidae** | Species *Emberiza rustica* |

Rustic Bunting

In summer, males are boldly marked black and white on the head, reddish above, and white below with a red-brown breast-band. Females and autumn males have a paler central crown stripe on a duller head; both sexes raise their crown feathers at times. Immatures are streaked yellow-brown, with a broad pale stripe over the eye, a white ear-spot edged black, and a cream band under the cheek. The rump is rust-brown; two thin white wingbars are usually obvious.

OCCURRENCE Breeds in Sweden, Finland, and Russia; rare migrant to E Europe and vagrant in W Europe; in wet forested areas; May to September.

VOICE Short, sharp, penetrating *tic* or *zit;* song short, rhythmic, rattling phrase.

MALE (AUTUMN)

rufous rump

| Length **12–13cm (4¾–5in)** | Wingspan **14–17cm (5½–6½in)** |

| Family **Emberizidae** | Species *Emberiza caesia* |

Cretzschmar's Bunting

This is very like an Ortolan Bunting (see p.434) but rust-red on the throat and moustache where the Ortolan is yellow. The head and chest are blue-grey, the back rufous-brown, and the underside a rich rust-orange. Females are duller; both sexes have a white eye-ring. The immature shows the eye-ring, black-streaked upperparts and broad, bright rusty edges to the tertials; it is a richer, more rufous bird than a young Ortolan.
OCCURRENCE Breeds in Greece, Crete, Turkey, and some Greek islands; very rare vagrant elsewhere.
VOICE Sharp, short *tsip* note, very like Ortolan Bunting's.

blue-grey head

streaked rufous-brown above

MALE

| Length **14–15cm (5½–6in)** | Wingspan **23–26cm (9–10in)** |

| Family **Emberizidae** | Species *Emberiza leucocephalos* |

Pine Bunting

Difficulties arise with intermediates between the Pine Bunting and Yellowhammer (see p.431) but most pure birds are identifiable. Males are easy, with their white crown and cheek, edged black and rufous; the back is rusty with black streaks and the underside whitish. Females are greyer, with a less marked head; immatures have white underparts streaked grey, a white spot on the ear coverts, a white eye-ring, and white feather edges on the wings (yellowish on Yellowhammer).
OCCURRENCE Vagrant in NW Europe, late autumn and winter, from Asia.
VOICE Metallic ticking notes, like Yellowhammer's.

MALE (WINTER) streaked, rusty back   white cheeks

| Length **16–17cm (6½in)** | Wingspan **26–30cm (10–12in)** |

| Family **Emberizidae** | Species *Emberiza melanocephala* |

Black-headed Bunting

This big bunting has easily identifiable males with a black head, yellow chin, rufous back, and yellow underside; females and immatures are difficult to distinguish from vagrant or escaped Red-headed Buntings. Females have large grey bills, hooded grey-brown heads, and pale yellow underparts, with some rufous above (unlike Red-headed); immatures are slightly more streaked, especially on the crown (more than Red-headed) and rather bright above (Red-headed typically dull).
OCCURRENCE Breeds in S Italy and Balkans; rare late summer vagrant in NW Europe.
VOICE Various metallic, ticking calls; fuller accelerating song.

black head

yellow underside

MALE

| Length **16–17cm (6½in)** | Wingspan **26–30cm (10–12in)** |

| Family **Emberizidae** | Species *Emberiza aureola* |

Yellow-breasted Bunting

The summer males of this small, neat bunting are distinctive, with a black and chestnut head, upperparts, and breast-band, against bright yellow on the chest and belly, and a white shoulder patch. In autumn, immatures are streaked brown, rather yellower or buff beneath, finely streaked, with two white wingbars; they have a pale central stripe on the top of the head, a broad pale stripe over each eye, a pale band under the cheek, and a pale spot at the rear of the dark-edged ear coverts.
OCCURRENCE Breeds from E Finland into Russia; rare migrant/vagrant in NW Europe in autumn.
VOICE Short, ticking *tsik* note.

pale stripe over eye

two wingbars

yellow or buff below

FEMALE

| Length **15–16cm (6–6½in)** | Wingspan **21–24cm (8½–9½in)** |

Family **Emberizidae**	Species *Pheucticus ludovicianus*

Rose-breasted Grosbeak

Very rare in Europe, this is an unmistakable bird by virtue of its large size for a finch, very large bill, and striking pattern. Males are black and white with red on the breast; immatures brown and streaked, males with red under the wings. These, and adult females, are boldly marked with a dark cap, very broad white band over the eye, dark cheeks, and white throat; two bars of white spots cross the wing. The breast is buff, streaked with wavy black lines.

OCCURRENCE Very rare vagrant in autumn in NW Europe, from North America.

VOICE Short, hard, *chik* call.

white spots on wings

MALE

Length **18–20cm (7–8in)**	Wingspan **30–32cm (12–12½in)**

Family **Emberizidae**	Species *Junco hyemalis*

Dark-eyed Junco

This small American bunting is sparrow-like, rather inconspicuous and quiet. Males are smoky grey, blacker around the face, with an oval white belly patch, white tail sides, and a pale pinkish bill. Females are tinged browner, especially on the wings, and a little sullied below (less pure white) but share the same basic pattern, which is unlike any European species.

OCCURRENCE Very rare vagrant, sometimes remaining all winter, in NW Europe, from North America.

VOICE Short, ticking note, sometimes repeated.

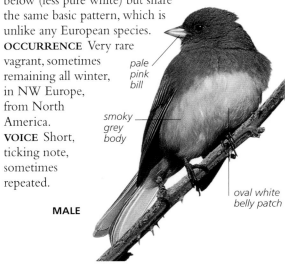

pale pink bill

smoky grey body

oval white belly patch

MALE

Length **14cm (5½in)**	Wingspan **25–30cm (10–12in)**

Family **Emberizidae**	Species *Zonotrichia albicollis*

White-throated Sparrow

American sparrows are more like European buntings. Adult White-throated Sparrows have a thin white crown stripe edged black, a broad white band over the eye (yellow at the front), grey cheeks, and a white chin patch. Immatures are duller, with a greyish and less clear-cut chin and throat. The body plumage, streaked brown above and grey beneath, as well as the general stance and behaviour, recall a Dunnock (see p.390).

OCCURRENCE Rare vagrant in NW Europe, mostly in spring, from North America.

VOICE Call a sharp, persistent, thin *zit*, and thicker *chink* notes.

white band over eye with yellow in front

streaked brown upperparts

ADULT

Length **15–17cm (6–6½in)**	Wingspan **20–25cm (8–10in)**

Family **Icteridae**	Species *Dolichonyx oryzivorus*

Bobolink

Bobolinks visiting Europe in autumn are rather yellowish birds, striped with black and cream; they are dumpy and heavy-bellied, with a narrow, tapered head and tail. The bill is sharply triangular, and the head shows a central yellow-buff crown stripe edged black and broad buff stripes over each eye. The back has two long straw-coloured lines; the underside is creamy with faint streaks only on the flanks. A short, narrow, spiky tail helps to separate it from a common bunting or sparrow.

OCCURRENCE Very rare vagrant in autumn in NW Europe, from North America.

VOICE Short, sharp *pink*.

black and cream stripes on upperparts

creamy below

long, narrow tail

ADULT (AUTUMN)

Length **16–18cm (6½–7in)**	Wingspan **30cm (12in)**

VAGRANTS

The list that follows consists of birds that occur only very rarely in Europe, known as vagrants or accidentals. It also looks a little further beyond Europe, to list those birds whose normal range is the Middle East and North Africa, so covering a total faunal area known as the Western Palearctic.

Vagrants to Europe arrive from Asia and North America (and fewer from Africa). Western Europe, especially the UK, is well placed to receive birds that are blown off course from eastern North America and cross the Atlantic. It was thought that such birds cross the ocean on board ship, but it is now accepted that even small birds can, with a following wind, survive a flight across the Atlantic, although they probably do not survive long afterwards. Larger species, however, such as some wildfowl, may live for years in Europe and a few (that have been trapped, ringed, and released in order to follow their movements) have even returned to North America in subsequent years. These are not, in any true sense, European birds, but are included here to complete the range of species that have been recorded. Many appear again, others may not: by their nature these "accidentals" are unpredictable.

Common Name	Scientific Name	Family/Scientific Name	Description
Wildfowl			
White-faced Whistling Duck	*Dendrocygna viduata*	Wildfowl/Anatidae	Large, noisy duck, vagrant in North Africa, from southern Africa
Lesser Whistling Duck	*Dendrocygna javanica*	Wildfowl/Anatidae	Small duck from Africa
Bar-headed Goose	*Anser indicus*	Wildfowl/Anatidae	Pale grey goose from Asia
Spur-winged Goose	*Plectropterus gambensis*	Wildfowl/Anatidae	Large goose from Africa
Cotton Pygmy-goose	*Nettapus coromandelianus*	Wildfowl/Anatidae	Small duck, vagrant in North Africa, from southern Africa
Baikal Teal	*Anas formosa*	Wildfowl/Anatidae	Colourful surface-feeding duck from Asia
Cape Teal	*Anas capensis*	Wildfowl/Anatidae	Surface-feeding duck, vagrant in North Africa, from southern Africa
Red-billed Teal	*Anas erythrorhyncha*	Wildfowl/Anatidae	Surface-feeding duck, vagrant in North Africa, from southern Africa
Cape Shoveler	*Anas smithii*	Wildfowl/Anatidae	Surface-feeding duck, vagrant in North Africa, from southern Africa
Southern Pochard	*Netta erythrophthalma*	Wildfowl/Anatidae	Diving duck, vagrant in North Africa, from southern Africa
Canvasback	*Aythya valisineria*	Wildfowl/Anatidae	Large, pale Pochard-like duck from North America
Redhead	*Aythya americana*	Wildfowl/Anatidae	Pochard-like diving duck, vagrant in UK, from North America
Gamebirds			
Caucasian Grouse	*Tetrao mlokosiewiczi*	Grouse/Tetraonidae	Sleek black grouse, in Middle East
Caspian Snowcock	*Tetraogallus caspius*	Gamebirds/Phasianidae	Large mountain grouse, in Middle East

Common Name	Scientific Name	Family/Scientific Name	Description
Gamebirds *continued*			
Caucasian Snowcock	*Tetraogallus caucasicus*	Gamebirds/Phasianidae	Large mountain grouse, in Middle East
Double-spurred Francolin	*Francolinus bicalcaratus*	Gamebirds/Phasianidae	Brown gamebird, in North Africa
Black Francolin	*Francolinus francolinus*	Gamebirds/Phasianidae	Blackish gamebird, rare in Middle East
Helmeted Guineafowl	*Numida meleagris*	Guineafowl/Numididae	Dark, white-spotted gamebird, in North Africa
Ostrich			
Ostrich	*Struthio camelus*	Ratites/Struthidae	Vagrant in North Africa, from southern Africa
Albatrosses			
Wandering Albatross	*Diomedea exulans*	Albatrosses/Diomedeidae	Large seabird from southern oceans
Yellow-nosed Albatross	*Diomedea chlororhynchos*	Albatrosses/Diomedeidae	Large seabird from southern oceans
Black-browed Albatross	*Diomedea melanophris*	Albatrosses/Diomedeidae	Long-winged seabird from South Atlantic
Shy Albatross	*Diomedea cauta*	Albatrosses/Diomedeidae	Large seabird from southern oceans
Petrels and Shearwaters			
Fea's Petrel	*Pterodroma feae*	Petrels and Shearwaters/ Procellariidae	Large, rare petrel from Madeira
Black-capped Petrel	*Pterodroma hasitata*	Petrels and Shearwaters/ Procellariidae	Large petrel from Caribbean
Atlantic Petrel	*Pterodroma incerta*	Petrels and Shearwaters/ Procellariidae	Large petrel from South Atlantic
Soft-plumaged Petrel	*Pterodroma mollis*	Petrels and Shearwaters/ Procellariidae	Large petrel from South Atlantic
Zino's Petrel	*Pterodroma madeira*	Petrels and Shearwaters/ Procellariidae	Large, rare petrel from Madeira
Bulwer's Petrel	Bulweria bulwerii	Petrels and Shearwaters/ Procellariidae	Dark, long-tailed petrel from mid-Atlantic islands
Streaked Shearwater	*Calonectris leucomelas*	Petrels and Shearwaters/ Procellariidae	Vagrant in Middle East, from tropical oceans
Flesh-footed Shearwater	*Puffinus carneipes*	Petrels and Shearwaters/ Procellariidae	All-dark shearwater from Indian Ocean
Audubon's Shearwater	*Puffinus lherminieri*	Petrels and Shearwaters/ Procellariidae	Small shearwater from Indian Ocean
Wedge-tailed Shearwater	*Puffinus pacificus*	Petrels and Shearwaters/ Procellariidae	Large dark shearwater from Indian Ocean
White-faced Storm-petrel	*Pelagodroma marina*	Petrels and Shearwaters/ Procellariidae	Small petrel from South Atlantic
Swinhoe's Storm Petrel	*Oceanodroma monorhis*	Petrels and Shearwaters/ Procellariidae	Small, dark, oceanic petrel from Pacific

Common Name	Scientific Name	Family/Scientific Name	Description
Gannets and Cormorants			
Cape Gannet	*Morus capensis*	Gannets/Sulidae	Black and white gannet from southern Africa
Masked Booby	*Sula dactylatra*	Gannets/Sulidae	Gannet-like seabird from tropical oceans
Brown Booby	*Sula leucogaster*	Gannets/Sulidae	Dark brown gannet, in Red Sea
Double-crested Cormorant	*Phalacrocorax auritus*	Cormorants/Phalacrocoracidae	Large black waterbird, vagrant in UK, from North America
Darters and Anhingas			
African Darter	*Anhinga rufa*	Darters and Anhingas/Anhingidae	Sharp-billed, cormorant-like waterbird, vagrant in North Africa, from southern Africa
Tropicbirds and Frigatebirds			
Red-billed Tropicbird	*Phaethon aethereus*	Tropicbirds/Phaethontidae	White seabird from tropical oceans
Lesser Frigatebird Tropicbird	*Fregata ariel*	Frigatebirds/Fregatidae	Large, fork-tailed seabird from tropical oceans
Bitterns and Herons			
American Bittern	*Botaurus lentiginosus*	Herons/Ardeidae	Stripe-necked bittern from North America
Least Bittern	*Ixobrychus exilis*	Herons/Ardeidae	Small bittern from North America
Schrenck's Bittern	*Ixobrychus eurhythmus*	Herons/Ardeidae	Small bittern from Asia
Dwarf Bittern	*Ixobrychus sturmii*	Herons/Ardeidae	Small bittern from Africa
Black-headed Heron	*Ardea melanocephala*	Herons/Ardeidae	Dry-ground heron, vagrant in North Africa, from southern Africa
Goliath Heron	*Ardea goliath*	Herons/Ardeidae	Very large heron, vagrant in Middle East, from Africa
Black Heron	*Egretta ardesiaca*	Herons/Ardeidae	Small dark heron from Africa
Mangrove Heron	*Butorides striatus*	Herons/Ardeidae	Small dark heron, in North Africa and Middle East
Snowy Egret	*Egretta thula*	Herons/Ardeidae	White egret from North America
Intermediate Egret	*Egretta intermedia*	Herons/Ardeidae	White egret from Africa, Asia
Storks			
Yellow-billed Stork	*Mycteria ibis*	Storks/Ciconiidae	White stork with red mask, vagrant in Middle East, from Africa
Ibises			
Bald Ibis	*Geronticus eremita*	Ibises/Threskiornithidae	All-dark ibis, rare breeder in North Africa

Common Name	Scientific Name	Family/Scientific Name	Description
Birds of Prey			
American Kestrel	*Falco sparverius*	Falcons/Falconidae	Tiny falcon from North America
Amur Falcon	*Falco amurensis*	Falcons/Falconidae	Red-footed falcon from Asia
Barbary Falcon	*Falco (peregrinus) pelegrinoides*	Falcons/Falconidae	Peregrine-like falcon, in Middle East and North Africa
Sooty Falcon	*Falco concolor*	Falcons/Falconidae	Large grey falcon from Middle East
Crested Honey Buzzard	*Pernis ptilorhyncus*	Hawks/Accipitridae	Medium-sized bird of prey, vagrant in Middle East, from Asia
American Swallow-tailed Kite	*Elanoides forficatus*	Hawks/Accipitridae	Fork-tailed kite from North America
Pallas's Fish Eagle	*Haliaeetus leucoryphus*	Hawks/Accipitridae	Huge eagle from Asia
Bald Eagle	*Haliaeetus leucocephalus*	Hawks/Accipitridae	Massive eagle from North America
Rüppell's Vulture	*Gyps rueppellii*	Hawks/Accipitridae	Distinctive vulture, vagrant in Middle East, from Africa
Bateleur	*Terathopius ecaudatus*	Hawks/Accipitridae	Acrobatic eagle, vagrant in Middle East, from Africa
Dark Chanting Goshawk	*Melierax metabates*	Hawks/Accipitridae	Large grey hawk, vagrant in Middle East, from East Africa
Shikra	*Accipiter badius*	Hawks/Accipitridae	Small pale sparrowhawk, in Middle East
Swainson's Buzzard	*Buteo swainsoni*	Hawks/Accipitridae	Large hawk from North America
Tawny Eagle	*Aquila rapax*	Hawks/Accipitridae	Big brown eagle, rare in North Africa
Verreaux's Eagle	*Aquila verreauxii*	Hawks/Accipitridae	Large black eagle, rare in North Africa
Rails, Crakes, and Coots			
Striped Crake	*Porzana marginalis*	Rails, Crakes, and Coots/Rallidae	Small pale crake, rare in North Africa
Sora	*Porzana carolina*	Rails, Crakes, and Coots/Rallidae	Dark crake from North America
Allen's Gallinule	*Porphyrula alleni*	Rails, Crakes, and Coots/Rallidae	Moorhen-like bird from Africa
American Purple Gallinule	*Porphyrula martinica*	Rails, Crakes, and Coots/Rallidae	Moorhen-like bird from North America
American Coot	*Fulica americana*	Rails, Crakes, and Coots/Rallidae	Coot from North America
Cranes			
Hooded Crane	*Grus monacha*	Cranes/Gruidae	Dark-headed crane from Asia
Siberian White Crane	*Grus leucogeranus*	Cranes/Gruidae	Large crane from Asia
Sandhill Crane	*Grus canadensis*	Cranes/Gruidae	Grey crane from North America

Common Name	Scientific Name	Family/Scientific Name	Description
Bustards			
Arabian Bustard	*Ardeotis arabs*	Bustards/Otididae	Very large, pale bustard, in Middle East
Waders			
Painted Snipe	*Rostratula benghalensis*	Waders/Scolopacidae	Snipe-like wader from Africa
Crab Plover	*Dromasar-deola*	Plovers/Charadriidae	Heavy-billed, black and white wader, in Middle East
Egyptian Plover	*Pluvianus aegyptius*	Plovers/Charadriidae	Small wader from Africa
Black-headed Lapwing	*Vanellus tectus*	Plovers/Charadriidae	Striking plover from Africa
Red-wattled Lapwing	*Vanellus indicus*	Plovers/Charadriidae	Eye-catching plover from Asia
Chestnut-banded Plover	*Charadrius pallidus*	Plovers/Charadriidae	Small sand plover from southern Africa
Three-banded Plover	*Charadrius tricollaris*	Plovers/Charadriidae	Small plover from Africa
Kittlitz's Plover	*Charadrius pecuarius*	Plovers/Charadriidae	Small, compact plover, vagrant in Middle East, from Africa
Short-billed Dowitcher	*Limnodromus griseus*	Waders/Scolopacidae	Long-billed wader from North America
Hudsonian Godwit	*Limosa haemastica*	Waders/Scolopacidae	Large, long-billed wader from North America
Eskimo Curlew	*Numenius borealis*	Waders/Scolopacidae	Medium-sized wader (possibly extinct) from North America
Little Curlew	*Numenius minutus*	Waders/Scolopacidae	Whimbrel-like wader from Asia
Grey-tailed Tattler	*Heteroscelus brevipes*	Waders/Scolopacidae	Medium to small grey wader from SE Asia
Willet	*Catoptrophorus semipalmatus*	Waders/Scolopacidae	Godwit-like wader from North America
Swinhoe's Snipe	*Gallinago megala*	Waders/Scolopacidae	Dark snipe from Asia
Pin-tailed Snipe	*Gallinago stenura*	Waders/Scolopacidae	Snipe-like bird from Asia
Auks			
Ancient Murrelet	*Synthliboramphus antiquus*	Auks/Alcidae	Small, Little Auk-like seabird from N Pacific
Parakeet Auklet	*Cyclorrhynchus psittacula*	Auks/Alcidae	Small auk-like seabird from N Pacific
Crested Auklet	*Aethia cristatella*	Auks/Alcidae	Small Puffin-like seabird from N Pacific
Tufted Puffin	*Lunda cirrhata*	Auks/Alcidae	Large dark puffin from N Pacific
Skuas, Terns, and Gulls			
South Polar Skua	*Stercorarius maccormicki*	Skuas/Stercoraridae	Large skua from southern oceans
White-cheeked Tern	*Sterna repressa*	Terns/Laridae	Black-capped dark tern, in Middle East

Common Name	Scientific Name	Family/Scientific Name	Description
Skuas, Terns, and Gulls *continued*			
Aleutian Tern	*Sterna aleutica*	Terns/Sternidae	Grey tern from Arctic Pacific
American Little Tern	*Sterna antillarum*	Terns/Sternidae	Little Tern from North America
Crested Tern	*Sterna bergii*	Terns/Sternidae	Large tern from Indian Ocean
Elegant Tern	Sterna elegans	Terns/Sternidae	Long-billed tern from Caribbean
Brown Noddy	*Anous stolidus*	Terns/Sternidae	Dark, tern-like seabird from tropical oceans
Brown-headed Gull	*Larus brunnicephalus*	Gulls/Laridae	Small gull, vagrant in Middle East, from Asia
Glaucous-winged Gull	*Larus glaucescens*	Gulls/Laridae	Large gull, vagrant in North Africa, from North America
Sooty Gull	*Larus hemprichii*	Gulls/Laridae	Large dusky gull, in Red Sea
Grey-headed Gull	*Larus cirrocephalus*	Gulls/Laridae	Vagrant in North Africa, from southern Africa
Skimmers			
African Skimmer	*Rynchops flavirostris*	Skimmers/Rynchopidae	Long-billed bird, vagrant in Middle East, from Africa
Sandgrouse			
Pallas's Sandgrouse	Syrrhaptes paradoxus	Sandgrouse/Pterocliidae	Partridge-like bird, vagrant in W Europe, from Asia
Lichtenstein's Sandgrouse	*Pterocles lichtensteinii*	Sandgrouse/Pteroclididae	Small, barred sandgrouse, in Middle East
Spotted Sandgrouse	*Pterocles senegallus*	Sandgrouse/Pteroclididae	Large pale sandgrouse, in Middle East
Crowned Sandgrouse	*Pterocles coronatus*	Sandgrouse/Pteroclididae	Pale sandgrouse, in Middle East
Chestnut-bellied Sandgrouse	*Pterocles exustus*	Sandgrouse/Pterocliidae	Large sandgrouse, in Middle East
Pigeons and Doves			
Yellow-eyed Dove	*Columba eversmanni*	Pigeons and Doves/ Columbidae	Small pigeon from Asia
Bolle's Pigeon	*Columba bollii*	Pigeons and Doves/ Columbidae	Dark pigeon, endemic to Canary Islands
Laurel Pigeon	*Columba junoniae*	Pigeons and Doves/ Columbidae	Dark pigeon, endemic to Canary Islands
Trocaz Pigeon	*Columba trocaz*	Pigeons and Doves/ Columbidae	Dark pigeon, endemic to Madeira
Oriental Turtle Dove	*Streptopelia orientalis*	Pigeons and Doves/ Columbidae	Dark dove from Asia
Laughing Dove	*Streptopelia senegalensis*	Pigeons and Doves/ Columbidae	Small dark dove, in Middle East
African Collared Dove	*Streptopelia roseogrisea*	Pigeons and Doves/ Columbidae	Brown and pinkish dove, vagrant in Middle East, from Africa

Common Name	Scientific Name	Family/Scientific Name	Description
Pigeons and Doves *continued*			
Namaqua Dove	*Oena capensis*	Pigeons and Doves/ Columbidae	Tiny, long-tailed dove, in Middle East
Mourning Dove	*Zenaida macroura*	Pigeons and Doves/ Columbidae	Sharp-tailed, Collared Dove-like species-from North America
Cuckoos and Coucals			
Didric Cuckoo	*Chrysococcyx caprius*	Cuckoos/Cuculidae	Green and white cuckoo, vagrant in Middle East, from Africa
Oriental Cuckoo	*Cuculus saturatus*	Cuckoos/Cuculidae	Small cuckoo from Asia
Black-billed Cuckoo	*Coccyzus erythrophthalmus*	Cuckoos/Cuculidae	Small cuckoo from North America
Senegal Coucal	*Centropus senegalensis*	Coucals/Centropodidae	Black-capped, rufous cuckoo, in North Africa
Owls			
Pale Scops Owl	*Otus brucei*	Owls/Strigidae	Small eared owl, rare in Middle East
Brown Fish Owl	*Ketupa zeylonensis*	Owls/Strigidae	Large, eared owl, in Middle East
Hume's Tawny Owl	*Strix butleri*	Owls/Strigidae	Small, pale desert owl, in North Africa and Middle East
Marsh Owl	*Asio capensis*	Owls/Strigidae	Small Short-eared Owl-like bird, in North Africa
Nightjars			
Nubian Nightjar	*Caprimulgus nubicus*	Nightjars/Caprimulgidae	Small nightjar, rare in Middle East
Egyptian Nightjar	*Caprimulgus aegyptius*	Nightjars/Caprimulgidae	Pale nightjar, rare in Middle East
Swifts			
White-throated Needletail	*Hirundapus caudacutus*	Swifts/Apodidae	Large swift from Asia
Chimney Swift	*Chaetura pelagica*	Swifts/Apodidae	Dumpy swift from North America
Plain Swift	*Apus unicolor*	Swifts/Apodidae	All-dark swift, in Canary Islands
Pacific Swift	*Apus pacificus*	Swifts/Apodidae	Large swift from Asia
Kingfishers			
Belted Kingfisher	*Ceryle alcyon*	Giant Kingfishers/ Cerylidae	Big kingfisher from North America
Pied Kingfisher	*Ceryle rudis*	Giant Kingfishers/ Cerylidae	Large, black and white kingfisher, in Middle East
White-throated Kingfisher	*Halcyon smyrnensis*	Halcyon Kingfishers/ Dacelonidae	Big, red-billed kingfisher, in Middle East
Bee-Eaters			
Blue-cheeked Bee-eater	*Merops persicus*	Bee-eaters/Meropidae	Greenish bee-eater from Middle East

Common Name	Scientific Name	Family/Scientific Name	Description
Bee-Eaters *continued*			
Little Green Bee-eater	*Merops orientalis*	Bee-eaters/Meropidae	Small, bright bee-eater, in Middle East
Rollers			
Indian Roller Middle	*Coracias benghalensis*	Rollers/Coracidae	Colourful roller, vagrant in East, from South Asia
Woodpeckers			
Northern Flicker	*Colaptes auratus*	Woodpeckers/Picidae	Colourful woodpecker from North America
Levaillant's Green Woodpecker	*Picus vaillantii*	Woodpeckers/Picidae	Large, scarce green woodpecker, in North Africa
Yellow-bellied Sapsucker	Sphyrapicus varius	Woodpeckers/Picidae	Small woodpecker from North America
Shrikes			
Black-crowned Tchagra	*Tchagra senegala*	Shrikes/Laniidae	Boldly patterned shrike, in North Africa
Brown Shrike	*Lanius cristatus*	Shrikes/Lanidae	Dull shrike from Asia
Long-tailed Shrike	*Lanius schach*	Shrikes/Lanidae	Brownish shrike from Asia
Crows			
Fan-tailed Raven	*Corvus rhipidurus*	Crows/Corvidae	Short-tailed raven, in North Africa and Middle East
Brown-necked Raven	*Corvus ruficollis*	Crows/Corvidae	Somewhat thinner-billed raven, in North Africa and Middle East
House Crow	*Corvus splendens*	Crows/Corvidae	Grey and black crow, introduced in Middle East from India
Daurian Jackdaw	*Corvus dauuricus*	Crows/Corvidae	Pied jackdaw from Asia
Larks			
Hoopoe Lark	*Alaemon alaudipes*	Larks/Alaudidae	Large pale lark, in Middle East
Thick-billed Lark	*Rhamphocoris clotbey*	Larks/Alaudidae	Heavy desert lark, in North Africa
Bimaculated Lark	*Melanocorypha bimaculata*	Larks/Alaudidae	Large lark, rare in Middle East
White-winged Lark	*Melanocorypha leucoptera*	Larks/Alaudidae	Large lark from Asia
Black Lark	*Melanocorypha yeltoniensis*	Larks/Alaudidae	Stocky lark from Asia
Hume's Short-toed Lark	*Calandrella acutirostris*	Larks/Alaudidae	Small lark, vagrant in Middle East, from Asia
Oriental Skylark	*Alauda gulgula*	Larks/Alaudidae	Plain lark from Middle East
Chestnut-headed Sparrow-lark	*Eremopterix signata*	Larks/Alaudidae	Small, sparrow-like lark from Africa
Black-crowned Sparrow-lark	*Eremopterix nigriceps*	Larks/Alaudidae	Small, finch-like lark, in Middle East

Common Name	Scientific Name	Family/Scientific Name	Description
Larks *continued*			
Dunn's Lark	*Eremalauda dunni*	Larks/Alaudidae	Small plain lark, rare in Middle East
Temminck's Lark	*Eremophila bilopha*	Larks/Alaudidae	Pale, horned, desert-living lark, in Middle East and North Africa
Bulbuls			
White-eared Bulbul	*Pycnonotus leucotis*	Bulbuls/Pycnonotidae	Black and brown bulbul, vagrant in Middle East, from Asia
Martins and Swallows			
Tree Swallow	*Tachycineta bicolor*	Swallows/Hirundinidae	Glossy green and white swallow from North America
Plain Martin	*Riparia paludicola*	Swallows/Hirundinidae	Small brown martin, in North Africa
Ethiopian Swallow	*Hirundo aethiopica*	Swallows/Hirundinidae	Long-tailed swallow from Africa
Wire-tailed Swallow	*Hirundo smithii*	Swallows/Hirundinidae	Slender swallow from Africa
Rock Martin	*Ptyonoprogne fuligula*	Swallows/Hirundinidae	Small, pale Crag Martin-like bird, in Middle East
Warblers and Allies			
Graceful Prinia	*Prinia gracilis*	Warblers/Sylviidae	Small, long-tailed warbler, in Middle East
Scrub Warbler	*Scotocerca inquieta*	Warblers/Sylviidae	Small, long-tailed warbler, in Middle East
Gray's Grasshopper Warbler	*Locustella fasciolata*	Warblers/Sylviidae	Shy warbler from Asia
Clamorous Reed Warbler	*Acrocephalus stentoreus*	Warblers/Sylviidae	Large reed warbler, in Middle East
Thick-billed Warbler	*Acrocephalus aedon*	Warblers/Sylviidae	Large reed warbler-like bird from Asia
Basra Reed Warbler	*Acrocephalus griseldis*	Warblers/Sylviidae	Rare reed warbler, in Middle East
Desert Warbler	*Sylvia nana*	Warblers/Sylviidae	Pale, sandy warbler from N Africa and Middle east
Eastern Crowned Leaf Warbler	*Phylloscopus coronatus*	Warblers/Sylviidae	Small leaf warbler from Asia
Tristram's Warbler	*Sylvia deserticola*	Warblers/Sylviidae	Small, Whitethroat-like warbler, in North Africa
Ménétrie's Warbler	*Sylvia mystacea*	Warblers/Sylviidae	Rare migrant warbler, in Middle East
Arabian Warbler	*Sylvia leucomelaena*	Warblers/Sylviidae	Dark-headed warbler, in Middle East
Two-barred Greenish Warbler	*Phylloscopus (trochiloides) plumbeitarsus*	Warblers/Sylviidae	Small warbler (like Greenish Warbler) from Asia
Green Warbler	*Phylloscopus (trochiloides) nitidus*	Warblers/Sylviidae	Small warbler from Asia
Plain Leaf Warbler	*Phylloscopus neglectus*	Warblers/Sylviidae	Small leaf warbler from Asia

Common Name	Scientific Name	Family/Scientific Name	Description
Warblers and Allies *continued*			
Ruby-crowned Kinglet	*Regulus calendula*	Kinglets/Regulidae	Tiny Goldcrest-like bird from North America
Waxwings			
Cedar Waxwing	*Bombycilla cedrorum*	Waxwings/Bombycillidae	Yellow-bellied waxwing, vagrant in UK, from North America
Grey Hypocolius	*Hypocolius ampelinus*	Waxwings/Bombycillidae	Shrike-like bird, in Middle East
Mockingbirds			
Northern Mockingbird	*Mimus polyglottus*	Mockingbirds/Mimidae	Thrush-like bird from North America
Brown Thrasher	*Toxostoma rufum*	Mockingbirds/Mimidae	Rusty, thrush-like bird from North America
Catbird	*Dumetella carolinensis*	Mockingbirds/Mimidae	Chat-like bird from North America
Nuthatches			
Red-breasted Nuthatch	*Sitta canadensis*	Nuthatches/Sittidae	Small, stripe-headed nuthatch from North America
Eastern Rock Nuthatch	*Sitta tephronota*	Nuthatches/Sittidae	Large nuthatch, in Middle East
Krüper's Nuthatch	*Sitta krueperi*	Nuthatches/Sittidae	Small, dark-capped nuthatch, in Middle East
Sunbirds			
Nile Valley Sunbird	*Anthreptes metallicus*	Sunbirds/Nectariniidae	Long-tailed nectar-eater, in Middle East
Palestine Sunbird	*Nectarinia osea*	Sunbirds/Nectariniidae	Tiny, dark nectar-eater, in Middle East
Starlings			
Tristram's Starling	*Onychognathus tristramii*	Starlings/Sturnidae	Dark, red-winged starling, in Middle East
Chats and Thrushes			
Varied Thrush	*Zoothera naevia*	Thrushes/Turdidae	Small thrush from W North America
Wood Thrush	*Hylocichla mustelina*	Thrushes/Turdidae	Small, spotted thrush from North America
Veery	*Catharus fuscescens*	Thrushes/Turdidae	Tiny spotted thrush from North America
Hermit Thrush	*Catharus guttatus*	Thrushes/Turdidae	Small spotted thrush from North America
Swainson's Thrush	*Catharus ustulatus*	Thrushes/Turdidae	Small spotted thrush from North America
Tickell's Thrush	*Turdus unicolor*	Thrushes/Turdidae	Colourful thrush from Asia
American Robin	*Turdus migratorius*	Thrushes/Turdidae	Large thrush from North America
Siberian Rubythroat	*Luscinia calliope*	Chats/Saxicoliidae	Colourful chat from Siberia

Common Name	Scientific Name	Family/Scientific Name	Description
Chats and Thrushes *continued*			
Siberian Blue Robin	*Luscinia cyane*	Chats/Saxicoliidae	Vagrant in NW Europe, from Siberia
White-throated Robin	*Irania gutturalis*	Chats/Saxicoliidae	Large greyish chat from Middle East
Eversmann's Redstart	*Phoenicurus erythronotus*	Chats/Saxicolidae	Large redstart from Asia
Moussier's Redstart	*Phoenicurus moussieri*	Chats/Saxicoliidae	Brightly patterned chat, in North Africa
Güldenstädt's Redstart	*Phoenicurus erythrogaster*	Chats/Saxicoliidae	Striking chat from Asia
Blackstart	*Cercomela melanura*	Chats/Saxicoliidae	Small, grey, black-tailed chat, in Middle East
Fuerteventura Stonechat	*Saxicola dacotiae*	Chats/Saxicoliidae	Small chat, endemic to Canary Islands
Pied Wheatear	*Oenanthe pleschanka*	Chats/Saxicoliidae	Small dark wheatear from Middle East and extreme E Europe
Desert Wheatear	*Oenanthe deserti*	Chats/Saxicoliidae	Black-tailed brownish wheatear from North Africa and Middle East
Finsch's Wheatear	*Oenanthe finschii*	Chats/Saxicoliidae	Large wheatear, in Middle East
White-crowned Wheatear	*Oenanthe leucopyga*	Chats/Saxicoliidae	Blackish wheatear from Middle East
Mourning Wheatear	*Oenanthe lugens*	Chats/Saxicoliidae	Piebald wheatear, in Middle East
Red-rumped Wheatear	*Oenanthe moesta*	Chats/Saxicoliidae	Large wheatear, in Middle East
Hooded Wheatear	*Oenanthe monacha*	Chats/Saxicoliidae	Large wheatear, in Middle East
Persian Wheatear	*Oenanthe xanthoprymna*	Chats/Saxicoliidae	Dark wheatear, in Middle East
Accentors			
Black-throated Accentor	*Prunella atrogularis*	Accentors/Prunellidae	Dunnock-like bird, vagrant in Middle East, from Siberia
Siberian Accentor	*Prunella montanella*	Accentors/Prunellidae	Dunnock-like bird from Siberia
Radde's Accentor	*Prunella ocularis*	Accentors/Prunellidae	Dunnock-like bird from Siberia
Flycatchers			
Brown Flycatcher	*Muscicapa dauurica*	Flycatchers/Muscicapidae	Plain flycatcher from Asia
Acadian Flycatcher	*Empidonax virescens*	Tyrant Flycatchers/Tyrannidae	Greenish flycatcher from North America
Eastern Phoebe	*Sayornis phoebe*	Tyrant Flycatchers/Tyrannidae	Small, dull flycatcher from North America
Babblers			
Fulvous Babbler	*Turdoides fulvus*	Babblers/Timaliidae	Thrush-like bird, in North Africa
Arabian Babbler	*Turdoides squamiceps*	Babblers/Timaliidae	Pale, thrush-like bird, in Middle East

Common Name	Scientific Name	Family/Scientific Name	Description
Sparrows			
Dead Sea Sparrow	*Passer moabiticus*	Sparrows//Passeridae	Small, colourful sparrow, in Middle East
Desert Sparrow	*Passer simplex*	Sparrows//Passeridae	Pale, dark-billed sparrow, in North Africa
Hill Sparrow	*Carpospiza brachydactyla*	Sparrows//Passeridae	Pale sparrow, in Middle East
Chestnut-shouldered Sparrow	*Gymornis xanthocollis*	Sparrows//Passeridae	Pale sparrow, in Middle East
Pipits			
Long-billed Pipit	*Anthus similis*	Pipits and wagtails/ Motacillidae	Large, pale, mountainside pipit, in Middle East
Buff-bellied Pipit	*Anthus rubescens*	Pipits and wagtails/ Motacillidae	Dark-legged pipit from North America
Berthelot's Pipit	*Anthus berthelotii*	Pipits and wagtails/ Motacillidae	Small pipit, in Canary Islands and Madeira
Finches			
Blue Chaffinch	*Fringilla teydea*	Finches/Fringillidae	Large blue finch, endemic to Canary Islands
Canary	*Serinus canaria*	Finches/Fringillidae	Greenish finch, endemic to Canary Islands
Red-fronted Serin	*Serinus pusillus*	Finches/Fringillidae	Small finch, in Middle East
Syrian Serin	*Serinus syriacus*	Finches/Fringillidae	Small, greenish, upland finch, in Middle East
Crimson-winged Finch	*Rhodopechys sanguinea*	Finches/Fringillidae	Thickset mountain finch, in Middle East
Mongolian Finch	*Bucanetes mongolicus*	Finches/Fringillidae	Large finch from Asia
Trumpeter Finch	*Bucanetes githagineus*	Finches/Fringillidae	Small, pale pinkish finch, in Middle East
Sinai Rosefinch	*Carpodacus synoicus*	Finches/Fringillidae	Pale, pink-tinged finch, in Middle East
Long-tailed Rosefinch	*Uragus sibiricus*	Finches/Fringillidae	Small finch from Asia
Desert Finch	*Rhodospiza obsoleta*	Finches/Fringillidae	Pale finch, in Middle East
Evening Grosbeak	*Hesperiphona vespertina*	Finches/Fringillidae	Large finch from North America
Tanagers			
Scarlet Tanager	Piranga olivacea	Tanagers/Thraupidae	Large finch from North America
Summer Tanager	Piranga rubra	Tanagers/Thraupidae	Finch-like bird from North America
Buntings			
Rufous-sided Towhee	*Pipilo erythrophthalmus*	Buntings/Emberizidae	Thickset finch-like bird from North America
Lark Sparrow	*Chondestes grammacus*	Buntings/Emberizidae	Streaky-headed bunting-like bird, vagrant in UK from North America

Common Name	Scientific Name	Family/Scientific Name	Description
Buntings *continued*			
Song Sparrow	*Melospiza melodia*	Buntings/Emberizidae	Streaky bunting from North America
Fox Sparrow	*Passerella iliaca*	Buntings/Emberizidae	Rufous bunting from North America
White-crowned Sparrow	*Zonotrichia leucophrys*	Buntings/Emberizidae	Sparrow-like bird from North America
Red-headed Bunting	*Emberiza bruniceps*	Buntings/Emberizidae	Yellowish bunting from Asia
Black-faced Bunting	*Emberiza spodocephala*	Buntings/Emberizidae	Dark bunting from Asia
Cinnamon-breasted Bunting	*Emberiza tahapisi*	Buntings/Emberizidae	Dark-coloured bunting from Africa
Grey-necked Bunting	*Emberiza buchanani*	Buntings/Emberizidae	Slender bunting from Asia
Yellow-browed Bunting	*Emberiza chrysophrys*	Buntings/Emberizidae	Small bunting from Asia
Pallas's Reed Bunting	*Emberiza pallasi*	Buntings/Emberizidae	Small bunting from Asia
Cinereous Bunting	*Emberiza cineracea*	Buntings/Emberizidae	Dull bunting from Asia
House Bunting	*Emberiza striolata*	Buntings/Emberizidae	Small bunting, in North Africa and Middle East
Black-faced Bunting	*Emberiza spodocephala*	Buntings/Emberizidae	Dark bunting from Asia
Savannah Sparrow	*Passerculus sandwichensis*	Buntings/Emberizidae	Sparrow-like bunting, vagrant in NW Europe, from North America
Indigo Bunting	*Passerina cyanea*	Buntings/Emberizidae	Dark bunting from North America
New World Orioles			
Brown-headed Cowbird	*Molothrus ater*	New World Orioles/Icteridae	Black bird, vagrant in NW Europe, from North America
Yellow-headed Blackbird	*Xanthocephalus xanthocephalus*	New World Orioles/Icteridae	Glossy black bird from North America
Northern Oriole	*Icterus galbula*	New World Orioles/Icteridae	Colourful oriole from North America
New World Warblers			
Ovenbird	*Seiurus aurocapillus*	New World Warblers/Parulidae	Woodland warbler from North America
Yellowthroat	*Geothlypsis trichas*	New World Warblers/Parulidae	Stocky warbler from North America
Northern Waterthrush	*Seiurus novaeboracensis*	New World Warblers/Parulidae	Streaked warbler from North America
Tennessee Warbler	*Vermivora peregrina*	New World Warblers/Parulidae	Plain warbler from North America
Blackburnian Warbler	*Dendroica fusca*	New World Warblers/Parulidae	Colourful warbler from North America
Yellow Warbler	*Dendroica petechia*	New World Warblers/Parulidae	Small warbler from North America
Yellow-throated Vireo	*Vireo flavifrons*	Vireos/Vireonidae	Small, warbler-like bird from North America
Philadelphia Vireo	*Vireo philadelphicus*	Vireos/Vireonidae	Small, warbler-like bird from North America

499

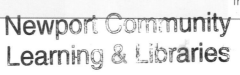

GLOSSARY

Many of the terms defined here are illustrated in the general introduction (pp. 8–53). For anatomical terms see also pp.10–11.

- **ADULT** A fully mature bird, able to breed, showing the final plumage pattern that no longer changes with age.
- **BARRED** With marks crossing the body, wing, or tail.
- **BROOD** Young produced from a single clutch of eggs incubated together.
- **CALL** Vocal sound often characteristic of a particular species, communicating a variety of messages.
- **COLONY** A group of nests of a highly social species, especially among seabirds but also others such as the Sand Martin and Rook.
- **COVERT** A small feather in a well-defined tract, on the wing or at the base of the tail, covering the base of the larger flight feathers.
- **CRYPTIC** Describes plumage pattern and colours that make a bird difficult to see in its favoured habitat.
- **DABBLE** To feed in shallow water, with rapid movements of the bill, sieving water through fine comb-like teeth to extract food.
- **DECLINING** Populations undergoing a steady decline over a number of years.
- **DIMORPHIC** Having two forms: sexually dimorphic means that the male and female of a species look different; otherwise indicates two colour forms.
- **DRUMMING** Sound made by woodpeckers with rapid beats of the bill against a hard object, or by a snipe, diving through the air with vibrating tail feathers.
- **EAR TUFT** A bunch of feathers on the head of an owl, capable of being raised as a visual signal and perhaps to assist camouflage.
- **ECLIPSE** The plumage of male ducks that is adopted during the summer, when they moult and become flightless for a short time.
- **ENDANGERED** Found in very small numbers, in a very small area or in a very restricted and declining habitat, so that the future security of the species is in doubt.
- **ESCAPEE** A bird that has escaped into the wild from a collection of some kind, such as a zoo or wildlife park.
- **EYE PATCH** An area of colour around the eye, often in the form of a "mask", broader than an eye-stripe.
- **EYE-RING** A more or less circular patch of colour, usually narrow and well-defined, around the eye.
- **EYE-STRIPE** A stripe of distinctive colour running in front of and behind the eye.

- **FAMILY** A category in classification, grouping species or genera that are closely related; ranked at a higher level than the genus.
- **FLIGHT FEATHER** Any one of the long feathers on the wing (primaries and secondaries).
- **FOREWING** The front part of a wing, including the outer primaries, primary coverts, and secondary coverts.
- **GAPE** A bird's mouth, or the angle at the base of the bill.
- **GENUS (_pl._ GENERA)** A category in classification: a group of closely related species, whose relationship is recognized by the same first name in the scientific terminology, e.g. _Larus_ in _Larus fuscus_.
- **HINDWING** The rear part of the wing, including the secondary feathers, especially when it has a distinctive colour or pattern.
- **HYBRID** The result of cross-breeding between two species; usually infertile. Rare in the wild.
- **IMMATURE** Not yet fully adult or able to breed; there may be several identifiable plumages during immaturity but many small birds are mature by the first spring after they have fledged.
- **INNER WING** The inner part of the wing, comprising the secondaries and rows of coverts (typically marginal, lesser, median, and greater coverts).
- **JUVENILE** A bird in its first plumage, that in which it makes its first flight, before its first moult in the autumn.
- **LEK** A gathering of birds at which males display communally, with mock fighting, while females choose which one to mate with.
- **LOCALIZED** More than 90 per cent of the population occurs at ten sites or less.
- **MOULT** The shedding and renewing of feathers in a systematic way; most birds have a partial moult and a complete moult each year.
- **MIGRANT** A species that spends part of the year in one geographical area and part in another, moving between the two on a regular basis. (See also p.26.)
- **ORDER** A category in classification: families grouped to indicate their close relationship or common ancestry; usually a more uncertain or speculative grouping than a family.
- **OUTER WING** The outer half of the wing, comprising the primaries, their coverts, and the alula, or bastard wing (the "thumb").
- **ORBITAL RING** A thin, bare, fleshy ring around the eye, sometimes with a distinctive colour.
- **PRIMARY** Any one of the long feathers, or quills, forming the tip and trailing edge of the outer wing, growing from the "hand".

- **RACE** _See_ Subspecies.
- **RARE** Found in small numbers or very low densities, although not necessarily at risk.
- **SCAPULAR** Any one of a group of feathers on the shoulder, forming a more or less oval patch each side of the back, at the base of the wing.
- **SECONDARY** Any one of the long flight feathers forming the trailing edge of the inner wing, growing from the ulna or "arm".
- **SECURE** The population is under no current threat.
- **SONG** Vocalization with character particular to the individual species, used to communicate a claim to a breeding territory and attract a mate.
- **SONG-FLIGHT** A special flight, often with a distinctive pattern, combined with a territorial song.
- **SPECIES** A group of living organisms, individuals of which can interbreed to produce fertile young, but do not normally breed, or cannot produce fertile young, with a different species.
- **SPECULUM** A colourful patch on a duck's hindwing, formed by the secondary feathers.
- **STREAKED** With small marks that run lengthwise along the body.
- **SUBSPECIES** A race; a recognizable group within a species, isolated geographically but able to interbreed with others of the same species.
- **SUPERCILIARY STRIPE** A stripe of colour running above the eye, like an eyebrow.
- **TERTIAL** Any one of a small group of feathers, sometimes long and obvious, at the base of the wing adjacent to the inner secondaries.
- **UNDERWING** The underside of a wing, usually visible only in flight or when a bird is preening.
- **UPPERWING** The upperside of the wing, clearly exposed in flight but often mostly hidden when the bird is perched.
- **VAGRANT** An individual bird that has strayed beyond the usual geographic range of its species.
- **VENT** The area of feathers between the legs and the undertail coverts.
- **VULNERABLE** Potentially at risk due to a dependence on a restricted habitat or range, or to small numbers.
- **WINGPIT** A group of feathers – the axillaries – located at the base of the underwing.
- **WINGBAR** A line of colour produced by a tract of feathers or feather tips, crossing the closed wing and running along the spread wing.
- **YOUNG** An imprecise term to describe immature birds; often meaning juveniles or nestlings.

INDEX

A

aberrations 19
Accentor, Alpine 391
accentors 368
Accipiter
  brevipes 449
  gentilis 150
  nisus 151
Accipitridae 136, 447
Accipitriformes 136
Acrocephalus
  agricola 477
  arundinaceus 353
  dumetorum 476
  melanopogon 349
  paludicola 476
  palustris 351
  schoenobaenus 350
  scirpaceus 352
Actitis
  hypoleucos 203
  macularia 461
Aegithalidae 304
Aegithalos caudatus 313
Aegypius monachus 143
Aix
  galericulata 443
  sponsa 443
Alauda arvensis 320
Alaudidae 314, 470
albatrosses 111
Alca torda 217
Alcedinidae 270
Alcedo atthis 271
Alcidae 214, 466
Alectoris
  barbara 451
  chukar 451
  graeca 451
  rufa 98
Alle alle 218
Alopochen aegyptiaca 442
Alpine Accentor 391
Alpine Chough 295
Alpine Swift 269
American Golden
    Plover 456
American Wigeon 443
Anas
  acuta 74
  americana 443

Anas cont.
  clypeata 76
  crecca 72
  crecca carolinensis 72
  discors 444
  penelope 70
  platyrhynchos 73
  querquedula 75
  rubripes 443
  strepera 71
Anatidae 55, 414
anatomy 10
Anser
  albifrons 64
  albifrons flavirostris 64
  anser 65
  brachyrhynchus 63
  caerulescens 441
  erythropus 442
  fabalis 62
  fabalis rossicus 62
Anseriformes 59
Anthropoides virgo 453
Anthus
  campestris 404
  cervinus 407
  godlewski 471
  gustavi 471
  hodgsoni 471
  petrosus 408
  pratensis 406
  richardi 471
  spinoletta 409
  trivialis 405
Apodidae 469, 491
Apodiformes 277
Apus
  affinis 493
  apus 282
  caffer 469
  melba 269
  pallidus 268
Aquatic Warbler 476
Aquila
  adalberti 447
  chrysaetos 154
  clanga 448
  heliaca 447
  nipalensis 448
  pomarina 448
Archaeopteryx 8
Arctic Redpoll 483

Arctic Skua 224
Arctic Warbler 478
Ardea
  cinerea 129
  purpurea 130
Ardeidae 136, 465
Ardeola ralloides 125
Arenaria interpres 192
Asio
  flammeus 263
  otus 262
Athene noctua 274
Audouin's Gull 487
Auk, Little 218
auks 214, 218, 466
Avocet 179
avocets 173
Aythya
  affinis 445
  collaris 444
  ferina 77
  fuligula 78
  marila 79
  nyroca 444
Azure Tit 480
Azure-winged Magpie
    482

B

Baillon's Crake 452
Baird's Sandpiper 458
Barbary Partridge 451
Bar-tailed Godwit 191
Barn Owl 256
Barnacle Goose 67
Barred Warbler 339
Barrow's Goldeneye 446
Bartramia longicauda 461
Bean Goose 62
Bearded Tit 312
Bee-eater 272
bee-eaters 270
Bewick's Swan 60
bill shape 22
binoculars 50, 51
BirdLife International
    53
bird-tables 49
birds of prey 136, 447
birdwatching 50
Bittern 122
  Little 123

bitterns 121
Black Duck 443
Black Grouse 95
Black Guillemot 216
Black Kite 140
Black Redstart 380
Black Stork 133
Black Tern 231
    White-winged 466
Black Vulture 143
Black Wheatear 388
Black Woodpecker 284
Black-backed Gull,
    Great 246
    Lesser 241
Black-bellied
    Sandgrouse 467
Black-headed Bunting
    485
Black-headed Gull 237
Black-necked Grebe 110
Black-shouldered Kite
    448
Black-tailed Godwit 190
Black-throated Diver
    104
Black-winged Pratincole
    454
Black-winged Stilt 178
Blackbird 371
Blackcap 337
Blackpoll Warbler 483
Blue Rock Thrush 383
Blue Tit 306
Blue-winged Teal 444
Bluetail, Red-flanked
    472
Bluethroat 379
Blyth's Pipit 471
Blyth's Reed Warbler
    476
Bobolink 486
Bombycilla garrulus 358
Bombycillidae 357
Bonaparte's Gull 463
Bonasa bonasia 92
Bonelli's Eagle 155
Bonelli's Warbler 333
Booted Eagle 156
Booted Warbler 478
Botaurus stellaris 122
Brambling 413

Branta
  bernicla 68
  bernicla hrota 68
  bernicla nigricans 68
  canadensis 66
  leucopsis 67
  ruficollis 442
breeding 15
Brent Goose 68
Bridled Tern 466
Broad-billed Sandpiper
  457
Brünnich's Guillemot
  466
Bubulcus ibis 126
Bucephala
  clangula 85
  islandica 446
Buff-breasted
  Sandpiper 459
Bullfinch 426
Bunting,
  Black-headed 485
  Cirl 432
  Corn 437
  Cretzschmar's 485
  Lapland 454
  Little 435
  Ortolan 434
  Pine 485
  Reed 436
  Rock 433
  Rustic 484
  Snow 429
  Yellow-breasted 485
buntings 428, 484
Burhinidae 173, 453
Burhinus
  oedicnemus 177
  senegalensis 453
Bush Robin, Rufous 472
Bustard,
  Great 172
  Little 171
bustards 163
Buteo
  buteo 152
  lagopus 153
  rufinus 449
Button-quail, Small 452
Buzzard 152
  Honey 139
  Long-legged 449
buzzards 136

C
Calandra Lark 315
Calandrella
  brachydactyla 316
  rufescens 470
Calcarius lapponicus 430
Calidris
  acuminata 460
  alba 197
  alpina 198
  alpina alpina 198
  alpina arctica 198
  alpina schinzii 198
  bairdii 458
  canutus 193
  ferruginea 195
  fuscicollis 458
  maritima 199
  mauri 459
  melanotos 460
  minuta 200
  minutilla 458
  pusilla 459
  ruficollis 459
  subminuta 457
  temminckii 196
  tenuirostris 458
calls 24
Calonectris diomedea 113
camera 50
camouflage 19
Canada Goose 66
Capercaillie 96
Caprimulgidae 254, 469
Caprimulgiformes 265
Caprimulgus
  europaeus 265
  ruficollis 469
Cardueline finches 410
Carduelis
  cannabina 419
  carduelis 417
  chloris 414
  flammea 421
  flammea flammea 421
  flavirostris 420
  hornemanni 483
  spinus 418
Carpodacus erythrinus 424
Carrion Crow 301
Caspian Plover 455
Caspian Tern 229
Catharus minimus 473

Cattle Egret 126
Cepphus grylle 216
Cercotrichas galactotes 472
Certhia
  brachydactyla 362
  familiaris 361
  familiaris familiaris 361
Certhiidae 357
Cetti's Warbler 332
Cettia cetti 332
Chaffinch 412
Charadriidae 173, 454
Charadriiformes 180
Charadrius
  alexandrinus 186
  asiatus 455
  dubius 184
  hiaticula 185
  leschenaultii 455
  mongolus 455
  morinellus 187
  semipalmatus 454
  vociferus 455
chats 368
Chersophilus duponti 470
Chiffchaff 335
Chlidonias
  hybridus 230
  leucopterus 466
  niger 231
Chough 294
  Alpine 367
Chrysolophus
  amherstiae 451
  pictus 452
Chukar 451
Ciconia
  ciconia 134
  nigra 133
Ciconiidae 131
Ciconiiformes 122
Cinclidae 357
Cinclus
  cinclus 367
  cinclus cinclus 367
  cinclus hibernicus 367
Circaetus gallicus 146
Circus
  aeruginosus 147
  cyaneus 148
  macrouros 473
  pygargus 149
Cirl Bunting 432
Cisticola juncidis 354

Citril Finch 416
Citrine Wagtail 472
Clamator glandarius 467
Clangula hyemalis 82
classifying birds 9
cliffs 34, 40
Coal Tit 304, 309
coasts 32, 34, 36
Coccothraustes
  coccothraustes 427
Coccyzus americanus 467
code of conduct 51
Collared Dove 251
Collared Flycatcher 480
  Semi-collared 305
Collared Pratincole 213
Columba
  livia 248
  oenas 249
  palumbus 250
Columbidae 247
Columbiformes 248
Common Gull 240
Common Sandpiper 203
Common Scoter 83
Common Tern 233
conservation 52
co-ordination 21
Coot 168
  Crested 453
coots 163, 453
Coracias garrulus 273
Coraciidae 270
Coraciiformes 271
Cormorant 72
  Pygmy 441
cormorants 117, 440
Corn Bunting 437
Corncrake 166
Corsican Nuthatch 481
Corvidae 291, 482
Corvus
  corax 303
  corone 301
  cornix 302
  frugilegus 300
  monedula 299
Cory's Shearwater 113
Coturnix coturnix 97
Courser, Cream-
  coloured 454
courtship 14
Crag Martin 324
crags 40

Crake,
  Baillon's 452
  Little 452
  Spotted 165
crakes 163, 452
Crane 170
  Demoiselle 453
cranes 170, 453
Cream-coloured
    Courser 454
Crested Coot 453
Crested Grebe,
    Great 107
Crested Lark 317
Crested Tern,
    Lesser 465
Crested Tit 308
Cretzschmar's Bunting
    485
Crex crex 166
Crossbill 422
  Parrot 423
  Scottish 484
  Two-barred 484
crossbills 410
Crow,
  Carrion 301
  Hooded 302
crows 357, 482
Cuckoo 253
  Great Spotted 467
  Yellow-billed 467
cuckoos 247, 467
Cuculidae 247, 467
Cuculiformes 253
Cuculus canorus 253
Curlew 189
Curlew Sandpiper 195
Cursorius cursor 454
Cyanopica cyanus 482
Cyanistes caeruleus 306
Cygnus
  columbianus 60
  columbianus
      columbianus 60
  cygnus 61
  olor 62
Cyprus Warbler 475
Cyprus Wheatear 473

D
Dabchick 101
dabbling 22
Dalmatian Pelican 440

Dark-eyed Junco 486
Dark-throated Thrush
    474
Dartford Warbler 342
dawn chorus 25
Delichon urbicum 326
Demoiselle Crane 453
Dendrocopos
  leucotos 470
  major 280
  medius 281
  minor 282
  syriacus 470
Dendroica
  coronata 483
  striata 483
Dipper 367
dippers 357
display 14
Diver,
  Black-throated 104
  Great Northern 105
  Red-throated 103
  White-billed 438
divers 101, 438
Dolichonyx oryzivorus 486
domestic pigeons 248
Dotterel 187
Dove,
  Collared 251
  Rock 248
  Stock 249
  Turtle 252
doves 247
Dowitcher,
  Long-billed 462
drumming 25
Dryocopus martius 284
Duck,
  Black 443
  Ferruginous 444
  Harlequin 446
  Long-tailed 82
  Marbled 468
  Ring-necked 468
  Ruddy 55
  Tufted 78
  White-headed 446
  Wood 443
ducks 55
Dunlin 198
Dunnock 368
duping 16
Dupont's Lark 470

Dusky Thrush 474
Dusky Warbler 478

E
Eagle,
  Bonelli's 155
  Booted 156
  Golden 154
  Imperial 447
  Lesser Spotted 448
  Short-toed 146
  Spanish Imperial
      447
  Spotted 448
  Steppe 448
  White-tailed 144
Eagle Owl 260
eagles 136
eggs 16
Egret,
  Cattle 126
  Great White 156
  Little 146
  Western Reef 441
egrets 121, 441
Egretta
  alba 156
  garzetta 127
  gularis 441
Egyptian Goose 442
Egyptian Vulture 145
Eider 80
  King 81
  Spectacled 445
  Steller's 445
Elanus caerulescens 448
Eleonora's Falcon
    450
Emberiza
  aureola 485
  caesia 485
  calandra 437
  cia 433
  cirlus 432
  citrinella 431
  hortulana 434
  leucocephalus 485
  melanocephala 485
  pusilla 435
  rustica 484
  schoeniclus 436
Emberizidae 428, 484
Enucleator,
  Pinicola 425

equipment 17
Eremophila alpestris 321
Erithacus rubecula 376
estuaries 32
evolution 8
Eyebrowed Thrush 474
extinction 9

F
Falco
  biarmicus 450
  cherrug 450
  columbarius 160
  eleonorae 450
  naumanni 158
  peregrinus 162
  rusticolus 450
  subbuteo 161
  tinnunculus 159
  vespertinus 449
Falcon,
  Eleonora's 450
  Gyr 450
  Red-footed 449
Falconidae 136, 449
Falconiformes 159
falcons 136, 425
Fan-tailed Warbler 354
farmland 44
feathers 11, 18
feeding 22
feral pigeon 248
Ferruginous Duck 444
Ficedula
  albicollis 480
  hypoleuca 393
  hypoleuca iberiae 393
  parva 480
  semitorquata 480
Fieldfare 372
Finch, Citril 416
finches 410, 483
Firecrest 356
Flamingo, Greater 132
flamingos 131
flight 20
Flycatcher,
  Collared 480
  Pied 393
  Red-breasted 480
  Semi-collared 480
  Spotted 392
flycatchers 328, 480
forest 46

Forster's Tern 466
Franklin's Gull 464
Fratercula arctica 215
Fringilla
   coelebs 412
   montifringilla 413
Fringillidae 410, 483
Fulica
   atra 168
   cristata 453
Fulmar 112
Fulmarus glacialis 112

G

Gadwall 71
Galerida cristata 317
Galeridae theklae 318
Galliformes 92
Gallinago
   gallinago 212
   media 462
Gallinula chloropus 167
Gallinule, Purple 453
gamebirds 90, 451
Gannet 118
gannets 117
Garden Warbler 338
gardens 48
Garganey 75
Garrulus glandarius 297
Gavia
   adamsii 438
   arctica 104
   immer 105
   stellata 103
Gaviidae 101, 438
Gaviiformes 103
geese 55
Glareola
   maldivarium 454
   nordmanni 454
   pratincola 213
Glareolidae 173, 454
Glaucidium passerinum 258
Glaucous Gull 245
Glossy Ibis 441
Godwit 173
   Bar-tailed 191
   Black-tailed 190
Goldcrest 355
Golden Eagle 154
Golden Oriole 286
Golden Pheasant
   452

Golden Plover 181
   American 456
   Pacific 456
Goldeneye 85
   Barrow's 446
Goldfinch 417
Goosander 88
Goose,
   Barnacle 67
   Bean 62
   Brent 68
   Canada 66
   Egyptian 442
   Greylag 65
   Lesser White-fronted
     442
   Pink-footed 63
   Red-breasted 442
   Snow 441
   White-fronted 64
Goshawk 150
Grasshopper Warbler 345
grassland 44
Great Black-backed
   Gull 246
Great Black-headed
   Gull 463
Great Bustard 172
Great Crested Grebe 107
Great Grey Owl 468
Great Grey Shrike 289
Great Knot 458
Great Northern Diver
   105
Great Reed Warbler 311
Great Shearwater 439
Great Skua 226
Great Snipe 462
Great Spotted Cuckoo
   467
Great Spotted
   Woodpecker 280
Great Tit 307
Great White Egret 128
Greater Flamingo 132
Greater Sand Plover 455
Greater Yellowlegs 460
Grebe,
   Black-necked 110
   Great Crested 107
   Little 106
   Pied-billed 438
   Red-necked 108
   Slavonian 109

grebes 101, 438
Green Sandpiper 204
Green Woodpecker 278
Greenfinch 414
greenfinches 394
Greenish Warbler 479
Greenshank 206
Grey Heron 129
Grey Owl, Great 468
Grey Phalarope 202
Grey Plover 182
Grey Shrike,
   Great 289
   Lesser 288
Grey Wagtail 402
Grey-cheeked Thrush
   473
Grey-headed
   Woodpecker 279
Greylag Goose 65
Griffon Vulture 142
Grosbeak,
   Pine 425
   Rose-breasted 486
Grouse 90
   Black 95
   Hazel 92
   Red 93
   Willow 93
Gruidae 163, 453
Gruiformes 166
Grus grus 170
Guillemot 219
   Black 216
   Brünnich's 466
Gull,
   Audouin's 463
   Black-headed 237
   Bonaparte's 463
   Common 240
   Franklin's 464
   Glaucous 245
   Great Black-backed
     246
   Great Black-headed
     463
   Herring 242
   Iceland 244
   Ivory 464
   Laughing 464
   Lesser Black-backed
     241
   Little 238
   Mediterranean 239

Gull cont.
   Ring-billed 463
   Ross's 465
   Sabine's 464
   Slender-billed 462
   Yellow-legged 243
Gull-billed Tern 228
gulls 214, 462
Gypaetus barbatus 447
Gyps fulvus 142
Gyr Falcon 450

H

Haematopodidae 173
Haematopus ostralegus 180
Haliaeetus albicilla 144
Harlequin Duck 446
Harrier,
   Hen 148
   Marsh 147
   Montagu's 149
   Pallid 449
harriers 136
hatching 16
Hawfinch 427
Hawk Owl 469
hawks 136
Hazel Grouse 92
health 42
Hen Harrier 148
Heron,
   Grey 129
   Night 124
   Purple 130
   Squacco 125
herons 121
Herring Gull 242
Hieraaetus
   fasciatus 155
   pennatus 156
Himantopus himantopus
   178
Hippolais
   caligata 478
   icterina 347
   languida 477
   olivetorum 477
   pallida 477
   polyglotta 348
hirundines 322
Hirundinidae 322
Hirundo
   daurica 327
   rustica 325

Histrionicus
    histrionicus 446
Hobby 161
Honey Buzzard 139
Hooded Crow 302
hooded gull 237, 238
Hoopoe 270, 274
House Martin 326
House Sparrow 395
Hume's Leaf Warbler 479
Hydrobates pelagicus 115
Hydrobatidae 111

I

Ibis, Glossy 441
ibises 441
Iceland Gull 244
Ichthyaetus,
    audouinii 293
    ichthyaetus 463
Icteridae 486
Icterine Warbler 347
Iduna caligata 478
Imperial Eagle 447
    Spanish 447
Infaustus,
    Perisoreus 293
Isabelline Shrike 482
Isabelline Wheatear 472
islands 34
Ivory Gull 464
Ixobrychus minutus 123

J

Jack Snipe 210
Jackdaw 299
Jay 297
    Siberian 293
Junco, Dark-eyed 486
Junco hyemalis 486
Jynx torquilla 277

K

Kentish Plover 186
Kestrel 159
    Lesser 158
Killdeer 455
King Eider 81
Kingfisher 271
kingfishers 270
Kite,
    Black 140
    Black-shouldered 448
    Red 141

kites 136
Kittiwake 214, 236
kleptoparasitism 23
Knot 193
    Great 458

L

Lady Amherst's
    Pheasant 451
Lagopus
    lagopus 93
    lagopus scoticus 93
    muta 94
lakes 31
Lammergeier 447
Lanceolated Warbler 476
Laniidae 285, 482
Lanius
    collurio 287
    excubitor 289
    excubitor algeriensis 289
    excubitor meridionalis 289
    excubitor pallidirostris
    289
    isabellinus 482
    minor 288
    nubicus 482
    senator 290
Lanner 450
Lapland Bunting 430
Lappet-faced Vulture 447
Lapwing 183
    Sociable 456
    Spur-winged 456
    White-tailed 457
Laridae 214, 462
Lark,
    Calandra 315
    Crested 317
    Dupont's 470
    Lesser Short-toed 470
    Short-toed 316
    Thekla 318
larks 314, 470
Larus
    argentatus 242
    argentatus argentatus 242
    cachinnans 243
    cachinnans
        cachinnans 243
    cachinnans michahellis
    243
    canus 240
    delawarensis 463

Larus cont.
    fuscus 241
    fuscus fuscus 241
    fuscus graellsii 241
    fuscus intermedius 241
    glaucoides 244
    hyperboreus 245
    marinus 246
    melanocephalus 239
    michahellis 243
    tridactyla 236
Laughing Gull 464
Leach's Petrel 116
Leaf Warbler, Hume's
    479
Least Sandpiper 458
lek 14
Lesser Black-backed
    Gull 241
Lesser Crested Tern
    465
Lesser Grey Shrike 288
Lesser Kestrel 158
Lesser Sand Plover 455
Lesser Scaup 445
Lesser Short-toed Lark
    470
Lesser Spotted Eagle 448
Lesser White-fronted
    Goose 442
Lesser Whitethroat 340
Lesser Yellowlegs 460
Leucophaeus
    atricilla 464
    pipixcan 464
Levant Sparrowhawk 449
life cycle 12
Limicola falcinellus 457
Limnodromus scolopaceus
    462
Limosa
    lapponica 191
    limosa 190
Linnet 419
linnets 394
Little Auk 218
Little Bittern 123
Little Bunting 435
Little Bustard 171
Little Crake 425
Little Egret 127
Little Grebe 106
Little Gull 238
Little Owl 259

Little Ringed Plover
    184
North Atlantic Little
    Shearwater 439
Little Stint 200
Little Swift 469
Little Tern 227
Locustella
    fluviatilis 476
    lanceolata 476
    luscinioides 346
    naevia 345
Locustella warblers 328
Long-billed Dowitcher
    462
Long-eared Owl 262
Long-legged Buzzard
    449
Long-tailed Duck 82
Long-tailed Skua 225
Long-tailed Tit 313
Long-toed Stint 457
Lophophanes cristatus 308
Loxia
    curvirostra 422
    leucoptera 484
    pytyopsittacus 423
    scotica 484
Lullula arborea 319
Luscinia
    luscinia 377
    megarhynchos 378
    svecica 379
    svecica cyanecula 379
    svecica magna 379
    svecica svecica 379
Lymnocryptes minimus 210

M

Madeiran Storm-petrel
    440
Magpie 296
    Azure-winged 482
magpies 291
Mallard 73
Mandarin 443
Manx Shearwater 114
Marbled Duck 444
markings 18
Marmaronetta
    angustirostris 468
Marmora's Warbler 475
Marsh Harrier 147
Marsh Sandpiper 207

marsh terns 220, 231
Marsh Tit 311
Marsh Warbler 351
Martin,
　Crag 324
　House 326
　Sand 323
martins 322
Masked Shrike 482
mating 15
Meadow Pipit 406
mechanical sounds 25
Mediterranean Gull
　239
Mediterranean scrub 42
Mediterranean
　Shearwater 439
Melanitta
　fusca 84
　nigra 83
　perspicillata 445
Melanocorypha calandra
　315
Melodious Warbler
　348
Merganser, Red-
　breasted 87
Mergus
　albellus 86
　merganser 88
　serrator 87
Merlin 160
Meropidae 270
Merops apiaster 272
Micropalama
　himantopus 461
migration 26
Milvus
　migrans 140
　milvus 141
Mistle Thrush 375
Montagu's Harrier 149
Monticola
　saxatilis 382
　solitarius 383
Montifringilla nivalis
　484
monogamy 15
Moorhen 167
moorland 38
Morus bassanus 118
Motacilla
　alba 403
　alba alba 403

Motacilla cont.
　cinerea 402
　citreola 472
　flava 401
　flava feldegg 401
　flava flava 401
Motacillidae 399, 471
moulting 19
mountains 38, 40
Moustached Warbler
　349
Muscicapa striata 392
Muscicapidae 370, 480
Mute Swan 59
mutual preening 15

N
natural selection 9
Naumann's Thrush 474
Neophron percnopterus
　145
Neornithes 8
nests 16
Netta rufina 446
Night Heron 124
Nightingale 378
　Thrush 377
nightingales 368
Nightjar 265
　Red-necked 469
nightjars 254, 469
North American
　warblers 483
Northern Diver, Great
　105
Nucifraga
　caryocatactes 298
　caryocatactes
　　macrothyncus 298
Numenius
　arquata 189
　phaeopus 188
Nuthatch 360
　Corsican 481
　Rock 481
nuthatches 357, 481
Nyctea scandiaca 468
Nycticorax nycticorax 124

O
Oceanites oceanicus 440
Oceanodroma
　leucorhoa 116
　castro 440

Oenanthe
　cypriaca 472
　hispanica 387
　hispanica hispanica 387
　hispanica melanoleuca
　　387
　isabellina 472
　leucura 388
　oenanthe 386
　oenanthe leucorhoa 386
Olivaceous Warbler 477
Olive-backed Pipit 471
Olive-tree Warbler 477
Onychoprion
　anaethetus 466
　fuscatus 465
Oriental Pratincole
　454
Oriole, Golden 286
orioles 285
Oriolidae 285
Oriolus oriolus 286
Orphean Warbler 474
Ortolan Bunting 434
Osprey 157
Otididae 169
Otis tarda 172
Otus scops 257
Ouzel, Ring 370
Owl,
　Barn 256
　Eagle 260
　Great Grey 468
　Hawk 469
　Little 259
　Long-eared 262
　Pygmy 258
　Scops 257
　Short-eared 263
　Snowy 468
　Tawny 261
　Tengmalm's 264
　Ural 468
owls 254, 468
Oxyura
　jamaicensis 89
　leucocephala 446
Oystercatcher 180

P
Pacific Golden Plover
　456
Paddyfield Warbler 477
Pagophila eburnea 464

pair bonding 14
Palearctic 28, 29
Pallas's Warbler 479
Pallid Harrier 449
Pallid Swift 268
Pandion haliaetus 157
Pandionidae 136
Panurus biarmicus 312
Parakeet, Ring-necked
　468
Paridae 304, 480
parks 48
Parrot Crossbill 423
parrots 468
partial migrants 27
Partridge,
　Barbary 451
　Grey 99
　Red-legged 98
　Rock 451
partridges 90
Parulidae 483
Parus
　cristatus 308
　cinctus 481
　cyanus 480
　lugubris 481
　major 307
Passer
　domesticus 395
　domesticus italiae 395
　hispaniolensis 396
　montanus 397
Passeridae 394, 484
Passeriformes 320
Passerines 270
Pechora Pipit 471
Pectoral Sandpiper 460
Pelecanidae 117, 440
Pelecaniformes 118
Pelecanus
　crispus 440
　onocrotalus 440
Pelican,
　Dalmatian 440
　White 440
pelicans 117, 440
Penduline Tit 305
Perdix perdix 99
Peregrine 162
Periparus
　ater 309
　ater ledouci 309
Perisoreus infaustus 293

Pernis apivorus 139
Petrel,
    Leach's 116
    Storm 115
petrels 111, 440
Petronia petronia 398
Phalacrocoracidae 117,
    441
Phalacrocorax
    aristotelis 120
    aristotelis desmaresti
    120
    carbo 119
    carbo sinensis 119
    pygmeus 441
Phalarope,
    Grey 202
    Red-necked 201
    Wilson's 462
Phalaropus
    fulicarius 202
    lobatus 201
    tricolor 462
Phasianidae 90, 451
Phasianus colchicus 100
Pheasant 100
    Golden 452
    Lady Amherst's 451
pheasants 90
Pheucticus ludovicianus
    486
Philomachus pugnax 194
Phoenicopteridae 131
Phoenicopteriformes
    132
Phoenicopterus ruber 132
Phoenicurus
    ochruros 380
    phoenicurus 381
Phylloscopus
    bonelli 333
    bonelli orientalis 333
    borealis 478
    collybita 335
    collybita tristis 335
    fuscatus 478
    humei 479
    inornatus 479
    proregulus 479
    sibilatrix 334
    swarzi 478
    trochiloides 479
    trochilus 336
Pica pica 296

Picidae 275, 470
Piciformes 284
Picoides tridactylus 283
Picus
    canus 279
    viridis 278
    viridis sharpei 278
Pied Flycatcher 393
pied flycatchers 328
Pied Wagtail 403
pied woodpeckers 281,
    282
Pied-billed Grebe 438
pigeons 247
    domestic 248
    feral 248
    racing 249
    town 248
Pin-tailed Sandgrouse
    467
Pine Bunting 485
Pine Grosbeak 425
Pinicola enucleator 425
Pink-footed Goose 63
Pintail 74
Pipit,
    Blyth's 471
    Meadow 406
    Olive-backed 471
    Pechora 471
    Red-throated 407
    Richard's 471
    Rock 408
    Tawny 404
    Tree 405
    Water 409
pipits 399, 471
Platalea leucorodia 135
Plectrophenax nivalis 429
Plegadis falcinellus 441
Plover,
    American Golden 456
    Caspian 455
    Goggle-eye 177
    Golden 181
    Greater Sand 455
    Grey 182
    Kentish 186
    Lesser Sand 455
    Little Ringed 184
    Pacific Golden 456
    Ringed 185
    Semipalmated 454
plovers 173

plumage 18
Pluvialis
    apricaria 181
    dominica 456
    fulva 456
    squatarola 182
Pochard 77
    Red-crested 446
Podiceps
    auritus 109
    cristatus 107
    grisegena 108
    nigricollis 110
Podicipedidae 101, 438
Podicipediformes 106
Podilymbus podiceps 438
Poecile
    montanus 310
    palustris 311
Polysticta stelleri 445
Pomarine Skua 223
Porphyrio porphyrio 453
Porzana
    parva 452
    porzana 165
    pusilla 452
Pratincole,
    Black-winged 454
    Collared 213
    Oriental 454
pratincoles 173
probing 22, 23
Procellariiformes 112
Procellariidae 111, 439
promiscuity 15
Prunella
    collaris 420
    modularis 390
Prunellidae 368
Psittacidae 468
Psittacula krameri 468
Ptarmigan 94
Pterocles
    alchata 467
    orientalis 467
Pteroclididae 467
Ptynoprogne rupestris 324
Puffin 215
Puffinus
    baroli 439
    gravis 439
    griseus 439
    puffinus 114
    yelkouan 439

Purple Gallinule 453
Purple Heron 130
Purple Sandpiper 199
Pygmy Cormorant 441
Pygmy Owl 258
Pyrrhocorax
    graculus 295
    pyrrhocorax 294
Pyrrhula pyrrhula 426

Q, R
Quail 97
radio tagging 53
Radde's Warbler 478
Rail, Water 164
rails 163
Rallidae 163, 452
Rallus aquaticus 164
rare species 438
Raven 303
Razorbill 217
Recurvirostra avosetta 179
Recurvirostridae 173
Red Kite 141
Red-backed Shrike 287
Red-breasted
    Flycatcher 480
Red-breasted Goose 442
Red-breasted Merganser
    87
Red-crested Pochard
    446
Red-eyed Vireo 483
Red-flanked Bluetail
    472
Red-footed Falcon 449
Red-legged Partridge 98
Red-necked Grebe 108
Red-necked
    Nightjar 469
Red-necked
    Phalarope 201
Red-necked Stint 459
Red-rumped
    Swallow 327
Red-throated Diver 103
Red-throated Pipit 407
Redpoll 421
    Arctic 483
Redshank 209
    Spotted 205
Redstart 374
    Black 366
Redwing 374

Reed Bunting 436
Reed Warbler 352
  Blyth's 476
  Great 353
Reef Egret, Western
  441
Regulus
  ignicapilla 356
  regulus 355
Remiz pendulinus 305
Remizidae 304
reservoirs 31
Rhodostethia rosea 465
Richard's Pipit 471
Ring Ouzel 370
Ring-billed Gull 463
Ring-necked Duck
  444
Ring-necked Parakeet
  468
Ringed Plover 185
  Little 184
Riparia riparia 323
River Warbler 476
Robin 376
  Rufous Bush 472
Rock Bunting 433
Rock Dove 248
Rock Nuthatch 481
Rock Partridge 451
Rock Pipit 408
Rock Sparrow 398
Rock Thrush 382
  Blue 383
rock thrushes 368
Roller 273
rollers 270
Rook 300
rooks 291
Rose-breasted
  Grosbeak 486
Rose-coloured Starling
  482
Roseate Tern 234
Rosefinch, Scarlet 424
Ross's Gull 465
Royal Tern 465
Ruddy Duck 89
Ruddy Shelduck 442
Ruff 194
Rufous Bush Robin
  472
Rüppell's Warbler 475
Rustic Bunting 484

S
Sabine's Gull 464
Saker 450
Sand Martin 323
Sand Plover,
  Greater 455
  Lesser 455
Sanderling 197
Sandgrouse 467
  Black-bellied 467
  Pin-tailed 467
Sandpiper,
  Baird's 458
  Broad-billed 457
  Buff-breasted 459
  Common 203
  Curlew 195
  Green 204
  Least 458
  Marsh 207
  Pectoral 460
  Purple 199
  Semipalmated 459
  Sharp-tailed 460
  Solitary 461
  Spotted 461
  Stilt 461
  Terek 457
  Upland 461
  Western 459
  White-rumped 458
  Wood 208
sandpipers 173
Sandwich Tern 232
Sardinian Warbler 344
Savi's Warbler 346
sawbills 87, 88
Saxicola
  rubetra 384
  torquatus 385
  torquata maura 398
Scarlet Rosefinch 424
Scaup 79
  Lesser 445
Scolopacidae 173, 457
Scolopax rusticola 211
Scops Owl 257
Scoter,
  Common 83
  Surf 445
  Velvet 84
Scottish Crossbill 484
sea 34, 35, 36

Sedge Warbler 350
Semi-collared
  Flycatcher 480
Semipalmated Plover 454
Semipalmated
  Sandpiper 459
Senegal Thick-knee 453
Serin 415
Serinus
  citrinella 416
  serinus 415
Shag 120
Sharp-tailed Sandpiper
  460
Shearwater,
  Cory's 113
  Great 439
  Little 439
  Manx 114
  Mediterranean 439
  Sooty 439
shearwaters 111, 439
Shelduck 69
  Ruddy 442
shelducks 55
Shorelark 321
Short-eared Owl
  263
Short-toed Eagle
  146
Short-toed Lark 316
  Lesser 470
Short-toed Treecreeper
  362
Shoveler 76
Shrike,
  Great Grey 289
  Isabelline 482
  Lesser Grey 288
  Masked 482
  Red-backed 287
  Southern Grey 289
  Woodchat 290
shrikes 285, 482
Siberian Jay 293
Siberian Thrush 473
Siberian Tit 481
Siskin 418
Sitta
  europaea 360
  neumayer 481
  whiteheadi 481
Sittidae 357, 481
skeleton 10

Skua,
  Arctic 224
  Great 226
  Long-tailed 225
  Pomarine 223
skuas 214
Skylark 320
Slavonian Grebe 109
Slender-billed Gull 462
Small Button-quail 452
Smew 86
Snipe 212
  Great 462
  Jack 210
Snow Bunting 429
Snow Goose 441
Snowfinch 484
Snowy Owl 468
Sociable Lapwing 456
Solitary Sandpiper 461
Somateria
  fischeri 445
  mollissima 80
  spectabilis 81
Sombre Tit 481
song 24
Song Thrush 373
Sooty Shearwater 439
Sooty Tern 465
Southern Grey Shrike
  289
Spanish Imperial Eagle
  447
Spanish Sparrow 396
Sparrow,
  House 395
  Rock 398
  Spanish 396
  Tree 397
  White-throated 486
Sparrowhawk 151
  Levant 449
sparrows 394
Spectabilis,
  Somateria 81
Spectacled Eider 445
Spectacled Warbler 475
Spoonbill 135
Spotless Starling 366
Spotted Crake 165
Spotted Cuckoo, Great
  467
Spotted Eagle 448
  Lesser 448

Spotted Flycatcher 392
Spotted Redshank 205
Spotted Sandpiper 461
spotted thrushes 368, 375
Spotted Woodpecker,
  Great 280
  Lesser 282
  Middle 281
spotted woodpeckers 275
Spur-winged Lapwing 456
Squacco Heron 125
Starling 365
  Rose-coloured 482
  Spotless 366
starlings 363, 482
Steller's Eider 445
Steppe Eagle 448
Stercorariidae 214
Stercorarius
  longicaudus 225
  parasiticus 224
  pomarinus 223
  skua 226
Sterna
  albifrons 227
  bengalensis 465
  caspia 229
  dougallii 234
  forsteri 466
  hirundo 233
  nilotica 228
  paradisaea 235
  sandvicensis 232
Sternidae 214, 465
Stilt, Black-winged 178
Stilt Sandpiper 461
stilts 173
Stint,
  Little 200
  Long-toed 457
  Red-necked 459
  Temminck's 196
Stock Dove 249
Stone-curlew 177
Stonechat 385
Stork,
  Black 133
  White 134
storks 131
Storm Petrel 115

Storm-petrel
  Madeiran 440
  Wilson's 440
Streptopelia
  decaocto 251
  turtur 252
Strigidae 254, 468
Strigiformes 261
Strix
  aluco 261
  nebulosa 468
  uralensis 468
Sturnidae 291, 482
Sturnus
  roseus 482
  unicolor 366
  vulgaris 365
Subalpine Warbler 343
Sulidae 117
Surf Scoter 445
Surnia ulula 469
Swallow 325
  Red-rumped 327
swallows 322
Swan,
  Bewick's 60
  Mute 59
  Whooper 61
swans 55
Swift 267
  Alpine 269
  Little 469
  Pallid 268
  White-rumped 469
swifts 266, 469
Sylvia
  atricapilla 337
  borin 338
  cantillans 343
  communis 341
  conspicillata 475
  curruca 340
  hortensis 474
  melanocephala 344
  melanothorax 475
  nisoria 339
  ruepelli 475
  sarda 475
Sylviidae 328, 474, 475
Syrian Woodpecker 470
syrinx 24

T

Tachybaptus ruficollis 106
Tadorna
  ferruginea 442
  tadorna 69
tail shape 20
Tarsiger cyanurus 472
Tawny Owl 261
Tawny Pipit 404
Teal 72
  Blue-winged 444
telescope 50, 51
Temminck's Stint 196
Tengmalm's Owl 264
Terek Sandpiper 457
Tern,
  Arctic 235
  Black 231
  Bridled 466
  Caspian 229
  Common 233
  Forster's 466
  Gull-billed 228
  Lesser Crested 465
  Little 227
  Roseate 234
  Royal 465
  Sandwich 232
  Sooty 465
  Whiskered 230
  White-winged Black 466
terns 214, 462
Tetrao
  tetrix 95
  urogallus 96
Tetraonidae 90, 92
Tetrax tetrax 171
Thekla Lark 318
Thick-knee, Senegal 453
Three-toed
  Woodpecker 283
Threskiornithidae 131, 441
Thrush,
  Blue Rock 383
  Dark-throated 474
  Dusky 474
  Eyebrowed 474
  Grey-cheeked 473

Thrush cont.
  Mistle 375
  Rock 382
  Siberian 473
  Song 373
  White's 473
thrushes 368, 472
Thrush Nightingale 377
Tichodroma muraria 359
Tichodromadidae 357
Timaliidae 304
Tit,
  Azure 480
  Bearded 312
  Blue 306
  Coal 309
  Crested 308
  Great 307
  Long-tailed 313
  Marsh 311
  Penduline 305
  Siberian 481
  Sombre 481
  Willow 310
tits and allies 304, 480
Torgos tracheliotus 447
town pigeon 248
towns 48
Tree Pipit 405
Tree Sparrow 397
Treecreeper 361
  Short-toed 362
treecreepers 357
Tringa
  erythropus 205
  flavipes 460
  glareola 208
  melanoleuca 460
  nebularia 206
  ochropus 204
  solitaria 461
  stagnatilis 207
  totanus 209
tripod 51
Troglodytes
  troglodytes 364
  troglodytes zetlandicus 364
Troglodytidae 363
Tryngites subruficollis 459
tubenose 111, 112
Tufted Duck 78
tundra 36
Turdidae 368, 472

Turdus
  iliacus 374
  merula 371
  naumanni 474
  obscurus 474
  philomelos 373
  pilaris 372
  ruficollis 474
  torquatus 370
  viscivorus 375
Turnix sylvatica 452
Turnicidae 452
Turnstone 192
Turtle Dove 252
Twite 420
Two-barred Crossbill 484
Tyto
  alba 256
  alba guttata 256
Tytonidae 254

U

upending 22
Upcher's Warbler 477
Upland Sandpiper 461
Upupa epops 274
Upupidae 270
Ural Owl 468
Uria
  aalge 219
  lomvia 466

V

vagrants 487
Vanellus
  gregarius 456
  leucurus 455
  spinosus 456
  vanellus 183
Velvet Scoter 84
Vireo, Red-eyed 483
Vireo olivaceus 483
Vireonidae 483
Vulture,
  Black 143
  Egyptian 145
  Griffon 142
  Lappet-faced 447
vultures 89

W

waders 173
Wagtail
  Citrine 472
  Grey 402
  Pied 403
  White 403
  Yellow 401
wagtails 399, 472
Wallcreeper 359
Warbler,
  Aquatic 476
  Arctic 478
  Barred 339
  Blackpoll 483
  Blyth's Reed 476
  Bonelli's 333
  Booted 478
  Cetti's 332
  Cyprus 475
  Dartford 342
  Dusky 478
  Eastern Bonelli's 333
  Fan-tailed 354
  Garden 338
  Grasshopper 345
  Great Reed 353
  Greenish 479
  Hume's Leaf 479
  Icterine 347
  Lanceolated 476
  Marmora's 475
  Marsh 351
  Melodious 348
  Moustached 349
  Olivaceous 477
  Olive-tree 477
  Orphean 474
  Paddyfield 477
  Pallas's 479
  Radde's 478
  Reed 352
  River 476
  Rüppell's 475
  Sardinian 344
  Savi's 346
  Sedge 350
  Spectacled 475
  Subalpine 343
  Upcher's 477
  Willow 336
  Wood 334
  Yellow-browed 479
  Yellow-rumped 483

warblers and allies 328, 474
Water Pipit 409
Water Rail 164
Waxwing 358
waxwings 357
Western Palearctic 28
Western Reef Egret 441
Western Sandpiper 459
wetlands 30
Wheatear 386
  Black 388
  Black-eared 387
  Cyprus 473
  Isabelline 472
wheatears 368
Whimbrel 188
Whinchat 384
Whiskered Tern 230
White Egret, Great 128
White Pelican 440
White Stork 134
White Wagtail 403
White's Thrush 473
White-backed Woodpecker 470
White-billed Diver 438
White-fronted Goose 64
  Lesser 442
White-headed Duck 446
white-headed gulls 241
White-rumped Sandpiper 458
White-rumped Swift 469
White-tailed Eagle 144
White-tailed Lapwing 457
White-throated Sparrow 486
Whitethroat 341
  Lesser 340
White-winged Black Tern 466
Whooper Swan 61
Wigeon 70
  American 443
wildfowl 55, 441
Willow Grouse 93
Willow Tit 310
Willow Warbler 336
Wilson's Phalarope 462
Wilson's Storm-petrel 440

wing shape 20
Wood Duck 443
Wood Sandpiper 208
Wood Warbler 334
Woodchat Shrike 290
Woodcock 211
woodland 46
Woodlark 319
Woodpecker,
  Black 284
  Great Spotted 280
  Green 278
  Grey-headed 279
  Lesser Spotted 282
  Middle Spotted 281
  Syrian 470
  Three-toed 283
  White-backed 470
woodpeckers 275, 470
Woodpigeon 250
Wren 364
wrens 363
Wryneck 275, 277

X

Xema sabini 464
Xenus cinereus 457

Y

Yellow Wagtail 401
Yellow-billed Cuckoo 467
Yellow-breasted Bunting 485
Yellow-browed Warbler 479
Yellow-legged Gull 243
Yellow-rumped Warbler 483
Yellowhammer 431
Yellowlegs,
  Greater 460
  Lesser 460

Z

Zitting Cisticola *see* Fan-tailed Warbler 354
Zonotrichia albicollis 486
Zoothera
  dauma 473
  sibrica 473

ACKNOWLEDGMENTS

THE AUTHOR would like to thank: the team at Dorling Kindersley for their hard work and patience; Marcella for her encouragement and forbearance at home; Chris Gomersall for his dedication in the pursuit of photographs; and Richard Thewlis for the research on the maps.

DORLING KINDERSLEY would like to thank: Sean O'Connor, Rachel Gibson, Kim Bryan, Simon Maugham, Peter Frances, and Rick Morris for getting the project started; Ira Pande, Atanu Raychaudhari, and Rimli Borooah for editorial assistance; Bryn Walls, Kirsten Cashman, Shuka Jain, Pallavi Narain, Elizabeth Thomas, Sukanto Bhattacharjya, and Suresh Kumar for design assistance; John Goldsmid and Umesh Aggarwal for DTP design assistance; Carolyn Clerkin for compiling the acknowledgments. We would also like to thank Roland Fiala for the use of his sound recordings and his help in putting together the audio CD. The publisher would like to thank the following for their kind permission to reproduce their photographs: (Key: a=above; c=centre; b=below; l=left; r=right; t=top).

Alamy Images: Buschkind 103bc, 104tr.

Andy and Gill Swash: 331br, 356tr, 410bc, 416tr.

Aquila Wildlife Images: Darren Frost 452bl; Hanne and Jens Erikson 138cla, 154tr; Mike Wilkes 414tr; Paul Harris 470tr.

Ardea London Ltd: Chris Knights 421cb; John Daniels 188tc; Peter Steyn 439tr, 452tr.

Arto Juvonen: 331tr, 351tr.

BBC Natural History Unit: Dietmar Nill 254tr, 286cr; Elio Della Ferrera 142cra; Hans Christoph Kappel 254br; Jose B Ruiz 254cr; Klaus Nigge 142crb; Rico & Ruiz 366tr, 469tl.

Richard Brooks: 180cra, 222tl, 243tr, 362crb, 427tr.

Laurie Campbell Photography: 275cra, 325cbl, 363bc, 363bl.

R.J. Chandler: *91cl*, 105cra, 97tr, 186tr, 182cra, 228tc, 390cla, 419cla, 454bl, 456bl, 457bl, 457br, 458bl, 459bl, 470bl.

Robin Chittenden: 11br, 18br, 88tr, *221bl*, 240cla, 268cca, *386ccb*, 335crb, 474tl, 478br, 483br, 484tr, 485br

Corbis: Eric and David Hosking *371tcr*.

David Cotteridge: 34br, 111cla, 115cra, 115crb, 127tc, 132cr, 66tc, 152ca, 95cl, 173bc, 175ca, 176cla, 176br, 178tc, 186tc, 213tr, 195tc, 200cla, 204tr, 226cb, 245cr, 219tr, 278cla, 281ccr, 275bl, 277tr, 322br, 327cra, 369tr, 369fcra, 379cra, 379bcr, 378tc, 380tla, 386cc, 388ccb, 382tr, 330tl, 339cra, 339ccb, 344tc, 329crb, 341tc, 333cr, 285tl, 396ccb, 400tr, 400br, 409cr, 408tc, 405cr, 405tr, 404cbr, 415ccb, 423ctr, 428crb, 434crb, 458tc, 437cr, 467br, 473br, 475tl, 475br, 477tr, 477bl, 481bl, 483tr, 485tr.

Dorling Kindersley: Natural History Museum, London 16cb (Ellipatical).

Dreamstime.com: 12qwerty 19tr, Menno67 19cl, Scooperdigital 139bc, 140tr, 141bl, 156bl, Tonybrindley 251bl, 252tr;

Goran Ekström: 269c, 408bcr, 425crb, 479bl.

Hanne and Jens Eriksen: 75ca, 155crb, 165cla, 202tr, 201cra, 285bl, 288tr, 382tc, 409cb, 411cla, 411tc, 419tr, 421tr, 428bl, 435tr, 434crb, 447tr, 447bl, 454tl, 456br, 457tr, 458tr, 458br, 462tl, 463br, 465bl, 468br, 471tl, 472br, 480bl, 482tl, 482br.

FLPA – Images of nature: 62bcl, 66cr, 169tr, 170cr, 178cla, 243tc, 276bc, 284tr, 318c, 329clb, 340cra, 350br, 313cr, 395bcr, 435ccb, 468tl; E & D Hosking 439br; E Coppola/A Petretti/Panda 450tl; Foto Natura Stock 113cra; Fritz Polking 260tc; H Hautala 460br; Hans Dieter Brandl 282cla, 470tl; John Holmes 474bl; Lee Rue 72cbr; M Melodia/Panda 113tr; Martin B Withers/FLPA 251cla; P Harris/Panda Photo 295cr; Panda Photo 113tc, 248cr; Peter Steyn 469bl; R Wilmshurst 57cl, 74ca; Richard Brooks 120cl, 322cb, 324tr, 369clb, 383tc, 383tr, 343ccb, 432cra, 432cr; Robin Chittenden 415tr; Roger Tidman/FLPA 133cl; S C Brown 230ccl; Silvestre 90bl, 92tr; Silvestris 330br, 349tc; Tony Hamblin 85crb; W S Clark 155tc, 449bl; W Wisniewski 109cl; Yossi Eshbol 119c, 123tr, 451tl.

Bob Glover: 222bc, 243cbl, 246tc, 241cla, 238tr, 227tc, 227ccb, 249cr, 249cb, 324clb, 408cb, 338ccb, 340cb, 419crb, 428tr.

Brian Small: 231tr.

Chris Gomersall Photography: 1, 4, 5, 8tr, 9bcla, 10cr, 18tc, 18trb, 18cl, 18cr, 21cl, 54c, 103ccla, 104cb, 101bc, 105tr, 105crb, 107cr, 111cr, 111crb, 112c, 114tr, 114ca, 118crb, 121fclb, 122tc, 122c, 124cr, 128c, 128tc, 129tc, 129bc,

134tr, 134cr, 55tr, 56c, 57, 57crb, 57fbl, 61crb, 61ccb, 59c, 66tr, 66cl, 67cr, 67bcr, 68clb, 69cr, 69bl, 73, 73c, 71cr, 70cb, 78cra, 78crb, 77cla, 77cra, 77crb, 80crb, 85clb, 88cr, 89bc, 137ca, 138clb, 138fcrb, 157tc, 157tr, 157cra, 142cr, 144ca, 144cb, 154tc, 154cb, 141tc, 141crb, 158cb, 158crb, 160tr, 160crb, 90tr, 91tr, 91c, 93crb, 94tr, 94crb, 96tr, 96bc, 95c, 98tr, 99c, 163tr, 163cr, 163bl, 166cr, 167tc, 167cra, 168crb, 169br, 172tr, 172cra, 172bc, 173tr, 173bl, 175bl, 176crb, 176bc, 180tr, 180crb, 179cra, 177tr, 177c, 181bc, 183cra, 198tc, 193crb, 199bcr, 192ccb, 203tc, 203cr, 208cr, 204crb, 209tr, 209cra, 209tcr, 206tr, 206cra, 206crb, 190tc, 190cla, 191cla, 189cla, 189ccb, 220tr, 221cla, 212tr, 214tr, 221cra, 221fcra, 221cr, 222tc, 237tc, 242tr, 236ccb, 232cb, 232cra, 235cra, 234cra, 234cbr, 214clb, 214br, 215cca, 216cb, 219tc, 219crb, 219trb, 217tr, 258cra, 247cl, 248tr, 250c, 251c, 262ccb, 263tc, 266tr, 268ccb, 267ccb, 271cra, 274tr, 274cra, 275bc, 278tr, 314fclb, 314clb, 319ccb, 316tr, 315tc, 315tc, 315ccb, 322crb, 325tc, 325cra, 399bc, 405cb, 403tr, 363clb, 368bc, 389crb, 364cla, 367br, 357fbl, 358tr, 358cla, 390ccb, 368cra, 369tl, 369cla, 376tc, 376tc, 376ccr, 378cra, 378cr, 380tc, 380tr, 384tc, 373tc, 374tr, 375ccb, 328tr, 342ccb, 350ccb, 350ca, 352cb, 306ccb, 308ccb, 291cra, 291cr, 291crb, 292tr, 292cl, 296tr, 297crb, 294tc, 300tc, 300cbr, 301tr, 301ccb, 303tcl, 365tc, 365cra, 394cl, 395tc, 395cra, 395ccb, 394br, 398tr, 398ccb, 412cb, 414ccb, 418crb, 418cca, 428clb, 431ccr, 432cr, 432tr, 432crb, 437cra, 439tl, 440bl, 446br, 453tl, 463tr, 465tr, 465br, 467bl.

Mark Hamblin: 9bca, 101clb, 103tr, 117bc, 119tc, 129cr, 57fcl, 57fcrl, 57br, 58cl, 61cra, 68c, 74tc, 76ca, 79ca, 82cr, 84tc, 85crb, 151tr, 163br, 168ca, 168cr, 173br, 179tc, 185cra, 182c, 203tr, 247br, 253tr, 254bl, 254br, 255tc, 261tr, 259tr, 255br, 256tr, 265cra, 270bl, 272tc, 272cbr, 275br, 276tl, 276tr, 280tr, 281tr, 281cca, 399clb, 401tc, 369crb, 389cla, 390tr, 381cla, 386cal, 385tr, 375cla, 389, 392crb, 392, 304cl, 304bc, 306tc, 308tc, 310tr, 291bl, 297tc, 302cr, 411bl, 412cra, 420cla, 426tc, 426tr, 426cla, 436tr, 436cra, 431cra.

Huttenmoser: 357bc, 360tr, 360cla, 359bl.

Pentti Johansson: 283cb.

Rob Jordan: 102b.

Arto Juvonen: 92cal, 279cb.

Steve Knell: 218crb.

Chris Knights: 60cr, 64cr, 87crb, 177tc, 177cr, 259tc, 323ccb, 372cb, 285br, 394tr, 413crb, 417crb.

Mike Lane: 8ra, 9tcb, 9bcl, 12bl, 13bl, 16tr, 19tc, 101cca, 101crb, 102ta, 102tc, 102tr, 102cl, 104ca, 106ca, 108tr, 107tr, 107cl, 107bc, 109tr, 111bl, 115cca, 117br, 118tr, 120tc, 120crb, 56tl, 56tc, 56tr, 57tr, 57cla, 57bl, 57fbr, 58cla, 58fbr, 64tc, 62tr, 63tc, 65c, 71c, 72tr, 79tr, 78tr, 82tc, 89cc, 138tc, 147cr, 147crb, 152tr, 90br, 91bc, 93cra, 94cla, 94ca, 97tc, 100tc, 163cl, 164tr, 168c, 174tl, 174tc, 174fbl, 174cl, 175tc, 175tr, 175clb, 175fbl, 180tc, 184tr, 185tc, 186ca, 186crb, 181tr, 183tr, 197tr, 175br, 197ca, 199tr, 194tr, 190tr, 188tr, 220bl, 221ca, 225tc, 225tr, 225crb, 237cla, 240c, 233tr, 234tr, 214cla, 214cra, 215tr, 216tr, 314bc, 320tr, 316tc, 323tc, 331tl, 331cr, 399cra, 403crb, 402tr, 368cb, 368crb, 368bl, 369fcla, 369br, 379tc, 388tr, 384cca, 385tc, 373tr, 372tr, 372cla, 371tr, 371cla, 337ccr, 329cra, 339tr, 344cra, 342cla, 350tc, 351tc, 353tr, 353ca, 353cr, 369cra, 377cb, 378tr, 378cr, 381tr, 388tc, 384ccb, 339tr, 337tr, 343ca, 353cla, 304tr, 310tc, 294tc, 365cra, 419bl, 417br, 427cl, 427tr, 460tl, 463tl, 463bl, 466tr, 466bl, 468tl, 468tr, 472tr, 475tr, 476tl, 482bl.

Gordon Langsbury: 104cr, 118cra, 126tc, 126cr, 125cr, 95cr, 205crb, 229cca, 291clb, 295tr, 430tr, 441tr, 441bl, 442tl, 442br, 449tl, 449tr, 453br, 454br, 461tr, 461br, 462tl.

Wayne Lankinen: 443bl, 443br.

Sampo Laukkanen: 302tcl.

Henry Lehto: 170tc, 416cla, 479tl.

Jari Peltomäki: 87tc, 188tc, 389br, 393tr, 411tr, 420tr.

Markus Varesvuo: 222clb, 235tc, 245tr, 254tr, 255cl, 260tr, 268tr, 269tr, 330tr, 345tr, 357fbr, 357br, 361tr, 362tc, 3698tc, 377car.

Tim Loseby: 58cr, 127tc, 67tc, 74cb, 83crb, 85tr, 207tr, 202crb, 222ca, 244tr, 404ca, 403tc, 384cla, 339cla, 389, 392, 398cla, 421tc, 416ccb, 424tc, 424cla, 424ccb, 434cra, 433cla, 433ccb, 477tl, 484br.

Tomi Muukkonen: 221fbl, 240tr, 264tc, 331bl, 354tr, 357tr, 359tr.

George McCarthy: 9bra, 9bc, 58tl, 58fbl, 120tr, 121crb, 121fcrb, 121fbl, 121bc, 123cr, 124tc, 124cl, 126tr, 125tr, 125cla, 128tc, 132bl, 69cra, 80tr, 80ca, 153tc, 93cb, 100tc, 100cra, 163c, 165tr, 175cb, 175crb, 176ca, 192tr, 205tr, 207cla, 191tr, 220crb, 224tr, 241crb, 235cla, 230tr, 254crb, 257tr, 282tc, 326ccb,